Diesel Engine Transient Operation

Constantine D. Rakopoulos
Evangelos G. Giakoumis

Diesel Engine Transient Operation

Principles of Operation and Simulation Analysis

Springer

Professor C.D. Rakopoulos
Dr. E.G. Giakoumis

National Technical University of Athens
School of Mechanical Engineering
9 Heroon Polytechniou St.
15780, Zografou Campus
Athens
Greece

ISBN 978-1-84882-374-7 e-ISBN 978-1-84882-375-4

DOI 10.1007/978-1-84882-375-4

A catalogue record for this book is available from the British Library

Library of Congress Control Number: PCN applied for

Cover image: Volvo D16E six-cylinder, variable geometry turbocharged and aftercooled truck diesel engine (Image provided by Volvo Trucks)

Cover design: eStudioCalamar, Figueres/Berlin

Printed on acid-free paper

9 8 7 6 5 4 3 2 1

springer.com

Preface

Traditionally, the study of internal combustion engines operation has focused on the steady-state performance. However, the daily driving schedule of automotive and truck engines is inherently related to unsteady conditions. In fact, only a very small portion of a vehicle's operating pattern is true steady-state, *e.g.,* when cruising on a motorway. Moreover, the most critical conditions encountered by industrial or marine engines are met during transients too. Unfortunately, the transient operation of turbocharged diesel engines has been associated with slow acceleration rate, hence poor driveability, and overshoot in particulate, gaseous and noise emissions. Despite the relatively large number of published papers, this very important subject has been treated in the past scarcely and only segmentally as regards reference books. Merely two chapters, one in the book *Turbocharging the Internal Combustion Engine* by N. Watson and M.S. Janota (McMillan Press, 1982) and another one written by D.E. Winterbone in the book *The Thermodynamics and Gas Dynamics of Internal Combustion Engines, Vol. II* edited by J.H. Horlock and D.E. Winterbone (Clarendon Press, 1986) are dedicated to transient operation. Both books, now out of print, were published a long time ago. Then, it seems reasonable to try to expand on these pioneering works, taking into account the recent technological advances and particularly the global concern about environmental pollution, which has intensified the research on transient (diesel) engine operation, typically through the Transient Cycles certification of new vehicles.

For a number of years now, the vast majority of diesel engines have been turbocharged and this trend is sure to continue. Although turbocharging the diesel engine is beneficial because it increases its (specific) brake power, and also because it provides better fuel economy and reduced CO_2 emissions, it is the turbocharged diesel engine that suffers way more than its naturally aspirated counterpart from poor transient response. This originates in what is known in the engine community as 'turbocharger lag', which is the key factor responsible for the slow speed response and heavy exhaust emissions. Consequently, the turbocharged diesel engine will be the focus of analysis in this book, with the behavior of its naturally aspirated counterpart highlighted only on specific aspects (*e.g.,* cold starting). As a matter of fact, the title of the book could well have read *Turbocharged Diesel Engine Transient Operation.*

Although there are many operating schedules experienced by diesel engines that can loosely be termed transient, we have focused on the most influential ones in terms of engine performance and exhaust emissions, namely load acceptance, acceleration and cold starting, as well as their combinations, most notably in the form of Transient Cycles.

Emphasis in the book is placed on the in-cylinder thermodynamic discrepancies, exhaust emissions and methods of improving transient response. However, it has been our intention to cover the subject from all relevant aspects; consequently, the interested reader will be able to find information on areas of 'lesser popularity' such as second-law (exergy, availability) analysis, compressor surging or crankshaft torsional deformation during transients. Although automotive applications are usually the main case studied, industrial or marine engines' transient response is dealt with too. Moreover, the analysis is thermodynamics rather than control oriented. Control matters are only briefly discussed, mainly in Chapter 6, where the effects of the various control strategies on the engine transient response improvement are pinpointed.

The book is organized as follows: in Chapter 1, an introduction to transient diesel engine operation is given, highlighting various typical load acceptance and acceleration schedules of both automotive and industrial/marine engines, and detailing the importance and different evolution pattern of transient operation compared with steady-state conditions. Chapter 2 describes the complex thermodynamic issues of transient operation (located in the fuel injection mechanism, heat transfer, combustion, air-supply and exhaust gas recirculation processes), starting, of course, with the discussion of the fundamental turbocharger lag problem. Chapter 3 focuses on the dynamic issues encountered during transients, namely friction overshoot, components stress and crankshaft torsional deformation. Chapter 4 discusses briefly the experimental procedure involved in transient diesel engine research, addressing in more detail the instantaneous particulate matter measurement techniques and the heat release analysis of transient pressure data. Chapter 5 deals with the very important aspect of exhaust emissions during transients, and Chapter 6 with the various methods developed over the years for reducing the turbocharger lag phase and improving transient response. Chapter 7 focuses on other aspects of transient conditions, namely cold starting, operation when the turbocharger compressor experiences surge and low-heat rejection engine operation. Chapter 8 provides an alternative coverage of the subject through the perspective of the second law of thermodynamics. Finally, Chapter 9 summarizes the various modeling approaches developed over the years for the simulation of transient operation.

This book is the outcome of many years of research on the subject and it is intended to serve as a reference for engineers and researchers but it should also be useful to (post-graduate) students as a supplementary text. It is expected that the reader is already familiar with (basic) aspects of internal combustion engine operation. Consequently, when dealing, for example, with the combustion development during transients, only a brief reminder is provided at the beginning of the section concerning some fundamental features of (steady-state) combustion, and afterwards we focus on the discrepancies and special behavior noticed during transients that diversify the operation from steady-state conditions. Wherever possible, we provide experimental results to support our analysis. There are a few points, however, where this was not feasible due to lack of relevant experimental work (*e.g.*, transient operation when the turbocharger compressor experiences surge or crankshaft transient torsional deformation).

At this point, we would like to express our thanks to the various publishers and companies, who have granted permission to reproduce figures and photos from their publications. In particular, the assistance of Messrs Jörg Albrecht of MAN Diesel SE, Elmar Gasse of Daimler AG, Chris Nickolaus of Cambustion Ltd, John Zambelis of Isuzu Motors Greece, Martin Stenbäck of Lysholm Technologies AB, Günther Krämer of BorgWarner Turbo Systems and Masayasu Kondo of Mitsubishi Heavy Industries Europe Ltd is greatly appreciated.

Finally, we would like to thank our families, and our colleagues and students at the National Technical University of Athens for their continuous support. Great thanks are due to the editorial staff at Springer London; their editorial and technical assistance has helped us enormously during the preparation of the book, and is thus deeply appreciated.

Prof.-Dr Constantine D. Rakopoulos
Lecturer Dr Evangelos G. Giakoumis

National Technical University of Athens
School of Mechanical Engineering
February 2009

Acknowledgements

The authors and the publisher wish to thank the following publishers and companies who have kindly given permission for the use of copyright material:

- Aachener Kolloquium
- ABB Turbo Systems Ltd
- American Association for Aerosol Research
- American Chemical Society
- American Society of Mechanical Engineers
- Audi AG
- BMW AG
- BorgWarner Turbo Systems
- Cambustion Ltd
- Conseil International des Machines à Combustion (CIMAC)
- Controlled Power Technologies Ltd
- Daimler AG
- Dekati Ltd
- Elsevier Limited
- General Motors Corporation
- GWV Fachverlage GmbH
- IEEE
- Inderscience Enterprises Ltd
- Isuzu Motors Ltd
- Lysholm Technologies AB, Sweden
- MAN Diesel SE
- Mitsubishi Heavy Industries Ltd
- Oxford University Press
- Professional Engineering Publishing
- PSA Peugeot Citroën
- Robert Bosch GmbH
- Russian Authors Society

- Scania CV AB
- SAE International
- Springer Science + Business Media
- The McGraw-Hill Companies
- Tognum AG
- Toyota Motor Europe
- Volkswagen AG
- Volvo Trucks

Contents

Notation

a	Specific availability/exergy (J/kg)
A	Availability/exergy (J), or surface area (m^2)
b	Piston acceleration (m/s^2)
B	Bearing loading (N)
c_p	Specific heat under constant pressure (J/kg K)
c_v	Specific heat under constant volume (J/kg K)
d	Diameter (m)
D	Cylinder bore (m)
D_p	Particle diameter (m)
E	Energy (J)
f	Friction coefficient (Ns/m), or frequency (Hz)
F	Force (N)
g	Specific Gibbs free enthalpy (J/kg), or gravitational acceleration (9.81 m/s^2)
G	Mass moment of inertia (kg m^2), or Gibbs free enthalpy (J)
h	Specific enthalpy (J/kg), or heat transfer coefficient (W/m^2K), or oil film thickness (μm)
i	Gear ratio
I	Irreversibility (J), or light intensity (W/m^2)
k	Crankshaft stiffness coefficient (N m/rad), or thermal conductivity (W/m K), or extinction coefficient (m^{-1}), or spring stiffness (N/m)
K	Whitehouse–Way combustion model preparation rate constant, or dissociation constant
L, ℓ	Length (m)
m, M	Mass (kg)
$\dot{m}$	Mass flow-rate (kg/s)
n	Number
N	Engine speed (rpm)
O	Opacity (%)
p	Pressure (Pa or bar)

Q	Heat (J)
r	Pressure ratio, or crank radius (m)
R_{mol}	Universal gas constant (8,314 J/kmol K)
R_s	Specific gas constant (J/kg K), or swirl ratio
Re	Reynolds number
s	Specific entropy (J/kg K)
S	Piston stroke (m), or entropy (J/K)
t	Time (s)
T	Absolute temperature (K)
u	Specific internal energy (J/kg), or speed (m/s)
U	Internal energy (J)
V	Volume (m^3), or vehicle velocity (kph or mph)
V_h	Cylinder swept volume (m^3)
W	Work (J)
$\dot{W}$	Power (J/s)
x	Mole fraction, or displacement (m)
y	Governor control lever position (mm)
z	Fuel pump rack position (mm), or injector nozzle holes

Greek symbols

α	Thermal diffusivity (m^2/s)
β	Connecting rod angle (deg or rad)
γ	Ratio of specific heat capacities c_p/c_v
Δp	Pressure drop (Pa or bar)
ε	Angular acceleration (s^{-2}), or second-law efficiency (%), or aftercooler effectiveness (%)
η	Efficiency (%)
θ	Bearing loading angle (deg or rad), or grade angle (deg or rad)
λ	Relative air–fuel ratio, or ratio of crank radius to connecting rod length
μ	Chemical potential (J/kg), or dynamic viscosity (N s/m^2)
ρ	Density (kg/m^3)
σ	Crankshaft stress (N/m^2)
τ	Torque (N m), or time constant (s)
τ_{id}	Ignition delay (ms)
φ	Crank angle (deg)
Φ	Fuel–air equivalence ratio
ψ	Flow availability/exergy (J/kg)
ω	Angular velocity (s^{-1})

Subscripts

o	Reference conditions
2p	Two-pulse (governor)
a	Ambient conditions, aerodynamic

AC	Aftercooler
b	Boost
bl	Blow-by
br	Brake, or break-up
c	Coolant
C	Compressor
ch	Chemical
cl	Clearance
coupl	Coupling
cv	Control volume
cyl	Cylinder
d	Duration, or downstream
D	Demand, or damping
dyn	Dynamometer
e	Engine
em	Exhaust manifold
eq	Equivalent
f	Fuel, or frontal area
fb	Fuel burning
fi	Fuel injected
fl	Flywheel
fr	Friction
fw	Flyweight
g	Gas
gov	Governor
gr	Gravitational, or grade
i	Any species, or injected
im	Inlet manifold
in	Inertia
ind	Indicated
ins	Insulation
irr	Irreversibilities
is	Isentropic
kin	Kinetic
l	Reciprocating masses
L	Loss, or load
m	Mechanical, or molar
N	Nozzle
p	Particle, or periodic
r	Rotating masses
rod	Connecting rod
s	Start, or swirl
S	Stiffness
sc	Scavenge
se	Sensing element
st	Steady-state
sup	Supporting

rest	Restoring
T	Turbine
TC	Turbocharger
thr	Thrust
tot	Total
tr	Traction, or trapped
u	Upstream
v	Vapor
V	Vehicle
var	Various
vol	Volumetric
w	Wall, or work
wgv	Waste-gate valve

Superscripts

o	True dead state
ch	Chemical
tm	Thermomechanical
w	Water

Abbreviations

°CA	Degrees crank angle
A/C	Aftercooler or aftercooled
AFR	Air–fuel ratio
BDC	Bottom dead center
bmep	Brake mean effective pressure (bar)
bsfc	Brake specific fuel consumption (g/kWh)
BSN	Bosch smoke number
CAN	Controller area network
CARB	California Air Resources Board
CI	Compression ignition
CNG	Compressed natural gas
CO	Carbon monoxide
CPC	Condensation particle counter
CVS	Constant volume sampling
DI	Direct injection
DMA	Differential mobility analyzer
DMS	Differential mobility spectrometer
DOC	Diesel oxidation catalyst
DPF	Diesel particulate filter
ECU	Engine control unit
EGR	Exhaust gas recirculation
ELPI™	Electrical low-pressure impactor
EM	Electric motor

EMG	Electric motor-generator
EoI	End of injection
EPA	Environmental Protection Agency
ESC	European Steady-state Cycle
ETC	European Transient Cycle
EU	European Union
EUDC	Extra Urban Driving Cycle
EVC	Exhaust valve closing (°CA)
EVO	Exhaust valve opening (°CA)
FGT	Fixed geometry turbine
FID	Flame ionization detector
fmep	Friction mean effective pressure (bar)
GVWR	Gross vehicle weight ratio
HC	Hydrocarbons
HEV	Hybrid-electric vehicle
HP	High-pressure
HRR	Heat release rate
HSDI	High speed direct injection
I	Intercooler
ICE	Internal combustion engine
IDI	Indirect injection
imep	Indicated mean effective pressure (bar)
isfc	Indicated specific fuel consumption (g/kWh)
ISG	Integrated starter-generator
IVC	Intake valve closing (°CA)
IVO	Intake valve opening (°CA)
LEV	Low emission vehicle
LFG	Landfill gas
LHR	Low-heat rejection
LHV	Lower heating value (kJ/kg)
LII	Laser induced incandescence
LP	Low-pressure
mph	Miles per hour
NEDC	New European Driving Cycle
NMHC	Non-methane hydrocarbons
NRTC	Non-road Transient Cycle
NYCC	New York City Cycle
OBD	On board diagnostics
PAHs	Polycyclic aromatic hydrocarbons
PFSS	Partial flow sampling system
P-I-D	Proportional-integral-differential
PM	Particulate matter
PRA	Piston ring assembly
PSZ	Plasma spray zirconia
RME	Rapeseed methyl ester
RoHR	Rate of heat release
SCR	Selective catalytic reduction

SMD	Sauter mean diameter (μm)
SMPS™	Scanning mobility particle sizer
SN	Silicon nitride
SOC	State of charge (%)
SOF	Soluble organic fraction
SoI	Start of injection
TDC	Top dead center
TEOM®	Tapered element oscillating microbalance
THC	Total hydrocarbons
VGT	Variable geometry turbine
WHTC	World-wide Harmonized Transient Cycle

1

Transient Operation Fundamentals

1.1 Introduction

The most attractive feature of the compression ignition (diesel) engine is its excellent fuel efficiency, which can surpass 40% in vehicular applications and even 50% in large, two-stroke units of marine propulsion or electrical generation. Consequently, vehicles equipped with diesel engines achieve much lower specific fuel consumption and reduced carbon dioxide emissions than their similarly rated spark ignition counterparts over the entire operating range, resulting also in considerable money savings over the vehicle's lifetime. Moreover, diesel engines are characterized by low sensitivity in terms of air–fuel ratio variations, absence of throttling, high torque and high tolerability in peak cylinder pressures and temperatures that favors the application of various supercharging schemes.

Table 1.1. Characteristics of diesel engines for heavy trucks from 1930 to 1996 (reprinted with permission from Hoepke [1])

Maximum Power (kW)	70–110	107–132	140–154	283–304	309–441
Maximum Torque (Nm)	520–1010	620–1050	680–760	1285–1300	1850–2700
Displacement Volume (L)	8.5–16.6	8.3–11.6	9.7–12.7	11.5–18.3	12.1–18.3
Year	1930s	1955/6	1960	*ca.* 1985	1996 (Euro II)

For a number of years now, the vast majority of compression ignition engines is turbocharged and aftercooled; Table 1.1 shows the remarkable increase in engine power of heavy trucks during the last 70 years. Quadrupling of power output has been achieved without significant increase in displacement volume and with only modest increase in engine speed (mainly owing to improvements in the combustion process and fuel quality).

Instead, the main portion of the increase in engine output has been the result of super- or turbocharging, which has enabled more powerful yet smaller and more compact engines to be fitted to vehicles (downsizing). Turbocharging the diesel engine is beneficial because it not only increases its (specific) brake power, but also because it provides better fuel economy and reduced CO_2 emissions – due to leaner operation, increased mechanical efficiency and positive pumping work – and, in some circumstances, reduced exhaust gas and noise emissions. As a result, the turbocharged diesel engine is nowadays the most preferred prime mover in medium and medium-large unit applications, *i.e.*, truck driving, railroad locomotives, non-road mobile machinery, ship propulsion, electrical generation (Figure 1.1). Moreover, it continuously increases its share in the highly competitive automotive market. In fact, diesel-engined vehicles have already ensured a market share that is on a par with gasoline-engined ones across much of the European Union, they are beginning to gain considerable attention in the US and this trend is sure to continue.[1]

During recent decades, the increasingly stringent exhaust emission regulations have dominated the (automotive) industry, and forced manufacturers to new developments. For diesel engines, the emphasis is on reducing emissions of nitrogen oxides (NO_x) and particulate matter (PM), due to the toxicity of the inhaled nanoparticles and because these pollutants are typically higher than those from equivalent rated, port-injected gasoline engines equipped with three-way catalysts. Unfortunately, there is a trade-off between NO_x and PM reduction, resulting in a complicated control strategy requiring complex after-treatment systems. Sophisticated, high-pressure common rail injection systems, exhaust gas recirculation (EGR), or selective catalytic reduction (SCR), multi-valve configurations with variable valve timing, variable geometry turbochargers (VGT), exhaust after-treatment systems with particulate traps or urea-based deNO_x are among the measures applied for reduction of pollutant emissions and fuel consumption. Moreover, carbon dioxide (CO_2) emissions are becoming increasingly important owing to their connection with global warming; limiting CO_2 production can be achieved, primarily, through improvements in fuel economy and use of biofuels. Today's diesel-engined automobiles not only demonstrate greater fuel efficiency than ever before, but they also achieve emission levels at least 50% lower than those of a few years ago.

Unsurprisingly, the various technological advances mentioned above have also led to a significant increase in the complexity and cost of the engine and its control system, and this trend is sure to continue. For example, the Euro 5 details require improvements in PM emissions of 80% compared with Euro 4 (see also Appendix A, Table A.1), which will reflect in an increased cost of the order of €400 per (light-duty) vehicle or up to €3000 per heavy-duty vehicle.

---

[1] According to the European Automobile Manufacturers' Association (ACEA), diesel-engined cars accounted for 53.3% of total new car registrations in the EU in 2007. By 2017, North America and Asia combined are predicted to account for 45% of global annual demand for diesel-engined light vehicles, compared with only 25% in 2007, according to JD Power and Associates, a global marketing information services firm.

Figure 1.1. Modern, turbocharged, four-stroke, DI, diesel engines. *Upper:* Six-cylinder, variable geometry turbocharged and aftercooled, truck diesel engine of 16.1 L displacement volume, with electronic unit injectors and SCR for reduced NO_x emissions, producing up to 485 kW (image provided by Volvo Trucks). *Middle:* Sequentially turbocharged, and aftercooled, 20-cylinder, marine diesel engine MTU 20V 4000 with four turbochargers, producing 4.3 MW (Copyright Tognum AG). *Lower:* Four-cylinder, two-stage turbocharged and aftercooled, passenger vehicle diesel engine of 2.0 L displacement volume with common rail injection system (2,000 bar), EGR and diesel particulate filter (courtesy of BMW AG)

The level of particulates are set at 5 mg/km for passenger vehicles and 0.03 g/kWh for heavy-duty engines, which effectively means that a DPF (diesel particulate filter) would be mandatory downstream of the oxidation catalyst; more injections will be required during a cycle, especially post-injections in order to help regeneration of the DPFs. Similar improvement will be required from Euro V to Euro VI level, since for heavy-duty engines a reduction of the order of 66% for both NO_x and PM emissions will most probably be legislated (equally strong are the improvements required by the US EPA and the Japan Ministry of Environment). More than in the past, combination of both internal engine measures and efficient exhaust after-treatment devices will be required. Since the number of engine and turbocharger controllable components rises continuously (see also Figure 1.3 later in the section), complicated and sophisticated control algorithms are needed to achieve optimum performance. It is universally accepted today that great benefits in the analysis and control of internal combustion engines will (and have actually) come from optimizing the interaction of all the associated engine sub-systems by detailed *simulation models* of the engine processes (Chapter 9). The latter will also pave the way for more advanced engine concepts and alternative combustion systems, such as homogeneous charge compression ignition (HCCI) and low-temperature combustion (LTC).

Traditionally, the study of internal combustion engines operation has focused on the steady-state performance, with minor, if any, attention paid to the unsteady-state or more accurately termed *transient* operation. However, the majority of daily driving schedule involves transient conditions. In fact, only a very small portion of a vehicle's operating pattern is true steady-state, *e.g.*, when cruising on a motorway. Historically, however, the research on transient diesel engine operation was initiated from the observation by engine manufacturers in the 1960s that when highly-rated, medium-speed diesel engines are employed in sudden 0–100% step load changes, severe difficulties are encountered, even leading to engine stall. In recent years, it is the global concern about environmental pollution that has intensified the respective studies; particulate, gaseous and noise emissions typically go way beyond their acceptable values following the extreme, non-linear and non-steady-state conditions experienced during dynamic engine operation. A few representative results follow: cold- or warm-start emissions from heavy-duty diesel engines have been found to exceed up to 15 times their steady-state values; 50% of NO_x emissions from automotive engines during the European Driving Cycle stem from periods of acceleration, whereas instantaneous particulate matter and NO_x emissions during load increase transients have been measured to be 1 to 2 orders of magnitude higher than their respective quasi-steady values (Figure 1.2 [2–5]).

The latter remark highlights in the most explicit way the considerably differentiated evolution pattern of transient response from the respective steady-state operation, and it will be discussed in detail in Chapters 2 and 5. Acknowledging the above-mentioned findings, various legislative Directives in the European Union, Japan and the USA, have drawn the attention of manufacturers and researchers all over the world to the transient operation of (diesel) engines, in the form of Transient Cycles Certification for new vehicles [6, 7].

Moreover, transient operation is inherently characterized by quick changes in the operating conditions, which can prove particularly demanding in terms of engine response and in the reliability of fuel pumps and governors/controllers, so that proper interconnection is required between engine, governor, fuel pump, turbocharger and load through an appropriate control strategy. The latter is required in order to avoid over- or under-speeding, overshoot of exhaust emissions or (significant) departure from the acceptable fuel consumption levels.

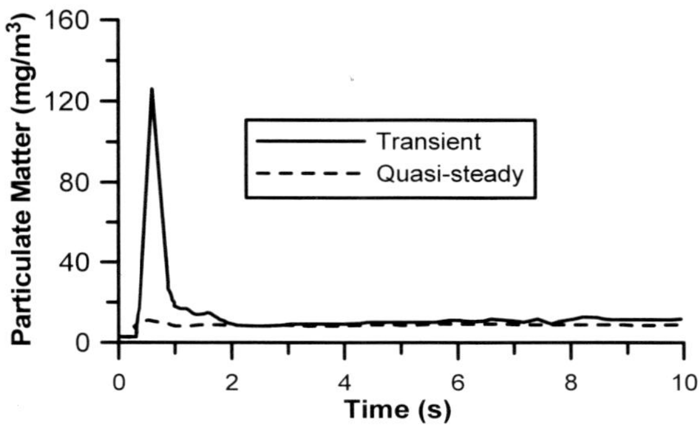

Figure 1.2. Transient *vs.* steady-state PM emissions during a typical load increase event of a turbocharged diesel engine (reprinted with permission from SAE Paper No. 2006-01-1151 [4], © 2006 SAE International)

Traditionally, the control optimization is undertaken during the design stage for steady-state operation, with the calibrated parameters, *e.g.*, injection strategy (pre-, main- and post-injection scheme, rate, timing and pressure of injection), VGT vanes position, boost pressure, EGR valve position *etc.*, stored in 3-D maps with respect to engine rotational speed and load/fueling. The main objective of the diesel engine control system is then to provide the required torque with minimum fuel consumption and exhaust emissions. During transients, the behavior of the engine control system depends upon the specific application (vehicle, ship propulsion, locomotive traction, electrical generation, *etc.*), which imposes different requirements, compared with steady-state operation, with 'driveability'[2] issues playing now a primary role. For example, in electrical generation applications, zero speed droop is required together with short recovery period after the new load has been applied for the base units, as well as rapid start-up for the stand-by ones, whereas in vehicles, fast acceleration is pursued combined with limited smoke emissions. In locomotive applications, the major challenge is to ensure that the electrical system and the diesel engine both function as a synchronized unit and undesirable phenomena such as under-speeding or excessive smoke emissions are avoided.

The comprehensive open- or closed-loop control systems employed gather the signals from the various sensors located on the engine, fuel pump and turbocharger (Figure 1.3), process them based on look-up tables (steady-state maps) or, better

[2] Thereinafter meaning rapid engine torque and speed response during acceleration of vehicular engines, as well as avoidance of over- or under-speeding (stall) and successful load acceptance with small speed droop and short recovery period for industrial/marine engines.

still, model-based control theory with, for example, feed-forward control, and eventually determine, via actuators, the optimum position of the various valves, vanes, *etc.* In order to realize such meaningful analysis of engine performance under transients, accurate analysis of the relevant experimental data is also needed that can only be accomplished using complicated and expensive test facilities equipped with electronically controlled dynamometers and accurate and fast response exhaust emission analyzers (Chapter 4).

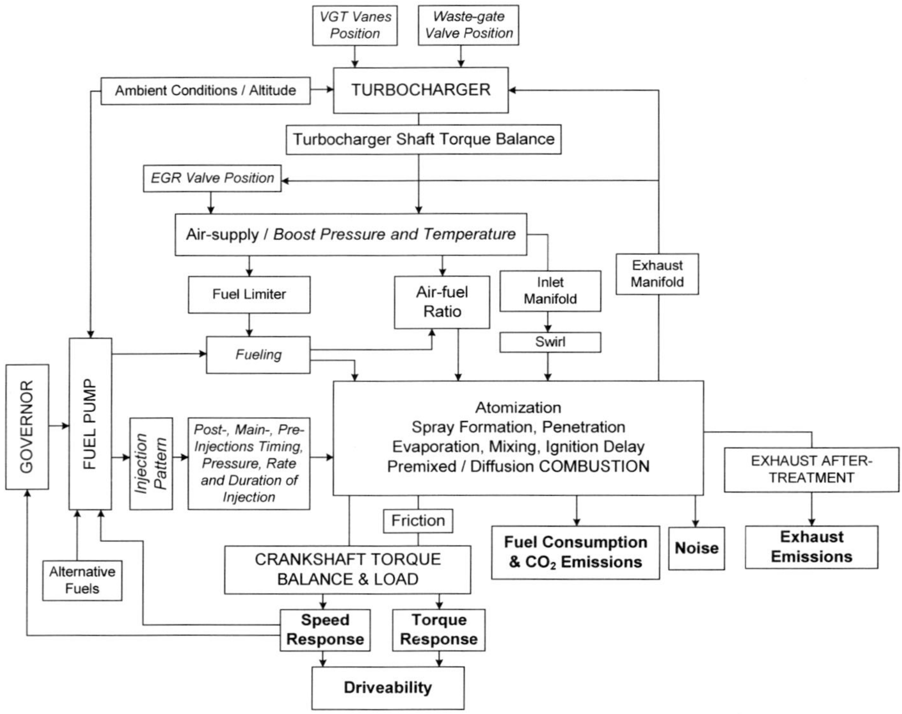

Figure 1.3. Simplified diagram showing some major air-supply and fueling controllable inputs (italics) and engine/vehicle outputs (bold), highlighting the complexity of a modern diesel engine powertrain

The fundamental aspect of transient condition lies in its operating discrepancies compared with steady-state operation (*i.e.,* operation at the same engine speed and fuel pump rack position of the respective transient cycles). Wheareas during (fully warmed-up) steady-state conditions, crankshaft rotational speed and fueling, hence all the other engine and turbocharger properties remain practically constant, during transient operation, both the engine speed and the amount of injected fuel change continuously. Consequently, the available exhaust gas energy varies, affecting turbine enthaply drop and, through the turbocharger shaft torque balance, the air-supply and boost pressure to the engine cylinders are influenced (Section 2.1). However, due to various dynamic, thermal and fluid delays in the system, mainly originating in the turbocharger principle of operation and its dymanic moment of inertia, combustion air-supply is delayed compared with fueling, eventually affecting torque build-up (driveability) and exhaust emissions. The most notable feature of transient turbocharged diesel engine operation that determines its whole operating profile is the *turbocharger lag*. Turbocharger lag is usually realized as slow response rate after a load or speed increase transient event as well as in the form of black smoke emissions coming out of the exhausts of diesel-engined vehicles. Unfortunately, there is a trade-off between exhaust smoke and engine response since the fuel–air ratio at which smoke becomes a problem is substantially lower than that at which maximum torque would be produced by the engine. During the early cycles of a transient event and until the air-supply and engine torque have been built-up efficiently, the following phenomena may be noticed that drastically differentiate the operation from steady-state conditions

- instantaneous torsional deformations in the driving system of the mechanical fuel injection pump leading to incomplete combustion;
- retardation of the dynamic injection timing;
- excursion of the fuel–air equivalence ratio to higher than stoichiometric values;
- increase in the average diameter of the fuel droplets leading to increased jet penetration;
- influence in mixture formation resulting in poorer mixing;
- increased ignition delay and hard combustion course;
- deflections of the crankshaft due to great accelerations/decelerations, resulting in an increase of mechanical friction;
- sharp thermal and mechanical gradients in the cylinder walls and valves;
- turbocharger compressor surging;
- increased (combustion) noise radiation;
- increased particulate and gaseous exhaust emissions; and
- engine stall, when an extreme load increase is abruptly applied.

The arguments introduced previously highlight, on the one hand, the importance of transient operation, and on the other hand, its complexity and its completely different development pattern compared with steady-state operation, requiring careful and systematic experimental and simulation analysis. Clearly, transient turbocharged operation cannot be considered a series of steady-state operating points, nor can the engine be assumed to behave in a quasi-steady manner during transients.

1.2 Typical Transient Operation Cases

There is a variety of operating conditions experienced by (diesel) engines that can be classified as 'transient'; these may last from a few seconds up to several minutes. In this book, the term *transient* will be used to describe any of the following three *forced* changes in load and/or fueling:

1. load acceptance (change) at constant governor setting, mainly experienced in industrial applications, *e.g.*, electrical generation, but

also observed in various propulsion applications, *e.g.*, when a vehicle climbs a hill;

2. change in the fuel pump rack position or step change in pedal ('throttle') position resulting in speed changes (acceleration) of vehicular, locomotive or marine engines;
3. cold or hot starting.

These are the most fundamental transient cases,[3] from which the first two will be primarily addressed in this book. Combination of the above three transient schedules results in

4. simultaneous speed and load changes;
5. whole vehicle/ship propulsion schedule, *e.g.*, gear shift or full ahead/full astern pattern;
6. Transient Cycles, which consist of all of the above mentioned transients.

A number of typical transient cases will be described in the following sections, emphasizing the different development pattern of the engine properties as well as the requirements of the engine control system. At the same time, the mechanism of turbocharger lag and its implications for the engine response will be introduced before the more in-depth analysis of Chapter 2.

1.2.1 Load Increase (Acceptance) Transient Event

A detailed set of engine and turbocharger properties responses during a typical load increase transient event initiated from a step load change from 10 to 85% of full engine load, at constant governor setting, is illustrated in Figure 1.4; it concerns a four-stroke, six-cylinder, moderately turbocharged and aftercooled, medium-high speed, industrial diesel engine, with mechanical fuel pump-governor, rated at 236 kW at 1500 rpm. Such load increase transients are typical of industrial applications, *e.g.*, electrical generation (zero final speed droop is required here as will be shown later in the section), pump driving, *etc.*, but they are also experienced in marine and non-road mobile engines (*e.g.*, agricultural or excavator). On the other hand, although automotive engines do encounter load acceptance transients, *e.g.*, when a vehicle climbs a hill, or when engaging the clutch after a gear change, these are not so pronounced as the ones presented in this section. Eleven typical engine and turbocharger variables are depicted in Figure 1.4 with respect to the engine cycles or time, namely, engine speed, crankshaft angular acceleration, load torque, fuel pump rack position, peak cylinder pressure, air–fuel equivalence ratio, soot density and NO concentration, compressor boost pressure, turbocharger speed and turbine inlet temperature.

At the initial conditions, the engine and load (resistance) torques are equal and the air–fuel ratio is relatively high due to the low loading. As soon as the new

[3] Engines or vehicles encounter several other operating conditions that can loosely be termed 'transient', such as cyclic variation, warm-up, particulate trap regeneration, *etc.* These are not considered in this book.

higher load is applied (this is accomplished in 1.3 s), there is a significant deficit in the net (engine minus load) torque, since the engine torque cannot instantly match its increased load counterpart, hence the engine speed drops. This is sensed by the governor sensing element, which, in turn, shifts the fuel pump rack towards a position of increased fueling. At the same time, the air–fuel ratio decreases since the air-supply cannot instantly match the higher fueling owing to the delayed response of the turbocharger in building-up the required delivery pressure. The instantaneous relative air–fuel ratio, therefore, may reach even lower than stoichiometric values leading to intolerable smoke emissions. At the same time, the increased gas temperatures owing to the low air–fuel ratios are also reflected into high NO_x emissions, with the amount of available oxygen playing here a significant role.

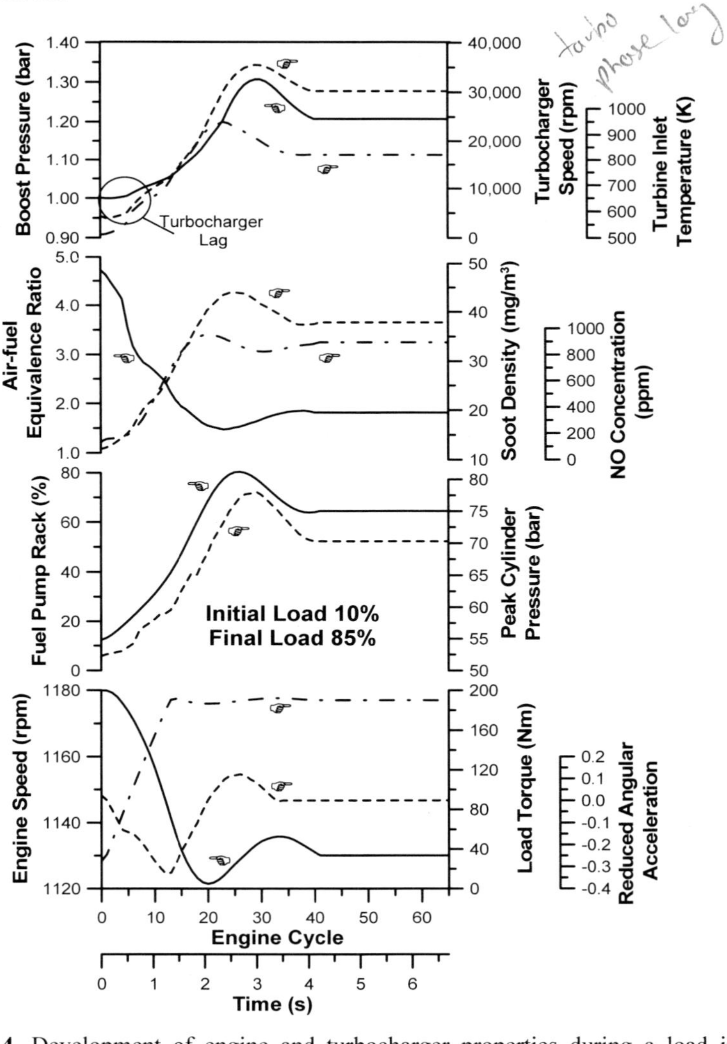

Figure 1.4. Development of engine and turbocharger properties during a load increase transient event of a four-stroke, moderately turbocharged and aftercooled diesel engine

Compressor boost pressure and turbocharger speed develop in a similar way, with the turbocharger lag being obvious during the first cycles of the transient event. Clearly, the increased exhaust gas power is not capable of instantly increasing the turbine power output, largely owing to the turbocharger inertia, so that the compressor operating point moves slowly towards the direction of increased boost pressure and air-mass flow-rate; during this period, the engine is practically running in naturally aspirated mode. During the early cycles (where the torque deficit is greatest), the highest value of crankshaft deceleration is experienced. The lowest engine speed or the maximum temporary speed droop (defined in Figure 1.5[4]), is observed for the present transient event, twenty cycles after the application of the load change. Because of the type of load involved (quadratic, as the particular engine is coupled to a hydraulic dynamometer), the drop in engine speed causes also a drop in resistance torque, resulting in quicker achievement of the final equilibrium between engine and load.

Final equilibrium, also termed recovery period, is defined in Figure 1.5 too; it concerns the time needed to reach the final (steady-state) engine speed. For the transient case demonstrated in Figure 1.4, this is achieved after almost 4 s (recall that the load application lasted for 1.3 s) or 40 cycles with a rather small final speed droop of the order of 47 rpm or 4% of the initial engine speed. For load changes of vehicular engines, much larger speed deviations are experienced since the speed governor has different, less tight characteristics (Section 3.2). The delays involved in the engine and turbocharger response build-up are best highlighted by the observed phase shift between the minima or maxima of the various parameters values during the transient test.

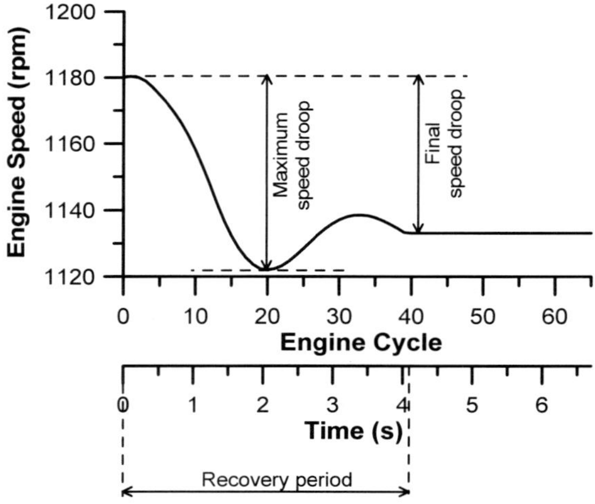

Figure 1.5. Definition of instantaneous maximum and final speed droop and recovery period for a load acceptance transient event

---

[4] In the literature, strict definition states that *droop* is the change in speed between no-load and full-load, while run-up is defined as the change in speed between full-load and no-load. Nonetheless, the majority of researchers in the field have used the term droop for intermediate load changes too and this approach will be followed in this book.

The engine of Figure 1.4 is fitted with a non-zero speed droop governor, a fact that led to positive final speed droop. What is more important is that it also possesses a high mass moment of inertia, which slowed down the development of the transient event and produced rather limited smoke emissions and crankshaft angular deceleration. The successful load acceptance was further supported by the speed-dependent resistance torque. The rather moderate rating of the engine (15 bar bmep) did not lead to complete deterioration of combustion during the early cycles as the turbocharger lag effects were rather limited. The particular engine makes use of the pulse turbocharging system with small exhaust pipes, and a twin entry turbine (each entry 'accepts' the exhaust gases from three cylinders, with minimum, if any, exhaust pressure waves overlap), a fact aiding fast response in contrast to constant pressure turbocharging. In the latter, usually met in large two-stroke (marine) engines, the large volume of the manifolds requires more time to be filled with the higher density gas, thus, delaying engine response even more (see also Figure 1.17 later in the chapter). Moreover, the imposed load change was not a very difficult task for the engine to cope with, as the final load was lower than the maximum acceptable and it was not applied instantly. For greater or more abrupt load changes engine stall might occur, especially if the engine is fitted with a fuel limiter.

This is actually the case illustrated in Figure 1.6 for a similarly rated, industrial, turbocharged diesel engine; here the load change was 0–100%. Since the fuel quantity was not allowed to exceed temporarily its steady-state full-load value, the engine was unable to handle the severe load increase owing to turbocharger response delay, and ultimately stalled. Usually, a fueling overload of about 10% (above steady-state values) is allowed in order for a turbocharged diesel engine to be able to withstand step load changes of the order of 0–100%. In vehicular applications, on the other hand, it is highly unlikely that the engine could stall because of a load increase, since load increases are much softer compared with industial engines. Instead, the problematic transient response would rather manifest itself as increased smoke emissions coming out of the exhaust pipe, owing to the mismatch between air-flow and fueling during the early cycles of the transient event.

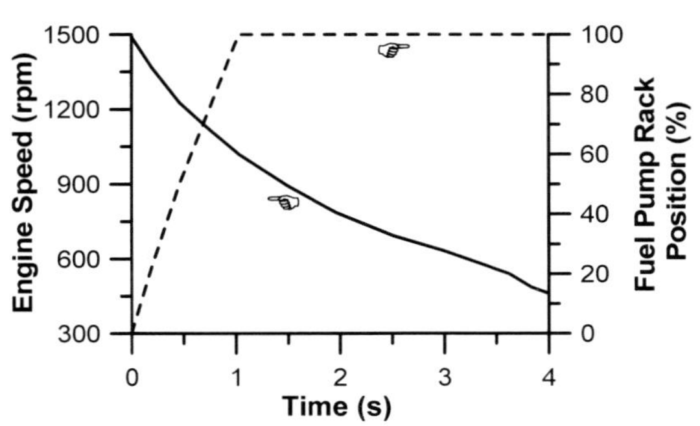

Figure 1.6. 0–100% step load increase transient event leading to engine stall

Summarizing the above remarks, the flow-chart of Figure 1.7 illustrates what happens in a turbocharged diesel engine after a new (higher) load has been applied.

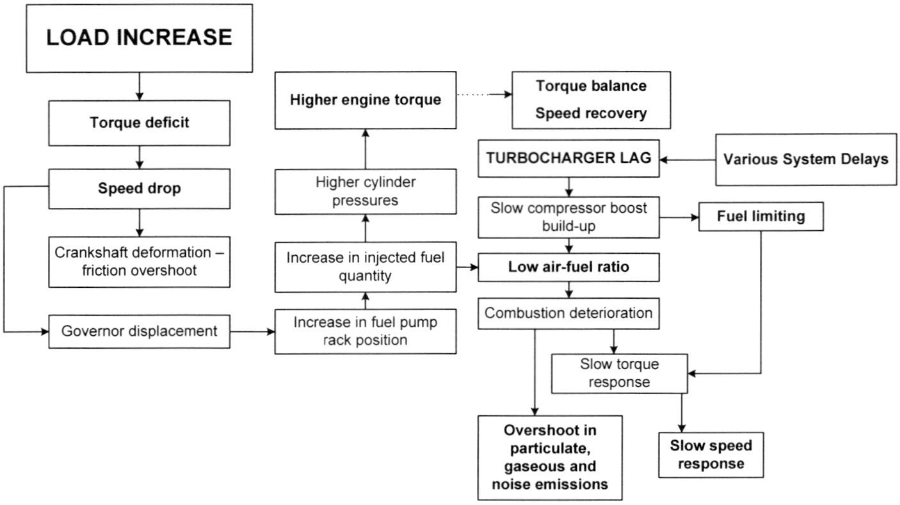

Figure 1.7. Series of events after a step load increase of a turbocharged diesel engine

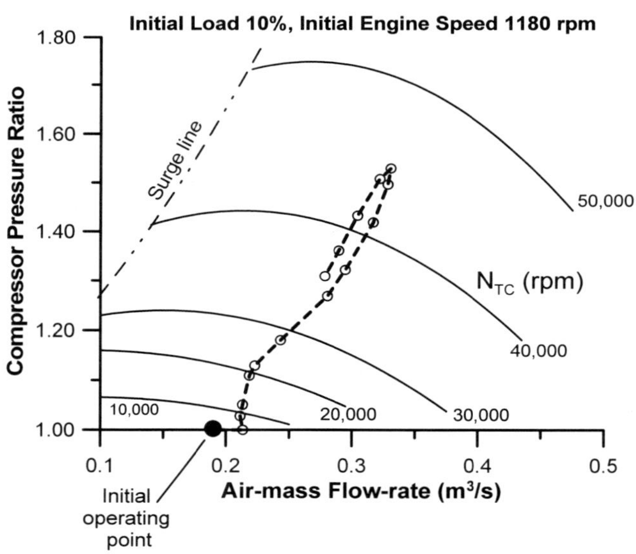

Figure 1.8. Development of engine transient response on the turbocharger compressor map

An alternative way of presenting a turbocharged diesel engine's transient response is by depicting it on the turbocharger compressor map, as is illustrated in Figure 1.8 for a 10–95% load increase transient event. This technique has the major disadvantage of ignoring the time scale involved; yet, it can prove quite useful when studying engine–compressor interdependence such as, for example, when the turbocharger compressor experiences surge or when the subject of concern is the matching between engine and turbocharger. Turbocharger lag is identified in Figure 1.8 too at the early stage of the event as the compressor delivery pressure builds-up slowly. As the initial engine speed and engine load are much lower than the maximum, a safe margin to the surge line is maintained throughout the whole transient test.

This was not the case, however, for the engine of Figure 1.9. A very fast load change schedule commencing from an initial load of almost 100% (*i.e.,* initial operating point very close to the surge line) resulted in the compressor operating point temporarily entering the unsteady region. Consequently, severe torsional loading of the turbocharger shaft was experienced owing to highly oscillatory and, for a few tenths of a second, even reversing flow (Section 7.2).

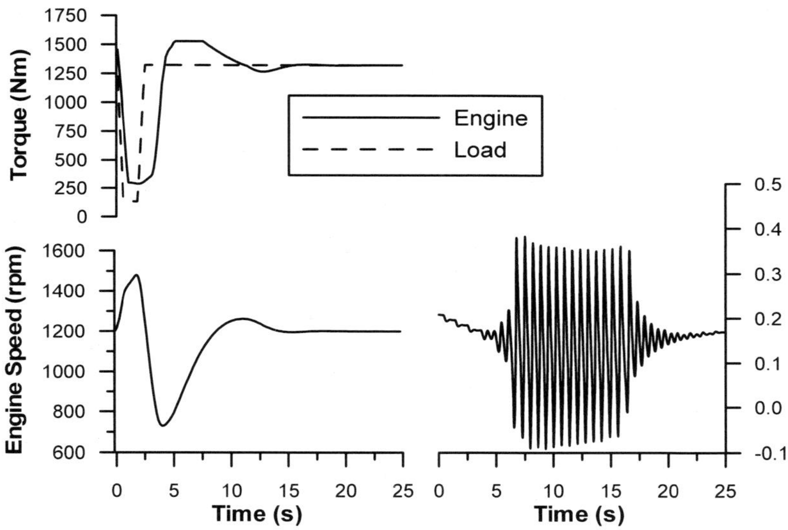

Figure 1.9. Prediction of engine transient behavior after a 96%–9%–87% load change schedule with turbocharger compressor experiencing surge (four-stroke, six-cylinder, turbocharged and aftercooled marine diesel engine)

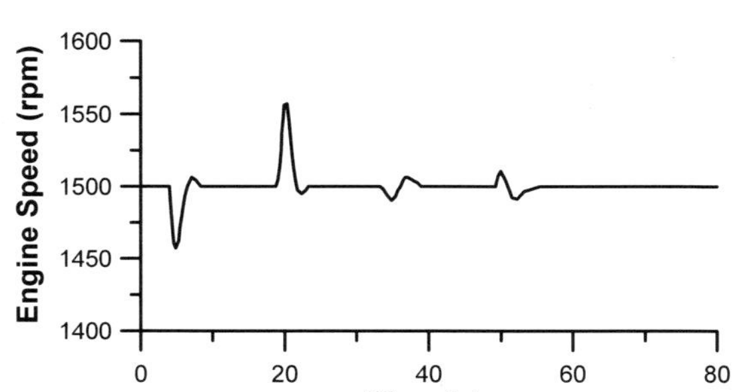

Figure 1.10. Series of load increases and decreases for an engine driving an electrical generator

A special, very challenging for the engine control system, case of load acceptance is when the engine is used for electrical generation. This is due to the particular response pattern required, *i.e.,* the engine speed must return to its initial value after the new load is applied with minimum instantaneous speed droop, while the recovery period must be maintained as short as possible. Electrical demand is, however, quite unpredictable and may involve large and abrupt (positive or negative) load changes. The difficult task of successful speed recovery is implemented by special isochronous governors. Figure 1.10 illustrates a series of load acceptance transient events, including both increases and decreases of load; recovery, in all four load changes, is accomplished within 2 to 3 s, whereas instantaneous speed droop, hence, voltage dip never exceeds 3.5%.

Another very critical schedule that may even lead to engine stall is marine engine maneuver from full ahead to full astern. Typically, the operator requires the propeller to be turning from the ahead to the astern position in not more than 10 s, leading to vessel reversal in approximately 30 to 40 s. In such a case, the engine speed usually falls at about idling level before the astern clutch is selected; the engine speed continues to fall before the clutch is fully engaged, a fact that may lead to stall since the turbocharger speed is very low too. Figure 1.11 demonstrates representative results from a full ahead to full astern maneuver of a medium-speed, turbocharged diesel-engined ship-handling tug [8]; the engine is coupled to the fixed pitch propeller through a reverse reduction gearbox. The procedure is as follows:

- The maneuver commences at point (1) with the vessel moving ahead at 6.5 knots. The control lever in this case was moved from the position of full ahead quickly into an astern position. The control system ensures that the engine response is as fast as possible consistent with adequate protection of the machinery. Coincident with the control lever movement the fuel to the engine is cut off by the governor, and the gearbox is selected to neutral position.
- The engine speed starts to fall immediately, and the propeller shaft brake is applied at point (2) stopping the propeller in approximately 3 s from initiation of the original signal.
- As the engine speed is reduced still further, the astern clutch is selected and signaled to engage at point (3).
- The propeller shaft brake is released before the clutch commences to drive, and due to forward way on the vessel, the propeller is induced to turn in the ahead direction (point 4).
- The clutch commences to engage at point (5).
- The clutch is completely engaged at point (6), but over this period, the torque in the system builds-up to a very high level due to the energy expended in stopping the propeller and re-accelerating it in the required direction. Due to the short time interval, this energy is provided by the rotating engine system and it can result in a severe drop in engine speed.

Whereas a naturally aspirated engine recovers quickly from such a situation, the highly-rated engine of Figure 1.11 suffers the conditions explained previously and due to the lack of sufficient combustion air the recovery is prolonged. Indeed, if precautions are not taken to limit the drop in speed, a complete stall can occur.

Once the turbocharger has commenced to accelerate, excess air is made available and the system accelerates to the new desired speed.

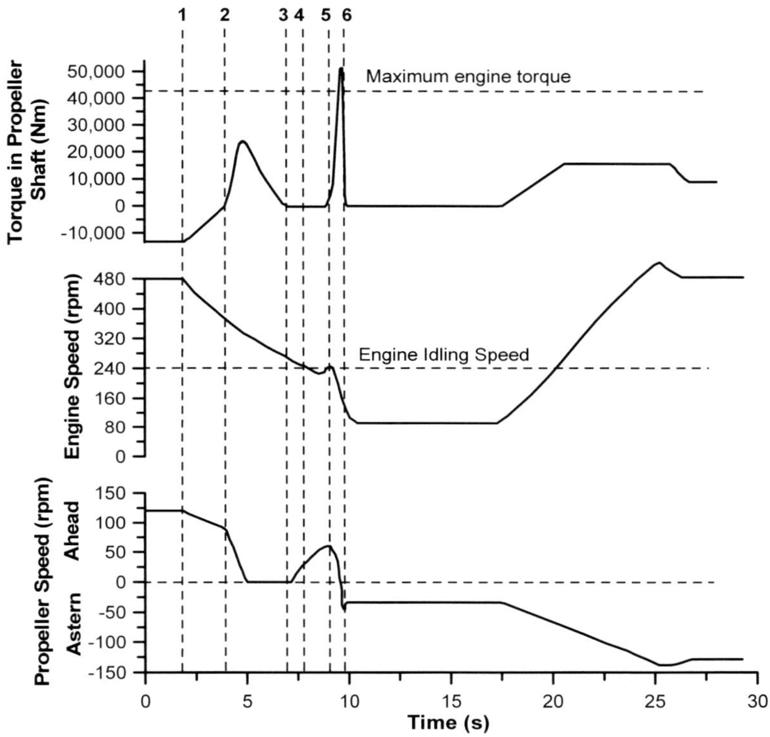

Figure 1.11. Transient response during full ahead to full astern maneuver of a ship-handling tug (four-stroke, medium-speed, turbocharged diesel engine – experimental data reprinted from Gillespie *et al.* [8], with permission from CIMAC)

Locomotive traction is a quite different propulsion mode that, nowadays almost universally, makes use of the diesel-electric configuration, *i.e.*, a diesel engine drives an alternator, which is connected to the wheels. The major sub-systems that comprise a typical diesel-electric locomotive are: 1. a two- or four-stroke, medium-speed, large bore, turbocharged diesel engine; 2. main generator; 3. electrical transmission consisting of traction-motor-generator and resistance dissipation grid; and 4. electrical control system incorporating an electro-hydraulic or electronic governor. The speed/power output of the engine is set at 8 or 9 discrete levels or modes (in addition to an idle speed setting), which are governed by throttle notches. Throttle notch No. 1 is the least powerful and No. 8 or 9 the most powerful one, typically corresponding to 900 rpm engine speed. The locomotive operator chooses the desired speed setting by means of the discretized throttle lever. The electrical system, which constitutes the engine load, is separately controlled to a specific maximum power output level at each throttle setting, chosen so as not to exceed an efficient engine output capability at the given speed setting. In addition to that, the electrical controller 'rate-limits' the power changes in traversing between throttle settings so as to match the engine's ability for simultaneous acceleration and production of shaft power.

Results of a special locomotive load change operation are demonstrated in Figure 1.12, showing a dead load pick-up (DLPU) test [9]. The DLPU test is performed in self-load test mode, with the voltage and current output of the main generator being dissipated in an electrically resistive grid. Initial conditions correspond to the maximum steady-state throttle notch: engine speed at 904 rpm, main generator output power equal to its maximum reaction power rating, and all other electrical control references at the maximum limit values. The generator field conductor (GFC) is then dropped out, which causes excitation current to the main generator to decay according to the main generator field time constant.

Generator output power, and therefore engine load, is reduced to near zero as the excitation current decays over the time the GFC is open. With control references remaining at their limits, picking up or closing the GFC at t=10 s results in a control response characterized by large, rapid increases in excitation current and generator output power. In this way, rapid transients in engine load are experienced. During GFC pick-up, the abrupt change in load causes engine speed to drop initially, resulting in a decrease in turbocharger speed. Turbocharger speed begins to rise between t = 10 to 15 s in response to higher fueling level and exhaust gas energy. An over-speed condition at t = 16 to 18 s results in a rapid decrease in fueling and therefore turbocharger speed. The near constant fueling rates that exist for t>20 s cause a gradual increase and leveling off of turbocharger speed up to its final steady-state value.

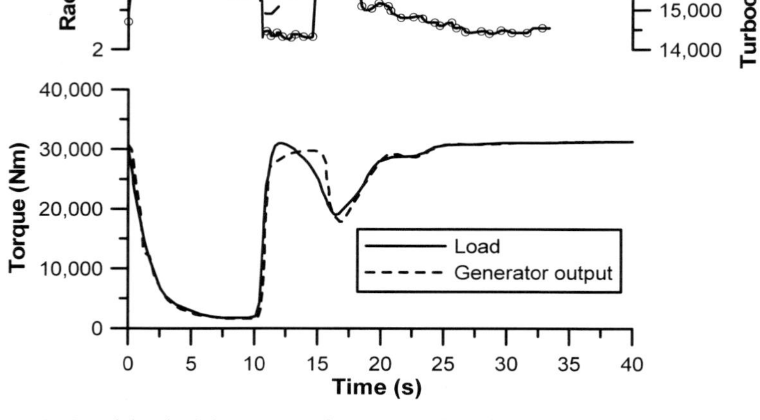

Figure 1.12. Dead load pick-up test of a two-stroke, eight-cylinder, turbocharged diesel-electric locomotive (adapted from Rackmil *et al.* [9])

1.2.2 Speed Increase (Acceleration) Transient Event

The fundamental difference between an acceleration and a load acceptance transient event lies in

- the load torque profile (demonstrated in Figure 1.13); and
- the speed governor behavior.

During load increase, resistance torque increases abruptly; the significant torque deficit that is established during the first cycles of the transient event is responsible for the initial drop in engine speed as well as for the response of the speed governor, engine fueling and torque up to the final equilibrium at the same or lower engine speed. On the other hand, during acceleration, load torque remains practically constant or increases moderately following engine/vehicle speed, while the governor reaches almost instantly its maximum fueling position, remaining there for most of the transient event; the fuel limiting function determining the exact fueling response. As is made obvious from Figure 1.13, it is at the early cycles that the greatest crankshaft acceleration (or deceleration) is established owing to the high surplus or deficit of torque, respectively; crankshaft acceleration depends, of course, on the magnitude of the speed or load increase, as well as on the specific speed/load change schedule. Side effects such as turbocharger lag or exhaust emissions overshoot are experienced in both transient cases, with the acceleration transient being strongly influenced by the initial engine loading.

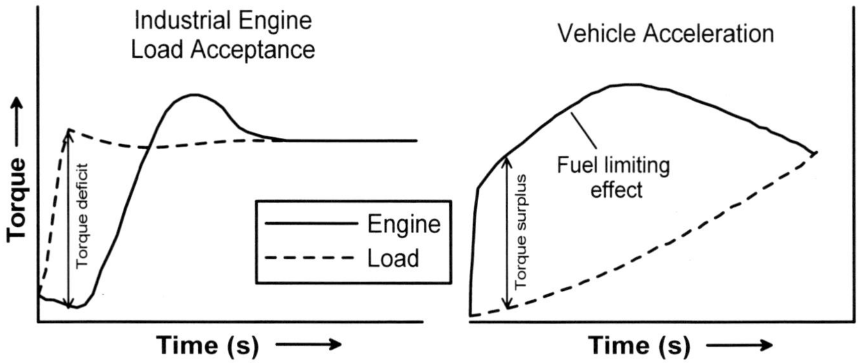

Figure 1.13. Comparison of torque development between typical load acceptance and acceleration transient events

A typical speed increase transient test will be analyzed with reference to Figure 1.14, which demonstrates vehicular engine acceleration from 2000 to 2800 rpm (speed change due to governor control lever or 'throttle' position movement, at constant engine load) for a six-cylinder, high-speed, turbocharged and aftercooled diesel engine without a fuel limiting module. It should be noted at this point that since the engines used for vehicular applications (particularly for passenger vehicles) typically have lower ratings than industrial engines, the implications raised by turbocharger lag are, usually, less dramatic. The low-load acceleration shown in Figure 1.14 is representative of passenger vehicles operating in low gear. At the initial, steady-state operating conditions, engine and load torque are in equilibrium. After the driver pushes the accelerator pedal, the governor control lever reaches almost immediately its maximum position corresponding to full-fueling. Since the resistance torque remains practically unaltered, the significant surplus of net (engine minus load) torque aids rapid build-up of boost pressure and engine speed unlike the load acceptance cases of the industrial engines of Figures 1.4 and 1.10. At the same time, significant increase in peak cylinder and exhaust manifold pressures and temperatures is noticed. Build-up of boost pressure is further supported by the relatively low inertia of the vehicular engine and its turbocharger as well as by the pulse turbocharging scheme involved; a waste-gate valve may be fitted internally or externally to the turbocharger in order to avoid over-boosting at high engine speeds.

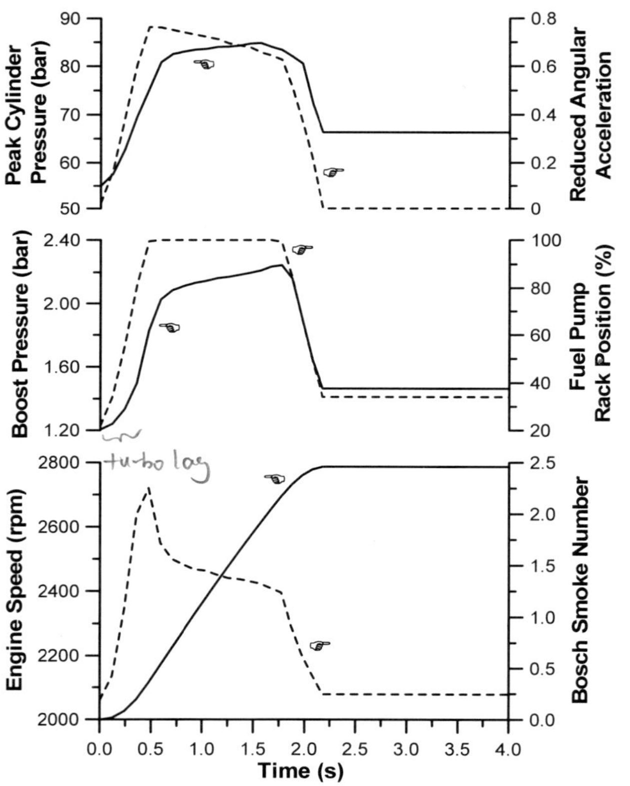

Figure 1.14. Low-load acceleration of a vehicular turbocharged diesel engine without fuel limiter

Despite the lower magnitude of turbocharger lag effects compared with a severe load acceptance case, high amounts of smoke emissions are noticed during the early cycles of the acceleration owing to the initially low air–fuel ratio as fueling has increased much more rapidly than air-supply. In fact, full-fueling has been established after 0.48 s, whereas boost pressure needs at least 0.6 s to assume an 'adequately' high value before peaking at 1.87 s. During much of this period the engine practically operates as naturally aspirated. Hence, the engine must be fitted with a fuel limiting device that determines fueling according to the current turbocharger compressor boost pressure and does not permit full-fuel flow, despite the fact that the governor has reached its maximum position; otherwise, smoke emissions would be intolerable. As will be discussed in Section 2.2.2, however, the fuel limiting module has an adverse effect on engine acceleration and driveability.

Later in the transient event, load torque increases gradually owing to the increase in engine speed, hence, aerodynamic resistance of the vehicle. As a result, the final operating condition at 2800 rpm, achieved after 2.2 s, corresponds to a higher loading than the initial one. Crankshaft angular acceleration assumes much higher values compared with the industrial engine load acceptance of Figure 1.4, owing to the much lower engine moment of inertia; the latter phenomenon unfavorably affects engine bearings stress and friction development. The engine of Figure 1.14 accelerated from a low initial load of almost 20% (*i.e.,* low engaged gear), which was a rather easy task as regards recovery period; much longer acceleration time would be required for speed increase transients in higher gear. Turbocharger inertia, match and configuration, valve timing and closing down of the EGR valve during transients are among the measures that influence the acceleration speed and smoke emissions profile of vehicular and truck engines; these will be discussed in Chapter 6.

A much more complex and demanding transient acceleration is experienced in the everyday gear change schedules of automotive and truck diesel-engined vehicles. Such a transient event with continuous up-gear changes, typical of acceleration when entering a highway, is illustrated in Figure 1.15 showing engine and vehicle speed development.

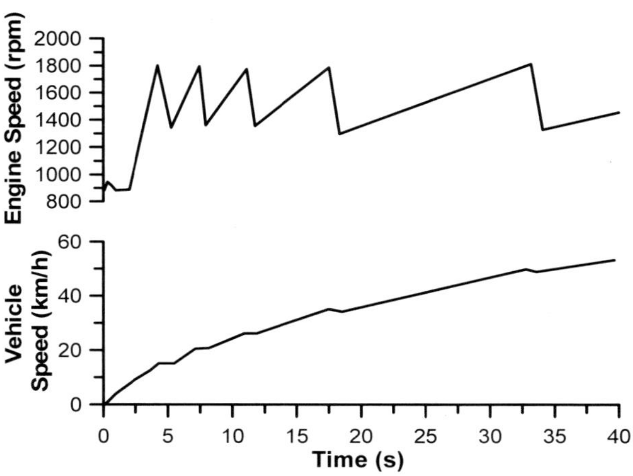

Figure 1.15. Development of engine and vehicle speed during up-gear change of a turbocharged truck (experimental results reprinted with permission from SAE Paper No. 770122 [10], © 1977 SAE International)

This is a more representative vehicular operating mode than the (easier in terms of laboratory reproduction) 'steady'-load acceleration of Figure 1.14. Increased amount of soot and NO_x emissions are noticed during this kind of vehicle operation too, originating in the deterioration of combustion, mainly, during the early cycles of the transient event (Section 2.3.2). The acceleration in first gear is the most critical part of the transient event since the turbocharger is called upon to accelerate from a very low speed, where the boost is practically zero. With the gradual increase in the selected gear, vehicle speed builds-up at a slower rate since the total (engine plus vehicle) inertia increases considerably.

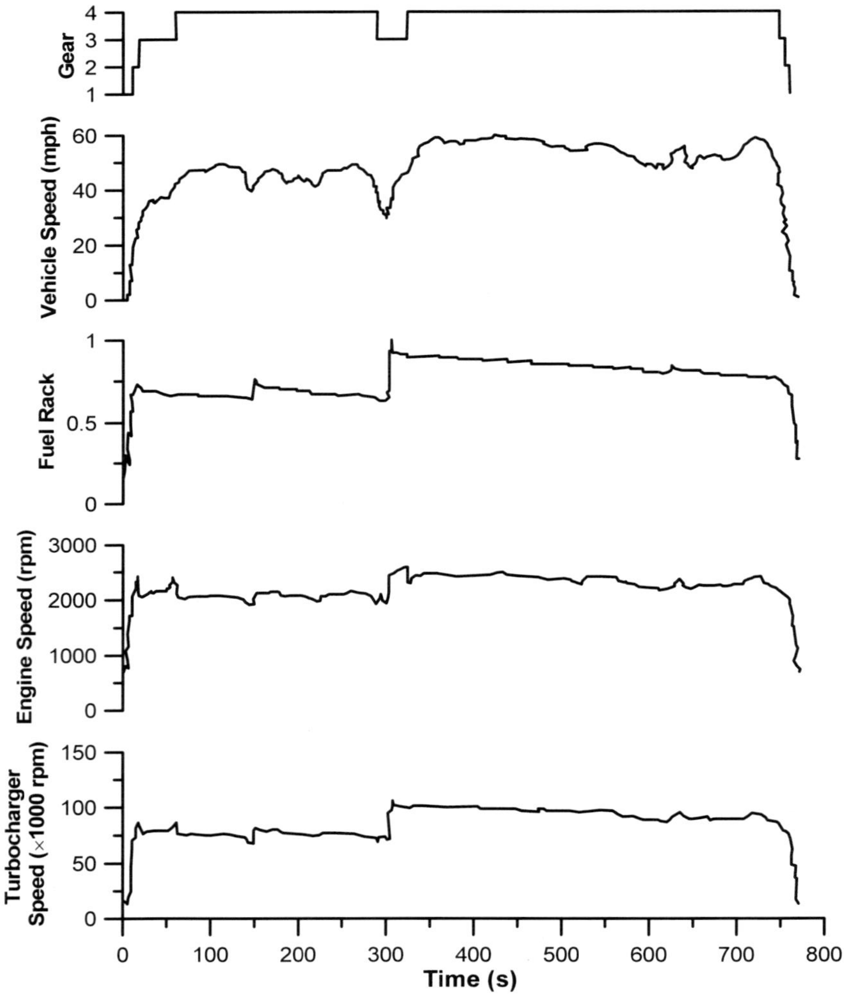

Figure 1.16. Histories of engine and vehicle parameters during the US Highway Driving Cycle for a 4×2 delivery truck (V-8, turbocharged and aftercooled diesel engine of 7.3 L displacement volume – simulation results reprinted with permission from SAE Paper No. 2001-01-1247 [11], © 2001 SAE International)

Starting from the 1980s new diesel-engined vehicles in the European Union, the US and Japan are tested for exhaust emissions using sophisticated transient tests (Transient Cycles), usually characterized by long duration and incorporating various speed and load changes under varying conditions; these simulate real driving patterns in urban and sub-urban conditions. Of particular importance here is the performance of the whole vehicle and not just of the engine during the Transient Cycle (Section 3.5), as it is illustrated in Figure 1.16 for a truck during the US Highway Driving Cycle. The engine is kept at relatively high loads, as indicated by driver demand or 'fuel rack' varying between roughly 0.6 and 1.0 during most of the cycle. Consequently, engine speed is also relatively high; it is close to 2000 rpm during the first 300 s of the cycle, while in the high speed part of the cycle it fluctuates between roughly 2300 and 2400 rpm.

The lowest sub-diagram of Figure 1.16 shows that turbocharger speed follows these trends, hence, during high speed driving, the rotor speed approaches 100,000 rpm. The most critical transient, besides the one at launch, occurs at 295 s into the cycle, when the high speed of the cycle is initiated. A variable geometry turbocharger or a hybrid-electric setup (Chapter 6) can help in achieving faster acceleration and lower exhaust gas emissions, although at the expense of a much more complex control system and/or space requirements and cost. Transient Cycles require highly sophisticated experimental facilities to be accurately reproduced, and complicated, time lengthy computer models for their simulation.

Acceleration is the most typical transient event experienced by passenger or truck engines due to the continuously changing driving operating conditions. Industrial, or non-road mobile machinery or diesel-electric locomotives undergo speed changes too, although with usually lower requirements as regards recovery period. For the latter engines, however, owing to the higher bmep ratings, the problem of smoke emissions during transients may be more acute, particularly if the engine is not equipped with a fuel limiter.

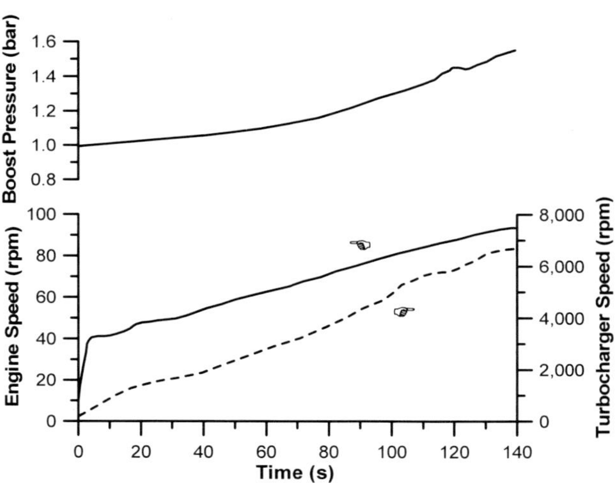

Figure 1.17. Acceleration of a low-speed, two-stroke, turbocharged, marine diesel engine of 11 MW power output (from Zinner [12], reprinted with kind permission from Springer Science+Business Media)

Ship engines are a further example of the need for successful acceleration; these may encounter both speed and extreme load increase (or decrease) transients. In Figure 1.17 an acceleration event is depicted for a low-speed, two-stroke, turbocharged, marine diesel engine directly connected to the propeller. Engine speed, turbocharger speed and boost pressure are presented in this figure, highlighting the different evolution pattern compared with the vehicle's acceleration of Figure 1.14, mainly in terms of recovery period. The latter assumes much higher values owing to the considerable engine and ship inertia as well as due to the constant pressure turbocharging scheme involved that damps down the exhaust gas pulses flowing out of the engine cylinders; fuel limiting effects contribute to the slow response too. Turbocharger lag is again obvious during the early stage of the transient event.

References

[1] Hoepke E. Hundred years of internal combustion engines in trucks. MTZ 1996;57:470–5 (in German).

[2] Wijetunge RS, Brace CJ, Hawley JG, Vaughan ND, Horrocks RW, Bird GL. Dynamic behavior of a high speed direct injection diesel engine. SAE Paper No. 1999-01-0829, 1999.

[3] Gullett BK, Touati A, Oudejans L, Ryan SP. Real-time emission characterization of organic air toxic pollutants during steady state and transient operation of a medium duty diesel engine. Atmos Environ 2006;40:4037–47.

[4] Hagena JR, Filipi ZS, Assanis DN. Transient diesel emissions: analysis of engine operation during a tip-in. SAE Paper No. 2006-01-1151, 2006.

[5] Rakopoulos CD, Giakoumis EG. Review of thermodynamic diesel engine simulations under transient operating conditions. SAE Paper No. 2006-01-0884, SAE Trans J Engines 2006;115:467–505.

[6] http://ec.europa.eu/enterprise/automotive/directives/vehicles/index.htm

[7] http://www.epa.gov/nvfel/testing/dynamometer.htm#engcycles

[8] Gillespie DA, Smith RN, Taylor PJ. Automatic control and associated problems with high pressure charged engines. Proceedings of the CIMAC 8th International Congress on Combustion Engines, Brussels, Belgium, 6–10 May 1968, Paper B.I5, pp. 1105–28.

[9] Rackmil CI, Blumberg PN, Becker DA, Schuller RR, Garvey DC. A dynamic model of a locomotive diesel engine and electrohydraulic governor. ASME Trans J Eng Gas Turbines Power 1988;110:405–14.

[10] Winterbone DE, Benson RS, Mortimer AG, Kenyon P, Stotter A. Transient response of turbocharged diesel engines. SAE Paper No. 770122, 1977.

[11] Filipi Z, Wang Y, Assanis D. Effect of variable geometry turbine (VGT) on diesel engine and vehicle system transient response. SAE Paper No. 2001-01-1247, 2001.

[12] Zinner K. Supercharging of internal combustion engines. Berlin Heidelberg New York: Springer-Verlag, 1978.

2

Thermodynamic Aspects of Transient Operation

The primary purpose of supercharging, nowadays universally employed in diesel engines, is the boost in power output without increase of speed. This is accomplished by increasing the engine air-supply (density) through compression in an external compressor, thereby facilitating increase in fueling. By far, the most popular and successful version of supercharging is the exhaust gas turbocharging. Here, the power required to drive the compressor is provided by a turbine that utilizes (part of) the available exhaust gas energy leaving the cylinders. When designing a turbocharged diesel engine, greatest care has to be given to the matching between engine and turbocharger and between fuel injection and air-supply/motion. These two interrelated optimizations are usually accomplished for steady-state operation but they are really put to the test during transients, where both the engine and the turbocharger are called upon to operate far from their design point conditions.

For a four-stroke engine, air-supply is primarily determined by rotational speed, delivery pressure and pressure difference between inlet and exhaust manifold during the valve overlap period; each of these three factors affects the transient breathing pattern of the engine too. Unfortunately, the flow characteristics of turbochargers are fundamentally different from those of reciprocating engines, a fact that leads to complex matching issues and problematic transient response. The performance of turbo-machines is very much dependent on the respective gas angles and it deteriorates when moving away from their design point. However, their use as superchargers is widely accepted due to their simplicity, small size and large flow capacity that enables significant increase in engine output, increase in the mechanical efficiency[1] but also due to their ability to operate under very high

[1] The fuel economy of a diesel engine is usually improved by turbocharging, since the mechanical losses do not increase in proportion to the increase in power output; this is not the case, however, for spark ignition engines, since turbocharging necessitates a reduction in compression ratio (hence indicated efficiency) in order to avoid knock.

mass flow-rates. As will be documented in the analysis of the next section, the key factors that primarily determine the transient torque and speed response of a turbocharged diesel engine is the engine air-supply and the EGR management (not fueling), being strongly influenced by the transient operation of the turbocharger; the latter affects unfavorably the whole system behavior and it is responsible for an overshoot in particulate, gaseous and noise emissions.

2.1 Turbocharger Lag and Transient Torque Pattern

The most notable feature of turbocharged diesel engine transient response that differentiates substantially the operation from steady-state conditions is the *turbocharger lag* or turbo-lag. Actually, there are various delays that affect the engine transient response, hence it may be more appropriate to speak of the whole system lag rather than only of the turbocharger; nonetheless, it is the latter's flow and dynamic inertia that contributes mostly to the total response delay. Turbocharger lag is caused because although the fuel pump responds rapidly to the increased fueling demand after a load or speed increase transient event, the compressor air-supply cannot match this higher fuel flow instantly but only after a number of engine cycles owing to the mechanical, thermal and fluid inertia of the whole (engine, manifolds and turbocharger) system. Consequently, without a properly matched transient fuel delivery on a *cycle-by-cycle basis,* the low trapped air-mass will initially lead to overshoot of the global fuel–air equivalence ratio. Turbocharger lag is pronounced with the continuous increase in engine rating; it affects primarily the engine torque pattern being realized as slow response after a load acceptance or acceleration as well as in the form of black smoke emissions coming out of the exhaust pipes of diesel-engined vehicles. Unfortunately, there is a trade-off between exhaust smoke and engine response since the fuel–air ratio at which smoke becomes a problem is substantially lower than that at which maximum torque would be produced by the engine.

After a ramp increase in engine load or fueling (see for example Figures 1.4 and 1.14), the governor forces an increase in the injected fuel quantity that is ultimately transformed into a boost of the exhaust gas energy. However, power production delays deteriorate this conversion; these are located in the fuel pump mechanism (Section 2.2) or in the in-cylinder processes (Section 2.3). For the former, mismatch in the (mechanical) fuel injection system is the key cause for the respective delay.[2] For the latter, the most notable delays emanate from the prolonged ignition delay period and the heat transfer to the cylinder and exhaust manifold walls; in fact, cylinder wall temperature adjusts to the new fueling conditions with a substantial delay, during which increased heat flux is observed from the gas to the walls. Although the detailed heat transfer mechanism during transients will be discussed in Section 2.3.1, it is important at this point to emphasize that this wall temperature adjustment lasts one order of magnitude

[2] A special behavior of the fuel injection system during transients, namely fuel limit is a consequence rather than a cause of the engine transient delay. In any case, fuel limit action further adds to the slow engine torque and speed build-up as will be discussed in Section 2.2.

longer than the respective transient event.[3] For the same reason, higher frictional losses are observed, when comparing transient to quasi-steady cycles, since the oil temperature corresponds to its initial steady-state value during the first cycles of the transient owing to the lubricating oil's thermal inertia (low oil temperature increases oil viscosity, hence friction losses) .

Other related power production delays concern the non-immediate change in bmep after the rack position has been altered, but only at the following injection for *each* cylinder, and the acceleration of the slider-crank mechanism rotating masses. The delay of the slider-crank mechanism kinematics observed after initiation of combustion is of the order of 90°CA (see Figure 2.1). Moreover, the increased gas enthalpy must 'wait' until the exhaust valve opens in order to move towards the exhaust manifold and turbine inlet. This further delays the energy effect on the turbine so that, in total, there is more than 200°CA delay between the first injection after a load or speed increase and the possible gains in turbine inlet pressure [1].

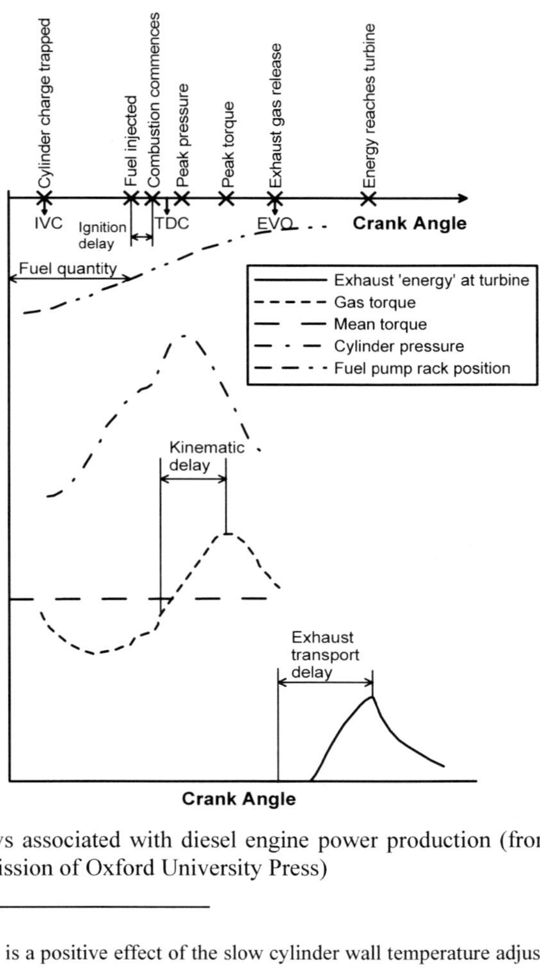

Figure 2.1. Delays associated with diesel engine power production (from Winterbone [1], reprinted by permission of Oxford University Press)

[3] Interestingly, there is a positive effect of the slow cylinder wall temperature adjustment since low wall temperatures favor higher volumetric efficiency of the incoming air-charge.

Further to the rather limited power production system delays, the energy transfer to the turbocharger is delayed by the fact that the nature of this transfer is via a continuous flow of relatively low density gas, whose temperature is reduced due to heat transfer to the 'colder' exhaust manifold walls. Consequently, only a portion of the increased exhaust gas power produced in the cylinder ultimately reaches the turbine. The above-mentioned process is strongly influenced by the exhaust gas turbocharging scheme involved, *i.e.*, pulse or constant pressure.

With constant pressure turbocharging (generally preferred for high engine power output and at medium to high load operation, see also Figure 6.31), the exhaust ports from all cylinders are connected to a single exhaust manifold of sufficiently large volume to smooth out the unsteady flow exiting from each cylinder. Although turbine efficiency is usually higher and construction simpler with one turbocharger unit, even for engines with a large number of cylinders, transient delay is more pronounced when a constant pressure turbocharging configuration is used (*e.g.*, in large two-stroke engines, as was the case depicted in Figure 1.17). The primary mechanism behind this behavior is the considerably longer time required to boost the pressure of the exhaust manifold, since its large plenum increases substantially its flow inertia, while the energy transfer is accomplished by mass flow only without any pressure wave influence. In other words, the high kinetic energy of the exhaust gases leaving the cylinder is not utilized. This delayed response of the exhaust manifold (Figure 2.2) may even lead to engine stall when a large load increase is applied, as is, for example, the case when maneuvering from full ahead to full astern. Nonetheless, the slower build-up of exhaust manifold pressure noticed in constant pressure marine (or generator) diesel engines does not constitute that much of a disadvantage, as these engines experience much fewer transients, mainly when maneuvering in port. During their propulsion in the open sea, which comprises by far the greatest part of their operating life, steady-state operation at high engine load is practically established. In any case, and owing to the vessel's large weight, engine response is always faster than the respective ship's one.

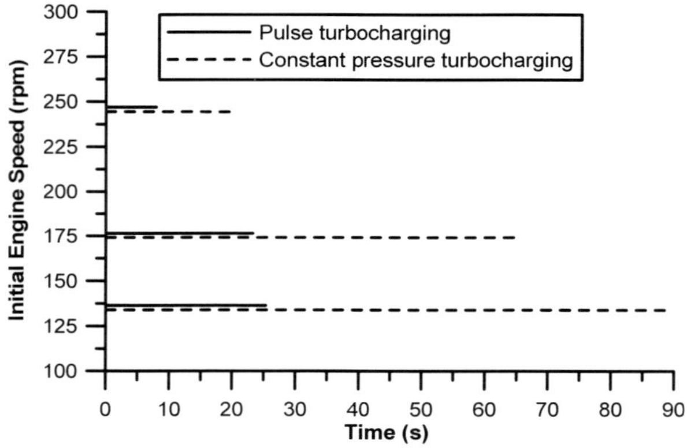

Figure 2.2. Comparison in the recovery period when accelerating from initial speed to full speed between pulse and constant pressure turbocharging (four-stroke, medium-speed diesel engine rated at 4,000 kW and 16 bar bmep at 400 rpm), (from Zinner [2], reprinted with kind permission from Springer Science+Business Media)

On the contrary, vehicular (diesel) engines make use of the pulse turbocharging scheme (patented by Alfred Büchi in 1925) although at the expense of lower turbine isentropic efficiency (recall that turbo-machines are better operated as steady flow rather than pulse wave devices). In general, manifold volume affects the amplitude of the pressure waves generated during the blow-down period of the exhaust process. The pressure waves travel at sonic velocity and their travel time in the exhaust manifold pipe is given by the following equation

$$\Delta\varphi = \frac{12\,L\,N}{\alpha} \tag{2.1}$$

where α is the sonic velocity, L the equivalent pipe length and N the engine speed. The wave amplitude is governed by the difference between instantaneous gas flow-rates in (from cylinders) and out (through cylinders), and manifold volume. Thus, exhaust manifold volume influences turbine expansion ratio and, hence, turbine power. Pulse turbocharging aids faster response of the turbine since the exhaust manifold volume is minimized, being divided into pipes of small cross section area, and the gas flow in the form of waves reaches the turbine entry immediately after the exhaust blow-down commences. Hence, in pulse turbocharging schemes, the exhaust gas energy that reaches the turbine is higher than with constant pressure turbocharging but the unsteadiness of the process is responsible for its rather inefficient exploitation.

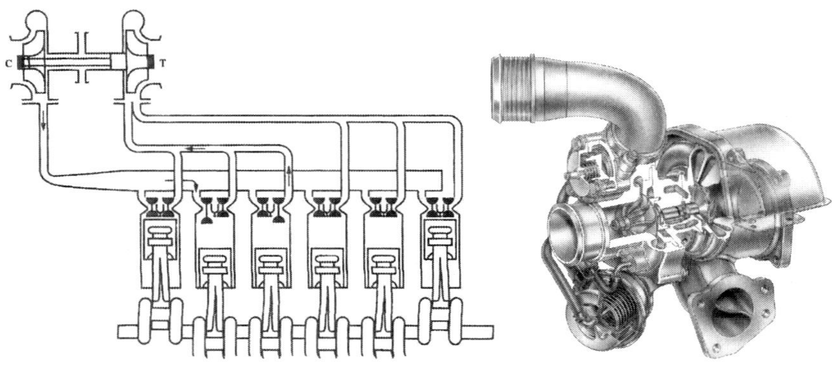

Figure 2.3. *Left:* Schematic diagram of a six-cylinder engine with pulse turbocharging configuration and twin-entry turbine. *Right:* Automotive turbocharger with twin-scroll turbine (courtesy of General Motors Corp)

In a pulse turbocharging configuration, use is made of turbines with multiple entries or scrolls, carefully chosen to serve specific cylinders in order to avoid interference between the traveling exhaust waves from the individual cylinders. Since the exhaust process duration in a four-stroke engine is of the order of 240°CA, it follows that not more than three cylinders should be connected to the same turbine entry. For example, cylinders No. 1, 2 and 3 on the one hand, and cylinders No. 4, 5 and 6 on the other, are served by different turbine entries in a four-stroke, six-cylinder engine having an ignition order of 1-5-3-6-2-4 and equal intervals of the order of 120°CA between successive firing cylinders (Figure 2.3).

Despite its advantages, pulse turbocharging is not universally applied; one such case is when a large number of engine cylinders, hence, turbine entries and manifold sections is involved. When the engine has an odd number of cylinders, pulse turbocharging is not a very attractive solution since, in this case, uneven exhaust pulses are generated in each section. This subject has been treated in detail by Zinner [2] and Watson and Janota [3], who also provide useful examples regarding cylinder number and disposition effects, mainly for steady-state operation.

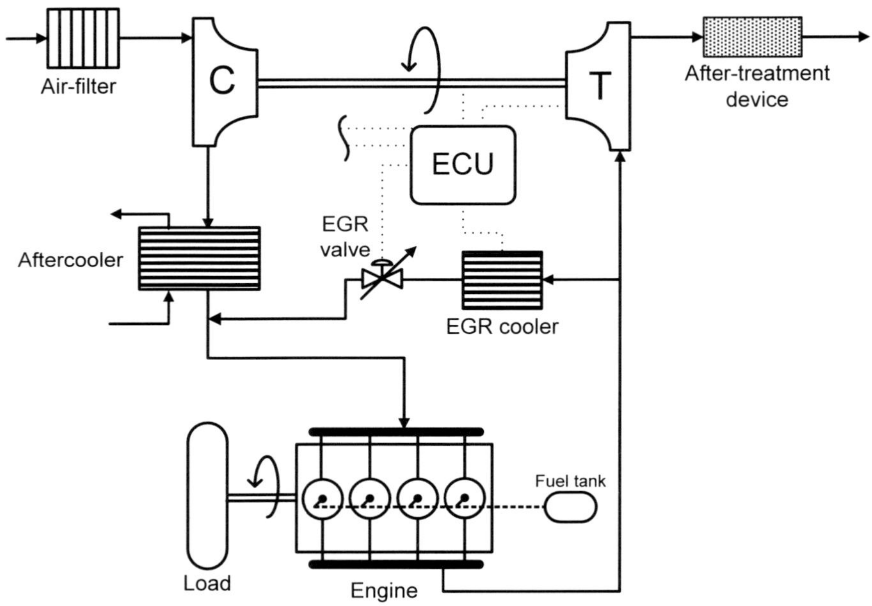

Figure 2.4. Schematic arrangement of current state of the art turbocharged diesel engine (C: compressor, T: turbine, ECU: engine control unit, EGR: exhaust gas recirculation; single line (—) indicates gas flow, double line (=) indicates mechanical connection, light dotted line (…) indicates signal and heavy dotted line indicates fuel flow)

By the time the exhaust gas reaches the turbine rotor, there are further, considerable delays induced by the turbocharger operation. In order to understand these delays, the basic operating pattern of the turbocharger needs first to be addressed. The main characteristic of a turbocharged engine is the lack of mechanical connection between turbocharger and engine crankshaft (Figure 2.4). Consequently, turbocharger and engine speeds are not mechanically but only indirectly interrelated. Turbocharger compressor boost pressure and air-mass flow-rate are typically interconnected via maps such as the one illustrated in Figure 2.5, showing also contours of constant isentropic efficiency and rotational speed.

The map of a radial turbocharger compressor (Figure 2.5) is characterized by:

- constant speed lines of variable slope; for radial compressors (typical in internal combustion engines applications) these are practically horizontal close to the surge line, but drop off rapidly;
- flow chokes at the right of the map when it reaches the sound velocity, typically at the inducer or at the entry to a vaneless diffuser or at the throat of a vaned diffuser;
- the pressure ratio is strongly dependent on rotational speed;
- there is an unstable region to the left of the surge line, in which operation is not viable (see also Section 7.2); and
- at high boost pressures, the speed range is usually narrow (a fact, which as will be discussed in Section 6.4.3, is related to application of two-stage units) and the isentropic efficiency low; both arguments holds true for turbocharger turbines too.

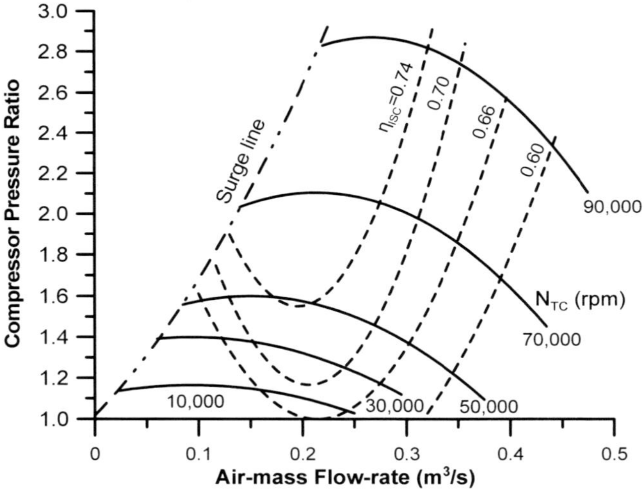

Figure 2.5. Typical map of aerodynamic type turbocharger compressor

It is made obvious then, that it is impossible for a radial compressor to achieve both high boost pressure and air-mass flow-rate when operating at low rotational speed. Instead, acceleration of the turbocharger to a higher speed is required, which, unfortunately, is delayed considerably by the turbocharger's mass moment of inertia. At steady-state conditions, the turbine power is equal to the sum of the compressor power and the losses in the turbocharger shaft bearings, so both compressor and turbine are in an equilibrium state. If the engine operating condition is changed after a ramp increase in load or fueling, the turbocharger needs to accelerate to a new operating point in order to provide the required higher boost. To achieve that, the turbine power must exceed the compressor power. For the turbocharger acceleration, the following differential equation describes the transient angular momentum balance on its shaft, based on Newton's second law of motion for rotational systems

$$\eta_{mTC}\tau_T - |\tau_C| = G_{TC}\frac{d\omega_{TC}}{dt} \tag{2.2}$$

where τ_C and τ_T are the instantaneous values for the compressor and turbine indicated torque, respectively, and η_{mTC} is the turbocharger mechanical efficiency. Changes in turbocharger rotating inertia G_{TC} have been found to have the strongest influence on its transient response, as will be described in Section 6.4.1. In fact, the inertia of the turbocharger disk is closely related to the fifth power of diameter, $G_{TC} \propto d_T^5$. This suggests that reducing turbine (and compressor) rotor diameter d_T will powerfully influence turbocharger and, consequently, engine speed response [4]; the latter remark emphasizes also the favorable use of more than one small turbocharger instead of a single (larger) unit.

At this point, it would be helpful to discuss briefly the merits of mechanical supercharging (Figure 2.6) with respect to transient delays. Contrary to the aerodynamic type compressors used in turbocharging systems, mechanical supercharging makes use of positive displacement compressors. The latter may be of piston type, usually installed in series with the exhaust gas turbocharger (this arrangement is met in large, two-stroke engines) or rotary type, *i.e.*, screw or Roots blower, common in smaller applications such as spark ignition engined passenger cars (lower sub-diagrams of Figure 2.6). Positive displacement compressors are characterized by relatively lower rotational speeds compared with exhaust gas turbochargers, whereas their flow characteristics resemble those of the reciprocating engine. In a mechanical supercharging configuration, a mechanical connection exists between the compressor and engine crankshaft that ultimately minimizes transient lag (at least during acceleration transients) ensuring much faster engine response. However, this is accomplished at the expense of higher system complexity and lower engine brake efficiency due to the mechanical losses associated with driving the compressor.

The key characteristics of positive displacement compressors can be summarized as follows, with reference also to Figure 2.7 [2]:

- steep speed lines, *i.e.*, small drop in flow-rate with increasing pressure ratio;
- the pressure ratio is largely independent of rotational speed, so that a large boost pressure can be obtained even with small flow-rates;
- there is no range of unstable operation, so that the compressor map area can be fully utilized; and
- the flow-rate is roughly proportional to speed, quite independent of pressure ratio.

There are two important parameters that favor engine transient response with such a supercharging scheme in contrast to exhaust gas turbocharging: the compressor speed is directly coupled to its engine counterpart via a fixed or variable gear ratio, hence compressor speed follows engine acceleration; on the other hand, high boost pressure can be achieved even at low (engine) speeds and air-mass flow-rates.

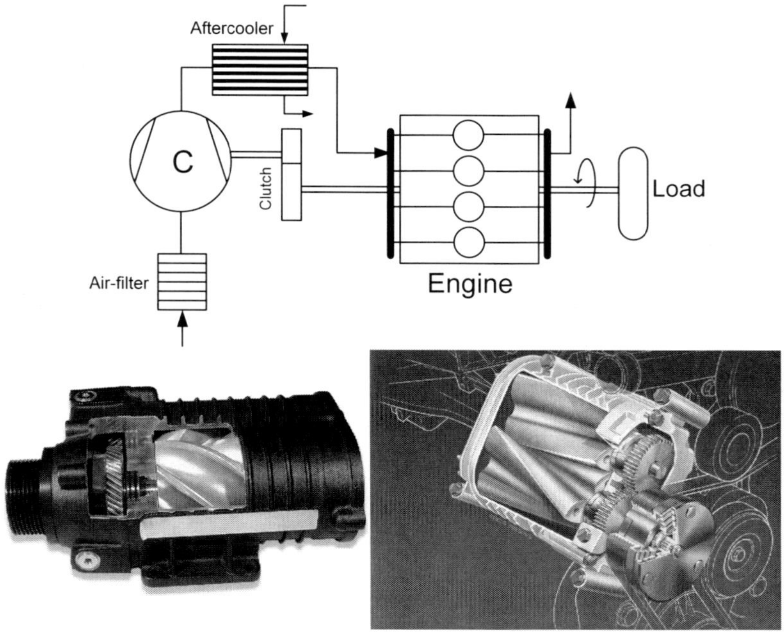

Figure 2.6. *Upper:* Schematic diagram of mechanically supercharged and aftercooled engine. *Lower left:* Mechanical, screw-type supercharger (courtesy of Lysholm® Technologies AB, Sweden). *Lower right:* Mechanical Roots blower installed on a Mercedes-Benz gasoline engine (source: Mercedes Benz)

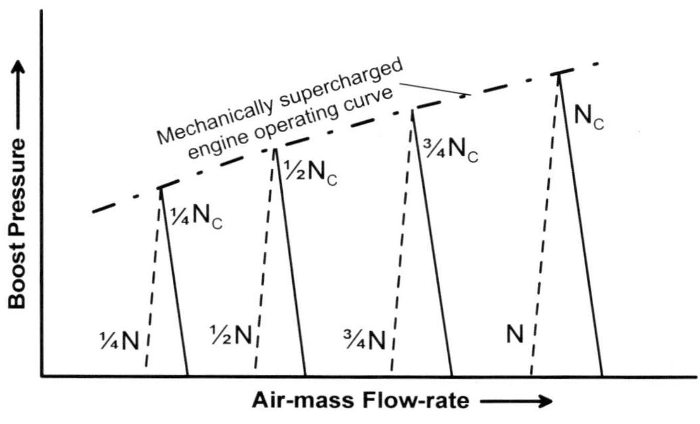

Figure 2.7. Flow map of a four-stroke supercharged engine with mechanically driven, positive displacement compressor (N corresponds to engine speed and N_C to compressor speed, solid line corresponds to positive displacement compressor curves and dashed line to four-stroke engine curves), (from Zinner [2], reprinted with kind permission from Springer Science+Business Media)

In addition to the delays discussed previously that concern the exhaust gas side, the inlet manifold contributes to the total system delay too via two mechanisms. The first is related to the EGR system and will be analyzed in more detail in Section 2.5. However, even at this point, it should be emphasized how the EGR management affects the system transient delay. In a diesel engine, recirculated exhaust gas replaces oxygen, hence during the turbocharger lag phase if the EGR rate is kept unaltered, an even lower amount of fresh air would hit the engine. Unsurprisingly, the main strategy in modern engines is shutting down the EGR valve during the first seconds of a transient event in order to aid faster build-up of combustion air-supply. The second mechanism originates from fluid mechanics grounds, namely the compressed air must first fill the inlet manifold volume before it affects the cylinder charge. Increased engine ratings generally result in higher inlet manifold volumes, hence slower response in the intake side. Similarly to its exhaust counterpart, the inlet manifold acts as a flow capacitor. This is documented in Figure 2.8, showing manifold pressure histories during the first second of a turbocharged diesel engine's load acceptance transient event.

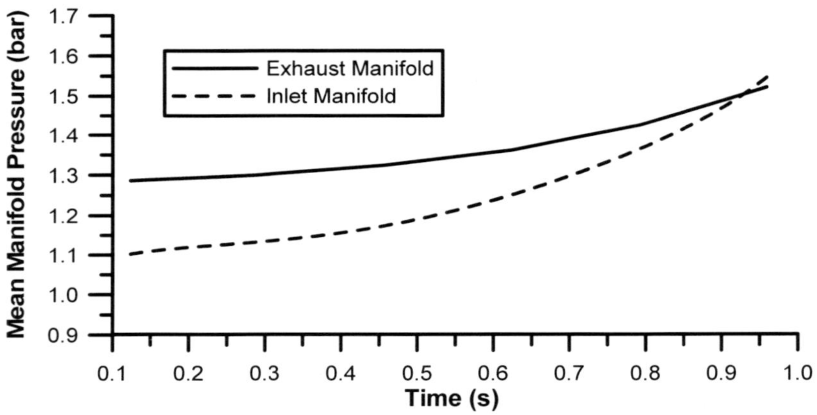

Figure 2.8. Manifold pressure histories during the first cycles of a 0–90% load increase transient event of a six-cylinder, DI, turbocharged diesel engine

Although the exhaust manifold pressure is increasing following the boost in fueling, the inertia of the turbocharger as well as the gas dynamics of filling the intake manifold cause a further lag on the intake side. With respect to the turbocharger compressor map curves (Figure 2.5), this lag concerns both boost pressure and air-mass flow-rate. Consequently, negative pressure difference is noticed for the first 0.95 s (roughly 15 cycles); on the other hand, during steady-state engine operation, a positive manifold pressure difference exists, *i.e.,* inlet manifold mean pressure is higher than its exhaust counterpart, a fact that ensures successful scavenging of the cylinder contents. It is, therefore, expected that during the turbocharger lag cycles and for the valve overlap period (which is particularly high in turbocharged engines to enhance their breathing and lower the gas temperatures at the turbine entry) scavenging is difficult, with substantial reverse flow occurring.

The latter phenomenon induces an even more unfavorable operating aspect in two-stroke engines, due to the absence of separate induction and exhaust strokes. This scavenging dysfunction is reflected into an increase of the residual gas fraction, reducing further the trapped air–fuel ratio. In modern automotive or truck diesel engines equipped with variable geometry turbochargers, the above phenomenon is further enhanced, since the VGT vanes are rapidly closed following initiation of the transient event, thus increasing the exhaust back-pressure even faster. Reduction of valve overlap or, better still, variable valve timing with valve overlap adjustment during transients could be a promising consideration in order to cope with these scavenging issues.

Until the point where the air-supply inside the cylinder has matched the increased quantity of injected fuel, the overall fuel–air equivalence ratio assumes higher than stoichiometric values resulting in excessive smoke emissions. The problem may manifest itself in a more prominent way depending on the specific application. For example, in vehicles, the matching between turbocharger and engine is usually established for steady-state conditions and for low specific fuel consumption. Thus, the turbocharger size is determined for high torque output, which usually leads to high moment of inertia and consequently slow air-charge response. This slow air-flow response cannot match the required fast fueling commands during transients and so it causes poor response and black smoke emissions from the vehicles' exhaust pipes. Summarizing, Figure 2.9 demonstrates the major causes of turbo-lag after a load or speed increase transient event.

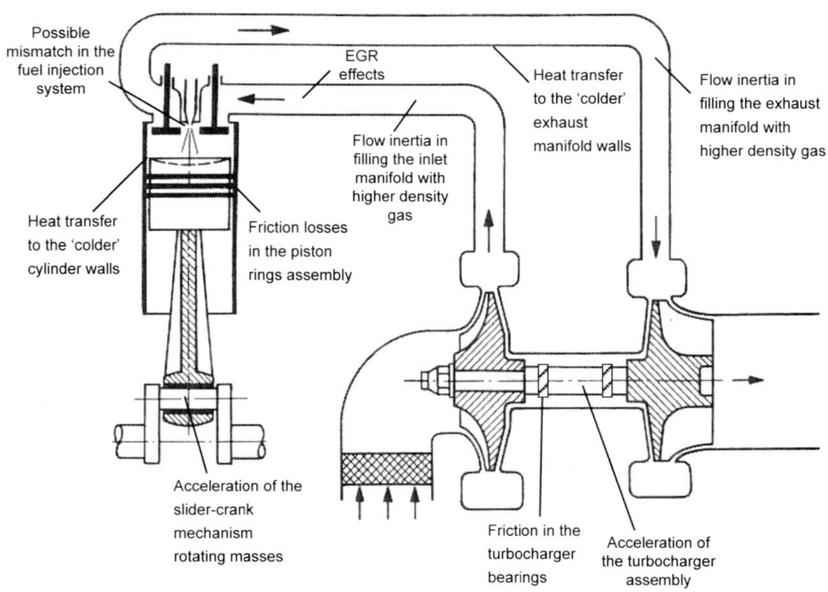

Figure 2.9. Schematic presentation of the major contributions to the system delay during transient response of a turbocharged diesel engine

The key parameter in the schematic arrangement of Figure 2.9 is the turbocharger mass moment of inertia, which is responsible for the biggest of the delays concerned. Unfortunately, turbocharger lag is more pronounced with increase in engine rating or the degree of turbocharging, which explains why it has become more prominent during recent years. Its most direct impact is on engine torque development and speed response.

An attempt to present the inherent difficulty of a turbocharged diesel engine to achieve fast response compared with a naturally aspirated engine, on a fundamental basis, was made by Watson [4]. He compared a naturally aspirated engine of 7 bar bmep with a turbocharged engine of the same size but twice this rating, *i.e.,* 14 bar bmep. Initially, the engines are running at no-load, no-speed conditions and they are rapidly accelerated to full-load, full-speed conditions. The governor responds to the demand and rapidly moves the fuel pump rack position to full-fuel delivery. Full-fueling for the turbocharged engine is twice that of the naturally aspirated, and it will need this fuel delivery to accept twice the load in the same timescale. Initially, however, the air-flow through the engines will be the same, since the turbocharger develops no boost at low-speed and low-load conditions. The need to accelerate the turbocharger rotor will then lead to slow boost pressure build-up. Thus, the fuel flow of the turbocharged engine is rapidly double that of the naturally aspirated, but the air-flow is pretty much the same. The maximum fuel delivery of the naturally aspirated engine will have been selected by the exhaust smoke-limited fuel–air ratio at steady-state full-load conditions. The fuel–air ratio developed instantly by the turbocharged engine will be twice its own fuel–air ratio when running steadily at full-load. In order to avoid excessive smoke emissions (well known dependence on fuel–air ratio), a fuel limiter (fueling determined according to boost pressure – see Section 2.2.2) would have to be installed, restricting further the engine torque developed during transients and leading to even poorer response. Figure 2.10 documents the above-mentioned remarks by highlighting

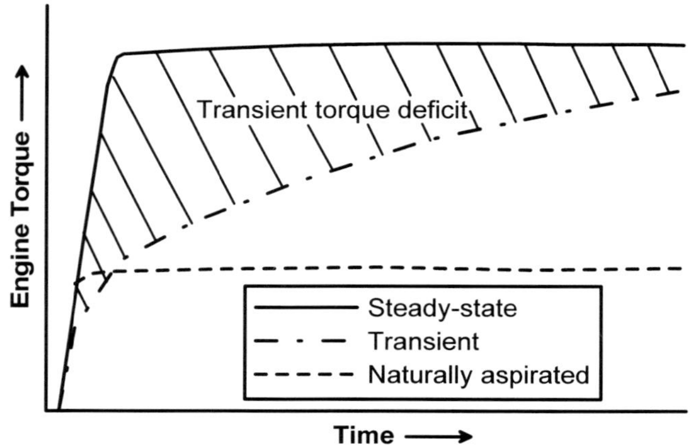

Figure 2.10. Comparison between transient and respective steady-state torque development during a load acceptance

- the *non-linearity* of turbocharged diesel engine transient torque build-up; and
- the substantial *torque deficit* during the first cycles of a load acceptance transient event when comparing the actual (transient) torque developed to the one under 'similar' steady-state operating conditions. Such torque deficit is observed during speed acceleration events too, as will be discussed later in the chapter.

Turbocharger lag and the subsequent fuel limit action are responsible for the turbocharged engine's response pattern illustrated in Figure 2.10. As was mentioned earlier, a key parameter that largely influences the magnitude of the transient torque deficit is the engine rating. The lower the rating (bmep) of the turbocharged diesel engine, the smaller the above mentioned discrepancies and the smaller the transient torque deficit. This can be best explained by the following case study [5].

Table 2.1 presents data of several truck engines, all rated at the same maximum power output of 325 kW at 2300 rpm, but with different numbers of cylinders and kinds of aspiration. In all engines, a limit of 135 bar peak cylinder pressure, 1000 K maximum turbine inlet temperature and 20.5 minimum air–fuel ratio at 1000 rpm were imposed; the latter requirement resulted in fuel limiting control for the four-cylinder engine below 1300 rpm and for the six-cylinder engine below 1500 rpm under steady-state operation, as well as for the V-8 engine under transient conditions.

Table 2.1. Data of engines considered for the analysis of Figure 2.11 (reprinted with permission from SAE Paper No. 840134 [5], © 1984 SAE International)

Cylinder number and disposition	V-16	V-12	V-8	In-line 6	In-line 4
Total displacement (L)	25	18.75	12.5	9.375	6.25
Bmep (bar)	6.78	9.04	13.56	18.08	27.1
Compression ratio	16	15	13.5	12	10
Aspiration	Naturally aspirated	Single-stage, pulse t/c	Single-stage, pulse t/c	Single-stage, pulse t/c	Two-stage, pulse t/c
Intercooling	–	no	yes	yes	yes
Boost pressure (bar)	(1.0)	1.57	2.0	2.9	4.3

Figure 2.11 provides a very instructive comparison between the selected truck diesel engines' acceleration, emphasizing the non-linearity of the turbocharged engines' transient behavior. In the case of the naturally aspirated V-16 engine, acceleration is almost linear since full-load torque is available almost instantly, while the steady-state torque curve (not shown) is relatively flat. Acceleration of

the in-line, six- and four-cylinder engines is highly non-linear, largely owing to turbocharger lag. The fuel limiter allows here more fuel to be injected as boost pressure builds up (recall that there is a 20.5 minimum air–fuel ratio limit). Consequently, the response time of the most highly-rated engine (in-line, four-cylinder, 27.1 bar bmep, two-stage turbocharged and aftercooled) is nearly 2.5 times that of the naturally aspirated, while at the same time achieving only 74% of the distance covered by the V-16 engine in this first gear acceleration! Times to intermediate speeds are even worse for the turbocharged engines compared with the naturally aspirated and are due to the non-linear acceleration. However, it should be pointed out that continued acceleration to higher gear would be less affected by the degree of turbocharging as, in that case, the turbocharger speed would not fall to a no-load value during gear change.

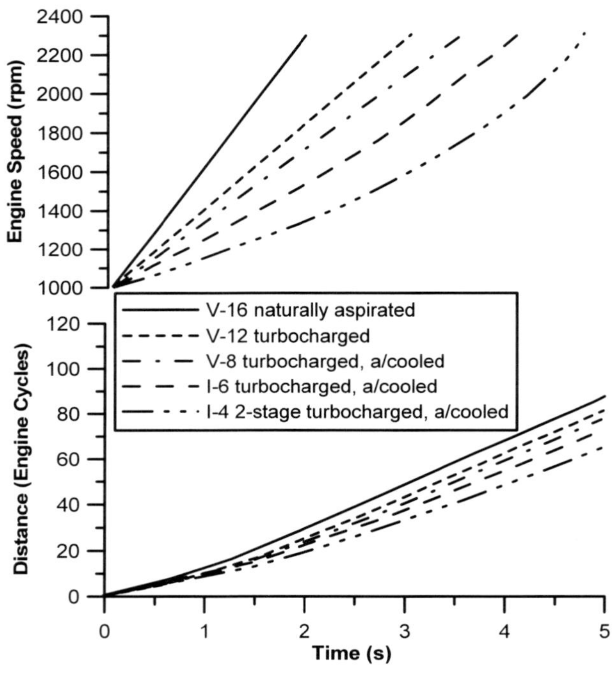

Figure 2.11. First gear acceleration from 1000 to 2300 rpm and distance covered (expressed in engine cycles) for the five truck engines of Table 2.1 possessing different kinds of aspiration and turbocharging options but rated at the same power output (simulation results reprinted with permission from SAE Paper No. 840134 [5], © 1984 SAE International)

The following interesting results were also revealed from the above analysis.

- The instantaneous engine bmep of the V-16 naturally aspirated engine equaled full-load, steady-state bmep after only 0.25 s (following governor response). This highlights the fast response a naturally aspirated engine can actually achieve. In fact, excluding starting, a naturally aspirated diesel engine can be assumed to behave in a quasi-steady manner during transients. On the other hand, the relative fuel–air equivalence ratio never exceeded 0.7 (20.7 air–fuel ratio).

- The acceleration of the V-12 engine was so slow that the boost pressure did not exceed ambient pressure even by the time the engine had reached the desired speed.
- For the in-line, six-cylinder engine, although the boost pressure reached 1.6 bar at maximum engine speed, most of this increase occurred during the last second of the transient event. The non-linearity of the boost pressure build-up was determined by the injected fuel quantity via the fuel limiter, and influenced the whole acceleration of the engine.
- For the four-cylinder, two-stage turbocharged engine, the high-pressure turbocharger accelerated reasonably well but the low-pressure unit provided no boost for the first three seconds.
- For all turbocharged engines, the transient bmep was very much less than the steady-state capability of the engine (*cf.* Figure 2.10).

Watson concluded that although the turbocharger lag did not actually get any worse with the increase in rating, it is its influence on engine response that becomes more and more serious.

Turbocharger lag effects are noticed during load or speed *decrease* transients too. Again, the air-charge cannot instantly match the new, decreased fueling rate. However, in this case, the fuel–air ratio temporarily assumes very low values, hence smoke, NO_x and CO emissions are not a problem; as will be discussed in Section 5.3, this does not hold true for HC emissions. On the other hand, for naturally aspirated diesel engines, all the system's delays summarized previously except for the predominant turbocharger lag and the interrelated delays of filling the exhaust and inlet manifold, do affect the engine transient response; their relative importance is, however, much lower owing to the considerably smaller engine ratings.

Turbocharged gasoline engines suffer from turbocharger lag too during transients. In fact, the situation can prove quite challenging (particularly so for carbureted engines, where fuel hold-up on the fuel-wetted surfaces must be avoided) owing to the wider speed, hence air-mass flow range of spark ignition engines that leads to much harder transients, *e.g.*, when accelerating from closed to wide open throttle. In this case, however, the problem manifests itself as poor driveability or slow acceleration rather than black smoke emissions. Typical results are illustrated in Figure 2.12 for a 2.8 L turbocharged SI engine low-load acceleration. During first gear acceleration, the engine behaves in a naturally aspirated mode for at least 1 s (up to 2200 rpm); things are quite improved at fourth gear acceleration, owing to the much higher turbocharger initial operating point. In any case, torque development is substantially delayed compared with steady-state operation (*cf.* Figure 2.10) affecting negatively vehicle driveability.

Turbocharger lag and the associated system delays are the main reason for the problematic transient response of turbocharged (diesel) engines, affecting unfavorably the combustion process and exhaust emissions; it is not surprising then that the majority of the analyses and research have focused on methods for improving this delay/response. Variable geometry turbines, air-injection in the compressor, retarded injection timing, electrically assisted turbocharging, variable valve timings are but a few measures that have been proposed as viable means for improving turbocharged diesel engine transient response; these and many other measures will be discussed in detail in Chapter 6.

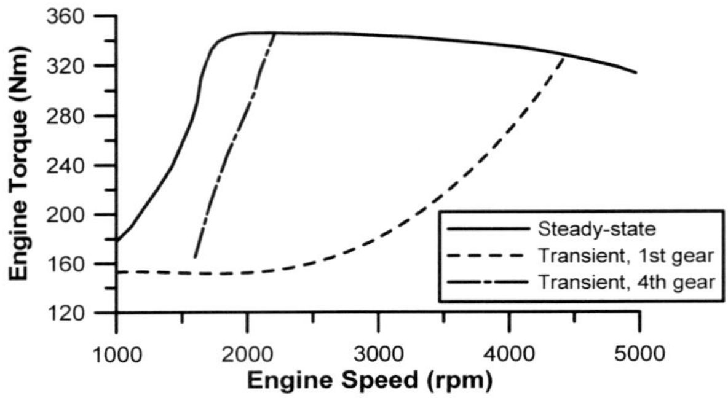

Figure 2.12. Comparison between transient and the respective steady-state torque development during acceleration in first and fourth gear for a 2.8 L displacement volume, turbocharged SI engine (from Heireth and Withalm [6], copyright VDI)

2.2 Fuel Injection

2.2.1 Mechanical Fuel Injection

One of the most important engine processes that influences combustion and thus heat release rate, exhaust emissions (mainly particulate matters) and transient response is the fuel injection system. As was discussed in the previous section, without a properly matched transient fuel delivery on a cycle-by-cycle basis the low in-cylinder trapped mass will initially result in overshoot of the global fuel–air equivalence ratio, leading eventually to intolerable smoke emissions. Hence, the dynamic characteristics of the injection system should be adjusted and matched to the current engine fueling requirements. The fuel injection mechanism operates in a highly transient manner even during steady-state engine operation (*e.g.*, at constant engine speed and control lever position).

In modern, high-pressure common rail injection systems governed by the engine ECU, the extremely high injection pressure poses one more significant challenge for the fuel injection dynamic response. Nonetheless, a relatively fast adaptation of the injection parameters to the incoming air-charge is usually achieved,[4] provided that the measurement of air-mass flow-rate is made promptly and accurately; this is not the case, however, in conventional mechanical injection systems. During transients, the continuously changing engine/pump speed, control

[4] At engine start-up, it usually takes 3 to 4 injections to establish line pressure in the high-pressure section of the injection system. Afterwards, stable, repeated injection is assured.

lever position and pump residual pressure result in the (mechanical) fuel pump experiencing a transient operation of its own. This is mainly the result of transient fluid mechanics of the fuel pump equipment and the dynamic response of the mechanical components of the fuel injection system (governor, needle lift and timing device). Consequently, various injection parameters such as line pressure, rate of injection, timing and duration of injection are influenced, with a subsequent differentiation in the amount and the profile of injected fuel per cylinder compared with the values under 'similar' steady-state conditions [7–10]. The latter behavior ultimately affects engine response.

Murayama *et al.* [7], by studying the accelerating behavior of a single-cylinder diesel engine fitted with mechanical fuel injection pump, found that

- owing to rapid and considerable changes in fueling, instantaneous torsional deformations in the driving system of the fuel pump take place, leading to incomplete combustion;
- a retardation of the dynamic injection timing takes place too, which was correlated to injected fuel-mass change via a linear relation (Figure 2.13);
- the shape of the fuel injection rate changes compared with the respective steady-state operation. A shortening of the throttling period at the early stage of fuel injection and the increase in fuel injection rate following throttling, are more pronounced under acceleration than under the corresponding steady-state conditions (Figure 2.14a). Later in the transient, the fuel injection rate curve assumes a more 'rectangular' shape. Moreover, during transients, injection starts later in the cycle compared with the respective steady-state operation; this originates in the lower values of the residual pressure of the fuel delivery system (Figure 2.14b).

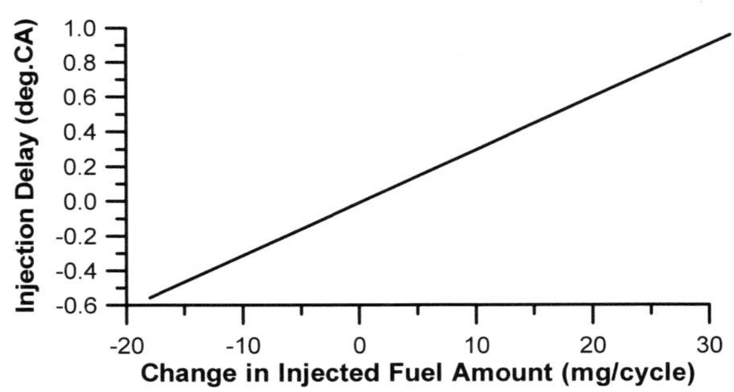

Figure 2.13. Change in the dynamic fuel injection timing due to the deformation of the injection pump driving system, resulting from changes in the amount of injected fuel (experimental results reprinted with permission from SAE Paper No. 800966 [7], © 1980 SAE International)

Although some of the above results were reached for naturally aspirated, indirect injection diesel engine operation, where the injection pressure are traditionally low, they have been confirmed for turbocharged DI engines too; in fact, even bigger discrepancies are expected for turbocharged diesel engines owing to the considerably higher difference between no-load and full-load fueling and to the much higher injection pressures involved.

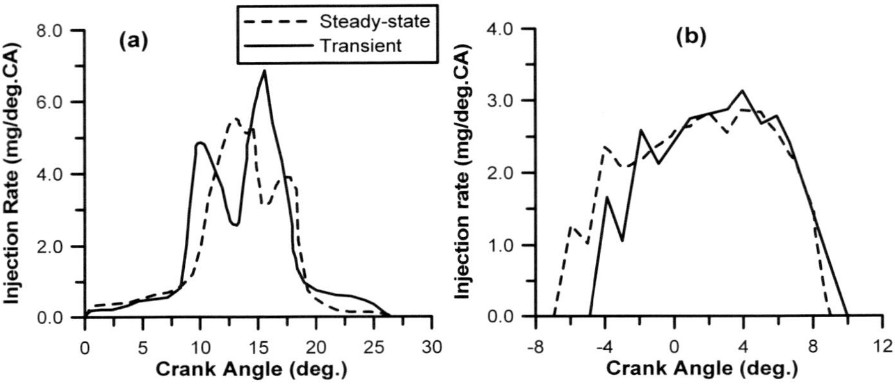

Figure 2.14. Change in fuel injection rate during an early cycle of an acceleration event compared with the respective steady-state operation. *a:* naturally aspirated engine (experimental results reprinted with permission from SAE Paper No. 800966 [7], © 1980 SAE International). *b:* turbocharged diesel engine (experimental results reprinted with permission from Harndorf and Kuhnt [11])

In order to focus on the dynamic injection characteristics, Arcoumanis and Baniasad [9] isolated the fuel injection system operation of a VE type distributor injection pump of an HSDI diesel engine and studied its transient response. Two dynamic operating conditions were examined mimicking 'real' engine transient operation, *i.e.,* 1. fuel pump setting (control lever) position varying from minimum to maximum in 13 ms at constant pump speed (termed 'fueling increase' transient), and 2. increasing pump speed at constant control lever position (termed 'acceleration' transient). From their analysis, the following were revealed.

During the fueling increase period at 2000 rpm engine speed (1000 rpm pump speed) 4 injection cycles were needed until the demanded fuel quantity is reached (Figure 2.15) (more cycles were needed for transients at higher engine/pump speeds). The transient injection rate signals (not shown in Figure 2.15) increased significantly in terms of the peak amplitude but only slightly in duration (*cf.* Figure 2.14), whereas the corresponding needle lift signal showed higher needle ascend velocities and longer signal durations due to the higher injection line pressures. Owing to the fluid and mechanical inertia of the system during these cycles, lower maximum needle lift, line pressure and injection rates were experienced compared with the demand conditions reached in cycle 4. The latter may result in poor fuel atomization and reduced spray penetration into the combustion chamber, as well as lower injected fuel quantities during the transient period in an operating engine. Actually, the injected fuel quantity was calculated to be 82, 86, 94 and 99%[5] of the demanded for the 4 transient cycles depicted in Figure 2.15.

However, in view of the slow build-up of air-supply to the engine cylinders, this gradual adjustment of fueling is favorable, as it prohibits very large fuel–air

[5] For a 'transient fueling' test at higher engine speed (2500 rpm), the respective percentages were 65, 73, 81, 86, 91, 97, 102 and 99% of the demanded fuel quantity. This leads to the conclusion that in terms of the injection cycle, the system has faster response at lower engine/pump speeds with the fuel quantity response deteriorating with increasing speed.

equivalence ratio overshoots during the first transient cycles. At the same time, residual pressure was, mainly, determined during the tail oscillation of the line pressure, and was dependent on the opening characteristics of the damping valve and the resultant reflected pressure.

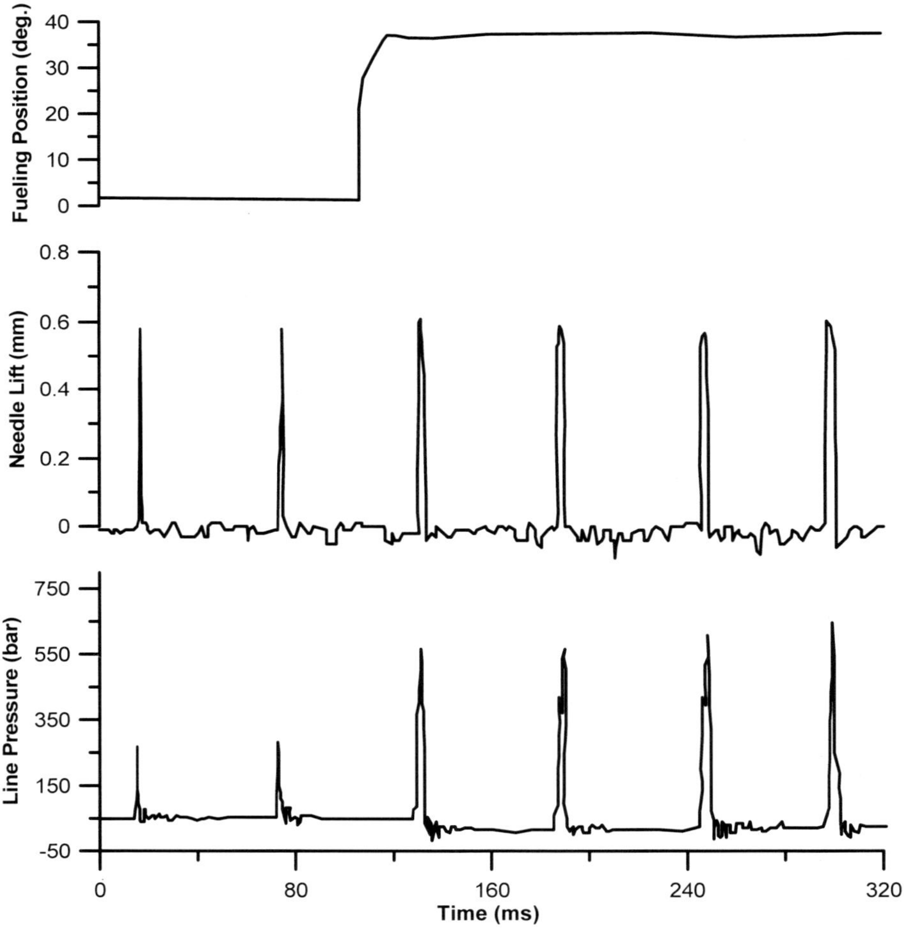

Figure 2.15. Response of speed control lever (fueling) position, needle lift and line pressure for a transient fueling test of a VE type distributor pump at constant speed (experimental results reprinted from Arcoumanis and Baniasad [9], with permission from Professional Engineering Publishing)

Catania *et al.* [10], studying a similar injection pump configuration, reached the conclusion too that the fuel injection system was capable of almost instantaneously adapting to the new stationary conditions in a few cycles, soon after the control lever was brought to its new steady-state position. They attributed the latter effect to the small sensitivity of the delivery pipe residual pressure to significant load variations and to the almost zero lasting cavitation effect on fluid properties at the start of the pump compression stroke. With the increase in pump speed, however, a slightly more intense cavitation[6] was observed, ultimately affecting the fuel fluid properties during the pump compression stroke.

During the increasing pump speed operation, residual pressures were found less than those at steady-state operation for lower pump speeds, but at higher pump speeds transient residual pressure exceeded its steady-state counterpart until final convergence towards the end of the transient event. At the start of the 'acceleration', transient injection advance timing suddenly dropped and recovered gradually with increasing pump speed (Figure 2.16 – upper sub-diagram). This trend in the transient injection timing, which was found independent of load, was attributed to: 1. advanced timing piston and spring mechanism response characteristics; and 2. fuel feeding pressure build-up in the pump housing cavity, being dependent on the pressure response of the fuel supply pump and the rate of pressure rise in the large pump cavity feeding chamber within the rather short transient time. The difference between transient and quasi-steady injection advance timing is one of the major contributors to the higher transient soot and NO_x emissions levels in diesel engines, to be discussed in Chapter 5.

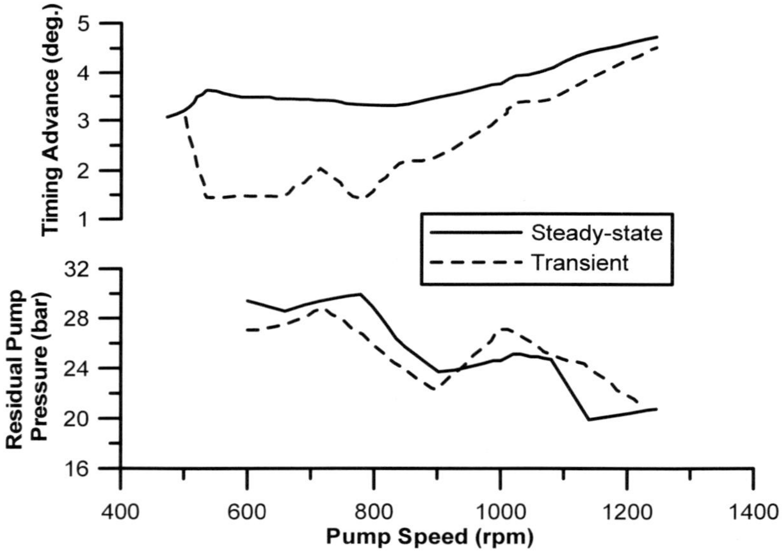

Figure 2.16. Comparison between transient and steady-state residual pressure and injection timing advance of a VE type distributor pump at constant control lever position (experimental results reprinted from Arcoumanis and Baniasad [9], with permission from Professional Engineering Publishing)

---

6 Cavitation in the pipe of a fuel injection system occurs when locally fuel pressure decreases below its vaporization value forming cavities; it is usually the more volatile components of diesel fuel that are prone to vaporization. Increased speeds and injection pressures as well as abrupt termination of injection are mainly responsible for creating vigorous wave actions inside the fuel pump mechanism leading to the above-mentioned local reduction of fuel pressure and thus formation of cavities. Vapor may subsequently condense inside the cavities leading to erosion and even fuel injection failure.

2.2.2 Fuel Limiter

Figure 2.17 illustrates a typical fueling/bmep *vs.* engine speed operating envelope for two turbocharged diesel engines in comparison with a naturally aspirated one. Irrespective of output, the problematic operation for turbocharged engines is located in the low-speed, medium-high load operating range, where the turbocharger compressor delivery pressure, hence, the air–fuel ratio are still low since the exhaust gas enthalpy is not yet capable of providing the required power to boost the compressor. Increasing the degree of turbocharging results in more abrupt low-speed air–fuel ratio curves, leading to even higher engine torque non-linearity. In the mid-range, engine output is, mainly, limited by mechanical considerations (high cylinder pressures) and in the high-speed range by turbocharger and engine over-speeding.

In view of the above, at low-speed, steady-state engine operation in order to avoid intolerable smoke emissions, special adjustment of the fueling system to the corresponding air-supply is required. This adjustment is usually realized via a fuel limiting function, which determines maximum fuel delivery according to the current turbocharger compressor boost pressure. Ideally, EGR effects should be taken into account too since the amount of recirculated exhaust gas lowers accordingly the fresh charge air–fuel ratio. As regards mechanical fuel pump/governor systems, the fuel limiting function can be accomplished by a damper system momentarily limiting fuel pump rack travel, or, more commonly, via a boost pressure sensitive device. In modern engines, the fuel limiting module is incorporated in the engine ECU together with the speed governor (*cf.* Figure 3.11b). The usual inputs to the fuel limiter are the compressor boost pressure and/or the air-mass flow-rate.

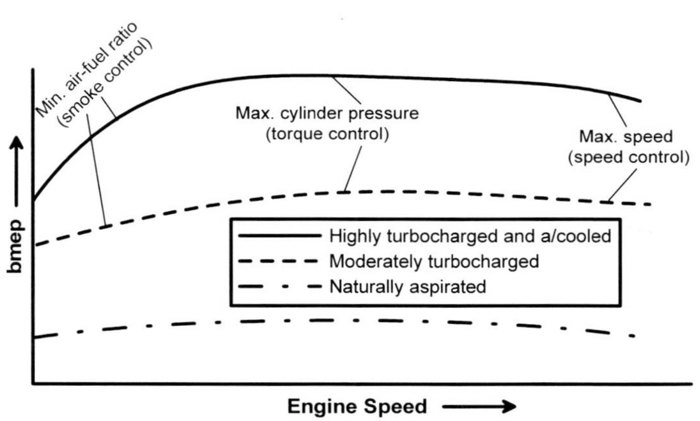

Figure 2.17. Envelope of maximum bmep *vs.* engine speed for steady-state operation of naturally aspirated and turbocharged diesel engines

Although a well designed fuel limiting device can actually prevent intolerable smoke emissions at steady-state engine operation, during turbocharger lag, much higher fuel–air equivalence ratios may be experienced. It is not surprising then that the steady-state matching often proves unsuccessful during transients. The most usual side effect is the transient torque deficit (*cf.* Figure 2.10) and the poorer engine response/acceleration; a very instructive comparative case study can be found in [4]. To describe the mechanism behind the trade-off between fast response and low smoke emissions, it is necessary to consider in more detail the way the fuel limiter operates.

The operating principles of a typical fuel limiter will be discussed with reference to Figure 2.18, which illustrates fueling and air-supply build-up during an acceleration transient event. Initially (point 1), the engine operates at or near idling conditions with the fuel–air equivalence ratio Φ_{idle} corresponding to fueling F_1 and air-supply A_1, both possessing very small values. Afterwards, the transient event commences, with the driver pushing the accelerator pedal to its maximum position. In case of a naturally aspirated engine, this would result to a direct increase in fueling (up to the corresponding final point). This is not the case, however, for the turbocharged diesel engine. Recall from the case study discussed in Section 2.1, that a turbocharged engine's maximum bmep is much higher than that of its naturally aspirated counterpart, hence maximum fueling corresponds to a considerably higher air-supply. However, owing to the various system delays analyzed in Section 2.1, air-supply build-up is, initially, very slow (1→3). In fact, for the first cycles after the acceleration commences, boost pressure and air-supply to the engine cylinders have not changed from their initial values.

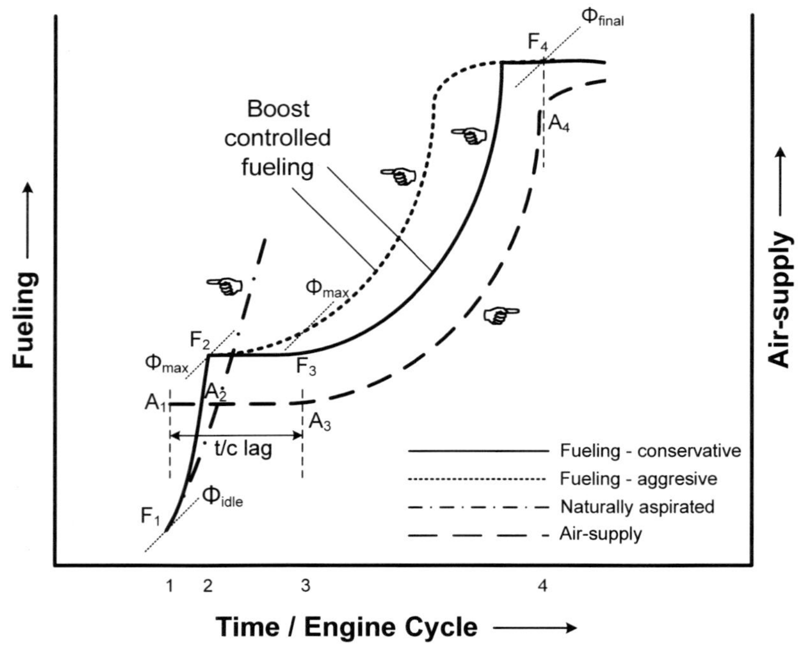

Figure 2.18. Typical fuel limit pattern of a turbocharged diesel engine acceleration event

During this period (turbocharger lag), the engine operates in a naturally aspirated mode. Because of this slow response, the fuel limiter allows, at first, only a modest increase in the injected fuel mass, namely from F_1 to F_2, to prevent excursion of the fuel–air equivalence ratio to very high values. At this point (2), maximum fuel–air ratio might be experienced. This fueling is then maintained, more or less, constant for the whole turbocharger lag period (up to point 3). As the acceleration event develops and the increased fueling produces exhaust gas of higher enthalpy, the turbocharger is gradually accelerated increasing the boost pressure and air-supply to the engine cylinders, curve A_3–A_4. During this period, fueling is steadily increasing too in proportion to air-supply, ultimately reaching point 4, which, usually, corresponds to the maximum fuel delivery for the current transient event. A critical parameter here, is the slope of curve F_3–F_4; the steeper the boost controlled fueling curve (dotted line in Figure 2.18), the faster the fueling response, hence engine torque build-up and speed response, but also the higher the fuel–air equivalence ratio and the respective soot emissions (most probably at point 3). In the case of fixed geometry turbines, air-supply at the final point 4 would correspond to the maximum value for the particular transient event. For variable geometry turbocharged engines, however, the gradual increase in exhaust manifold temperature will lead to a further increase in the turbine back-pressure, hence a slight increase of turbocharger speed will be accomplished, possibly differentiating the final conditions after several seconds [12,13]. In conclusion, fueling during a typical acceleration event is accomplished in three distinct phases as follows

- During the first phase (1→2 in Figure 2.18), the mass of injected fuel is rapidly increased to a value limited by the available air-flow owing to boost pressure lack. In this phase, fueling has increased at a much sharper rate than air-flow.
- During the second phase 2→3, fueling is practically maintained constant owing to the respective delay in the build-up of boost pressure during the turbocharger lag period.
- During the third phase 3→4, fueling is controlled by the engine ECU and it basically follows air-supply pressure history until the final equilibrium. A steeper 3→4 fueling curve will result in faster fueling increase and speed response, coupled, however, with excursion to high fuel–air equivalence ratios and soot emissions.

The above-described fuel limiting action primarily aims at limiting intolerable smoke emissions, which would otherwise be experienced if fueling were determined irrespective of air-supply. Nonetheless, the slow movement of the fuel rack towards full-fueling comes at a cost, namely slow engine torque build-up and speed recovery. The latter can be best documented with reference to Figures 2.19 and 2.20. Figure 2.19 focuses on the torque development during vehicle acceleration in fourth gear and can be studied in conjunction with the remarks regarding Figure 2.18.

Initially (point A at 1250 rpm) and following the driver's demand, the transient engine torque increased sharply owing to the rapid movement of the governor to its maximum value (point B). However, the air-supply could not match the fuel flow and the engine practically operated as naturally aspirated. The slow build-up of boost pressure was sensed by the fuel pump mechanism, which in turn limited the amount of injected fuel according to the current compressor boost pressure. Consequently, transient torque developed at a much slower rate compared with the steady-state curve (1300–2750 rpm), ultimately reaching point C_t, which corresponds to lower torque and at a higher engine speed than the respective steady-state point C_{st}. As a result, emission of excessive smoke has been prohibited but at the expense of poorer response. For naturally aspirated engines, the difference between points C_t and C_{st} would be much smaller, since the greatest of the above mentioned delays concerns the turbine acceleration [14].

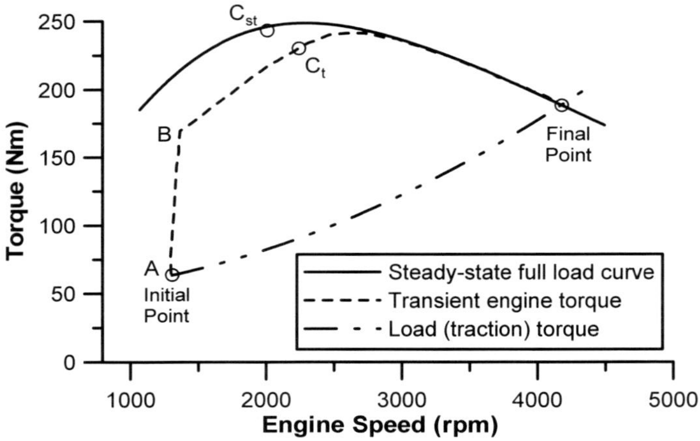

Figure 2.19. Curves of steady-state full-load torque, transient engine torque and load (traction) torque for vehicle acceleration in high gear, showing the effect of fuel limiting function (reprinted with permission from Woschni *et al.* [14])

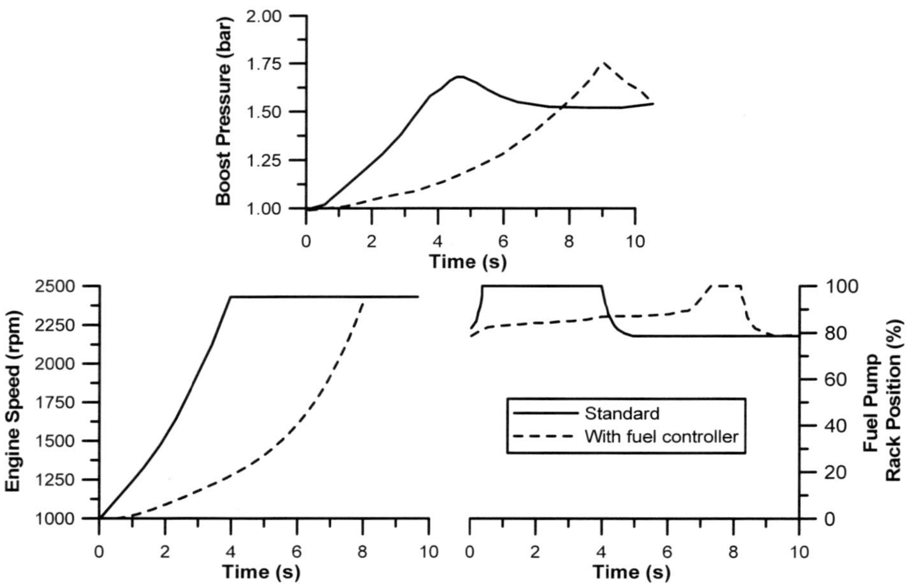

Figure 2.20. Acceleration of a turbocharged truck diesel engine against steady load with and without fuel limiting function (experimental data reprinted with permission from SAE Paper No. 810338 [4], © 1981 SAE International)

The effect of fuel limiting function on engine speed response is quantified in Figure 2.20, where the acceleration of a turbocharged truck engine of modest rating (10 bar maximum bmep) is compared with and without a system for controlling maximum fuel delivery as a function of boost pressure. Limiting maximum fuel delivery by 20% at zero boost is capable of reducing the total amount of emitted smoke owing to the fuel pump rack moving slowly towards full-fueling. However, the response time of the engine and turbocharger properties is doubled resulting in much poorer acceleration.

Further limitation of the fuel flow may prove critical for the engine's capability to recover high load changes leading even to engine stall, especially when the load application is very fast. This is, particularly, the case for industrial or marine applications, where the engine ratings as well as the load changes experienced are quite large. In view of the above, it seems reasonable to permit at least a 10% over-fueling margin for industrial engines that, furthermore, do not encounter transients that often, compared with vehicular applications. When the engine accelerates from an initial high load, smoke reduction from fuel limit action is generally negligible; on the other hand, the advantages of (slight) over-fueling with respect to successful engine speed recovery are considerable. Of course, the case is quite different when automotive engines are involved. The use of fuel limiters is here compulsory in order to conform with the stringent emission regulations. Nonetheless, other measures would have to be employed in order to deal with the root of the problem, *i.e.,* minimize turbocharger lag, hence simultaneously enhancing vehicle drive-ability and lowering emissions (Chapter 6).

2.3 In-cylinder Processes

2.3.1 Heat Transfer

At the onset of a transient event, cylinder wall temperature corresponds to the steady-state conditions, being mainly determined by engine loading/fueling. The following equation describes the respective heat flux from the gas to the cylinder wall

$$\frac{dQ_L}{dt} = A\, h_g (T_g - T_w) \qquad (2.3)$$

where A is the respective surface area, h_g the heat transfer coefficient and T_g, T_w the instantaneous gas and wall temperatures, respectively. During the early cycles of the transient event, the gradual increase in fueling is transformed into higher gas temperatures T_g and heat transfer coefficients h_g. Consequently, higher heat transfer rate dQ_L/dt to the walls is noticed, since the cylinder wall temperature T_w cannot adjust instantly to the new fueling conditions but only after a time delay owing to the wall's thermal inertia. The latter behavior is one of the contributors to the engine response delay as was discussed in Section 2.1 and identified in

Figure 2.9. Later in the section, it will be shown that this time delay can be quite extensive, typically one order of magnitude greater than the actual duration of the load or speed increase transient event or the time needed for the engine speed to reach its new final value. Among other things, this behavior affects also the evolution of (transient) combustion noise.

In order to evaluate the wall's thermal inertia, which determines the wall's temperature response, a heat transfer scheme, based on electrical circuit analogy, can be applied. This models the temperature distribution from the gas to the cylinder wall up to the coolant. By so doing, the cylinder wall thickness, thermal conductivity and thermal diffusivity are taken into account. For the heat transfer inside the cylinder wall, the one-dimensional unsteady heat conduction equation reads

$$\frac{\partial T}{\partial t} = \alpha_w \frac{\partial^2 T}{\partial x^2} \tag{2.4}$$

with $\alpha_w = k_w/(\rho_w c_w)$ the wall thermal diffusivity, ρ_w the wall density and c_w its specific heat capacity. Applying the boundary conditions to all wall sides (gas side and coolant side) of a four-stroke diesel engine, the following equation is valid with reference to Figure 2.21:

$$\frac{1}{4\pi} \int_0^{4\pi} \frac{dQ_L}{d\varphi} d\varphi = A \frac{k_w}{L_w} \left(\overline{T}_{w,g} - \overline{T}_{w,c} \right) = A h_c \left(\overline{T}_{w,c} - T_c \right) \tag{2.5}$$

where L_w is the cylinder wall thickness with k_w its thermal conductivity h_c the heat transfer coefficient from the external wall side (respective temperature $\overline{T}_{w,c}$) to the coolant; the overbar denotes mean temperatures over an engine cycle.

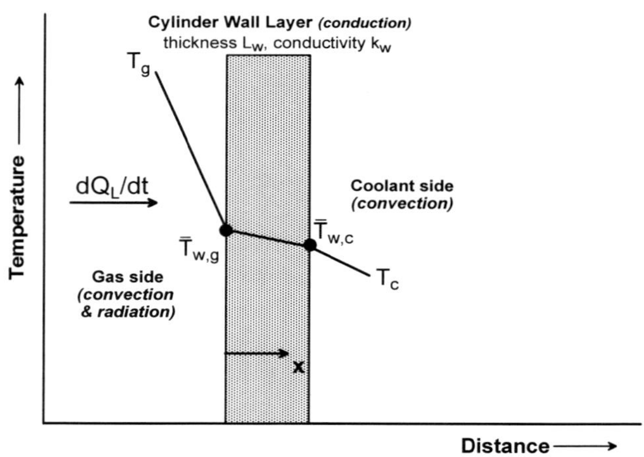

Figure 2.21. Schematic arrangement of cylinder wall – coolant heat transfer scheme

Representative results quantifying the influence of the above mentioned wall thermal balance are reproduced in Figures 2.22 and 2.23, showing cylinder wall and cylinder liner thermal equilibrium procedure after a load increase and a speed increase transient event, respectively. Measurements were accomplished using special 'J'-type thermocouples, which comprised of two thin insulated wires each having a 0.003" diameter. The two wires were fixed inside a hollow quartz tube of 1.75 mm outside diameter with special high temperature cement and fastened inside their position on the cylinder head or liner with the aid of the same material. Point TC1 in Figures 2.22 and 2.23 refers to the thermocouple installed on the cylinder head between two fins, in contact with the outside metal surface, TC2 refers to the thermocouple installed on the cylinder head interior at a distance of 0.6 mm from the fire deck, while TC3 thermocouple was installed at the valve bridge interior at a distance of 10 mm from the fire deck. In a similar way, the cylinder liner thermocouples were installed at a distance of 5 mm from the gas side surface, at two opposite positions across the cylinder bore on the valve plain, one on the exhaust valve side (TC4) and the other on the inlet valve side (TC5).

As is depicted in Figure 2.22, the steepest slope is observed for the valve bridge thermocouple (TC3), while the most moderate response is at the outside surface of the cylinder head point (TC1). As expected, the valve bridge being one of the most sensitive areas on the cylinder head, suffers from thermal distortion caused by sharp temperature gradients during a transient event (thermal shock). The important finding from Figure 2.22 is that even 250 s after initiation of the load increase, the temperatures have still not reached their final steady-state values. Hence, during the few seconds of the transient event (up to the point of engine speed recovery) transient heat loss to the cylinder walls is higher than under the respective steady-state conditions. For turbocharged engines, this behavior would lower the available exhaust gas energy to the turbine and further contribute to the turbocharger lag delay. For the cylinder liner alone, and due to the air-cooling of the engine, the temperature response (TC4 and TC5) is rather unstable although the main finding of slow thermal equilibrium is confirmed.

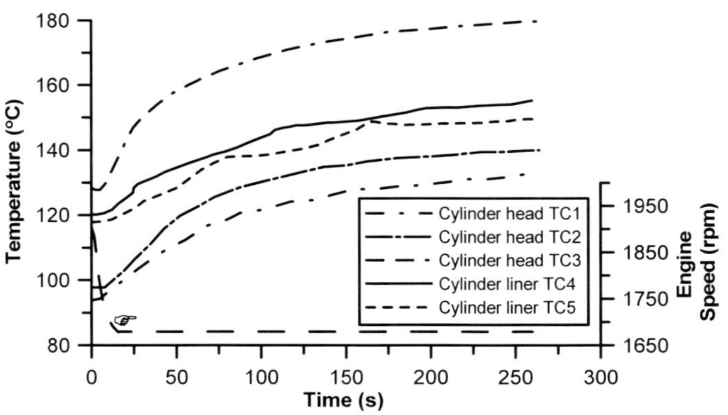

Figure 2.22. Experimental temperature profiles *vs.* time during a 12–70% load increase transient event of a single-cylinder, air-cooled, naturally aspirated diesel engine [15]

Similar observations are valid for the case of speed increase transients, as is illustrated in Figure 2.23. Temperature levels, for all positions studied, present smaller differences when comparing initial and the final operating conditions.

Although fueling reached 100% for a few seconds during the initial phase of the acceleration, the high thermal inertia of the cylinder structure delayed significantly the wall temperature response, hence the highest temperatures observed actually correspond to the final fueling of 30% rather than to the instantaneously maximum 100%. Again, the steepest thermal gradient is noticed in the valve bridge area. Contrary to the (usually much more demanding) load acceptance case, thermal equilibrium is achieved now after 120 s for all positions examined on the cylinder head and liner.[7]

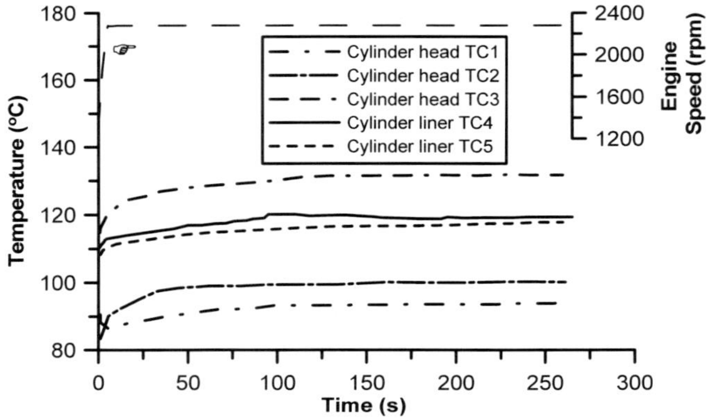

Figure 2.23. Experimental temperature profiles *vs.* time during a low–medium load, 1300–2300 rpm acceleration event of a single-cylinder, air-cooled, naturally aspirated diesel engine [15]

A similarly slow, but not that long in duration, response is observed for the heat transfer rate to the exhaust manifold walls (Figure 2.24). Evaluation of the exhaust gas temperature depends strongly on the location of the thermocouple, which for the analysis of Figure 2.24 was located at the exhaust manifold at an approximate distance of 0.2 m from the corresponding valve.

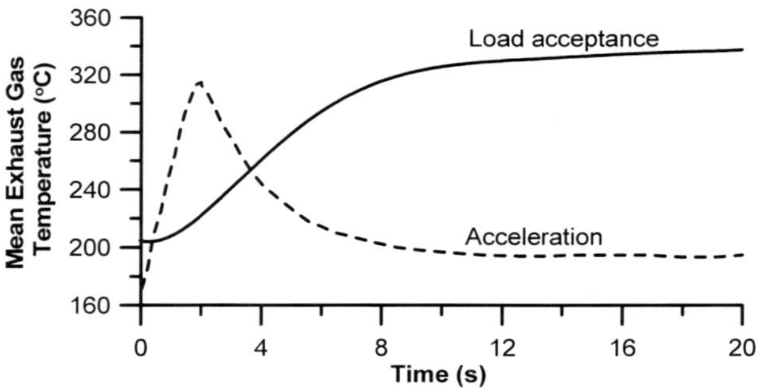

Figure 2.24. Experimental mean exhaust gas temperature *vs.* time for the load acceptance and acceleration transient events of Figures 2.22 and 2.23 [15]

---

[7] The initial drop in TC1 temperature is caused by the increase in engine speed at the first five seconds of the acceleration, which increases the air velocity through the fins and the respective heat transfer coefficient with a simultaneous decrease in air temperature.

Further aspects of transient heat transfer phenomena will be discussed in Section 5.7 concerning evolution of combustion noise, and in Section 7.3 concerning low-heat rejection engine operation.

2.3.2 Combustion

Diesel engine combustion comprises a variety of physical and chemical processes being typically characterized by a high degree of non-uniformity. In a diesel engine, the overall air–fuel ratio is always weak of stoichiometric, in order to achieve complete combustion; this is a consequence of the rather limited time available for mixture preparation. The fuel is injected in liquid form as nearly conical jets (sprays) into the engine cylinder, under a very high-pressure difference between injection line and cylinder gas. However, before it can be burned, atomization and spray penetration into the compressed air must be established that will enable heating, evaporation and mixing of the fuel with a sufficient quantity of air entrained from the surrounding air via a diffusion process. At the same time, pre-flame oxidation and localized ignition occur that prepare chemically the fuel–air mixture for burning by decomposing the heavier hydrocarbons into lighter components, forming combustion radicals and facilitating pre-ignition chemical reactions. The above physical and chemical processes are referred to collectively as 'preparation', with the temperature and pressure of the compressed air prior to fuel injection playing here an important role and air-swirl promoting the fuel–air mixing rate. It is the droplets at the outer edge that evaporate first creating a fuel–air vapor film around the still liquid cone jet. The fuel–air equivalence ratio is highest on the centerline of the jet, decreasing to zero at the boundaries (Figure 2.25). The sprays eventually hit the cylinder wall or piston and mix together.

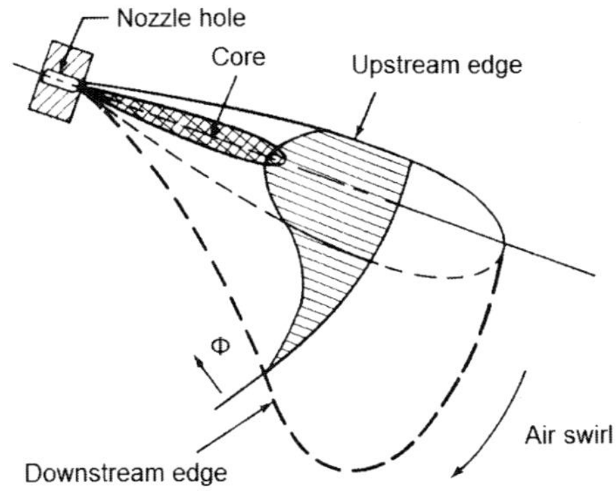

Figure 2.25. Schematic of fuel spray injected radially outward from the chamber axis into swirling air-flow; shape of fuel–air equivalence ratio distribution within jet is indicated (from Heywood [16], © 1988, reproduced with permission of the McGraw-Hill Companies)

The prepared fuel–air mixture is then under the control of a chemical process and may be burned (reacted) at a rate that can be calculated by a chemical kinetics equation. Diesel combustion in conventional type DI or IDI engines is primarily influenced by the inducted air properties (density, swirl, temperature) as well as by the injection system characteristics (number and pattern of injections, injection pressure and rate, fuel cetane number, number and diameter of nozzle holes). A variety of other parameters such as compression ratio, cylinder wall temperature, intake system design, valve configuration, combustion chamber geometry, *etc.*, affect strongly the whole process development. Summarizing, for cylinders with single injection strategy, diesel engine combustion comprises the following stages:

- ignition delay period, where the injected fuel is physically and chemically prepared and mixed with air;
- a rapid premixed burning phase of the already prepared mixture during the previous ignition delay period; this stage is usually characterized by a high rate of gas pressure increase and is, mainly, responsible for radiation of combustion noise; and
- a slower, mixing- or diffusion-controlled burning phase, where the burning rate is not governed by chemical kinetics but, rather, by the fuel injection and subsequent turbulent mixing rates.

Application of multi-hole injectors, pre- and pilot injections, exhaust gas recirculation, swirl, and the use of bowl-in pistons complicate the development and modeling of combustion even more. As a result, advanced multi-dimensional simulation models as well as optical engine cylinders are required for complete understanding of the heterogeneous in nature in-cylinder processes in a compression ignition engine; for transient operation studies, the use of advanced multi-dimensional smulation models is, at the moment, prohibited due to the need for simulating a large number of engine cycles up to the final equilibrium conditions. A thorough review of diesel engine combustion principles during steady-state operation can be found, for example, in Heywood [16] and Benson and Whitehouse [17]. In this book, only the differentiations that are experienced during transients compared with the respective steady-steady operation will be discussed.[8] These are primarily the result of 1. the dynamic response characteristics of the fuel injection system, 2. the different quantity and pressure of air-supply owing to the previously discussed turbocharger lag (Section 2.1), which affects directly fuel–air ratio, heat release rate and exhaust emissions, and 3. the lower cylinder wall temperature compared with the respective steady-state operation as discussed in the previous section.

Figure 2.26 shows data for a turbocharged DI diesel engine's response to a step increase in engine load. Cycles No. 6–11 of the transient event have been isolated and the development of engine and exhaust manifold pressures, turbine inlet temperature, engine speed, fueling, mean boost pressure, and relative air–fuel ratio are illustrated with respect to crank angle or time.

---

[8] By the term *respective* steady-state operation, it is meant operation at the same engine speed and fuel pump rack position as the corresponding transient cycles; however, as was discussed in Section 2.1, differentiation in air-mass flow-rate, hence, fuel–air equivalence ratio exists that affects accordingly the engine processes.

It is obvious how the increased fueling and boost pressure are gradually transformed into much higher cylinder gas pressures, hence exhaust manifold pressures and temperatures (not shown on Figure 2.26); maximum cylinder pressure increases from 57.5 bar (cycle No. 6) to 83.8 bar (cycle No.11). This is a 45% increase in peak cylinder pressure resulting in an equally high rate of bearings loading (Section 3.1) and engine construction stress. Relative air–fuel ratio assumes very low values during the depicted cycles, which is expected to increase smoke emissions.

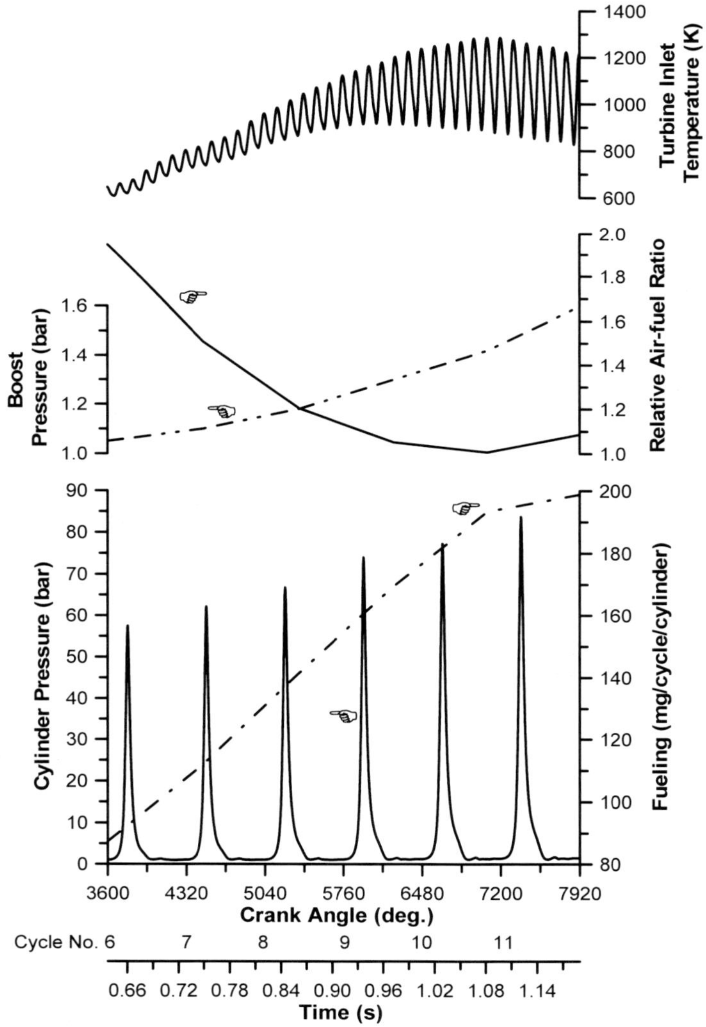

Figure 2.26. Development of various engine properties during cycles No. 6–11 after a 10–90% step increase in engine load (experimental and simulation results for a DI, turbocharged and aftercooled diesel engine with pulse turbocharging configuration and mechanical fuel injection pump without fuel limiter)

During the early transient cycles after a load or speed increase, the higher pressure, larger volume fuel jets are injected into an environment that is practically unchanged from the condition before the transient commenced, thus the higher-momentum fuel jet is not accompanied by enhanced gas motion. The rate of mixture preparation is, therefore, reduced and the heterogeneity of the mixture is increased. Figure 2.27 illustrates in the most explicit way the difference between steady-state and transient spray formation for a typical acceleration transient event compared with the respective steady-state conditions. Clearly, the rate of fuel vaporization during transients is much lower, being responsible for worse mixture preparation and higher rate of wall impingement [11,13].

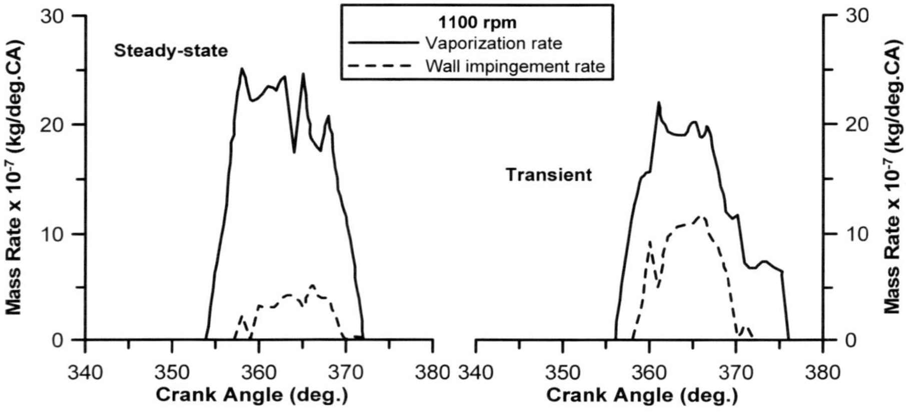

Figure 2.27. Comparison of the behavior of a spray jet during steady-state and transient conditions at the 1100 rpm operating point (experimental results reprinted with permission from Harndorf and Kuhnt [11])

In general, the following off-design phenomena are experienced during combustion of the early transient cycles

1. an influence in mixture formation due to air-deficiency (for the increased fueling) caused by turbocharger lag;
2. a change (increase) in the average fuel droplet diameter caused by the lower density and swirl compared with the respective quasi-steady conditions, which leads to increased jet penetration; over-penetration results in impingement of liquid fuel on the cooler cylinder surfaces, lowering mixing rate;
3. the lower end gas and wall temperatures in combination with a higher amount of end gas result in an increased ignition delay and hard combustion course, during the early cycles, where the turbocharger lag is more pronounced.

Winterbone and Tennant [18] worked on a six-cylinder, turbocharged diesel engine and analyzed pressure–crank angle data after load increase transients, using the Whitehouse–Way combustion model [19] (see also Section 9.3.2.3). This model concerns a two-part expression; specifically for the preparation rate P (kg/°CA), it holds

$$P = K \cdot m_{fi}^{1-x} \cdot m_u^x \cdot p_{O_2}^y \tag{2.6}$$

with p_{O_2} the partial pressure of oxygen in the mixture, $m_{fi} = \int (dm_{fi}/d\varphi)d\varphi$ the mass of injected fuel up to the crank angle φ considered, with injection rate $dm_{fi}/d\varphi$ given in Section 9.8, $m_u = m_{fi} - \int P d\varphi$ the mass of unprepared fuel, and K, x and y are constants derived after calibration against experimental data. This relation was initially derived on the assumption that the rate of preparation is proportional to the surface area of all the droplets having uniform diameter, while proportionality with respect to the droplets diameter has also been proposed. A variable value of the exponent x makes allowance for this fact and for the usually existing non-uniform droplets' diameter distribution. The last term on the right hand side of Equation 2.6 allows for the effect of oxygen availability on mixing. Winterbone and Tennant concluded from their experimental analysis of transient pressure diagrams that the combustion process was somehow deteriorated after a load increase. Further to the discrepancies discussed in the previous page:

4. Reduction of the injected fuel preparation rate constant 'K' of the Whitehouse–Way model (Equation 2.6) was observed, indicating that, other things being equal, the preparation proceeds more slowly. This can be attributed to the different evolution profile of injection rate (Figure 2.14), particularly to the shortening of the throttling period at the early stage of fuel injection that affects the premixed combustion phase.
5. A reduction in the exponent 'x' was also detected, which was attributed to poorer mixing owing, probably, to over-penetration resulting in deposition of the fuel on the piston (this can be attributed to the dynamic response of the fuel injection system as was discussed in Section 2.2.1). The reduction of exponent 'x' is then responsible for a reduction of the mass transfer rate between fuel and air and, hence, slowing down of the whole combustion process.

Figure 2.28 expands on the in-cylinder properties evolution of Figure 2.26, going one step further and focusing on the rate of heat release (RoHR) development during the same cycles of the 10–90% load increase. What is obvious during the early cycles of the transient event is a quite sharp rate of premixed heat release (for cycles No. 6 and 7); even in modern common rail systems with pilot injection, the above phenomenon cannot be completely eliminated.

As the boost pressure and injected fuel quantity increase, the peak cylinder pressures increase as well (Figure 2.26), while injection is advanced in order to allow optimal combustion of the larger fuel quantity. Hence, for the rest of the cycles depicted in Figure 2.28, a lower initial HRR peak and a higher duration of diffusion combustion are noticed, mainly, attributed to the slight increase in cylinder wall temperature as the transient event develops, and to the increased quantity of injected fuel.

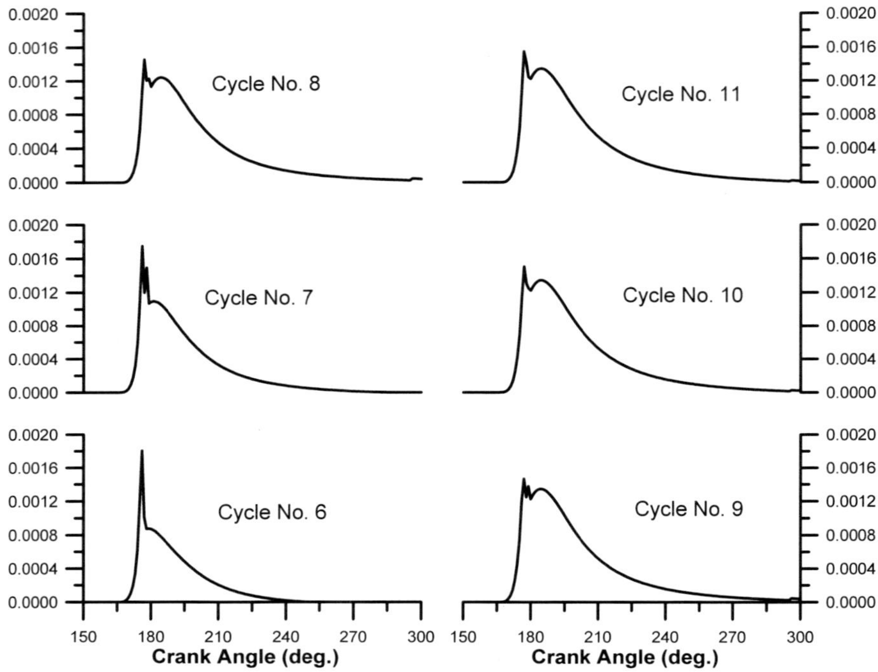

Figure 2.28. Rate of heat release (kJ/deg. CA) diagrams for cycles No. 6–11 of the 10–90% load increase transient event of Figure 2.26

The influential parameter that is responsible for the heat release behavior in Figure 2.28 is the ignition lag during the transient response. The ignition delay period is composed of a physical delay, incorporating liquid fuel atomization, vaporization and mixing, coupled with a simultaneous chemical delay process; the latter is the result of pre-combustion (decomposition and oxidation) reactions in the fuel–air mixture. It is the mixture preparation that is mainly responsible for the physical delay prior to ignition, with the air-charge pressure and temperature playing a primary role. Ignition delay has a direct impact on the intensity of heat release immediately after mixture auto-ignition. A relative reduction in the in-cylinder gas density due to turbocharger lag may cause less vigorous spray break-up and fuel–air mixing, thus resulting in increased ignition delay period and particulate matter emissions; measurements have documented this behavior, as is shown in Figure 2.28 and further quantified in Figure 2.29.

Consequently, the synergistic effect of higher fueling with still low wall temperatures and charge air pressure during turbocharger lag, ultimately, results in

- longer ignition delay period during the early cycles of the transient event; in fact, as is illustrated in Figure 2.29, the transient ignition lag is still higher than the respective steady-state conditions would dictate, even after 3 s through the specific 850–2000 rpm acceleration transient event, although a strong decreasing trend is established after the first cycles; and

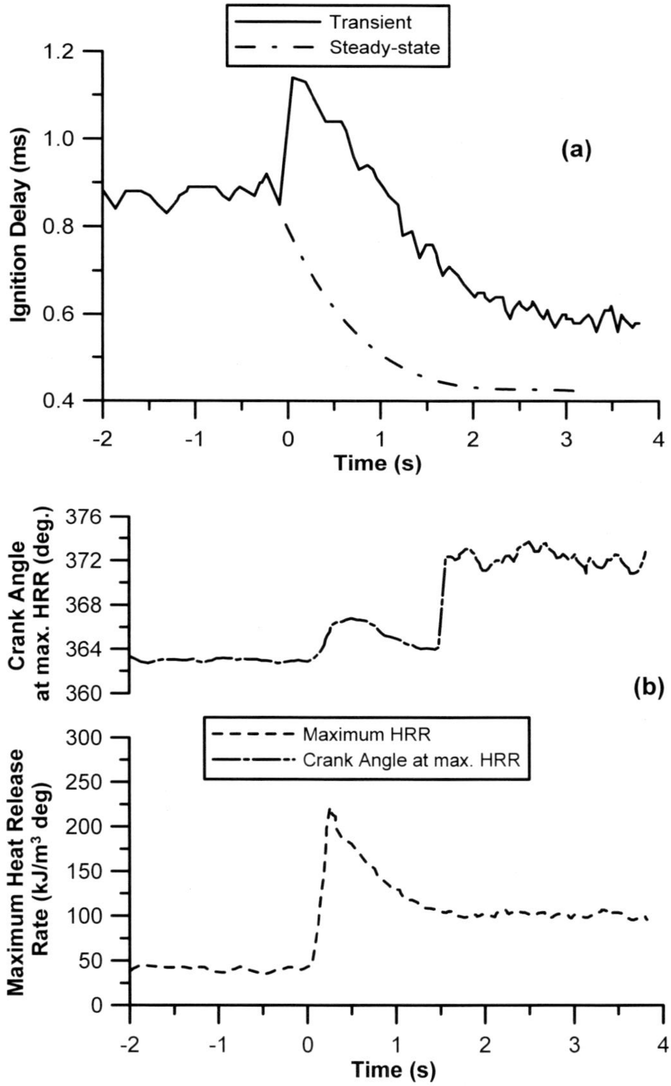

Figure 2.29. *a:* Comparison between steady-state and transient ignition delay; *b:* maximum heat release rate and crank angle at maximum HRR during an acceleration transient event (experimental results reprinted with permission from Harndorf and Kuhnt [11])

- significant and abrupt initial peak in the amount of premixed heat release.[9] In general, the peak magnitude of the heat release rate during the premixed combustion phase directly correlates with the magnitude of the maximum

[9] The duration of premixed combustion can be defined as the crank angle duration between the start of combustion and the minimum heat release before the start of diffusion combustion. Likewise, the duration of diffused combustion can be defined as the period from the end of the premixed phase up to the time where heat release crosses the zero line.

pressure rise. Therefore, the fact that during transients, turbocharger lag promotes longer ignition delay and a higher premixed combustion rate is reflected into an increase in the radiated (combustion) noise and smoke emissions as will be discussed in detail in Chapter 5.

Later in the transient, the distribution profile changes with the premixed peak being much smaller and the diffusion-burning phase becoming predominant; actually, diffusion burn fraction has been calculated to be at least 90% of the total during the late cycles of a transient event. The latter behavior is responsible for the gradual shift of the point of maximum HRR later in the cycle (upper sub-diagram of Figure 2.29b).

Ignition delay is typically defined experimentally as the time between start of (dynamic) injection and start of combustion; the former is determined as the start of injector needle movement, whereas the latter can be defined as the start of cylinder gas pressure rise (first or second derivative of gas pressure signal). Traditionally, constant volume vessels or bombs have been used for ignition delay measurements, ignoring the effect of piston movement. Steady-state ignition delay data are usually correlated by Arrhenius type equations of the form

$$\tau_{id} = C_1 \, p_g^{-C_2} \exp\left(\frac{E_a}{R_{mol} T_g}\right) \tag{2.7}$$

where τ_{id} is the ignition delay time in ms, p_g and T_g are the integrated mean gas pressure and temperature during the ignition delay, E_a is an apparent activation energy for the fuel auto-ignition process, R_{mol} is the universal gas constant, and C_1, C_2 are constants depending on the fuel and the injection characteristics. Based on the fuel–air equivalence ratio Φ discrepancies during transients compared with steady-state operation, Assanis *et al.* [20] proposed an updated version of Equation 2.7 valid under transient conditions

$$\tau_{id} = C_1 \, p_g^{-C_2} \, \Phi^{-C_3} \exp\left(\frac{E_a}{R_{mol} T_g}\right) \tag{2.8}$$

By including the global equivalence ratio dependence, it was postulated that Φ is actually a measure of the probability of finding local pockets of fuel–air mixture within flammability limits for auto-ignition sites to be promoted. The adjustable constants in Equation 2.8 were fitted in order to minimize the least-square error between measured and correlated ignition delay, as well as to ensure that the latter would tend to zero at extremely high load. The ignition delay data were correlated as a function of the overall equivalence ratio and the mean pressure and temperature over the ignition delay interval. This ignition delay formula gave satisfactory results, mainly for medium to high engine speeds.

Summarizing the previous remarks, Figure 2.30 illustrates the various discrepancies/anomalies encountered during the early seconds of a turbocharged diesel engine load increase or acceleration transient event.

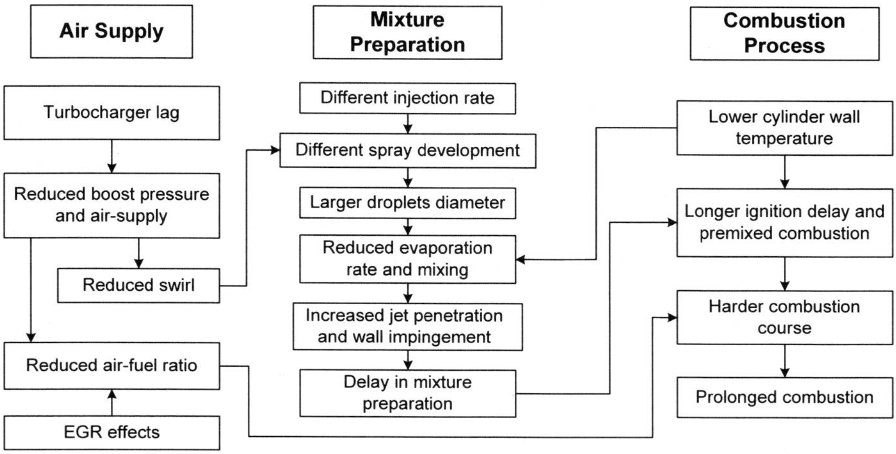

Figure 2.30. In-cylinder discrepancies during the early cycles of a turbocharged diesel engine transient event (adapted from Harndorf and Kuhnt [11])

With the gradual acceleration of the turbocharger, hence build-up of air-supply and delivery pressure (typically, after 15–30 cycles for an HSDI engine), combustion improves and the previously discussed anomalies diminish. At this point, (slight) retardation of the injection timing might be imposed by the engine ECU in order to control the already high gas pressures.

2.4 Variable Geometry Turbine

One of the main causes for the unsatisfactory transient response of turbocharged diesel engines lies in the fact that the matching between engine and turbocharger cannot be optimum throughout the whole engine operating range. Whereas internal combustion engines are designed to operate at wide air-mass flow range, the performance of turbo-machines is very much dependent on the respective gas angles and it deteriorates when moving away from their design point. Automotive diesel engines employ the pulse turbocharging configuration. Consequently, the turbocharger turbine is called upon to operate under non-steady flow conditions, with a two or even three scroll arrangement creating different instantaneous flow regimes in each sector. During an engine cycle, the variation in air-mass flow-rate is quite large, sometimes ranging from zero flow to choke; this variation depends on the number of engine cylinders and their disposition, degree of turbocharging, exhaust manifold geometry, *etc*. Since the instantaneous mass flow reaching the turbine blades varies substantially, the resultant gas angles will definitely depart from their design values, creating various losses, namely incidence and boundary layer separation. Moreover, the turbine practically operates in a highly transient manner even at steady-state engine operation, since the instantaneously high gas flow tends to accelerate it, whereas reduction in the air-flow produces the opposite effect (Figure 2.31). From the above remarks, it is made clear that since the flow and operating characteristics of a turbocharger (turbine) are fundamentally different from those of the reciprocating engine, complex matching issues arise leading also to problematic transient response.

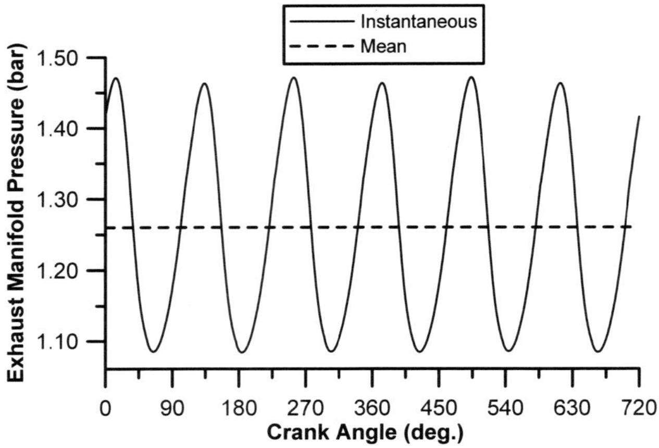

Figure 2.31. Instantaneous and mean exhaust manifold pressure during steady-state engine operation of a six-cylinder, pulse turbocharged diesel engine (simulation results)

Usually, engine operation is optimized at or around its design point/conditions. For example, an engine driving an electrical generator for base load operation will be optimized at full-load conditions and the turbine nozzle area will be typically selected for optimum performance at this operating point. When moving away from the design point however, *i.e.,* at part load operation, the turbine nozzle area will be too large for the reduced gas flow and temperature, reducing accordingly the turbine available energy. As a result of this behavior, shortage of air-supply to the engine cylinders will be established (particularly so if the EGR rate is kept at a high level for successful NO_x emissions control), as the power produced by the turbine will be reduced at a much higher rate.

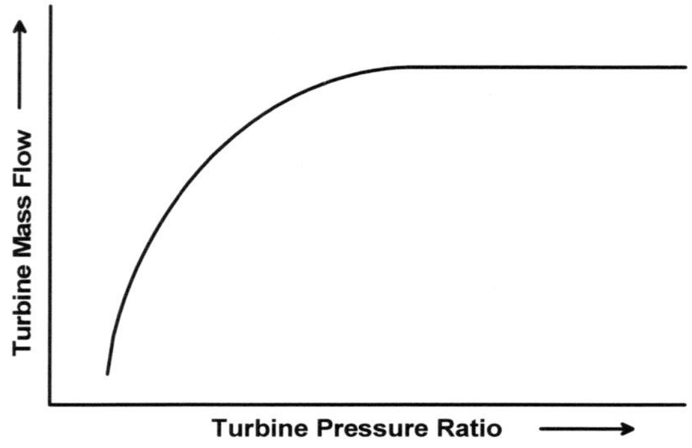

Figure 2.32. Typical curve of axial turbine expansion ratio *vs.* mass flow

This will deteriorate engine performance at steady-state conditions by lowering the allowable amount of injected fuel for smoke-free operation, thereby reducing bmep. Consequently, transient response will suffer too. The previously mentioned disproportionality emanates from the typical nozzle-type relation between turbine expansion ratio and mass flow that is depicted in Figure 2.32. Actually, it has been argued [5] that a linear increase in turbine expansion ratio with volumetric flow-rate would be ideal, but this cannot be realized with a conventional exhaust gas turbocharger of fixed geometry. Obviously, in order to accommodate the different operating requirements induced at part load operation, a reduction of turbine nozzle area would be needed.

Figure 2.33. *Upper left:* Variable geometry turbocharger with electric vane actuation for automotive use (courtesy of Toyota Motor Europe). *Upper right:* Cutaway of automotive turbocharger showing the swing blade angle mechanism (courtesy of Audi AG). *Lower left:* VGT arrangement for radial turbine, featuring nozzle rings with adjustable vanes (courtesy of MAN Diesel); *Lower right:* Adjustable turbine nozzle blades/variable turbine geometry of an ABB TPS57 turbocharger; the clearances for the movable nozzle blades are reduced by springs that push the blades against the opposing casing wall (courtesy of ABB Turbo Systems Ltd)

On the other hand, optimizing engine performance for low-load operation (*i.e.*, by selecting a smaller turbocharger frame) would result in poorer efficiency overall as well as small speed range, over-boosting and turbine flow choking at high engine speeds and loads. It follows then that, ideally, a continuously variable turbine nozzle area would be needed to adjust the turbocharger characteristics to the specific engine operating point (preferably approaching the linear expansion ratio *vs*. mass flow-rate relation), hence maximize delivery pressure and air-supply to the engine cylinders. This holds particularly true for automotive applications, which are characterized by wide speed and load operating range as well as by frequent transient events; another potential application is high bmep engines, where the increased rating enhances the previously discussed matching discrepancies between no-load and full-load conditions. The adjustment of the turbine nozzle area can be realized through a variable geometry turbine (VGT, Figure 2.33) that alters the whole turbine map by pivoting the swing blade angles or by keeping the nozzle blades fixed and changing the inlet area by moving the nozzle sidewalls. A variable geometry turbine can be considered, therefore, as a series of different turbine frames incorporated in a single unit, with variable flow characteristics but constant moment of inertia.

Variable geometry turbines are quite flexible in their operation and can prove beneficial in terms of engine performance at both steady-state and transient conditions; on the contrary, during engine cold starting, the gains obtained are much smaller. During steady-state operation, a small turbine nozzle area will increase the overall boost pressure level (this is primarily desirable at low-load conditions), establishing a less abrupt low-speed torque curve (Figure 2.34). By so doing, the low-speed, smoke-limited operating range of a turbocharged diesel engine can be significantly reduced. The overall higher air-mass flow-rate and air–fuel ratio will reduce smoke emissions, and the high exhaust gas pressure will ensure that a positive pressure difference exists between exhaust and inlet manifold supporting exhaust gas recirculation. When the engine speed and load are high enough, the blades are opened (nozzle area is increased) reducing gas velocities and preventing the excessive increase in boost pressure, which may harm the engine and lead to turbocharger over-speeding.

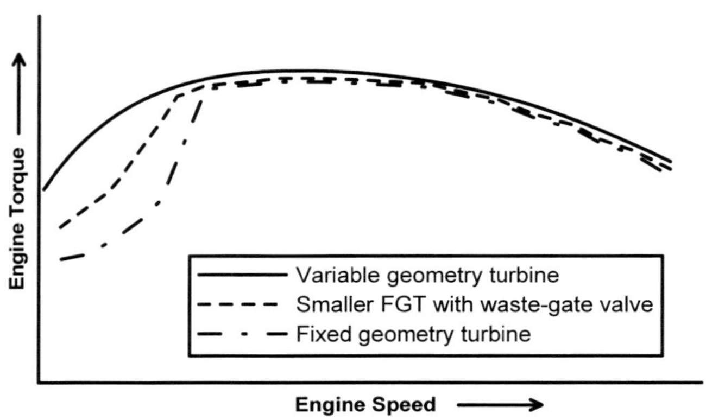

Figure 2.34. Improvement in low-speed torque from VGT operation compared with conventional (fixed geometry – (FGT)) or waste-gated turbocharging

A serious drawback that accompanies variable geometry turbine operation is the fact that the increased exhaust manifold pressure compared with its inlet manifold counterpart leads to back-flow of exhaust gas into the engine cylinders during the valve overlap period requiring a differentiated valve timing schedule. Moreover, pumping work is increased, worsening engine efficiency. An additional drawback is the, usually, lower turbine efficiency at the small nozzle areas established during low-load/speed operation.

An important aspect of VGT configuration is the need for a much more comprehensive and lengthy matching procedure compared with fixed geometry units; this has to take into account both steady-state and transient requirements for the various nozzle areas over the complete operating range, accounting also for EGR effects. Furthermore, a sophisticated closed-loop control system (see Appendix B for some fundamental aspects of control theory) that determines nozzle area according to the operating conditions (Figure 2.35) is needed [21–26]. This typically uses boost-pressure as the measured feedback output, as is demonstrated in Figure 2.36. At any operating point, the engine ECU uses the rotational speed and fueling values to calculate the desired boost pressure based on steady-state optimized maps. The aim of the controller is then to keep this pressure as close as possible to the mapped values applying, for example, P-I-D control, by relaying to the appropriate VGT vane position. A very instructive analysis in terms of VGT P-I-D control can be found in [27].

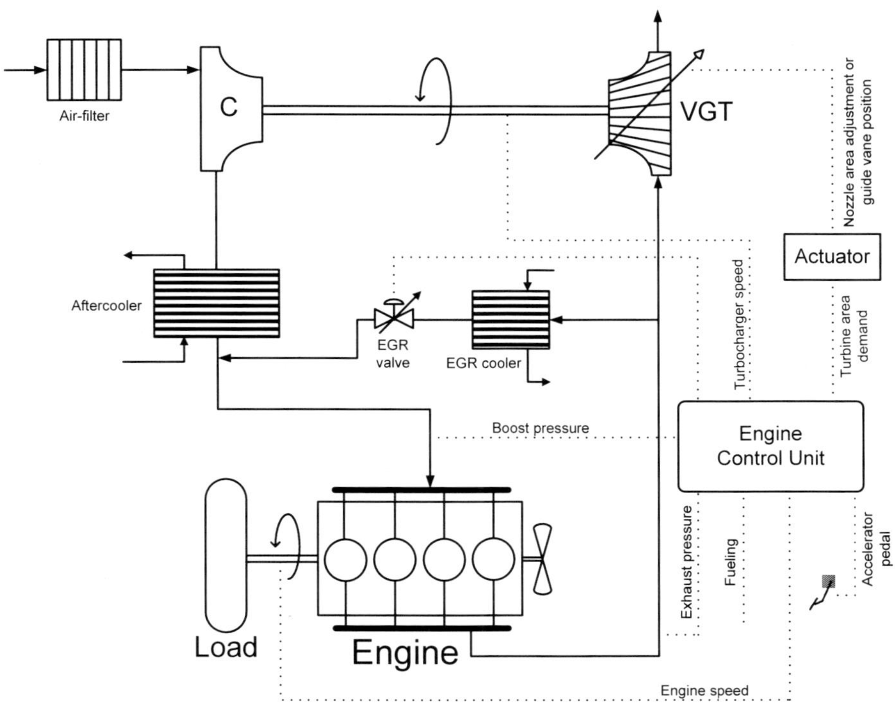

Figure 2.35. Schematic arrangement and simplified control diagram of variable geometry turbocharged diesel engine with exhaust gas recirculation

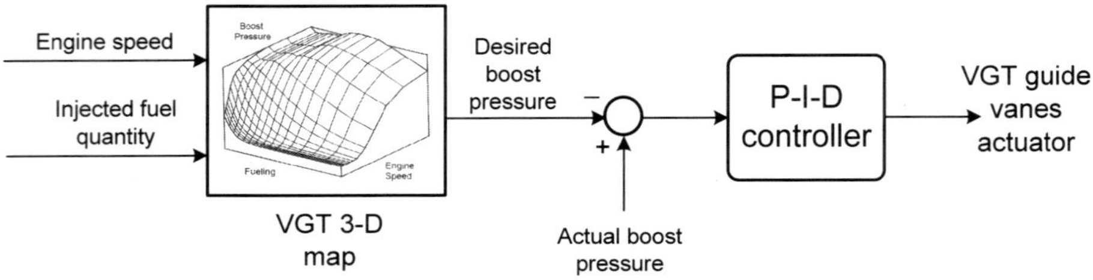

Figure 2.36. Simplified diagram of boost-pressure P-I-D, closed-loop controlled VGT system

Obviously, the response of the engine is strongly influenced by the specific VGT control system gain. Figure 2.37 illustrates how a relatively low gain, in a closed-loop strategy, results in smooth but rather long recovery period, in contrast to a high gain setup that responds almost instantly but leads to 'hunting' due to over-compensation. The latter behavior is typical when conventional P-I control algorithms are applied based on steady-state maps that often prove unable to handle the system constraints and delays. An optimization procedure is required taking into account both driveability and emission effects.

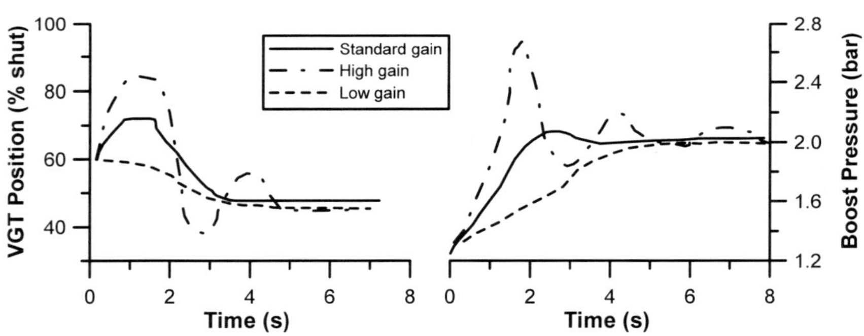

Figure 2.37. Effect of VGT gain on engine transient response at constant engine speed (simulation results adapted from Dekker and Sturm [28])

Historically, VGT control systems in low-bmep engines aimed to keep the VGT at its minimum area, increasing back-pressure and therefore turbine power at all times until the target boost pressure is reached. The VGT then opens to hold the boost at that value. Alternatively, speed dependent control strategies have been employed with smaller nozzle areas assigned the lower the engine speed. In recent years, increased global attention regarding exhaust emissions and the evolution of electronics have led to alternative VGT control strategies that are more suited to high-bmep engines; these usually focus on minimization of fuel consumption or exhaust emissions (among other things through accurate EGR control) or a combination of the above, depending on the specific engine operating conditions.

A possibly differentiated strategy is applied during transients, primarily aiming at turbocharger lag minimization. During transient operation (either load or speed increase) the main strategy of the VGT is the reduction of the nozzle area by closing down the vanes so as to increase rapidly the back-pressure and enthalpy drop across the turbine, thereby boosting the compressor operating point (see also the conceptual diagram of Figure 2.38). By so doing, fast increase of engine air-supply is established, minimizing fuel limiting function and improving driveability and emissions. As soon as the air-supply to the engine has been built-up, the VGT vanes are gradually opened in order to prevent over-boosting. The positive effects of VGT during transients are, therefore, the result of improvement in flow characteristics and air-supply, whereas the dynamics (inertia) of the turbocharger are not altered.

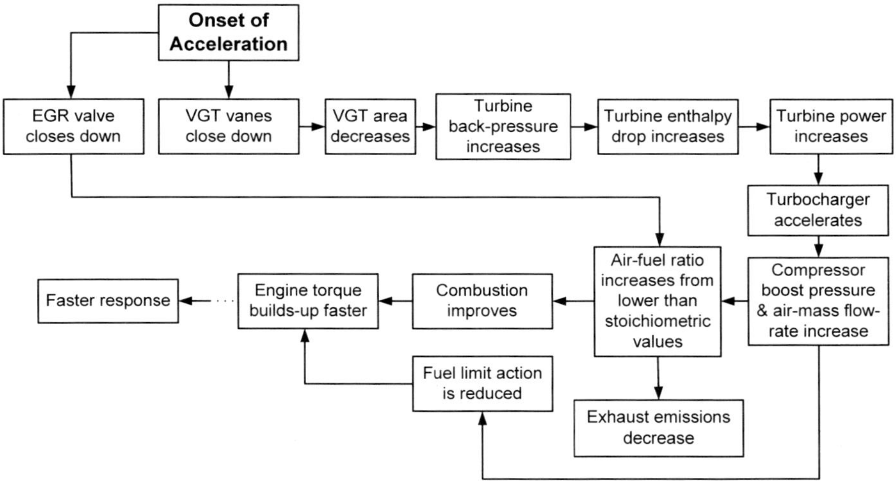

Figure 2.38. Series of events after onset of acceleration for a VGT diesel engine

Figure 2.39 describes in an explicit way the previously discussed operating mode of a variable geometry turbocharged engine during a transient event at constant engine speed. As soon as the driver applies the new fueling (corresponding to almost 27 mg/stroke), the air–fuel ratio decreases since the air-supply cannot instantly match the higher fueling due to the various system delays discussed in Section 2.1. The abrupt increase in fueling causes a step increase in desired boost pressure, hence a large error exists between demand and actual boost pressure. The VGT controller, responds to this error by closing down the VGT vanes. As a result, exhaust manifold pressure increases very fast, increasing accordingly the turbine expansion ratio and power. The increased power delivered to the compressor boosts its operating point and it is subsequently transformed into higher air-supply. Consequently, air–fuel ratio recovers much faster than with fixed geometry units and smoke emissions (not shown in Figure 2.39) are limited. Nonetheless, a brief peak in soot emissions cannot be utterly prohibited as the minimum value of AFR at approximately t = 0.2 s suggests. At about t = 1.5 s, VGT vanes are beginning to open in order to avoid over-boosting of the engine. What is missing from this figure is an indication of the engine efficiency during the transient event. This has surely decreased due to the elevated pumping losses following the increased exhaust back-pressure. We should keep in mind though that it is very difficult to achieve optimization in both fast response and low fuel consumption for each turbine nozzle area. In any case, this disadvantage of VGT operation is more than offset during transients from the faster turbocharger and engine speed response (see also Section 6.4.4 for comparative VGT *vs.* fixed geometry turbine results).

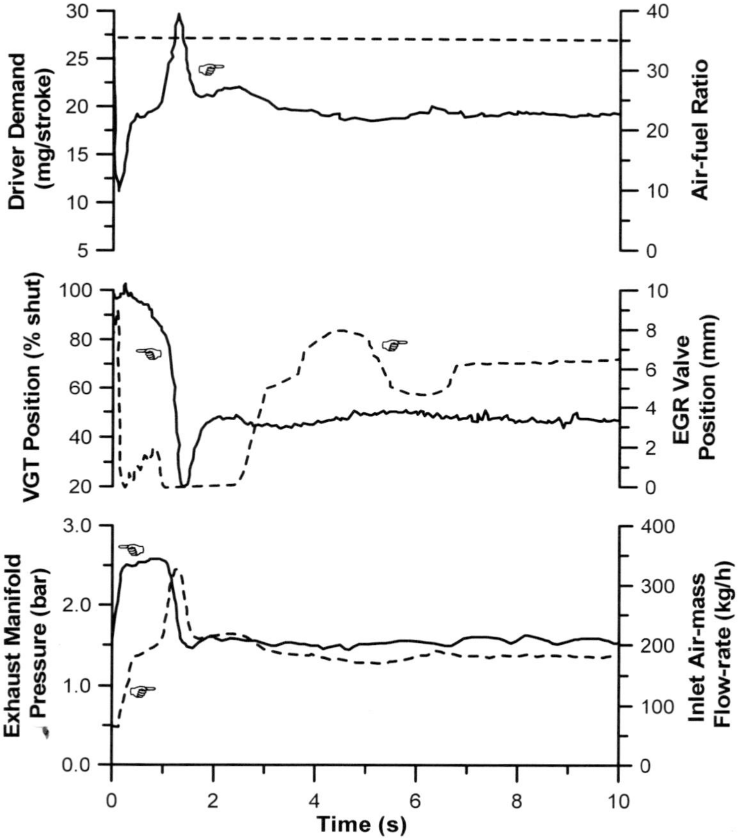

Figure 2.39. Engine and turbocharger properties development after a 10–170 Nm step load increase at 2500 rpm of a VGT engine (four-stroke, four-cylinder, 2 L displacement volume, HSDI, automotive diesel engine – experimental results reprinted with permission from SAE Paper No. 1999-01-0829 [23], © 1999 SAE International)

A case in point is that of pneumatically actuated VGT configurations. As the lower sub-diagram of Figure 2.39 suggests, a dynamic problem may arise, namely large overshoots in boost pressure and air-flow are experienced owing to hysteresis in the VGT vane mechanism. The vacuum at the vane actuator reduces steadily but the response of the vanes themselves is non-linear, which leads to a period of stagnation followed by a sudden flip open. The stagnation causes an overshoot in pressure and air-flow, then the rapid opening leads to undershoot, settling only after several seconds.

2.5 Exhaust Gas Recirculation

The increasingly stringent emission regulations have rendered exhaust gas recirculation (EGR) quite popular in recent years as a successful means for reducing NO_x emissions from both spark and compression ignition engines. Introduction of cooled (exhaust) gas into the combustion chamber (see also the conceptual diagram of Figure 2.40), results in dilution of the air-charge by replacing O_2 with the non-reacting CO_2. Consequently, both the specific heat capacity of the in-cylinder gas mixture and the peak flame temperatures of the cycle are reduced. As a result, NO_x emissions are reduced too, aided by the lower oxygen availability. However, smoke emissions increase since the soot oxidation process is diminished; the same holds true for HC and CO emissions.

The EGR system may be high pressure or low pressure. In the former case (Figure 2.40), the exhaust gas is drawn from upstream of the turbocharger turbine and is diverted back to the inlet side (typically downstream of the aftercooler); the recirculated exhaust gas flow is established from the pressure difference between the higher-pressure exhausted gas from the cylinders and the lower-pressure inducted air-supply. Since in this case, both EGR and turbine flow are driven by the exhaust gas, and, further, EGR flow reduces the amount of exhaust gas to be expanded in the turbine, there is clearly a strong relationship between the two flows that needs to be taken into account at the design stage in terms of optimum matching between engine and turbocharger (speed limits, peak cylinder pressure and surge avoidance). Nowadays, exhaust gas recirculation in automotive diesel engines is usually combined with a variable geometry turbine arrangement. In this case, the pressure difference between exhaust and inlet side for safe EGR flow can be determined by manipulating the VGT guide vanes; such a coupled EGR-VGT control system is discussed by Wijetunge *et al.* [29]. In general, interactions between EGR and other sub-systems, notably VGT and fuel injection, have been documented to have significant effects on the engine dynamic behavior, as will be discussed later in the section.

A low-pressure EGR unit, on the other hand, draws exhaust gas from downstream of the turbine or the DPF and diverts it to a point upstream of the turbocharger compressor; its main advantages are the somewhat cooled exhaust gas as well as the prevention of fouling of the EGR components, typical for the HP units exposed to the un-processed exhaust gases.

Besides, an EGR system can be internal or external; in the former case, the exhaust gas is re-routed back into the inlet manifold at the end of the exhaust stroke, for example through variable valve phasing. Most commonly, however, external EGR systems are employed, which have the added advantage of cooling down the exhaust, thus preventing decrease of the engine volumetric efficiency.

A typical, high-pressure, external EGR system, represented schematically in Figure 2.40, comprises the following components:

- EGR valve; this can be a continuously variable unit incorporating electric or pneumatic (or both) actuation;
- EGR cooler for cooling down the exhaust stream;
- EGR venturi mixer, where the exhaust gas is mixed with the incoming air-charge; such devices have been also applied for promoting gas flow from the exhaust to the inlet side;
- temperature and pressure sensors in the exhaust and inlet streams, and/or air-mass flow measurement, typically via hot wire anemometer; and
- comprehensive open- or closed-loop control system; this gathers the signals from the various sensors including engine and turbocharger, processes them based on look-up tables (steady-state maps) or model-based control theory, most probably applying correction factors to account for transient effects, and eventually determines the amount of exhaust gas to be recirculated through the EGR valve, *i.e.,* it determines the EGR valve position.

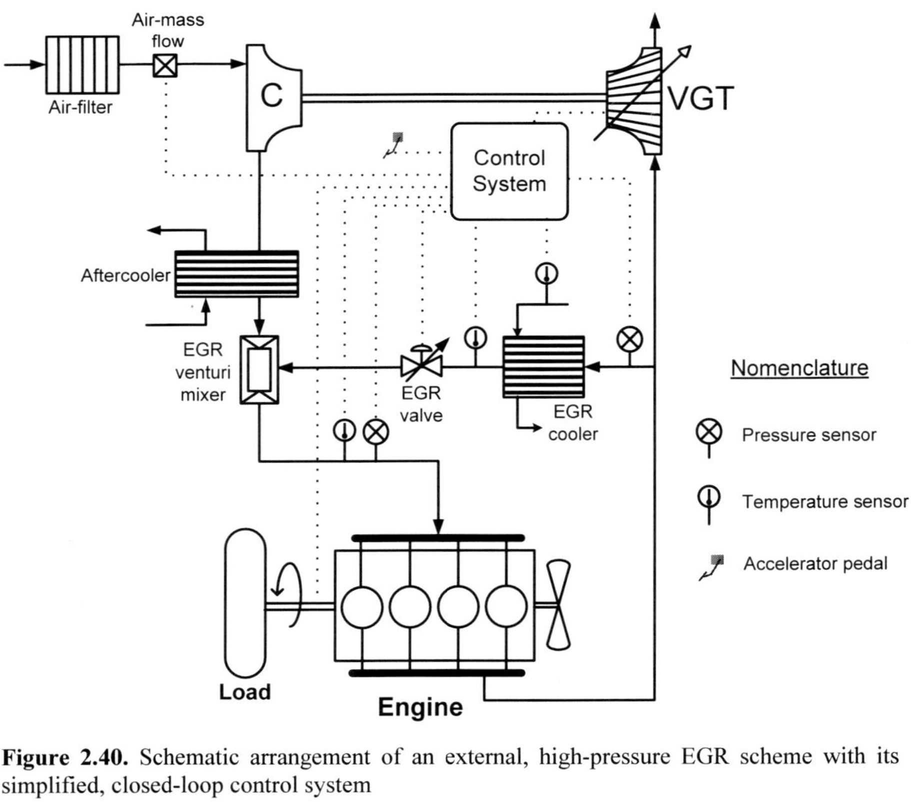

Figure 2.40. Schematic arrangement of an external, high-pressure EGR scheme with its simplified, closed-loop control system

In an open-loop EGR control system, the air–fuel ratio with the best NO_x reduction under the boundary condition of no smoke increase is mapped over injected fuel quantity and engine rotational speed; compensation is applied for high altitude operation, cold starting conditions (through the cooling water temperature) and transient conditions through a correction coefficient. However, during transients, feedback (closed-loop) systems are better suited to prevent excursion of the fuel–air equivalence ratio to high values and overshoot in smoke emissions. Guzzella and Amstutz [30] describe such a closed-loop EGR control system (Figure 2.41), where the measured air-mass flow-rate is used as the feedback property. Unlike in SI engines, recirculated exhaust gas in a diesel engine replaces oxygen, hence promotes enrichment of the mixture. During normal EGR mode, the intake manifold may be filled with 20 or 30% recirculated gas. If a rapid acceleration occurs, further deterioration of the air–fuel ratio and smoke emissions would be established, since the slow turbocharger response would delay increase of air-supply, whereas the smoke maps based on measured boost will fail because they assume that the mass-flow to the engine is all air. The closed-loop, feed-forward control system of Figure 2.41 describes how the net air-supply to the engine cylinders is estimated based on both EGR rate and total mass-flow to the engine (Equation 9.9) by using the measurement of the air-mass flow through the compressor. Based on the estimated effective air-mass flow to the engine, the maximum smoke-free, fuel injected quantity is then determined.

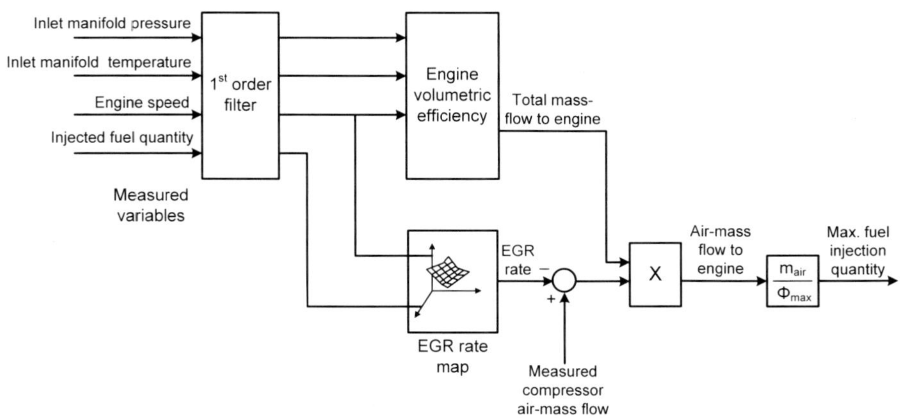

Figure 2.41. EGR calculation during transient operation (reprinted from Guzzella and Amstutz [30], © 1998 IEEE)

However, EGR flow is still largely controlled based on steady-state philosophy, *i.e.*, the EGR valve is kept open throughout the transient event. This eventually results in an extensive fuel limiting action (to avoid smoke emissions), delaying torque response and directly affecting driveability, up to the point where the turbocharger compressor boost pressure and the air-supply have been built-up satisfactorily. Since the slow turbocharger response induces quite different operating requirements during transients compared with steady-state operation, it is doubtful that the above EGR control strategy can prove successful overall. This will be shown by the following comparative case study.

Figure 2.42 illustrates the response of a turbocharged diesel engine during a load acceptance transient event at constant engine speed by comparing two different EGR strategies; in the first case, the EGR valve remains open during the transient event (most probably incorporating a corrective function through the measured compressor air-supply, such as the one described in Figure 2.41, to avoid excessive smoke emissions), whereas in the second case, the engine ECU closes down the EGR valve during the first seconds of the transient event.

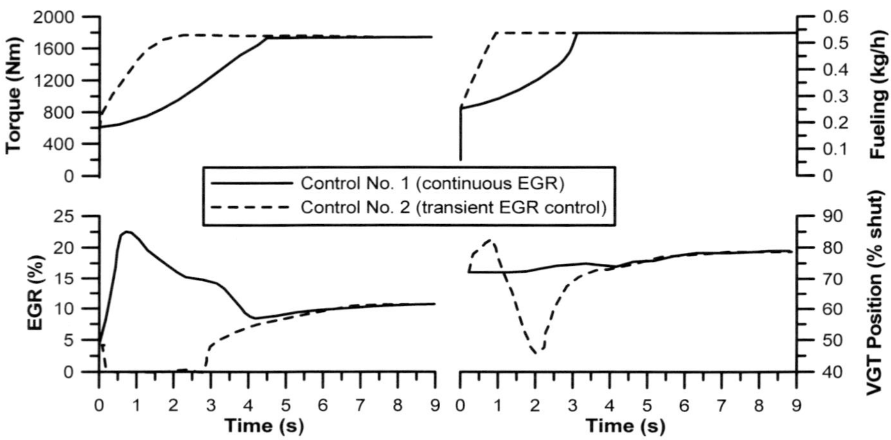

Figure 2.42. Engine response with continuous *vs.* transient controlled EGR valve during a load acceptance transient event at 1500 rpm (12 L displacement volume, VGT diesel engine, rated at 315 kW at 2000 rpm – simulation results adapted from Dekker and Sturm [31])

During the short period of time (3 s) where the EGR valve is closed down, the VGT guide vanes, being boost pressure loop-controlled, are directed to a more closed position in order to help build-up of turbine enthalpy drop. This ultimately results in much faster rise of turbocharger speed, compressor boost pressure (not shown in Figure 2.42), fueling and engine torque response (upper sub-diagram of Figure 2.42), hence driveability. The main mechanism behind the need for a drastic differentiation of the EGR strategy during transients is the deficiency of combustion air through the turbocharger lag phase. Even in an engine without EGR, the slow increase of boost pressure and air-supply during the early transient cycles results in a fuel limiting action to prevent excessive smoke emissions. Recall that recirculated exhaust gas in a diesel engine replaces oxygen promoting enrichment of the mixture. In an engine with EGR, if the strategy of Figure 2.41 (*i.e.,* continuous EGR) were followed, overshoot in smoke emissions would be prevented, but since the EGR valve is kept open, this would be accomplished only with an even more extensive fuel limitation, and driveability would suffer. Instead, the main engine control strategy in modern turbocharged engines aims at shutting down the EGR valve as soon as a load or speed transient is detected (detection can be established through the accelerator pedal signal in Figure 2.40). By so doing, the primary gain is that the EGR-caused fueling restriction is diminished and engine torque can be built-up faster (still slower though than a naturally aspirated engine)[10] with a penalty in nitrogen oxide emissions paid during this period, as will be discussed in Section 5.2. At the same time, closing of the variable geometry turbine vanes is initiated for better boost pressure build-up.

Returning to Figure 2.39 (in Section 2.4), we can now expand on the previously discussed remarks by highlighting specific aspects of the ECU control behavior in terms of coupled EGR and VGT operation during a load acceptance transient event. Initially, the EGR valve in Figure 2.39 is open, and then at the start of the transient it shuts rapidly in an attempt to maintain an adequate air–fuel ratio. Immediately after the valve closure, however, there will still be exhaust gas residuals present in the inlet manifold[11] – owing to the relatively large residual volume between the EGR valve and the combustion chamber – particularly so if the exhaust manifold pressure rises sharply, while the valve is closing due to the flow restriction of the closed VGT. The peak in exhaust manifold pressure following the closing of the VGT vanes causes a brief high EGR flow into the inlet manifold (the VGT control responds faster than the EGR valve closing), which in turn results in very low transient air–fuel ratios. Of particular note, here, is the brief opening of the EGR valve 3 s after initiation of the transient event; this is the result of the specific control strategy being derived directly from steady-state ECU maps. At the start of the transient event, the boost error exceeds acceptable limits and this in itself disables exhaust gas recirculation; however, the ECU still looks up the new EGR valve position demand based on the current fueling and speed conditions. As soon as the boost pressure error falls to within acceptable limits, control is handed back to the EGR map-based system, which in turn opens the valve because in the steady-state condition EGR is scheduled for the current engine speed and fueling [23].

An important aspect of the exhaust gas recirculation scheme applied is the magnitude of pressure difference between exhaust and inlet manifold. Obviously, the higher this difference (*e.g.,* by closing down the vanes in variable geometry turbine engines) the higher the exhaust gas flow through the EGR cooler, albeit, at the expense of higher pumping losses. An interesting comparison during both steady-state and transient conditions was conducted by Weber *et al.* [32] and a representative result from this work is reproduced in Figure 2.43. The low-pressure EGR unit (*i.e.,* exhaust gas drawn downstream of the turbine) managed to achieve the target brake mean effective pressure almost 0.9 s (or 40%) earlier than its high-pressure counterpart. It has been argued, however, that at higher power output, the high-pressure EGR loop can prove more efficient, mainly, in terms of fuel consumption.

---

[10] Ironically, owing to the usually slow response of the EGR valve, the valve area restriction causes during the early cycles of a transient event an increase in the pressure difference, hence increase in the amount of recirculated exhaust gas.

[11] This exhaust gas residual may distribute unevenly to the engine cylinders resulting in cylinder-to-cylinder variation in terms of NO_x emissions.

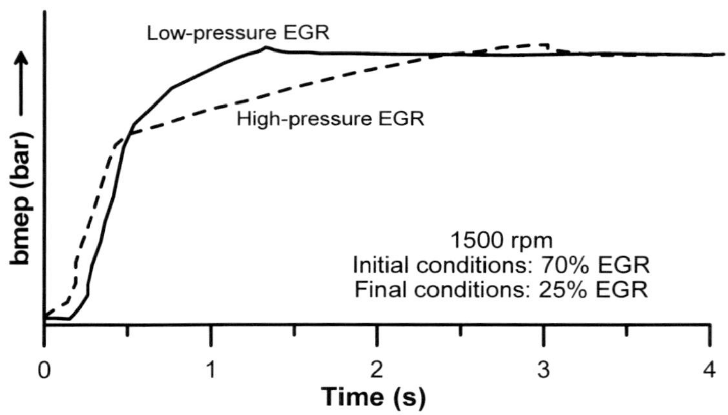

Figure 2.43. Effect of EGR strategy on transient performance of a VGT engine at 1500 rpm (simulation results adapted from Weber *et al.* [32])

Further aspects of exhaust gas recirculation and variable geometry turbine effects on transient operation will be discussed in Section 5.2, concerning NO_x emissions at the expense of smoke, and in Section 6.4.4, regarding engine and vehicle response throughout a Transient Cycle.

References

[1] Winterbone DE. Transient Performance. In: Horlock JH, Winterbone DE (eds). The thermodynamics and gas dynamics of internal combustion engines, Vol. II. Oxford: Clarendon Press, 1986; pp. 1148–212.

[2] Zinner K. Supercharging of internal combustion engines. Berlin Heidelberg New York: Springer-Verlag, 1978.

[3] Watson N, Janota MS. Turbocharging the internal combustion engine. London: MacMillan, 1982.

[4] Watson N. Transient performance simulation and analysis of turbocharged diesel engines. SAE Paper No. 810338, 1981.

[5] Watson N. Eliminating rating effects on turbocharged diesel response. SAE Paper No. 840134, 1984.

[6] Heireth H, Withalm G. The transient behavior of turbocharged SI engine. 18th FISITA Congress, Hamburg 1980 (in German).

[7] Murayama T, Miyamoto N, Tsuda T, Suzuki M, Hasegawa S. Combustion behaviors under accelerating operation of an IDI diesel engine. SAE Paper No. 800966, 1980.

[8] Takaishi T, Tateishi M, Kunimoto E. Dynamic response of fuel injection system of diesel engines. Institution of Mechanical Engineers, Seminar on 'Engine Transient Performance', Nov. 1990, pp. 5–12.

[9] Arcoumanis C, Baniasad MS. Analysis of transient response of diesel fuel injection systems. Institution of Mechanical Engineers, Seminar on 'Diesel Fuel Injection Systems', 28–29 Sept 1995, pp. 99–113.

[10] Catania AE, Dongiovanni C, Mittica A, Negri C, Spessa E. Study of automotive diesel injection-system dynamics under control. SAE Paper No. 962020, 1996.

[11] Harndorf H, Kuhnt H-W. Improvement of transient behavior of turbocharged diesel engines through additional air injection in the turbocharger. MTZ 1995;56:20–8 (in German).

[12] Benajes J, Luján JM, Bermúdez V, Serrano JR. Modelling of turbocharged diesel engines in transient operation. Part 1: insight into the relevant physical phenomena. Proc Inst Mech Eng, Part D, J Automobile Eng 2002;216:431–41.

[13] Assanis DN, Filipi, ZS, Fiveland SB, Syrimis M. A methodology for cycle-by-cycle transient heat release analysis in a turbocharged direct injection diesel engine. SAE Paper No. 2000-01-1185, 2000.

[14] Woschni G, Doll M, Spindler W. Simulation of the stationary and transient performance of small high speed car diesel engines. MTZ 1991;52:468–77 (in German).

[15] Rakopoulos CD, Hountalas DT, Mavropoulos GC, Giakoumis EG. An integrated transient analysis simulation model applied in thermal loading calculations of an air-cooled diesel engine under variable speed and load conditions. SAE Paper No. 970634, SAE Trans, J Engines 1997;106:923–39.

[16] Heywood JB. Internal combustion engine fundamentals. New York: McGraw-Hill, 1988.

[17] Benson RS, Whitehouse ND. Internal combustion engines. Oxford: Pergamon Press, 1979.

[18] Winterbone DE, Tennant DWH. The variation of friction and combustion rates during diesel engine transients. SAE Paper No. 810339, 1981.

[19] Whitehouse ND, Way RGB. Rate of heat release in diesel engines and its correlation with fuel injection data. Proc Inst Mech Eng, Part 3J 1969-70;184:17–27.

[20] Assanis DN, Filipi ZS, Fiveland SB, Syrimis M. A predictive ignition delay correlation under steady-state and transient operation of a direct injection diesel engine. ASME Trans, J Eng Gas Turbines Power 2003;125:450–7.

[21] Watson N, Banisoleiman K. A variable-geometry turbocharger control system for high output diesel engines. SAE Paper No. 880118, 1988.

[22] Pilley AD, Noble AD, Beaumont AJ, Needham JR, Porter BC. Optimization of heavy-duty diesel engine transient emissions by advanced control of a variable geometry turbocharger. SAE Paper No. 890395, 1989.

[23] Wijetunge RS, Brace CJ, Hawley JG, Vaughan ND, Horrocks RW, Bird GL. Dynamic behaviour of a high speed direct injection diesel engine. SAE Paper No. 1999-01-0829, 1999.

[24] Filipi Z, Wang Y, Assanis D. Effect of variable geometry turbine (VGT) on diesel engine and vehicle system transient response. SAE Paper No. 2001-01-1247, 2001.

[25] Bartsch P, Prenninger P, Allmer I. Transient performance optimization of turbocharged engines by means of gas exchange simulations. Institution of Mechanical Engineers, International Conference on 'Turbocharging and Air Management Systems', Paper C554/016, London, 1998, pp. 237–251.

[26] Winkler N, Ångström H-K. Simulations and measurements of a two-stage turbocharged heavy-duty diesel engine including EGR in transient operation. SAE Paper No. 2008-01-0539, 2008.

[27] Brace CJ, Cox A, Hawley JG, Vaughan ND, Wallace FW, Horrocks RW, Bird GL. Transient investigation of two variable geometry turbochargers for passenger vehicle diesel engines. SAE Paper No. 1999-01-1241, 1999.

[28] Dekker JH, Sturm WL. Model based development of engine control algorithms. Institution of Mechanical Engineers, 3rd International Conference on 'Computers in Reciprocating Engines and Gas Turbines', Paper C499/021, 1996, pp. 163–72.

[29] Wijetunge RS, Hawley JG, Vaughan ND. Application of alternative EGR and VGT strategies to a diesel engine. SAE Paper No. 2004-01-0899, 2004.
[30] Guzzella L, Amstutz A. Control of diesel engines. IEEE Control Syst 1998;18:53–71.
[31] Dekker HJ, Sturm WL. Simulation and control of a HD diesel engine equipped with new EGR technology. SAE Paper No. 960871, 1996.
[32] Weber O, Joergl V, Shutty J, Keller P. Future breathing system requirements for clean diesel engines. Aachener Kolloquium, Fahrzeug- and Motorentechnik, Aachen, Germany, 2005.

3

Dynamics

The research on diesel engine transient operation has primarily focused on the relevant thermodynamic processes because of their direct impact on heat release and consequently performance and pollutants emissions. On the other hand, issues concerning (engine) dynamics are often over-simplified. Such notable dynamic issues that contribute to the non-linearity and complexity of diesel engine transient operation are

- development of the various forces of the slider-crank mechanism that may lead to considerable stress of the engine bearings;
- crankshaft torsional deformation resulting from the different magnitude of instantaneous torque values induced by the engine and load;
- governor clutch movement determining the actual fuel pump rack position;
- development of engine friction; and
- other dynamic issues induced by the whole vehicle propulsion.

3.1 Engine Dynamics

3.1.1 Kinematics and Forces of the Slider-crank Mechanism

At each instant of time, the displacement of the piston from the top dead center (TDC) position is given by the following equation with reference to Figure 3.1

$$x(\varphi) = r\left(1-\cos\varphi\right) + L_{rod}\left(1-\sqrt{1-\lambda^2\sin^2\varphi}\right) \tag{3.1}$$

where $r=S/2$ is the crank radius and $\lambda=r/L_{rod}$ with L_{rod} the connecting rod length.

Differentiating the above equation with respect to time, we get the instantaneous piston velocity

$$u_{pist}(\varphi) = \omega r \sin\varphi \left(1 + \frac{\lambda \cos\varphi}{\sqrt{1-\lambda^2 \sin^2\varphi}} \right) \tag{3.2}$$

and differentiating once again with respect to time, we get the instantaneous piston (linear) acceleration

$$b(\varphi) = \omega^2 r \left(\cos\varphi + \lambda \frac{\cos 2\varphi + \lambda^2 \sin^4\varphi}{(1-\lambda^2 \sin^2\varphi)^{3/2}} + \frac{1}{\omega^2} \varepsilon \frac{u_{pist}(\varphi)}{r\omega} \right) \tag{3.3}$$

The last term on the right-hand side of Equation 3.3 accounts for the influence of the crank's angular acceleration $\varepsilon = d\omega/dt$ on the piston linear acceleration. Further, the connecting rod angle β is given by

$$\sin\beta = \lambda \sin\varphi \Leftrightarrow \beta = \cos^{-1}\left(\sqrt{1-\lambda^2 \sin^2\varphi} \right) \tag{3.4}$$

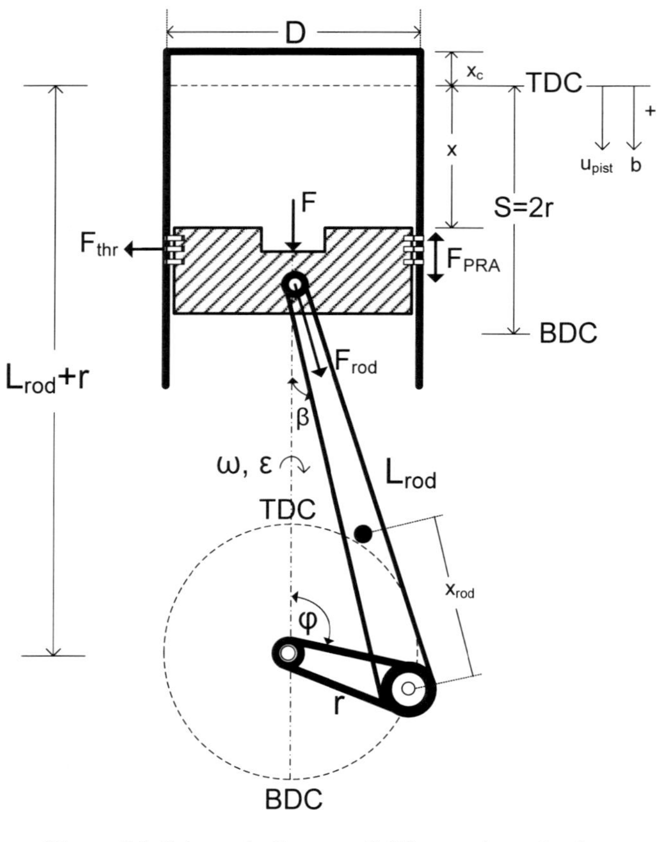

Figure 3.1. Schematic diagram of slider-crank mechanism

Total force acting on the piston comprises of thermodynamic and inertia terms

$$F(\varphi) = F_g(\varphi) + F_{in}(\varphi) = p_g(\varphi)A_{pist} - m_l b(\varphi) = p_g(\varphi)A_{pist} - m_l r\omega^2 \left(b(\varphi)/r\omega^2\right) \quad (3.5)$$

This force propagates into the thrust force $F_{thr}(\varphi) = F(\varphi)\tan\beta$ and the force in the direction of the connecting rod $F_{rod}(\varphi) = F(\varphi)/\cos\beta$. Also, A_{pist} is the piston surface area = $\pi D^2/4$. Piston rings assembly friction force F_{PRA} in Figure 3.1 will be discussed in Section 3.3.

The reciprocating mass m_l in Equation 3.5 is given by

$$m_l = m_{pist} + m_{rod.l} \quad (3.6a)$$

whereas for the rotating masses

$$m_r = m_{rod.r} + m_{crank.eq} \quad (3.6b)$$

In the above equations, m_{pist} is the mass of the piston assembly (piston, rings, wrist pin) and $m_{crank.eq}$ is the equivalent rotating mass of the crank with reference to the crankpin.

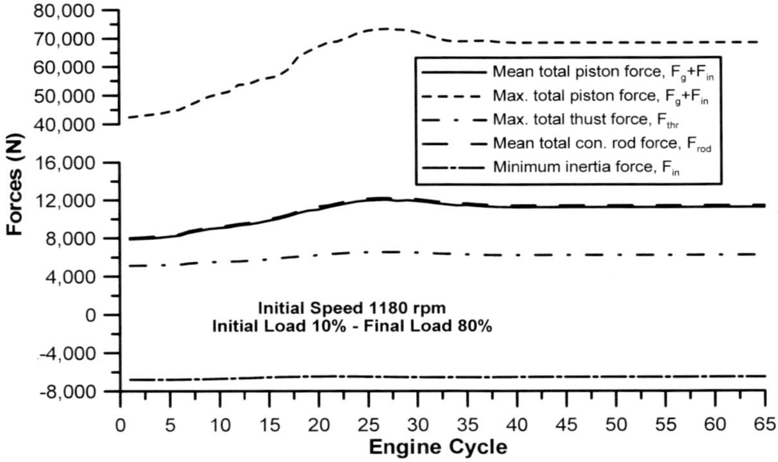

Figure 3.2. Mean and maximum, over each engine cycle, slider-crank mechanism forces during a 10–80% load increase transient event ('total' denotes the sum of gas and inertia forces contribution)

The connecting rod is usually assumed equivalent to two masses, one reciprocating with the piston assembly $m_{rod.l}$ and the other rotating with the crank $m_{rod.r}$. The following two equations apply with reference to Figure 3.1:

$$m_{rod} = m_{rod.r} + m_{rod.l} \quad (3.7a)$$

$$m_{rod.l}\,(L_{rod} - x_{rod}) = m_{rod.r}\,x_{rod} \tag{3.7b}$$

with x_{rod} the distance between the connecting rod's center of gravity and its big end bearing and m_{rod} the connecting rod mass. It follows then that $m_{rod.r}x_{rod}^2 + m_{rod.l}(L_{rod} - x_{rod})^2 \neq G_{rod}$ with G_{rod} the real connecting rod mass moment of inertia; the difference is, however, usually negligible.

Figure 3.2 illustrates the development of some mean and maximum, over each engine cycle, slider-crank mechanism forces during a 10–80% load increase transient event of a turbocharged diesel engine. Clearly, thrust forces are always of much less importance owing to the small connecting rod angles involved, and the same holds true for inertia forces (notice the negative sign, actually denoting the minimum rather than the maximum value of this force). The latter are not so pronounced for the particular operating case depicted in Figure 3.2 owing to the relatively low engine speed (recall that inertia forces depend on the engine speed squared).

For the loading of the slider-crank mechanism bearings, the following equations can be developed with reference to Figure 3.3:

$$B_{0x} = -m_{rod.l} \cdot b\sin\beta \quad , \quad B_{0y} = -\frac{F(\varphi)}{\cos\beta} - m_{rod.l} \cdot b\cos\beta \tag{3.8}$$

for the connecting rod small end bearing,

$$B_{1x} = -m_{rod.r} \cdot u_{pist} \cdot \omega \cdot \cos\beta \quad , \quad B_{1y} = \frac{F(\varphi)}{\cos\beta} - m_{rod.r} \cdot r\omega^2 \cos(\varphi + \beta) \tag{3.9}$$

for the connecting rod big end bearing,

$$B_{2x} = \frac{F(\varphi)}{\cos\beta}\sin(\varphi + \beta) \quad , \quad B_{2y} = -\frac{F(\varphi)}{\cos\beta}\cos(\varphi + \beta) + m_{rod.r} \cdot r\omega^2 \tag{3.10}$$

for the crank pin (single cylinder contribution),

$$B_{3x} = -\frac{1}{2}\frac{F(\varphi)}{\cos\beta}\sin(\varphi + \beta) \quad , \quad B_{3y} = \frac{1}{2}\left[\frac{F(\varphi)}{\cos\beta}\cos(\varphi + \beta) - m_r r\omega^2\right] \tag{3.11}$$

for the crank journal (single cylinder contribution), and

$$\begin{aligned} B_{4x} &= \frac{1}{2}\left[F(\varphi)\tan\beta + m_r r\omega^2 \sin(\varphi)\right] \\ B_{4y} &= \frac{1}{2}\left[-F(\varphi) + m_r r\omega^2 \cos(\varphi)\right] \end{aligned} \tag{3.12}$$

for the main crankshaft bearing (single cylinder contribution).

The corresponding total bearing force is then given for each case by

$$B_i = \sqrt{B_{ix}^2 + B_{iy}^2} \tag{3.13}$$

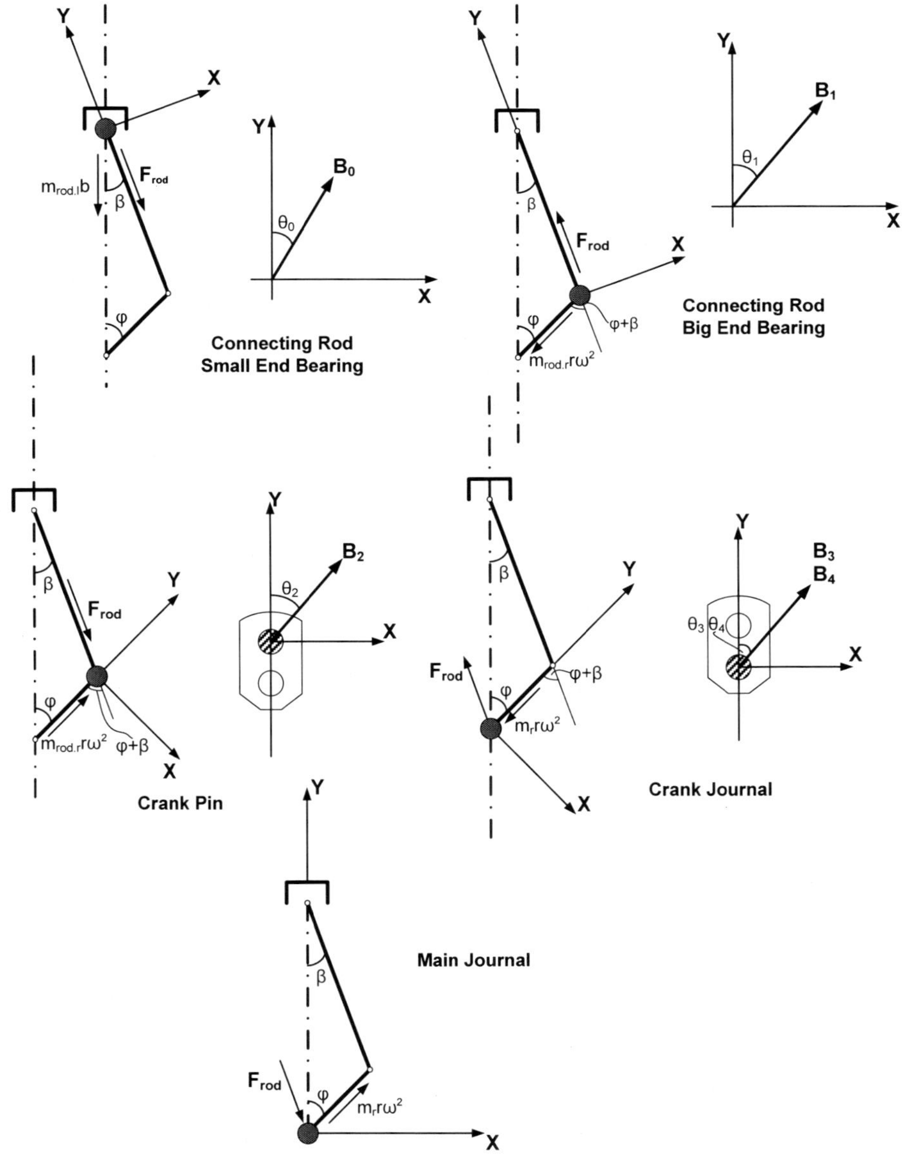

Figure 3.3. Forces distribution for bearing loadings evaluation [1]

and the angle θ, shown in Figure 3.3 is calculated from

$$\theta_i = \tan^{-1}\left(\frac{B_{ix}}{B_{iy}}\right) \tag{3.14}$$

with i = 0…4 according to the bearing under study.

Figure 3.4 illustrates the development of maximum bearing loadings during the same 10–80% transient step load increase. The response of piston force ($F_g(\varphi) = p_g(\varphi)\, A_{pist}$), engine speed and fueling are also depicted in the same figure for comparison purposes.

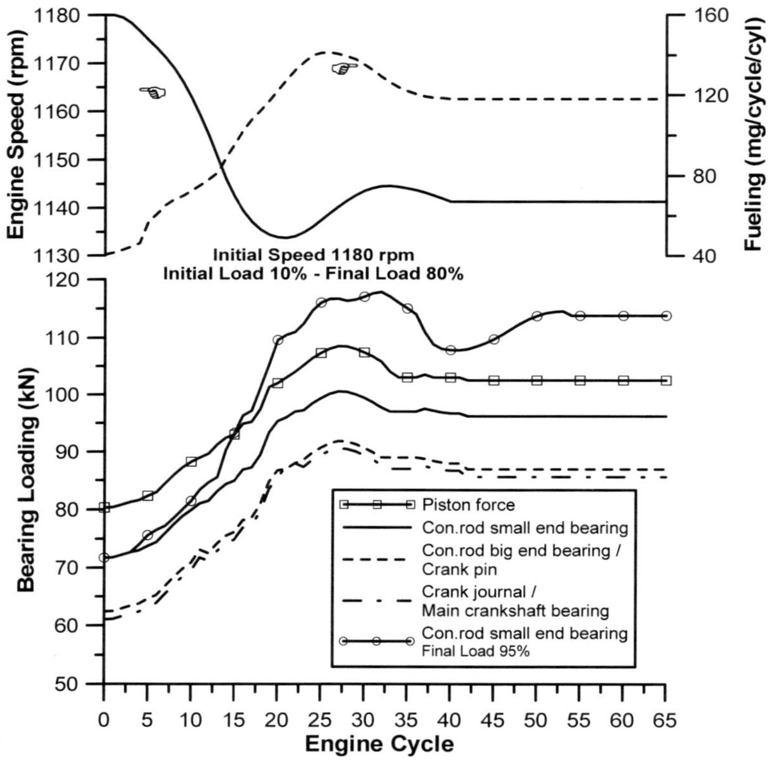

Figure 3.4. Maximum bearing loadings *vs.* engine cycle for a 10–80% load increase transient event

The main findings derived from Figure 3.4 are summarized below.

- Maximum values, over each engine cycle, of all bearing loadings develop during a transient event in a way that closely follows the fueling and the peak cylinder pressure profiles owing to the dominance of gas pressures on the B_i terms.

- Turbocharger lag effect is obvious during the first cycles of the transient event.
- The connecting rod's big end bearing loading equals the crank pin's one, whereas the crank journal's loading equals the main crankshaft bearing's one.
- Loading gradually decreases as we move downwards the kinematics mechanism, *i.e.,* from the piston towards the crankshaft, as the initial (gas) force is 'propagated'.
- Contribution of inertia forces is minimal (as regards maximum (absolute) values only), primarily due to the low engine rotational speed (*cf.* the ω^2 term in Equations 3.9–3.12).

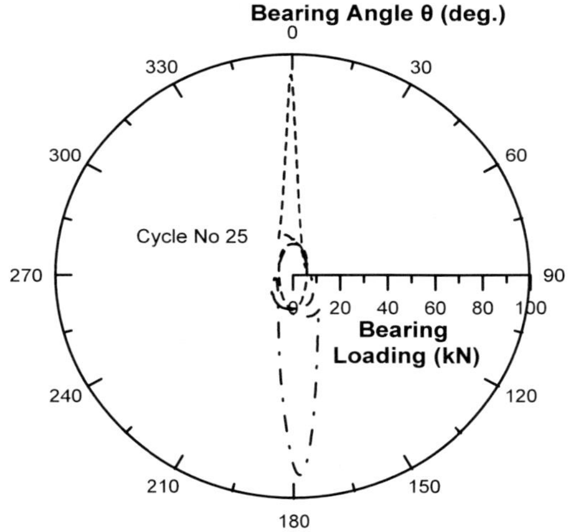

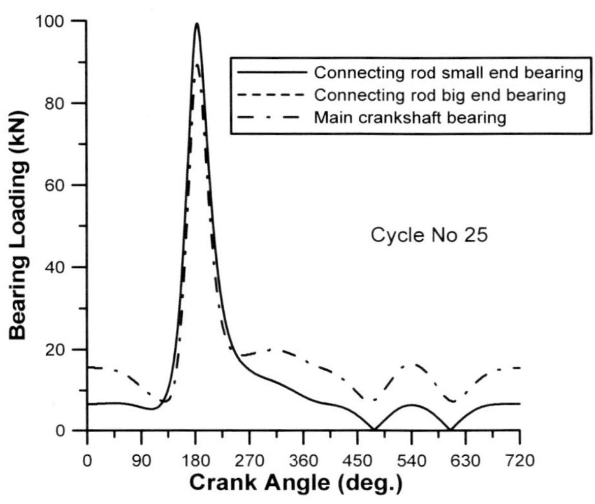

Figure 3.5. Development of connecting rod and main crankshaft bearing loadings during an intermediate cycle of a 10–80% load increase transient event [1]

- The crank angle where the maximum loading is observed ranges for all bearings and for all transient cycles from 1–4°CA after firing TDC, as is also the case with the peak cylinder pressure for the particular engine.
- As regards loading rate, the maximum increase is observed between cycles 10–30 where the main part of the increase in fueling occurs; it is of the order of 16 kN/s for the specific 10–80% load increase transient.

Figure 3.5 focuses on an intermediate cycle of the transient event (cycle No. 25, where maximum fueling occurs), illustrating the in-cycle build-up of the connecting rod and main crankshaft bearing loadings. The contribution of gas force is evident during the compression and expansion strokes; however, it is the inertia force $F_{in}(\varphi)$ that defines the loading profile during most of the open part of the cycle, where the cylinder pressure assumes lower values. In the upper diagram of Figure 3.5 the above data are illustrated on a polar diagram, which is a more common way of depiction in terms of this kind of analysis.

Lee *et al.* [2] investigated experimentally the thermal and load stress during various transient schedules (starting, shutting down, changing load at constant engine speed); consistent with the results presented in the previous figures, they concluded that although the force/loading levels increase significantly with increasing load, the actual values of the mechanical stress during transients are only marginally higher than those corresponding to the respective steady-state conditions.

3.1.2 Crankshaft Torque Balance

In the general case, the crankshaft is assumed to be a flexible, elastic body that may deform during engine operation; the lumped mass model of Figure 3.6 can be employed. This illustrates a condensed (crank)shaft model, *i.e.*, rigid enough between the cylinders and elastic between flywheel and load. The elastic crankshaft rotary motion is excited by the gas and inertia force fluctuation. The rotary motion of the shaft is expressed by the following two differential equations describing its angular momentum balance based on Newton's second law of motion:[1]

$$\tau_e(\varphi) - \tau_{fr}(\varphi) - \tau_S - \tau_D = G_e \frac{d\omega}{dt} + \frac{1}{2}\frac{\partial G_e(\varphi)}{\partial \varphi}\left(\frac{d\varphi}{dt}\right)^2 \tag{3.15a}$$

$$\tau_S + \tau_D - \tau_L(\varphi_L) = G_L \frac{d\omega_L}{dt} \tag{3.15b}$$

In Equations 3.15, G_e is the engine (see next section) and G_L is the load (resistance) mass moment of inertia, respectively.

[1] Sir Isaac Newton's (1642–1727) formulation of second law of motion for rotational systems states that the sum of torques acting on a body about a given axis is equal to the product of its rotational moment of inertia and the rotational acceleration about the axis.

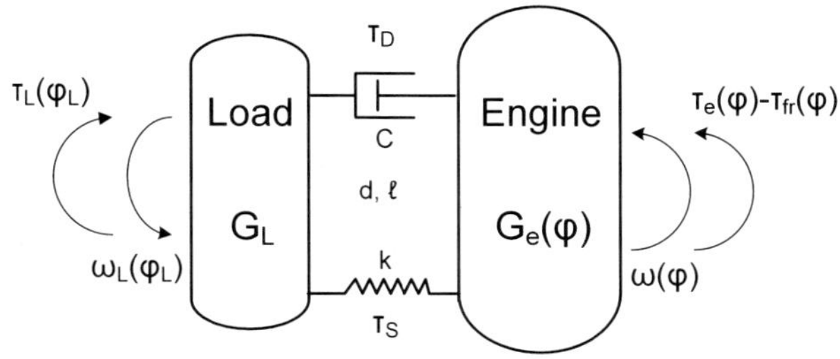

Figure 3.6. Schematic arrangement of the engine-load dynamic system for crankshaft torque balance analysis

Also, $\omega = d\varphi/dt$ is the engine angular velocity and $\omega_L = d\varphi_L/dt$ is its load counterpart. Further, $\tau_{fr}(\varphi)$ is the total (from all cylinders) friction torque and $\tau_e(\varphi) = \sum_{i=1}^{n_{cyl}} \tau_{ei}(\varphi)$ denotes the total engine indicated torque that includes the contribution of gas, inertia and (the negligible) gravitational forces. Individual cylinder indicated torque $\tau_{ei}(\varphi)$ is given explicitly by

$$\tau_e(\varphi) = \overbrace{\tau_g(\varphi)}^{\text{Gas}} + \overbrace{\tau_{in}(\varphi)}^{\text{Inertia}} + \overbrace{\tau_{gr}(\varphi)}^{\text{Gravitational}} =$$

$$\left[\overbrace{\left(p_g(\varphi)\, A_{pist} \frac{u_{pist}(\varphi)}{r\omega}\right)}^{\text{Gas}} - \overbrace{\left(m_l\, b(\varphi) \frac{u_{pist}(\varphi)}{r\omega}\right)}^{\text{Inertia}} + \overbrace{\left(m_l\, g \frac{u_{pist}(\varphi)}{r\omega} + m_r\, g \sin\varphi\right)}^{\text{Gravitational}}\right] r \tag{3.16}$$

with $p_g(\varphi)$ the instantaneous cylinder gas pressure and g the gravitational acceleration.[2]

For the torsional stiffness and damping torques we have, respectively,

$$\tau_S = k\,(\varphi - \varphi_L) \tag{3.17}$$

$$\tau_D = C\,(\omega - \omega_L) \tag{3.18}$$

with k the shaft stiffness coefficient, and C the damping factor (Figure 3.6).

In most of the cases, however, the crankshaft can be assumed short and rigid enough, hence, $\varphi \equiv \varphi_L$ and $\omega \equiv \omega_L$. Then, by adding both sides of Equations 3.15, these are replaced by the following much simpler equation:

$$\tau_e(\varphi) - \tau_{fr}(\varphi) - \tau_L(\varphi) = G_{tot} \frac{d\omega}{dt} \tag{3.19}$$

---

[2] For the calculation of total engine torque at the flywheel end, the crank angle difference between successive firing cylinders is required in order to formulate n_{cyl} equations similar to 3.16.

where G_{tot} represents the total mass moment of inertia of the engine–load configuration reduced to the crankshaft axis.

For the resistance torque term, τ_L in Equations 3.15b and 3.19, the following relation can be applied (for small speed changes):

$$\tau_L(\omega_L) = C_1 + C_2 \omega_L^{C_3} \tag{3.20}$$

- for a linear load-type (electric brake, generator) C_3=1;
- for a quadratic load-type (hydraulic brake, fixed pitch propeller, vehicle aerodynamic resistance) C_3=2; and
- C_1 is the speed-independent load term (*e.g.*, road slope – see Section 3.5).

An even more detailed angular momentum balance can be formulated by taking into account all possible crankshaft deformations between pulley, each cylinder (or crank) of a multi-cylinder diesel engine, flywheel and load. This eventually leads to a system of multiple differential equations (similar to those given in Equations 3.15), depending on the number of cylinders (cranks) of the engine in question, as it is depicted in Figure 3.7 for a six-cylinder engine.

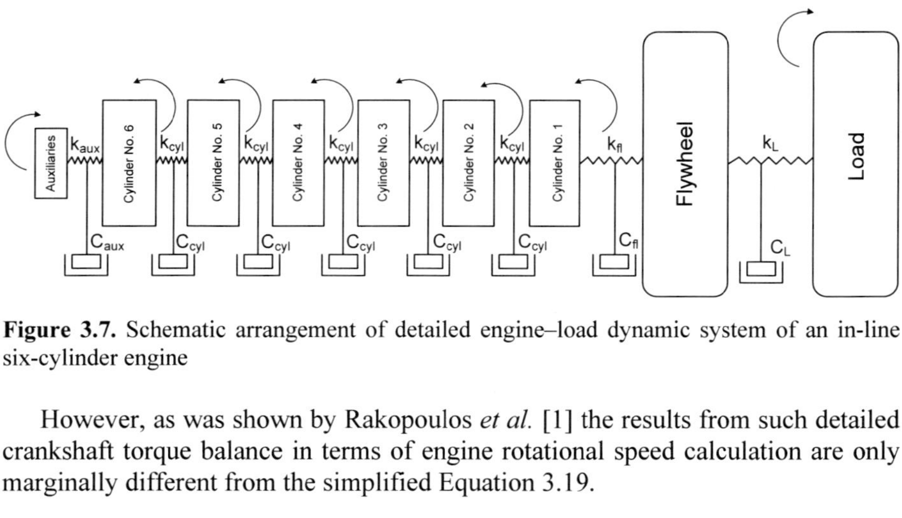

Figure 3.7. Schematic arrangement of detailed engine–load dynamic system of an in-line six-cylinder engine

However, as was shown by Rakopoulos *et al.* [1] the results from such detailed crankshaft torque balance in terms of engine rotational speed calculation are only marginally different from the simplified Equation 3.19.

3.1.3 Mass Moments of Inertia

The total engine mass moment of inertia G_e needed in the crankshaft torque balance analysis of the previous section is calculated from the following equation

$$G_e = G_{fl} + G_{coupl} + n_{cyl}(G_r + G_l) + G_{var} \tag{3.21a}$$

Here, the total engine inertia includes flywheel G_{fl} and, if applicable, elastic coupling G_{coupl} contribution, with n_{cyl} the number of engine cylinders and G_{var} other engine-based terms such as gears, pulleys *etc.;* G_l is the varying, according to crank angle, reciprocating masses inertia reduced to the crankshaft axis, given by

$$G_l = m_l r^2 \left(\frac{u_{pist}(\varphi)}{r\omega} \right)^2 \tag{3.21b}$$

and G_r is the rotating masses moment of inertia, which according to Figure 3.8 is computed by

$$G_r = \overbrace{G_w + \left(G_{Kr} + m_K r^2\right) + 2G_B}^{\text{Crank assembly}} + G_{rod.r} =$$

$$\underbrace{\frac{\pi}{32}\rho d_w^4 \ell_w}_{\text{Crank journal}} + \underbrace{\left(\frac{\pi}{32}\rho d_K^4 \ell_K \left(d_K^2 + 8r^2\right)\right)}_{\text{Crank pin}} + \underbrace{2G_B}_{\text{Crank cheeks}} + \underbrace{m_{rod.r} r^2}_{\substack{\text{Connecting rod} \\ \text{big end bearing}}} \approx m_r r^2 \tag{3.21c}$$

with m_r defined in Equation 3.6b, and ρ the material density. Ideally, the counterweights moment of inertia should be taken into account in Equation 3.21c.

Further, by substituting the $u_{pist}(\varphi)$ term from Equation 3.2, into Equation 3.21b, we get

$$\bar{G}_l = \left(\frac{1}{2} + \frac{\lambda^2}{8} + \ldots\right) m_l r^2 \approx \frac{1}{2} m_l r^2 \tag{3.21d}$$

which can be used as a very good approximation for the mean, over an engine cycle, reciprocating masses moment of inertia.

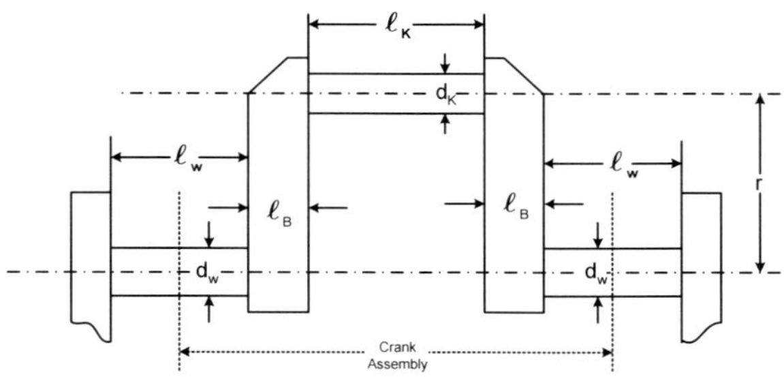

Figure 3.8. Schematic arrangement of crank for calculation of rotating masses moment of inertia

3.2 Governor

3.2.1 Governor Fundamentals

Unlike spark ignition engines, the diesel engine's fuel pump curves are characterized by increasing values with increasing engine speed. Even a slight increase in speed from the equilibrium conditions leads to increase in the injected fuel quantity, hence engine torque. At idling conditions (without speed regulation), engine speed would either increase constantly or decrease to stall. Likewise, after cold starting and while the engine operates at idling conditions, engine speed would constantly rise due to the gradual decrease in resistance and friction torque during warm-up. Hence, the diesel engine, being inherently unstable, particularly at low loads and speeds, requires some kind of feedback control in order to maintain engine speed. This control is realized through the speed governor that determines fueling consistent with current engine operating conditions and requested changes in desired engine speed.

In order to understand the mechanism behind the governor control we need to look back into the angular momentum balance of the engine crankshaft (Equation 3.19). Since friction and resistance torque are, mainly, dependent on engine speed, and, in general, $\tau_e \propto \eta_e \dot{m}_{fi}$, with η_e the engine brake efficiency, it is made obvious that in order to manipulate speed the only engine variable that can be externally regulated is the injected fuel quantity $\dot{m}_{fi}$; this regulation is achieved through the speed governor. Engine speed regulation is accomplished via negative feedback control, *i.e.*, decrease in engine speed causes increase in fueling. Figure 3.9 depicts the block diagram of the basic governor action.

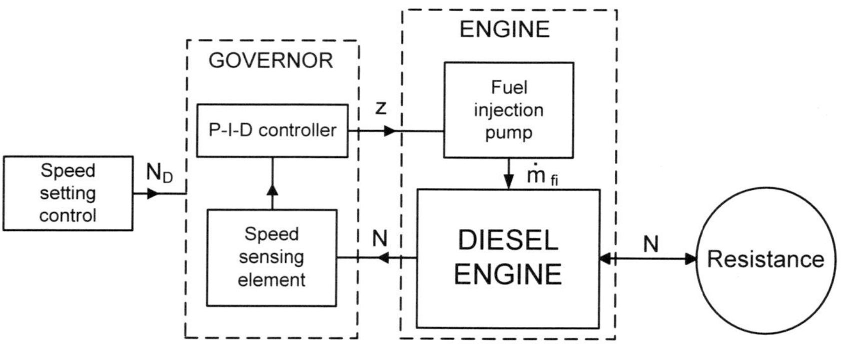

Figure 3.9. Block diagram of typical speed governor operation (N: actual engine speed, N_D: demand speed, $\dot{m}_{fi}$: injected fuel quantity, z: fuel pump rack position)

The desired or demand engine speed N_D (as dictated, for example, by the position of the accelerator pedal in passenger vehicles or as set by the governor control lever in industrial engines) and the actual engine speed N are the two basic inputs to the governor control loop. Any difference in the values of these two signals (*e.g.*, decrease in actual engine speed due to sudden load application) results in displacement of the governor sensing element; the latter alters the fuel pump rack position z, therefore the injected fuel quantity $\dot{m}_{fi}$, and engine torque τ_e in order to establish a new equilibrium between engine and resistance (load) at a different or even the same rotational speed.

Governors are either direct or indirect acting. Both types consist of a sensing element for the sensing of speed change; the indirect acting governors incorporate also a servomechanism that is connected to the fuel pump rack, correcting the fueling. The sensing elements are usually mechanical (with flyweights) or electrical/electronic. Servomechanism feedback may be unity or vanishing; the vanishing feedback indirect acting governor achieves constant governing (zero speed droop after a load change) and is termed isochronous. On the other hand, direct acting governors are characterized by permanent speed droop according to the applied load change. Direct acting governors apply proportional (P) control, meaning that the change in injected fuel quantity is determined by the difference between actual and desired speed (power amplification). Isochronous governors incorporate proportional plus integral (P-I) control. Finally, a governor may apply differential control (P-I-D controller), in which case the fuel pump rack movement is also affected by the rate of change of engine speed. In mechanical fuel injection systems, the governor is a separate device, usually attached to the fuel pump casing (Figure 3.10), whereas in modern, common rail injection systems, the governor function is integrated in the engine control unit.

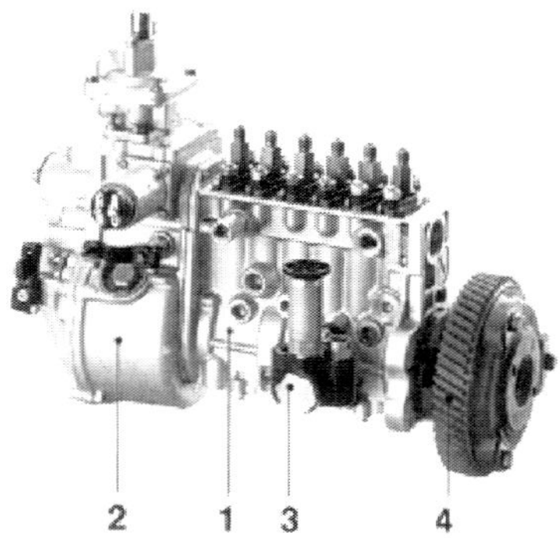

Figure 3.10. Mechanical in-line fuel pump incorporating a speed governor. 1: fuel injection pump; 2: governor; 3: fuel-supply pump; 4: timing device (reprinted with permission from Bosch [3])

Figure 3.11a describes in more detail the control loop of a conventional mechanical governor fitted on a vehicle engine; it emphasizes on its auxiliary functions, *i.e.*, fuel limiting (identified through the input signal of turbocharger compressor boost pressure, as was discussed in Section 2.2.2) as well as load regulation, starting requirements or altitude compensation (atmospheric pressure signal). Figure 3.11b illustrates the much more complex loop of a modern electronic governor integrated in the engine ECU receiving several signals from various sensors and processing them in a micro-processor.

If the governor regulates speed at a unique operating condition, for example idling, or maximum speed, it is called one-speed governor. A common type of automobile governor regulates speed at two distinct speeds, *i.e.*, idle and maximum speed and is termed a minimum/maximum governor. In Figure 3.12 this is accomplished via two springs of different stiffness, whereas in the middle of the operating range, engine output is directly affected by the accelerator pedal position, enhancing driveability. When the governor regulates speed over the whole engine operating range, it is termed all-speed or variable speed.

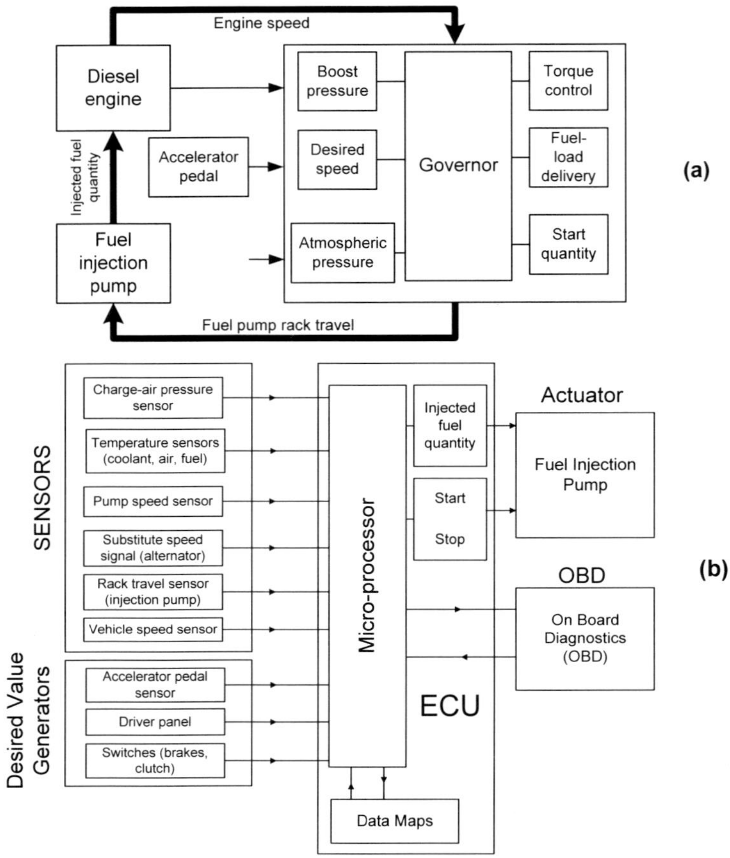

Figure 3.11. *a:* Mechanical governor closed-loop control of a vehicle engine. *b:* Electronic governor closed-loop control of a modern passenger vehicle engine (reprinted with permission from Bosch [3])

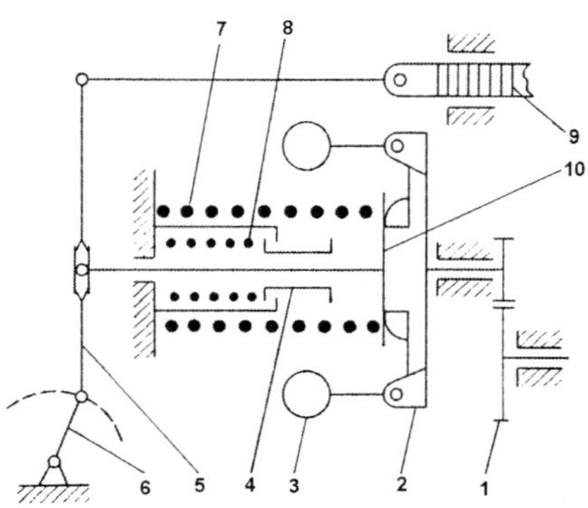

Figure 3.12. Min-max governor (1: step-up gear, 2: cross-piece, 3: flyweight, 4: support bushing, 5: lever, 6: control lever, 7: external spring, 8: internal spring, 9: fuel rack, 10: clutch) (from Krutov [4], reprinted with permission from the Russian Authors Society)

The characteristics of the governor depend upon the specific application. For example, the vanishing feedback indirect (isochronous) governor is ideal for electrical generation applications, where strict maintenance of engine rotational speed is desired irrespective of engine load (see also Figure 1.10). On the contrary, vehicular applications require less tight governing, otherwise the driveability of the vehicle would be impossible. Agricultural tractors, road-sweeping machines, ships *etc.*, use variable speed governors, which regulate speed over the whole engine operating range; for example in ships, a variable speed governor maintains each defined cruising speed.

In Figure 3.13, which illustrates a typical variable speed governor, the springs are incorporated into the flyweights; the latter move outward within the specified adjustment range as long as speed increases. Each control lever position is allocated a given speed at which regulation begins. Control lever movement is transmitted through the two piece linkage lever and the guide block to the variable fulcrum lever, and thus to the control rack.

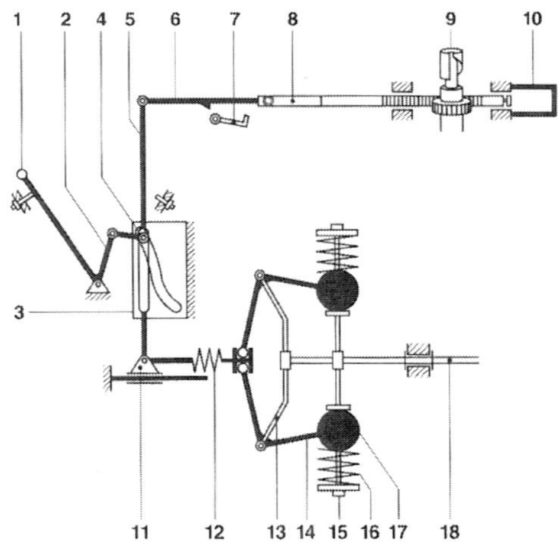

Figure 3.13. Variable speed governor (1: control lever, 2: linkage lever, 3: plate cam, 4: guide block, 5: fulcrum lever, 6: link fork, 7: full-load stop (automatic), 8: control rack, 9: pump plunger, 10: start quantity stop, 11: slider, 12: sliding bolt with drag spring, 13: governor hub, 14: bell crank, 15: adjustment nut, 16: governor spring, 17: flyweight, 18: camshaft) (reprinted with permission from Bosch [3])

Typical steady-state curves of a variable speed governor are illustrated in Figure 3.14. In addition to the low- and high-idle (maximum) speeds, this governor regulates intermediate speeds independent of engine load; the speed is set at the governor control lever. With reference to Figure 3.14, the engine starts with the start quantity determined through the automatic starting rack travel. During warm-up, low-idle speed finally levels off at point L. Full-load regulation follows the full-load curve, and torque control takes place between speeds N_1 and N_2 until speed regulation break-away starts at maximum full-load speed following the line from N_{vo} (maximum full-load speed) to N_{lo} (maximum speed). The remaining curves show the break-away characteristics for the intermediate speeds, whereby the increase in speed droop for decreasing engine speed is evident [3].

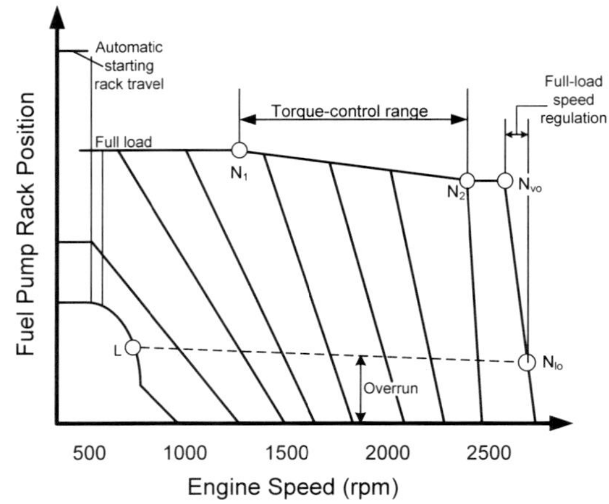

Figure 3.14. Steady-state curves of a mechanical variable-speed governor (reprinted with permission from Bosch [3])

Krutov [4], Welbourn *et al.* [5], Webb and Janota [6] and Catania *et al.* [7] discuss various issues concerning governor application and control under steady-state and transient conditions applying transfer function analysis. They also propose equations for modeling the various governor types, as they will be described briefly in the next sub-section. Rackmil and co-workers [8] discuss in detail the operation of an electro-hydraulic governor and its auxiliary functions (buffer system by-pass, compensation cut-off, load regulation and fuel limiting) for transient operation of a two-stroke, locomotive diesel engine applying quasi-linear engine modeling.

3.2.2 Governor Equations

A simple mechanical sensing element, following the Watt principle,[3] is illustrated in Figure 3.15. Here, the centrifugal force of the rotating flyweights (2) is opposed by the spring (9) force. In this sensing element the acting spring is located far from the flyweights and opposes their centrifugal force through the lever (10) motion.

The supporting, *i.e.*, centrifugal force of the flyweights for the mechanical sensing element of Figure 3.15 can be expressed by the following relation [4]:

$$F_{sup} = (a_{se} + b_{se} z) N_{gov}^2 \tag{3.22}$$

[3] James Watt (1736–1819) designed in 1787 a centrifugal governor to control the speed of a steam engine. Significant theoretical studies on governing were conducted during the second half of the 19th century by C. W. Siemens in 1853 and 1866, and James Clerk Maxwell (1831–1879); the latter presented in 1868 before the Royal Society of UK his classic work 'On governors' dealing with the equations of motion of a closed-loop system, such as the steam engine fitted with a Siemens governor.

where $a_{se} = (nm)_{fw} \ell \cdot r_o / d$ and $b_{se} = (nm)_{fw} \cdot (\ell / d)^2$ with n, m the number and mass of flyweights, respectively, d, ℓ, r_o are lengths as shown in Figure 3.15, z is the current displacement of clutch (8) ranging from 0 to z_{max} and $N_{gov}=i_{gov}$ N, with i_{gov} the transmission ratio between crankshaft and governor axis (typical value 0.5).

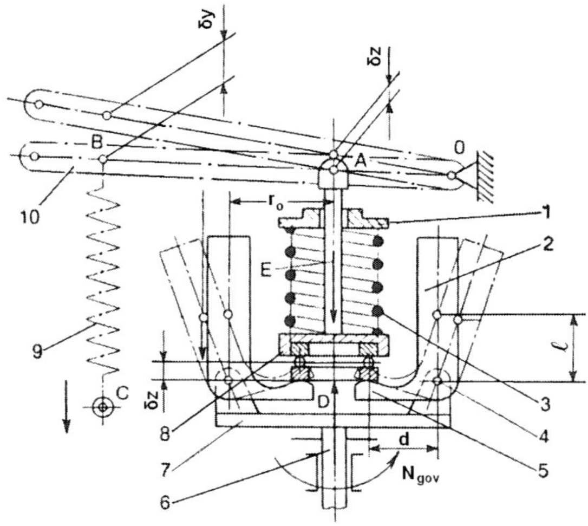

Figure 3.15. Mechanical speed sensing element (1: spring thrust plate, 2: flyweights, 3: spring alligned with clutch, 4: weight pivot, 5: lug, 6: governor shaft, 7: cross-piece, 8: clutch, 9: spring remote from clutch, 10: lever) (from Krutov [4], reprinted with permission from the Russian Authors Society)

The spring (9) restoring force is expressed as follows

$$F_{rest} = k\, i_{tr}\, (y + i_{tr} z) \tag{3.23}$$

where y is the (constant) prior spring strain (a measure of 'throttle position', often referred to as the 'governor setting') made possible either by moving thrust plate (1) or point C of spring (9), k is the spring stiffness (constant or varying with engine speed for conical springs with square-law characteristics) and $i_{tr}=\delta y_B/\delta z$ is the transmission ratio of the mechanism coupling the clutch (8) with the spring (9) by means of lever OB. For $i_{tr}=1 \Rightarrow \delta y_B \equiv \delta z$, which means that the restoring spring deformation is equal to the displacement z of the clutch (8). At steady-state conditions the above mentioned two forces are in equilibrium, so that it holds (neglecting the weight of ballarms and clutch) that

$$F_{sup} = F_{rest} \Leftrightarrow z = \frac{a_{se}\, i_{gov}^2\, N_{gov}^2 - k\, i_{tr}\, y}{k\, i_{tr}^2 - b_{se}\, i_{gov}^2\, N^2} \tag{3.24a}$$

For an electrical sensing element the supporting force is electromagnetic; a core of known reluctance and cross section area plays the role of the flyweights. A voltage is applied at its ends proportional to the current angular velocity ω of the engine. Speed changes are transformed into voltage changes and the core pushes the spring or is being pushed by it. Equation 3.22 holds true again, but now a_{se} is a function of core reluctance, cross section area and gain factor between engine speed and induced voltage at the core edges; b_{se} on the other hand equals zero. By altering the prior spring strain y, one manipulates the sensing element equilibrium curve (3.24a), the initial operating point and, thus, the whole engine speed response.

As with almost all aspects of transient operation, the governor steady-state curves (Figure 3.14 or Equation 3.24a) differentiate during a load or speed change. During transient operation, Newton's second law of motion for the mechanical sensing element of Figure 3.15 states

$$\underbrace{F_{sup}}_{\text{Supporting force}} + \underbrace{F_{2p}}_{\text{2 pulse term}} - \underbrace{F_{rest}}_{\text{Spring force}} \underbrace{- f_{gov}\frac{dz}{dt}}_{\text{Friction}} = \underbrace{m_{gov}\frac{d^2z}{dt^2}}_{\text{Acceleration}} \tag{3.24b}$$

where f_{gov} is the friction coefficient in Ns/m of the governor clutch (mainly a function of its speed) and m_{gov} is the mass of clutch and flywheels.

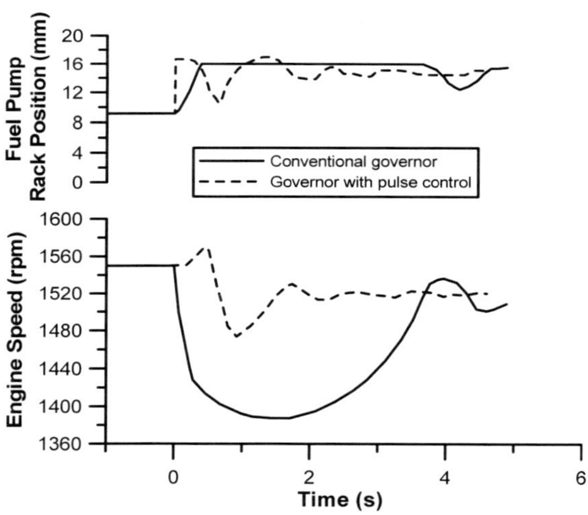

Figure 3.16. Effect of pulse governor (differential control) on the transient performance of a 12 cylinder turbocharged diesel engine driving an electrical generator (reprinted with permission from Zellbeck and Woschni [9])

The term F_{2p} (=$A_{2p}\varepsilon\, i_{gov}$ N) exists only for a *two-pulse* sensing element, *i.e.*, one which responds not only to the angular velocity variations but also to the crankshaft angular acceleration ones (this kind of feedback was defined in Section 3.2.1 as differential control). It can, therefore, prove very effective for improving engine maximum speed droop, although it may lead to an unstable recovery (Figure 3.16). At steady-state conditions it behaves as an ordinary mechanical sensing element being governed by the same equilibrium Equation 3.24a, though at transient conditions the second term on the left-hand side of Equation 3.24b appears (since angular acceleration $\varepsilon \neq 0$); term A_{2p} depends on the design of the ballhead and its mass moment of inertia.

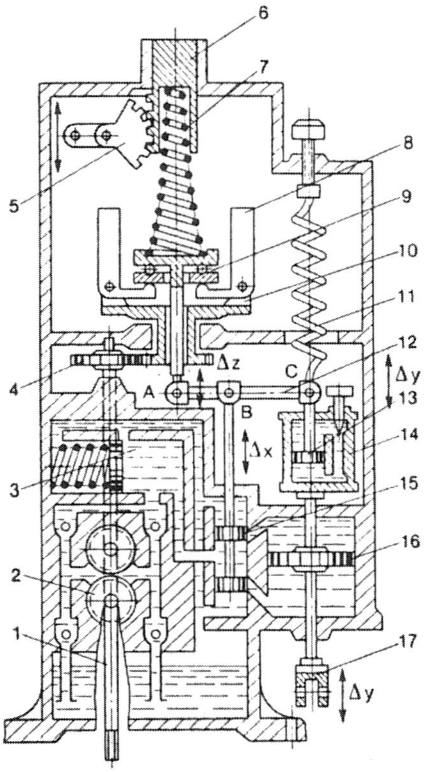

Figure 3.17. Isochronous governor with P-I feedback (1: driving shaft, 2: oil pump, 3: oil pressure accumulator, 4: driving gear, 5: spring prior strain variator, 6: spring upper support, 7: spring, 8: flyweight, 9: clutch, 10: cross-piece, 11: P-I element spring, 12: feedback lever, 13: P-I element piston, 14: P-I element, 15: control valve, 16: servopiston, 17: rod) (from Krutov [4], reprinted with permission from the Russian Authors Society)

Indirect acting governors, whether of unity or vanishing feedback, almost exclusively make use of hydraulic servomotors as power elements and are usually equipped with mechanical sensing elements. In cases where a governor must ensure a strictly constant engine speed irrespective of load, vanishing feedback is applied, which is a proportional plus integral (P-I) controller (Figure 3.17 with $i_{tr}=1$ for the sensing element). Depending on the load applied, piston (16) and rod (17), coupled with the rack, occupy distinct positions under the steady-state conditions, whereas point C is moved to its initial position by means of spring (11) when the transient terminates. Since point B of lever (12) occupies one and always the same position under all steady-state conditions, point A, coupled with clutch (9) of the sensing element, must also occupy one and the same position under all steady-state conditions. This may happen only at one and the same angular velocity, set by support (6), which occupies a permanent position. Thus, after a load change, restoration of speed is achieved to its initial value [4].

Rakopoulos *et al.* [10] based on the above analysis of Krutov [4], modeled several governor sensing elements and feedbacks; they coupled the analysis with a comprehensive, filling and emptying transient simulation code for a single cylinder, naturally aspirated diesel engine operating under load increase transients. Their analysis revealed the facts detailed below.

- Geometrical and technical characteristics of the governor affect the engine response seriously in terms of speed droop, recovery period and whole speed response profile. The impact of each term, however, was sometimes conflicting when comparing speed droop and recovery period.
- Speed droop is favored when the governor is equipped with a mechanical sensing element rather than an electrical one (when both possessing the same equilibrium curves and initial operating points), while the recovery period worsens. On the other hand, two-pulse sensing elements prove even more satisfactory due to their dependence on the angular acceleration term, which is of great importance during transient operation, but may lead to slight instability at the final operating condition, as was documented in Figure 3.16.
- Instantaneous maximum speed droop improves, *i.e.*, it is lower, with higher amplification in the servomechanism, though at the expense of final speed instability; it also improves with higher values of the number and mass of flyweights $(nm)_{fw}$, sensing element spring stiffness k and sensing element spring prior tension y, which lead to initial conditions close to the maximum spring deformation.
- The recovery period deteriorates with smaller values of the mass and number of flyweights $(nm)_{fw}$, low values of P-I element spring stiffness (or higher values of P-I element spring friction coefficients for the vanishing feedback governor) and extremely high amplification, though the trend is not always monotonic.
- The servopiston surface area, the density of the oil used and the differential pressure of the servomechanism all have a marginal effect on the transient response of the engine.

3.3 Friction

3.3.1 Friction Fundamentals

Frictional energy losses inside an internal combustion engine arise from the shearing of oil films between the various working surfaces, *e.g.*, between piston rings and cylinder liner or inside the journal bearings; friction is, thus, interrelated with lubrication. Friction energy is ultimately removed as wasted heat by the cooling system of the engine. The interest in determining friction losses in engines has increased recently owing to the global requirements for lower fuel consumption and decreased CO_2 emissions, but also for increased engine durability. A detailed review of steady-state engine friction can be found in Ferguson and Kirkpatrick [11] and Heywood [12]. In this book, only some basic aspects will be discussed that deal with the peculiarities of friction development during transient operation.

An important aspect of friction theory is the mode of lubrication, which, as shown in Figure 3.18, can be hydrodynamic, mixed or boundary. In hydrodynamic friction, the surfaces are separated by a liquid film, minimizing the respective wear. This is the desired lubrication mode during engine operation. As pressure increases or speed decreases, oil film thins out to the point where its thickness is comparable in size to the surface irregularities. This is the mixed lubrication regime. With further increase in load or decrease in speed, the boundary layer regime is reached, which, for internal combustion engine applications, is experienced, for example, around dead centers for piston rings. In Figure 3.18, the Stribeck variable is equal to $\mu N/p$, with μ the lubricant dynamic viscosity, N the relative speed between the two surfaces, and p the normal stress. The ordinate in Figure 3.18 is the friction coefficient, defined as the ratio of shear or tangential stress to normal stress acting on a surface.

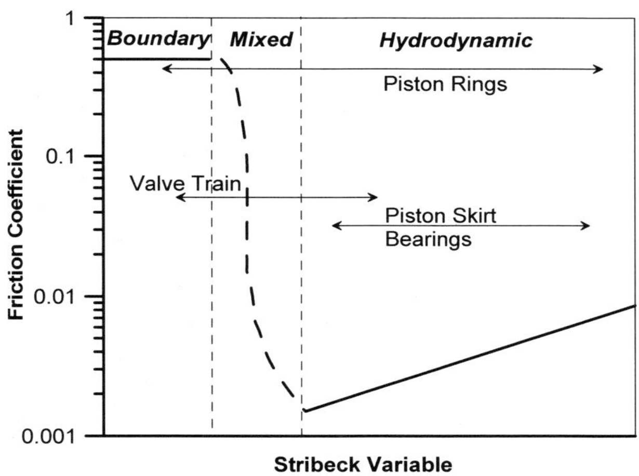

Figure 3.18. Stribeck diagram of friction coefficients for various internal combustion engine components

Engine friction losses comprise of: 1. mechanical friction and 2. accessories work (oil-, water-, fuel pumps, alternators, gears, *etc.*). Following Gish *et al.* [13] in this book, pumping losses are accounted for in the engine bmep by adding both closed and open parts' work and not in the frictional losses.

Mechanical friction occurs in

- piston assembly, consisting of compression and oil rings, and piston skirt;
- valve train; and
- loaded (main crankshaft and connecting rod) bearings.

Around 40–55% of total friction losses in a diesel engine are attributed to the piston assembly friction, 20–30% to loaded bearings, 10–15% to valve train, whereas 15–25% is due to engine auxiliary losses. Piston assembly is dominated by the rings friction contribution. Ring assembly lubrication is, mainly, hydrodynamic, except near dead centers where the oil film breaks down and mixed or even boundary (at firing TDC) lubrication is established. Piston skirt friction losses are generally much lower since the skirt is separated from the cylinder liner by a relatively thick oil film and the gas pressure load has already been largely carried out by the rings; lubrication is here hydrodynamic. Lubrication in the valve train is both boundary and mixed, with the respective frictional losses occurring at the cam-tappet interface, camshaft bearings, valve guides and seals, and at the rocker arm–fulcrum contact. Hydrodynamic lubrication is established in loaded bearings, with mixed conditions assumed around firing TDC, where loading increases considerably. Finally, auxiliary mechanical losses are, mainly, proportional to engine speed.

Ultimately, engine friction reduces the work produced by the thermodynamic gas processes inside the cylinder, and can therefore be calculated from the difference between indicated and brake work as it is described in the following equation on a fmep (friction mean effective pressure) basis

$$\text{fmep} = \text{imep} - \text{bmep} \tag{3.25}$$

For the calculation of engine friction, the following three approaches have been applied [14–17]:

1. global evaluation of mean, over the engine cycle, fmep equations;
2. evaluation of separate mean fmep equations for each component (bearings, valve train, *etc.*) mentioned above; and
3. evaluation of instantaneous friction components contribution at each degree crank angle.

In the first approach, which is the most frequently applied, use is made of the following general type equation:

$$\text{fmep} = \alpha + \beta\, p_{max} + \gamma \bar{u}_{pist} \tag{3.26}$$

where p_{max} is the peak cylinder pressure, $\bar{u}_{pist} = S \cdot N / 30$ is the mean piston velocity, and α, β and γ are constants specific for each engine that are derived after calibration against experimental data at steady-state conditions.

However, it is a well-known fact that friction torque varies significantly during an engine cycle (Figure 3.19); its magnitude compared with brake torque is not negligible, particularly at low loads where the most demanding transient events commence. Modeling of friction is, however, difficult due to the interchanging nature of lubrication and the large number of components that cannot easily be isolated, experimentally investigated, and studied separately even at steady-state conditions.

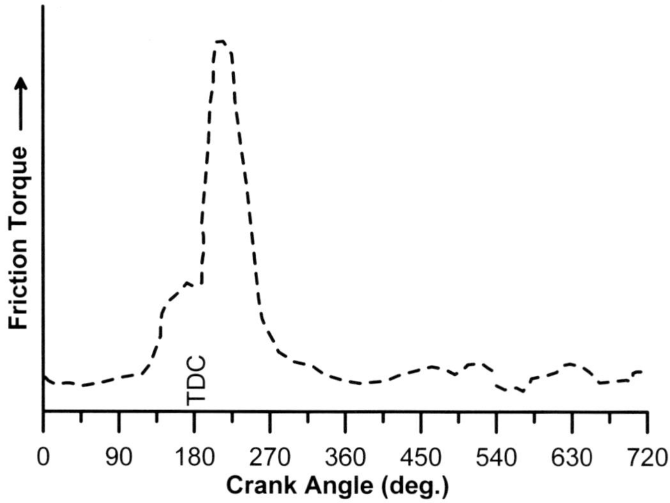

Figure 3.19. Instantaneous friction torque during an engine cycle

3.3.2 Development of Friction Torque during Transients

Many techniques of varying accuracy and complexity have been employed so far for measuring mean friction torque over an engine cycle at steady-state conditions. The most usual ones are: motoring method (here, with partial engine disassembly the distribution of individual components friction can be also estimated), Willans line (extrapolation of bmep *vs.* fuel flow lines to the point of zero fueling), Morse test (one cylinder in a multi-cylinder engine is not fired, reducing accordingly the engine total power) and indicator diagram method (subtraction of bmep, as found from dynamometer measurements, from imep, as computed from cylinder pressure measurements). For the continuously variable speed and fueling conditions, however, of a transient event, these techniques cannot be applied. Instead, instantaneous, total (from all components) transient friction torque can be measured, without need for engine modification, using

- a torque-meter (transducer) between engine and load for the instantaneous resistance torque measurement;
- a piezo-electric transducer for in-cylinder pressure measurement;
- an electromagnetic pick-up or shaft encoder for capturing instantaneous engine speed;
- a comprehensive storing and accurate data analyzing system;
- solution of Equation 3.19 for the unknown friction torque term, using the above-mentioned signals after they have been carefully treated and phased, *i.e.*,

$$\tau_{fr}(\varphi) = \tau_e(\varphi) - \tau_L(\varphi) - G_{tot}\frac{d\omega}{dt} \quad (3.27)$$

Solution of Equation 3.27 can be employed for friction torque evaluation at each degree crank angle during steady-state conditions too, as was the case in Figure 3.19.

Since the technical difficulties involved are considerable, unsurprisingly there is a scarcity of experimental results regarding development of (components) friction torque during transients. Figure 3.20, taken from the work of Winterbone and Tennant [18], shows the variation of mean values over an engine cycle for all torque terms of Equation 3.27 during a load acceptance transient event of a six-cylinder, turbocharged, truck diesel engine applying the above-mentioned experimental procedure.

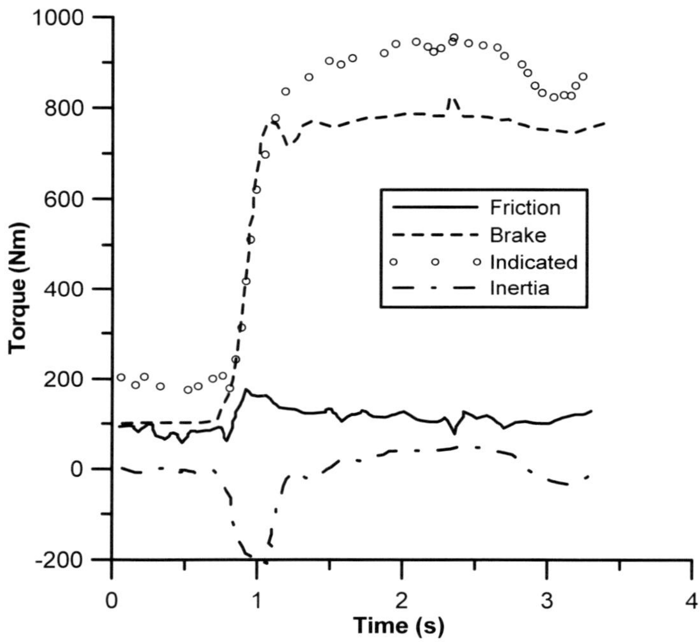

Figure 3.20. Development of mean, over each engine cycle, transient torque terms during acceleration of a turbocharged diesel engine (experimental results reprinted with permission from SAE Paper No. 810339 [18], © 1981 SAE International)

The interesting finding from Figure 3.20 is that the friction torque level overshoots the final steady-state value as the load is applied; this overshoot was estimated between 50 and 100% of the final level of friction. Winterbone and Tennant [18] concluded that during transients, friction is characterized by non-steady-state behavior, differentiating engine response and performance when compared with the corresponding steady-state values predicted by Equation 3.26. Consequently, they proposed that friction torque should generally be overestimated by some percentage during the transient event to account for the peculiarities of transient operation.

The postulated reasons for the apparent increase in friction during transients are

- during the first cycles of the transient event, oil temperature is still low (lower than at the corresponding steady-state operation for the same fueling and engine speed) resulting in higher friction losses compared with steady-state conditions (*cf.* Figure 9.10 for the effects of oil temperature on viscosity, hence friction loss);
- crankshaft deflection is not the same as for the respective steady-state conditions; and
- crankshaft is distorted due to the additional torque being transmitted during both acceleration and deceleration (see also Section 3.4), and also due to the release and absorption of angular momentum stored in the crankshaft mechanism.

Figure 3.21 illustrates the development of the instantaneous upper piston ring friction coefficient during an early cycle of a load increase transient of a turbocharged diesel engine; it refers to simulated results from an experimentally validated transient diesel engine simulation code [1, 19] incorporating a detailed (on a degree crank angle basis) friction model. It is observed that the period of mixed lubrication lasts longer around firing TDC than around the other three dead centers in the cycle. This is due to the increased piston ring friction force F_{PRA} resulting from increased gas pressures during the closed part of the cycle. The important finding from Figure 3.21 is that the friction coefficient development (profile *and* absolute values) remains practically unaltered for each cycle of the transient event (only the duration of mixed lubrication around firing TDC may increase slightly with increased loading/fueling); the maximum value of friction coefficient is determined by the condition of the wall material boundary lubrication. Thus, the values of the corresponding mean or maximum – over an engine cycle – piston rings friction force or torque during the transient event are determined solely by the mean or maximum gas pressures, respectively.

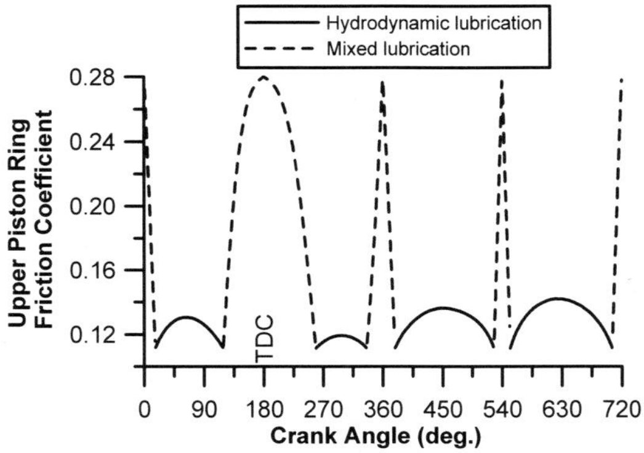

Figure 3.21. Development of upper piston ring friction coefficient during an early cycle of a 10–75% load increase transient event (simulation results)

Figure 3.22 focuses on the development of the reduced transient 'fmep' term, which can be identified as follows:

$$(\mathrm{fmep})_{\mathrm{trans}} = \frac{2\pi \int_0^{720} \tau_{\mathrm{fr}}(\varphi)\,d\varphi}{A_{\mathrm{pist}}\, r} \tag{3.28}$$

This fmep response arises from the synergistic effect of gas pressure (through fuel pump rack position response) and engine speed development. Bmep development for cylinder No. 1 of the six-cylinder engine under study is also provided on this figure for comparison purposes.

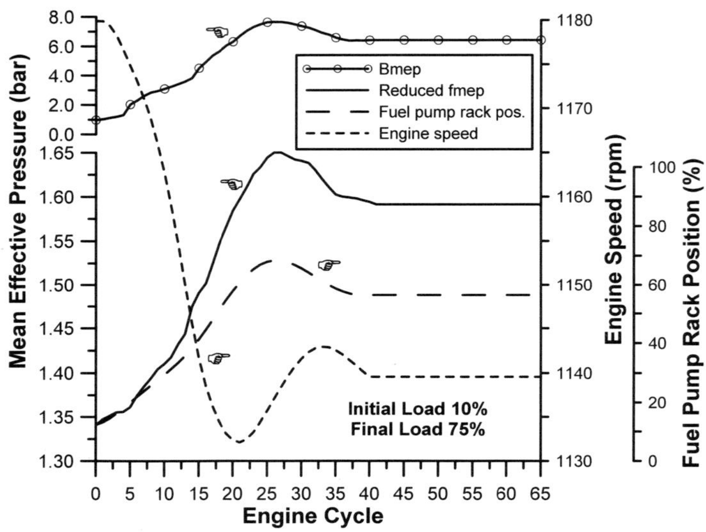

Figure 3.22. 'Reduced' fmep development *vs.* engine cycle for a 10–75% load increase transient event

Loaded bearings and auxiliaries are mostly dependent on engine speed, whereas the impact of gas pressure is dominant for the piston rings assembly. This is highlighted in Figure 3.23, which shows the development of the various friction components' reduced fmep for the same transient load increase of Figure 3.22. The dominance of the piston rings assembly is obvious for the present engine (five piston rings in total, of which the two oil rings each have 6 mm width). An increasing loading, *i.e.,* increased fueling, and hence gas pressures throughout the transient event are responsible for the increase of piston rings friction values. The primary mechanism here is the increase of piston rings and skirt friction forces, although the corresponding values of the Stribeck diagram and friction coefficients behave in an opposite way. On the other hand, piston rings friction force generally decreases with a decrease in engine speed, due to the lower values of mean piston speed as the load increase develops. For the examined transient event, however, the effect of engine speed is modest, owing to the small speed droop observed. All the other friction terms have smaller contribution to the total transient fmep, especially at the low engine speed under study. This holds, particularly, for the loaded bearings term, which is heavily dependent on engine speed. Only the auxiliaries' friction exhibits a markedly increasing trend during the transient event, due to the significant increase in injected fuel quantity from cycle 15 onwards.

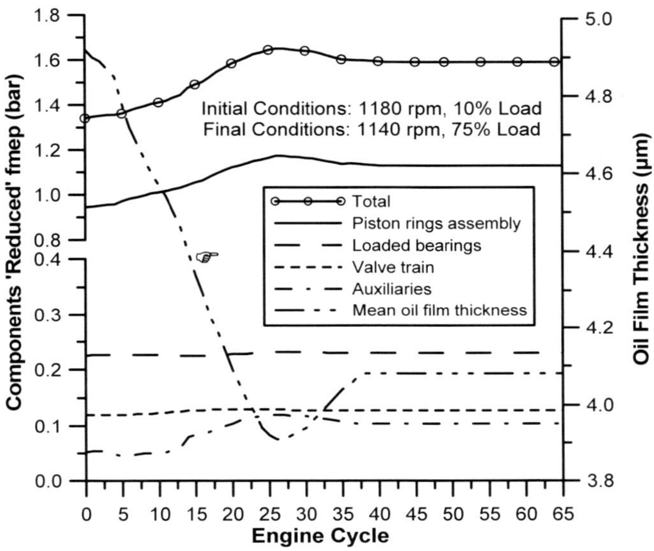

Figure 3.23. 'Reduced' components fmep and mean oil film thickness *vs.* engine cycle for a 10–75% load increase transient event

3.4 Crankshaft Torsional Deformation

An important aspect of internal combustion engine operation is that instantaneous engine torque fluctuates significantly during an engine cycle even under steady-state conditions. The main mechanism behind this is the cyclic nature of gas pressures and inertia reciprocating forces. This fluctuation may be of considerable magnitude, particularly during turbocharged diesel engine operation, where the cylinder pressure assumes very high values. On the other hand, resistance (load) torque remains practically constant during a cycle, owing to the adequately low non-uniformity of rotation; the latter being, mainly, determined by the flywheel mass moment of inertia. As a result, a significant fluctuation occurs in the instantaneous net (engine minus load) torque that eventually leads to cyclic speed irregularities, twists between individual cranks of a multi-cylinder engine and, finally, torsional (angular) deformation of the whole elastic crankshaft affecting also the bearing oil film thickness. Crankshaft deformation is further enhanced during transient operation owing to the dynamic instability induced by the considerable deficit of torque, during the early cycles of the transient event after a new, increased, load has been applied, or due to the considerable surplus of torque after abrupt fueling increase. With reference to Equations 3.15, the instantaneous torsional deformation of the crankshaft due to the torque difference between engine and load is defined as (φ–φ_L). Figure 3.24 illustrates the development of crankshaft torsional deformation during typical, steady-state operation of a single-cylinder diesel engine. This figure should be studied in conjunction with Figure 3.25 that depicts the development of all torque contributors during the same engine cycle [20].

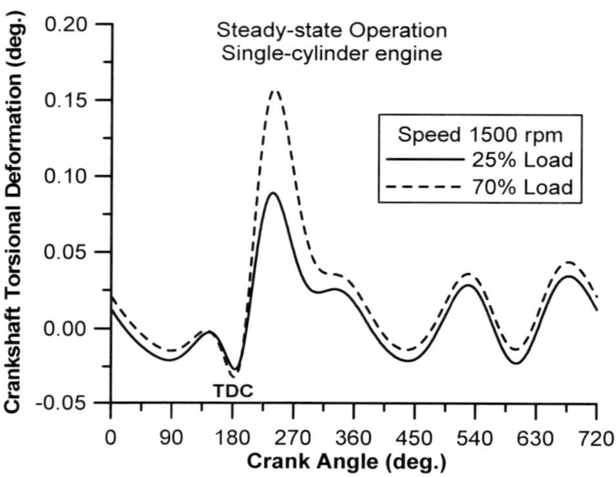

Figure 3.24. Development of crankshaft torsional deformation during steady-state, single-cylinder operation for two different engine loads (simulation results)

During compression (0–180°CA), a deficit of gas torque is observed that leads to engine speed decrease (upper sub-diagram of Figure 3.25) and to the 'negative' crankshaft deformation shown in Figure 3.24; torque, speed and deformation all peak around 'hot' TDC. After the start of combustion, there is a surplus of torque, as now the engine enters the power producing phase of operation. Consequently, 'positive' deformation is established, while the instantaneous engine speed increases. This lasts for the whole expansion stroke. The considerably higher amount of engine torque produced during expansion leads to the greater local peak in crankshaft deformation, *i.e.*, for the present engine 0.16° occurring at about 55°CA after 'hot' TDC, compared with the local minimum of –0.03° at 180°CA (70% load operation in Figure 3.24).

The main mechanism behind the crankshaft torsional deformation profile, over an engine cycle, is clearly the gas torque because of its direct impact on the total engine torque; in fact, Chen and Chen [21] concluded that torsional angle amplitude is, practically, a linear function of the corresponding gas torque. Closer examination of Figure 3.24 reveals that inertia torque influence is also present, mainly during the open part of the cycle where the cylinder pressure is low, as well as during the second half of compression. For the present engine the inertia contribution is rather small, due to the low engine speed (recall from Equation 3.5 that inertia forces vary according to the square of the engine speed).

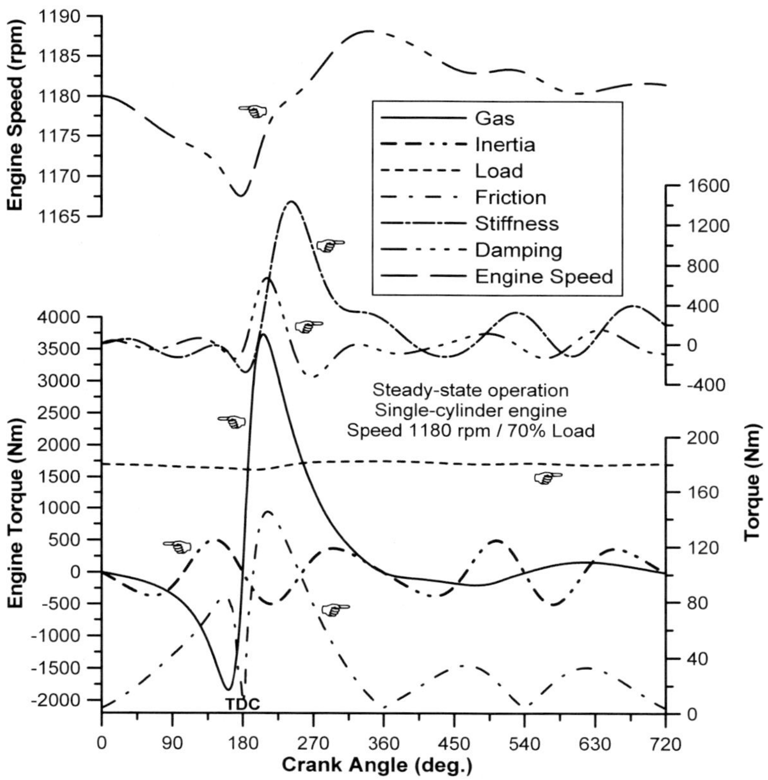

Figure 3.25. Engine speed and various torque contributors build-up during steady-state, single-cylinder engine operation (simulation results)

An interesting finding is that stiffness and damping instantaneous torques may indeed reach high values during an engine cycle. The profile of stiffness torque follows the torsional deformation as dictated by Equation 3.17. The profile of damping torque is indicative of the instantaneous difference between engine and load speeds ($\omega - \omega_L$). Owing to the 'internal inertia' of these 'processes', the respective torque peaks observed in the center sub-diagram of Figure 3.25 may be delayed compared with the initiating gas torque or cylinder pressure. However, despite their relatively high instantaneous values, both stiffness and damping torque mean (over the engine cycle) values are negligible compared with their gas or load counterparts.

The torsional deformation for the 25% load is also depicted in Figure 3.24 for comparison purposes. Here, lower maxima of crankshaft deformation are, overall, observed. This was intuitively expected because of the lower values of fueling and consequently engine gas torque produced during this cycle (inertia forces retain the same values as in the 70% load case, since engine speed is the same). Likewise, smaller deformations are experienced in naturally aspirated diesel or in spark ignition engines, where cylinder pressures are much lower. In the latter cases, a greater influence of inertia torque is also expected, particularly for small SI (automotive) engines operating at high rotational speeds.

Figure 3.26 expands the previous findings to the transient operation illustrating the development of the maximum and the mean, over each engine cycle, deformation and stress for a typical 10–80% load increase transient event of a turbocharged diesel engine.

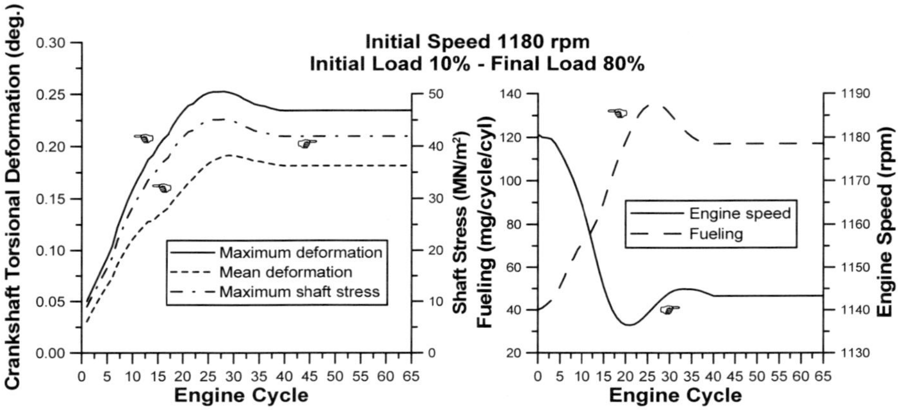

Figure 3.26. Development of the maximum and mean, over an engine cycle, crankshaft torsional deformation and stress during a 10–80% load increase transient event (simulation results)

Initially, the deformation is negligible due to the low engine load (*cf.* Figure 3.24). As the governor responds to the drop in engine speed caused by the abrupt load increase, fueling increases too (right side sub-diagram of Figure 3.26) leading to higher gas pressures and torques throughout the cycle; this results in greater maximum and mean, over the engine cycle, deformations. It is important to note that the instantaneous maximum deformation is considerably higher (up to 50% for the specific engine transient) than the respective mean value in the same cycle; this fact justifies the analysis on a degree crank angle basis, in order to be able to estimate the 'true' maximum stress (left sub-diagram of Figure 3.26) that is instantaneously experienced by the (crank)shaft, *i.e.*,

$$\sigma_{max} = \frac{\Theta}{2}\frac{d}{\ell}\Delta\varphi_{max} = \frac{\Theta}{2}\frac{d}{\ell}(\varphi_e - \varphi_L)_{max} \tag{3.29}$$

where Θ is the shear modulus and d, ℓ are the shaft diameter and length, respectively, between engine flywheel and load (shown in Figure 3.6).

It should be pointed out here that the evolution of transient maximum or mean, over the engine cycle, deformation and, hence, stress develop in a different way compared with the corresponding steady-state engine operation points (*i.e.*, at the same engine speed and fuel pump rack position). The difference is, mainly, attributed to

- the different air–fuel equivalence ratios experienced during transients owing to the turbocharger lag, which significantly affects the air-mass flow-rate, particularly during the early cycles of the transient event; and
- the transient operation of the fuel pump that differentiates from the steady-state fuel pump curves.

The above remarks are expanded in Figure 3.27, by showing the 'wave' of crankshaft deformation build-up for several cycles of a similar transient event. What is interesting here is that the increase in loading/fueling during the transient event leads also to greater in-cycle deformation fluctuations; the respective deformations between individual cylinders are, as expected, much smaller owing to the higher stiffness involved.

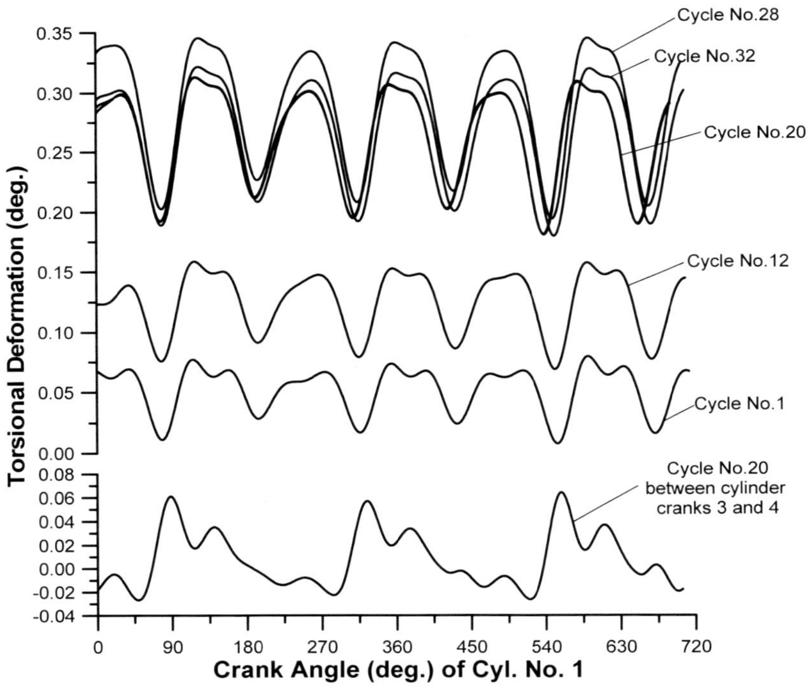

Figure 3.27. Development of torsional deformation between engine and load (upper sub-diagram) and between cylinders 3 and 4 (lower sub-diagram) during various cycles of a six-cylinder engine load increase transient event (simulation results)

3.5 Introduction to Vehicle Dynamics

Application of Transient Cycles for exhaust emission certification of passenger, light and heavy-duty, diesel-engined vehicles has pointed out the importance of studying the transient operation of the whole vehicle and not just of the engine. To this aim, all vehicle's sub-systems need to be properly investigated; the interconnection between engine, transmission and suspension/tire/road system are of particular importance in determining performance, fuel consumption and exhaust emissions. The latter interconnection induces a much higher degree of complication in the relevant studies (particularly if some kind of hybridization scheme is involved – Section 6.5.4), which need to be extended accordingly in order to account for the various dynamic and control issues incurred.

3.5.1 Simplified Analysis

Figure 3.28 illustrates a simplified engine powertrain module, consisting of the engine, (manual) transmission, final drive-train and wheels; this can be usually considered as an adequate first approximation for engine–vehicle powertrain studies.

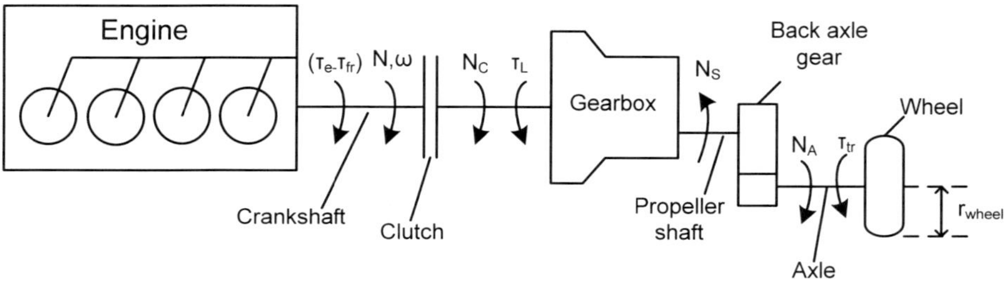

Figure 3.28. Schematic arrangement of simplified engine–gearbox–wheel drivetrain

With reference to Figure 3.28, Newton's second law of motion, expressed on the crankshaft axis, states [22–24]

$$\left[\tau_e(\varphi) - \tau_{fr}(\varphi)\right] - \frac{\tau_L}{\eta_{gear}} = \tau_A = G_V \frac{d\omega}{dt} \tag{3.30}$$

The above relation holds true when the clutch is engaged, *i.e.*, speed N_C equals engine rotational speed N. For the case where the gears are not engaged ($N_C \neq N$), Equation 3.30 is transformed into

$$\frac{\tau_L}{\eta_{gear}} + G_V \frac{d\omega}{dt} = 0 \tag{3.30a}$$

In Equation 3.30, engine brake torque $(\tau_e - \tau_{fr})$ is absorbed by the resistances to the vehicle's motion τ_L or by the vehicle acceleration τ_A. Further, we can assume that the various gears (clutch, gearbox, back-axle) absorb another portion of the engine brake torque owing to viscous and mechanical losses; the latter is defined through total gear efficiency η_{gear}. Also, N_S is the speed of the propeller shaft (if applicable), N_A the wheel rotational speed, i_b is the back-axle ratio and i_g the engaged ratio in the gearbox; the latter ratios are defined as follows with reference to Figure 3.28

$$i_g = \frac{N_C}{N_S} \text{ and } i_b = \frac{N_S}{N_A} \tag{3.31}$$

Instantaneous speed reached by the vehicle is then given by

$$V(km/h) = 2\pi r_{wheel} \frac{N_C(rpm)}{i_b i_g} \cdot 60 \times 10^{-3} \tag{3.32}$$

The main concern in this kind of study, apart from the correct engine torque evaluation, is the accurate estimation of the various resistances to the vehicle's motion. The three forces, which a vehicle experiences on the road (a fourth term is also present in the more general case where the vehicle tows a trailer), are acceleration dependent inertia force, velocity dependent resistive force, and grade dependent force. Total traction force F_{tr} is given by

$$F_{tr} = F_a + F_r + F_{gr} \tag{3.33}$$

The various terms in Equation 3.33 are defined as follows:

- The aerodynamic force F_a is given by

$$F_a = \frac{1}{2} \rho_\alpha c_d A_f V^2 \tag{3.34a}$$

with A_f the vehicle frontal area, ρ_α the air density and c_d the vehicle aerodynamic resistance (drag) coefficient.

- The rolling resistance force F_r is given by

$$F_r = m_V (f + c_{tr} V) \tag{3.34b}$$

The latter incorporates the tire rolling resistance (through friction coefficient f), the tire deformation, the friction in the wheel bearing and any other friction elements that are a function of the vehicle motion (defined through vehicle speed V); m_V is the loaded mass of the vehicle.

- The resistance due to road grade is given by

$$F_{gr} = m_V g \sin\theta \tag{3.34c}$$

Grade is defined as the 'rise' over the 'run', being the tangent of the grade angle θ; grade may reach up to 4% in main roads and 10–12% in secondary roads.

For the resistance (traction) torque it holds that $\tau_{tr} = F_{tr}\, r_{wheel}$. This traction torque τ_{tr} needs to be reduced to the crankshaft axis in order to be used in Equation 3.30

$$\tau_L = \tau_{tr}\left(\frac{1}{i_b}\right)\left(\frac{1}{i_g}\right) \tag{3.35}$$

Finally, the vehicle's total moment of inertia, reduced to the crankshaft axis, is given by

$$G_V = m_V\, r_{wheel}^2\left(\frac{1}{i_b}\right)^2\left(\frac{1}{i_g}\right)^2 + G_e + G_{other} \tag{3.36}$$

with G_e the engine moment of inertia as discussed in Section 3.1.3 and G_{other} the inertia of wheels, tires, *etc.* (when reduced to the crankshaft axis). Both G_e and G_{other} terms are usually very small compared with their vehicle mass counterpart, particularly for large vehicles.

Using the previous relations, engine acceleration $d\omega/dt$ from Equation 3.30 is finally given by

$$\varepsilon = \frac{d\omega}{dt} = \frac{\left(\tau_e - \tau_{fr}\right) - \dfrac{\left(\frac{1}{2}\rho_\alpha\, c_d\, A_f V^2 + m_V\left(f + c_{tr} V\right) + m_V\, g\, \sin\theta\right)\left(\dfrac{1}{i_b\, i_g}\right) r_{wheel}}{\eta_{gear}}}{m_V\, r_{wheel}^2\left(\dfrac{1}{i_b}\right)^2\left(\dfrac{1}{i_g}\right)^2 + G_e + G_{var}} \tag{3.37}$$

It is obvious from the above analysis that the lower the engaged gear (this means high i_g), the smaller the total vehicle moment of inertia G_V, resulting in quicker acceleration according to Equation 3.37. On the other hand, the higher the engaged gear, the lower the acceleration rate, but the higher the vehicle speed as calculated from Equation 3.32. The latter remarks can be validated by a numerical example. Typical first gear reduction in a gearbox is around 9 for a 30 tn, 350 kW truck; combined with i_b=5.5 reduction ratio in the rear axle and 0.55 m wheel radius produces a total vehicle mass moment of inertia (excluding engine inertia G_e) $G_V^{1_{st}\,gear} = 30{,}000\cdot(0.55)^2\left(1/9\right)^2\left(1/5.5\right)^2 = 3.70$ kg m^2. The latter is of the same order of magnitude as the engine moment of inertia. For the ninth gear, however, the respective inertia is $G_V^{9_{th}\,gear} = 30{,}000\cdot(0.55)^2\left(1/0.8\right)^2\left(1/5.5\right)^2 = 468$ kg m^2, or almost 130 times the inertia of the first gear.

Various driving modes can be studied using the above-mentioned analysis, *e.g.*, starting the vehicle from rest, shifting into higher or lower gear, climbing a hill, *etc.* Winterbone *et al.* [22] discuss some of them; for example, shift into higher gear comprises of

- clutch disengagement and isolation of the engine from the vehicle after the maximum engine speed is achieved; the gear lever is moved to the next higher gear;
- the fuel pump rack is set to its minimum position (when the driver removes his foot from the accelerator pedal). Consequently, the engine decelerates until it reaches a speed equal to the current vehicle speed in the higher gear selected; the vehicle continues to roll, driven by the inertia of the system; and
- the clutch is engaged; the driver fully depresses the accelerator pedal and the rack moves to its maximum position; the engine and vehicle accelerate up to the maximum engine speed.

Figure 3.29 is a typical result of an upward gear change schedule commencing from standstill; it refers to a six-cylinder, turbocharged diesel engine of 11.32 L displacement volume, rated at 175 kW at 1800 rpm, mounted on a 32 t truck with 6.51 m^2 frontal area [22].

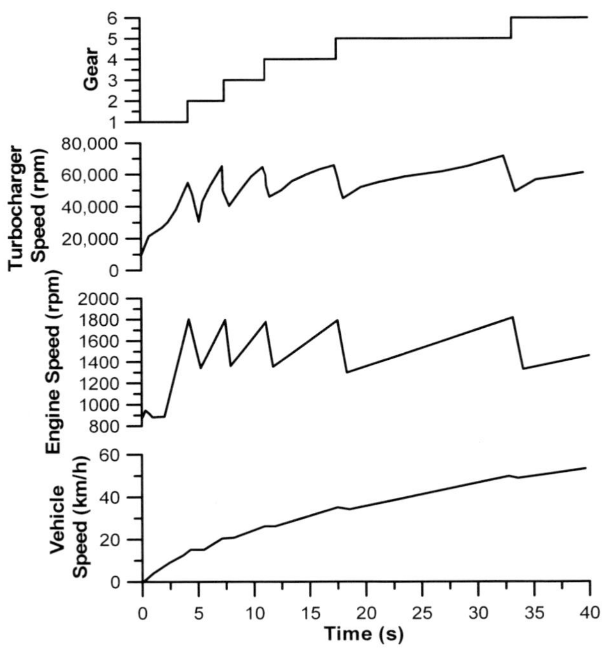

Figure 3.29. Up-gear change schedule of a turbocharged truck (experimental results reprinted with permission from SAE Paper No. 770122 [22], © 1977 SAE International)

The impact of turbocharger lag is here less prominent compared with load acceptance cases of the same engine. This is due to the fact that the acceleration of the engine helps the build-up of boost pressure. The first 5 s are the most critical part of the transient; this interval is influenced by turbocharger lag and system dynamics significantly more than the rest of the test, since the initial turbocharger speed and available exhaust gas enthalpy are minimal. As a result, smoke emissions are expected to peak during this period too. As the engaged gear increases, so does the inertia of the whole vehicle (Equation 3.36) leading to slower acceleration at higher engine and vehicle speeds.

Interestingly, the study of a similarly rated six-cylinder, two-stroke, turbocharged (with an additional Roots blower prior to the turbocharger compressor) diesel engine of 9 L displacement volume, having 37% more torque than its four-stroke counterpart of Figure 3.29, revealed 43% higher acceleration rate when both engines are installed in the same truck; the comparison concerned sixth gear acceleration [25]. The better response of the two-stroke engined truck was attributed to its Roots blower; the latter's steep pressure ratio *vs.* mass flow characteristics, typical of a positive displacement device (*cf.* Figures 2.5 and 2.7), resulted in an almost immediate rise in engine boost pressure, diminishing turbocharger lag and therefore aiding faster engine and vehicle acceleration.

3.5.2 Detailed Vehicle Dynamics Study

The analysis of Section 3.5.1 does not take into account special vehicle dynamic issues, *e.g.,* suspension effects, multiple driving axles, *etc*. Consequently, the study is engine rather than vehicle oriented. For the latter, more detailed analysis is needed treating the vehicle as separate lumped masses; the lumped mass representing the body is the 'sprung mass', and the wheels and axle masses supported by the tires are denoted as 'unsprung masses'. Such powertrain analyses incorporating detailed vehicle dynamics can be used [24, 26, 27]

- to obtain general vehicle performance data;
- for fuel consumption and exhaust emissions prediction during real driving conditions (gear shift, Transient Cycles);
- as hardware-in-the-loop development tools. An example is in the development of an electronic control module for a powertrain component. Typically, the control module may be simulated in software, and its performance can be assessed by coupling this simulation to the simulated component, *e.g.,* an engine and an engine control module system simulation;
- as vehicle simulators;
- as design tools to assist the design engineer early in the vehicle development process;
- for control system design; and
- for enhancing vehicle driveability.

Owing to the obvious experimental difficulties involved, detailed vehicular transient performance studies have been carried out, mainly, on a simulation basis [28–31] based on a modular modeling philosophy. In the following paragraphs, the analysis will be primarily based on the work conducted by Assanis and co-workers [29, 31]. Although this is not the place to discuss in detail vehicular dynamic issues, a few representative arguments will be made to aid the analysis of the following chapters.

A detailed vehicle dynamic simulation approach includes

1. the engine module describing the in-cylinder and manifolds thermodynamic processes;
2. the driveline module consisting of the torque converter, transmission, propeller shafts, differential, and drive shafts; this provides the connection between the engine and the vehicle dynamics module (see Figure 3.30);
3. the vehicle dynamics;
4. traction resistance as already described in Section 3.5.1; and
5. appropriate integration of all the above-mentioned modules.

A typical powertrain system of a 4×2 truck (*i.e.,* vehicle having two axles, the front-one steering and the rear-one driving) is shown in Figure 3.30.

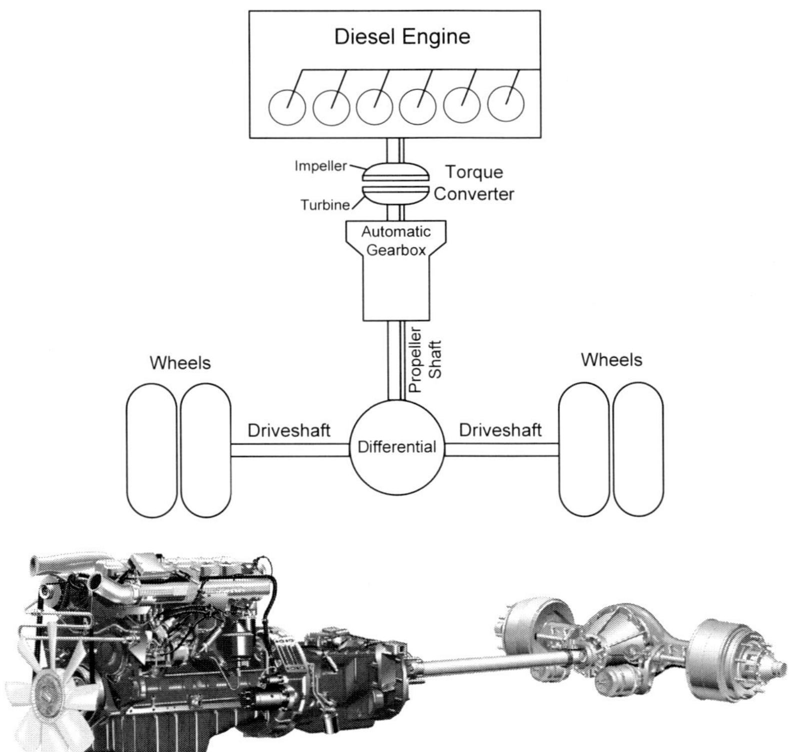

Figure 3.30. *Upper:* Schematic diagram of truck drivertrain with automatic transmission. *Lower:* Drivetrain of a 9 L displacement volume diesel-engined truck (courtesy of Scania CV AB, Illustration: Semcon Informatic Graphic Solutions)

The (turbocharged diesel) engine is connected to the torque converter, whose output shaft is then coupled to the automatic (or automated manual, or continuously variable) gearbox, propeller shaft, differential and two driveshafts, coupling the differential with the driven wheels. For manual gearbox, the engine is directly connected to the transmission via the clutch.

The torque converter input shaft, on the one end, and the wheels, on the other end, are the connecting points for the engine and the vehicle dynamics modules, respectively. The torque converter is usually a fluid coupling, utilizing hydrodynamic principles to amplify the torque input to the transmission at the expense of speed. It consists of an impeller, stator (reactor), and turbine (runner). The impeller is connected to the engine output shaft, and the turbine to the transmission input shaft. The stator is connected to the torque converter housing via a one-way clutch. The presence and arrangement of the stator cause the torque converter to act as a torque multiplication device when operating at low speed ratios, and as an approximately direct-drive fluid coupling at higher speed ratios. The speed ratio is defined as the ratio of turbine speed to impeller speed. The torque converter's output shaft is connected to the automatic transmission with planetary gears. The transmission output torque is multiplied by the transfer case gear ratio to determine the torque transmitted to the differential. The latter determines the acceleration of gears based on propeller shaft torques and inertias. On the one end, the instantaneous shaft speed is determined by the differential output; on the other end, the shaft speed is equal to the wheel speed.

Vehicle dynamics describe the motion of the selected rigid bodies (wheels, axles, suspension and vehicle body) that are allowed to move in space in the longitudinal and heave direction, subject to forces/moments and rigid constraints. A single degree of freedom can be selected in which the vehicle mass is assumed lumped at the center of gravity; such approach is best suited for vehicle acceleration on flat roads. Complexity can be enhanced with more degrees of freedom, as more severe excitations (road roughness, steering, braking, *etc.*) are introduced. This is essential for the investigation of vehicle–powertrain interactions during extreme transients that induce significant pitch motion. The dynamic behavior of sprung and unsprung masses components are coupled through the road/tire interaction. The input to the system is the road profile. The torque from the driveline is applied to the wheel hub, and the available traction force to accelerate the vehicle is affected by the wheel slip. The traction force increases linearly with the wheel slip and it saturates when it reaches a value that is equal to the tire normal force multiplied by the road/tire friction coefficient μ.

Figure 3.31 illustrates the *powertrain integration methodology* of the detailed dynamic vehicle analysis, focusing on the key parameters of each module. One of the main difficulties encountered in such studies is the appropriate link between the engine simulation module with its multi-body vehicle dynamics counterpart. The difficulty arises because of the different time steps required for the solution of the differential equations in each module. The engine module requires a time step of the order of 1°CA, which is much too small for the vehicle dynamics.

The goal of integration is to find the optimum operating mode for the whole powertrain, not just separate parts of it. A major enabling technology in the area of powertrain system integration has been the introduction of CAN (Controller Area Network) bus or similar digital communication links, now almost universally fitted to the vehicle ECUs. Complex time critical control tasks can be distributed among a group of controllers co-operating via such a bus. Various configurations may be employed, and a supervisory controller could govern the separate engine and transmission control systems.

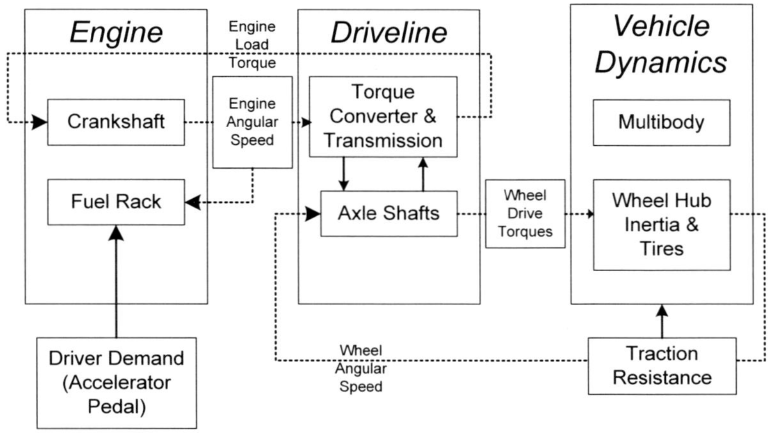

Figure 3.31. Powertrain system integration methodology (adapted from Assanis *et al.* [29])

A typical result from a detailed vehicular modeling approach was reproduced in Figure 1.16 concerning a turbocharged 4×2 truck during the US Highway Driving Cycle.

References

[1] Rakopoulos CD, Giakoumis EG, Dimaratos AM. Evaluation of various dynamic issues during transient operation of turbocharged diesel engine with special reference to friction development. SAE Paper No. 2007-01-0136, 2007.

[2] Lee KS, Assanis DN, Lee J, Chun KM. Measurements and predictions of steady-state and transient stress distributions in a diesel engine cylinder head. SAE Paper No. 1999-01-0973, 1999.

[3] Diesel fuel injection. Stuttgart: Robert Bosch GmbH, 1994.

[4] Krutov VI. Automatic control of internal combustion engines. Moscow: MIR Publishers, 1987.

[5] Welbourn DB, Roberts DK, Fuller RA. Governing of compression-ignition oil engines. Proc Inst Mech Eng 1959;173:575–91.

[6] Webb CR, Janota MS. Governors with load sensing. Proc Inst Mech Eng 1969-70;184:161–74.

[7] Catania AE, Dongiovanni C, Mittica A, Negri C, Spessa E. Study of automotive diesel injection-system dynamics under control. SAE Paper No. 962020, 1996.

[8] Rackmil CI, Blumberg PN, Becker DA, Schuller RR, Garvey DC. A dynamic model of a locomotive diesel engine and electrohydraulic governor. ASME Trans, J Eng Gas Turbines Power 1988;110:405–14.

[9] Zellbeck H, Woschni G. Prediction of the transient response of turbocharged diesel engines. MTZ 1983;44:81–6 (in German).

[10] Rakopoulos CD, Giakoumis EG, Hountalas DT. A simulation analysis of the effect of governor technical characteristics and type on the transient performance of a naturally aspirated IDI diesel engine. SAE Paper No. 970633, SAE Trans, J Engines 1997;106:905–22.

[11] Ferguson CR, Kirkpatrick AT. Internal combustion engines: applied thermosciences, 2nd Edition. New York: Wiley, 2001.
[12] Heywood JB. Internal combustion engine fundamentals. New York: McGraw-Hill, 1988.
[13] Gish RE, McCullough JD, Retzloff JB, Mueller HT. Determination of true engine friction. SAE Paper No. 580063, 1958.
[14] Ciulli E. A review of internal combustion engine losses, pt. 2: studies for global evaluations. Proc Inst Mech Eng, Part D, J Automobile Eng 1993;207:229–40.
[15] Chen SK, Flynn PF. Development of a single cylinder compression ignition research engine. SAE Paper No. 650733, 1965.
[16] Millington BW, Hartles ER. Frictional losses in diesel engines. SAE Paper No. 680590, 1968.
[17] Rezeka SF, Henein NA. A new approach to evaluate instantaneous friction and its components in internal combustion engines. SAE Paper No. 840179, 1984.
[18] Winterbone DE, Tennant DWH. The variation of friction and combustion rates during diesel engine transients. SAE Paper No. 810339, 1981.
[19] Rakopoulos CD, Giakoumis EG. Prediction of friction development during transient diesel engine operation using a detailed model. Int J Vehicle Design 2007;44:143–66.
[20] Giakoumis EG, Rakopoulos CD, Dimaratos AM. Study of crankshaft torsional deformation under steady-state and transient operation of turbocharged diesel engines. Proc Inst Mech Eng, Part K, J Multi-body Dynam, 2008;222:17–30.
[21] Chen SK, Chen S. Engine diagnostics by dynamic shaft measurement: a progress report. SAE Paper No. 932412, 1993.
[22] Winterbone DE, Benson RS, Mortimer AG, Kenyon P, Stotter A. Transient response of turbocharged diesel engines. SAE Paper No. 770122, 1977.
[23] Winterbone DE. Transient Performance. In: Horlock JH, Winterbone DE (eds). The thermodynamics and gas dynamics of internal combustion engines, Vol. II. Oxford: Clarendon Press, 1986;1148–212.
[24] Gillespie TD. Fundamentals of vehicle dynamics. Warrendale PA: SAE International, 1992.
[25] Winterbone DE, Loo WY. A dynamic simulation of a two-stroke turbocharged diesel engine. SAE Paper No. 810337, 1981.
[26] Rubin ZJ, Munns SA, Moskwa JJ. The development of vehicular powertrain system modeling methodologies: philosophy and implementation. SAE Paper No. 971089, 1997.
[27] Jennings MJ, Blumberg PN, Amann RW. A dynamic simulation of the Detroit diesel electronic control system in heavy duty truck powertrains. SAE Paper No. 861959, 1986.
[28] Fluga EC. Modeling of the complete vehicle powertrain using ENTERPRISE. SAE Paper No. 931179, 1993.
[29] Assanis D, Bryzik W, Chalhoub N, Filipi Z, Henein N, Jung D *et al.* Integration and use of diesel engine driveline and vehicle dynamics models for heavy-duty truck simulation. SAE Paper No. 1999-01-0970, 1999.
[30] Assanis D, Filipi Z, Gravante S, Grohnke D, Gui X, Louca L *et al.* Validation and use of SIMULINK integrated, high fidelity, engine-in-vehicle simulation of the international class VI truck. SAE Paper No. 2000-01-0288, 2000.
[31] Filipi Z, Wang Y, Assanis DN. Effect of variable geometry turbine (VGT) on diesel engine and vehicle system transient response. SAE Paper No. 2001-01-1247, 2001.

4

Experimental Measurements

4.1 Introduction: Steady-state Test Bed Review

Steady-state testing involves measuring and analyzing performance at discrete points within the engine speed and load operating range using an appropriately instrumented experimental test bed; this subject has been treated in detail by Zhao and Ladommatos [1] and partly by Stone [2]. Before data are logged, the engine is allowed to settle at each operating point. A basic engine steady-state test cell, usually comprising of a single-cylinder reciprocator, typically consists of the following devices or measurement points:

- dynamometer (brake) for monitoring and control of the power output; the latter is established through measurement of the load absorbed by the torque reaction on the dynamometer casing. The torque reaction can be electromagnetic or electric or hydraulic. In the former case, reference is made of an eddie-current dynamometer applying electromagnetic braking with eddie currents dissipated by a thin rotor into a flow of cooling water. An electric dynamometer can be of the DC or AC type with varying field strength; it can act either as a generator absorbing the engine power or as a motor for starting the engine and for motoring tests in order to evaluate the engine mechanical losses. Quite frequently, water brakes are used, which consist of a vaned rotor turning adjacent to a pair of vaned stators and with sluice gates separating the stators from the rotor, thus controlling the absorbed load. More advanced hydraulic dynamometers such as hydrostatic or variable fill are often applied;
- gravimetric tank to time the average consumption of a fixed volume, or a mass flow-meter for measuring the fuel consumption rate;
- air-flow meter for measuring the inlet air-supply, *e.g.*, by connecting the air intake to a large rigid box with an orifice at its inlet or via a viscous flow-meter or via positive displacement (Roots type) flow-meter or via hot-wire anemometer;

- for engine rotational speed evaluation a shaft (optical) encoder is usually employed that generates an output signal in the form of a series of clock pulses. Many researchers apply an electro-magnetic pick-up TDC marker installed on the flywheel housing opposite to its tooth gears, in such a way to produce a pulse signal indicating the presence of a tooth on the flywheel ring gear;
- flush-mounted (water-cooled) piezo-electric transducer, appropriately calibrated and coupled to a charge amplifier for continuously monitoring the cylinder pressure; in this case, an assumption has to be made regarding the minimum pressure recorded in the cycle, *e.g.,* being equal to the inlet manifold pressure. The start of combustion can then be calculated from the abrupt increase of this transducer's signal in the vicinity of 'firing' TDC. Alternatively, a piezo-resistive transducer can be applied;
- strain gauge or piezo-electric transducer for capturing the injector fuel line pressure (static injection timing is calculated from the abrupt increase of this transducer's signal);
- exhaust gas analyzers fitted through a diluted heated line (except for CO) into the exhaust pipe; most usually, these are of the chemiluminescence type for nitrogen oxides emissions (a technique depending on the emission of light), flame ionization detector (FID) for unburned hydrocarbons, infrared radiation absorption for carbon monoxide and dioxide, light beam obscuration or fouling of a filter paper for smoke emissions and integrated filter measurement for particulates mass (a filter collects the particulates, increasing accordingly its weight);
- 'J' or 'K'-type thermocouples for measuring the temperature of the exhaust gas after the cylinder;
- engine oil and cooling water temperature control;
- storage oscilloscope for an initial assess of the in-cylinder and fuel line pressure signals;
- software for storing the captured data on a computer;
- control of the data acquisition system by a functional computer code capable of processing the stored data.

Depending upon the type of engine and/or specific experimental investigation, the following may also be part of a more sophisticated steady-state engine test cell:

- injector needle lift measurement via gap sensor or via variable inductance system or via Hall effect transducer in order to detect dynamic injection timing;
- turbocharger compressor boost pressure manometer;
- thermometers for compressor inlet and exit temperatures and aftercooler exit temperature;
- turbocharger rotational speed measurement, *e.g.,* by using a stroboscope or optical sensor; for the latter, a laser beam is targeted on the compressor impeller, where it is reflected once per revolution from a reflecting marker;
- 'J' or 'K'-type thermocouples for measuring the temperature of the exhaust gas downstream of the turbine, as well as fast thermocouple with lower

than 10 μs response time, coupled to the appropriate amplifier, for cylinder surface temperature measurement;
- combustion noise-meter (a semi-anechoic test cell is required for overall engine noise emission measurements).

4.2 Transient Experimental Test Bed

Transient testing is much more demanding primarily in terms of *fast* acquisition and control; it requires complex, sophisticated and expensive facilities, particularly if Transient Cycles analysis is involved for the whole vehicle performance evaluation. Whereas for steady-state operation, fuel- or air-flow are measured for a relatively long period (of the order of a few minutes) and then their values are reduced to one cycle on a time-averaged basis, during transients, measurements must be undertaken at each engine cycle, a fact that poses a significant challenge for the capacitance of the measuring device (see also Section 4.2.3); for example at 1200 rpm, a four-stroke engine cycle lasts 0.1 s, which means that the fuel- or air-flow meter must be appropriately fast to capture the respective peaks. Moreover, transient test procedures require both fast response of the applied loading as well as tight set point holding. The measurements are undertaken after the steady-state evaluation of the engine and its sub-systems has been carried out; measurements at very low steady-state loads are usually required prior to transient experimentation, since the most demanding transient tests commence from a low or even zero initial load, where the engine and turbocharger behavior are often problematic and cannot be extrapolated from other measured data. Further, in order to evaluate the actual engine performance, transient test beds usually incorporate multi-cylinder, turbocharged engines so as to take into account the discrepancies induced by the turbocharger, exhaust manifold volume and cylinder interconnection and interdependence in terms of fuel injection and air-supply. In order to quantify the dynamic characteristics of the engine in relation to response, performance, emissions and fuel consumption, the respective instrumentation differs from its steady-state counterpart in as much as it requires (see also Figures 4.1 and 4.2)

- electronically closed-loop controlled, fast response dynamometer, discussed in more detail in Section 4.2.1, or at least a strain-gauge torque-meter in the case of mechanical (non-regulated) engine brakes for continuous monitoring of the load torque;
- fast response fuel flow-meter, *e.g.,* of the gravimetric balance type. Alternatively, the fueling signal from the engine ECU can be used provided that the appropriate software is available for translating and storing the ECU signals. Another system capable of transient fuel flow measurements applies a hydraulic equivalent of the electrical Wheatstone bridge. For mechanical fuel injection systems, a fuel pump rack position measurement is used, *e.g.,* by applying an inductive type displacement transducer mounted adjacent to the fuel pump with its axis parallel to the fuel rack;

- pressure transducer and temperature thermocouple connected to the appropriate amplifiers for continuously measuring the turbocharger compressor boost pressure and temperature;

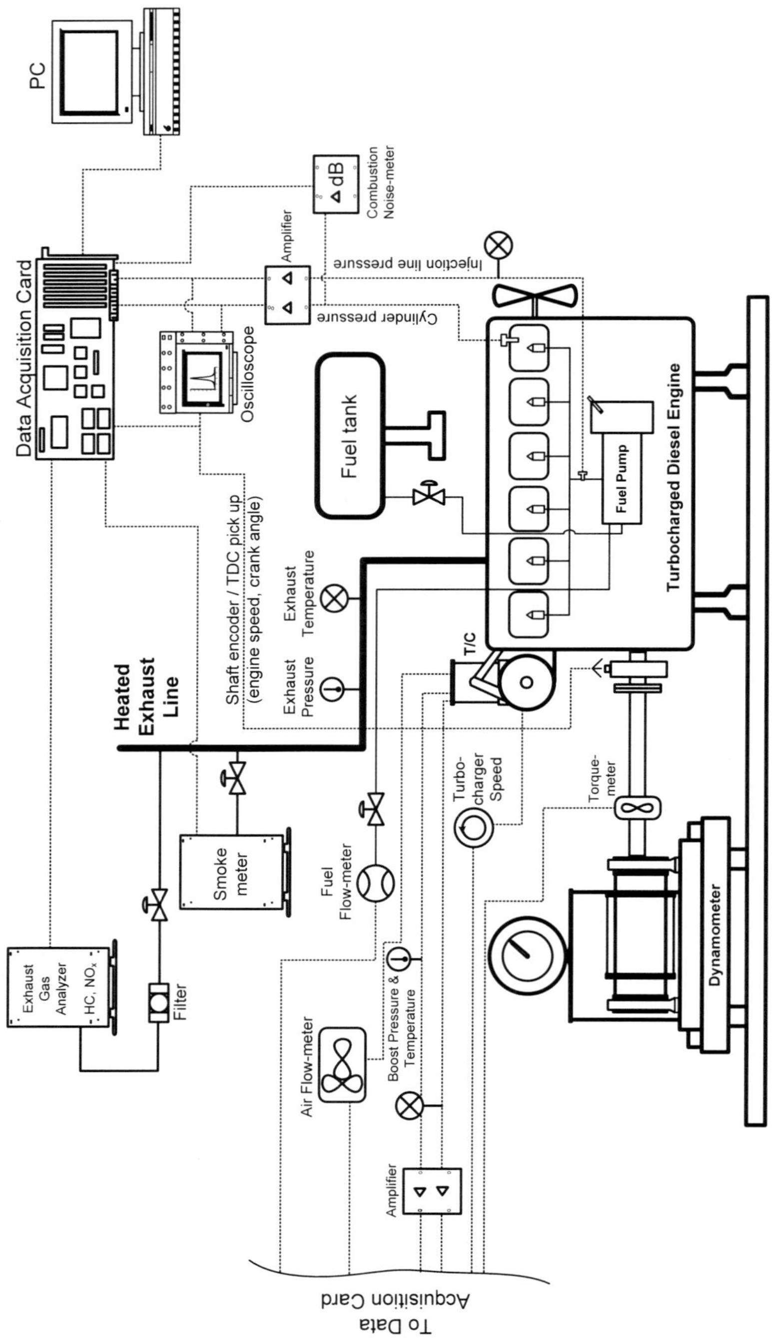

Figure 4.1. Schematic arrangement of a test bed for transient turbocharged diesel engine experimentation with electronically controlled dynamometer, identifying various measurement points and connections (dotted line denotes data signal)

- fast response exhaust emissions analyzers discussed in more detail in Section 4.2.2;
- fast response air-flow meter needed for accurate heat release analysis. This can be, for example, a hot-wire anemometer consisting of a very thin (platinum) wire heated to a temperature above that of the air-supply to the engine; a relationship between flow-rate and wire resistance can be applied based on the fact that the incoming air-flow imposes a cooling effect on the heated wire depending on the flow velocity, thus altering the wire's temperature, and so its electrical resistance. Hot-wire anemometers are characterized by high frequency response, but they may require frequent calibration in case of dust accumulated on the wire sensor;
- high frequency, *e.g.*, 100 kHz, 16 bit, data acquisition card capable of simultaneously storing all recorded channels at a sampling rate corresponding to 1 degree crank angle or even less in order to accurately phase the various recorded signals to the instantaneous engine crankshaft position;
- large computer storage space for storing the data of hundreds or even thousands of cycles;
- a possible cylinder-by-cylinder measurement since, particularly during the early cycles of the transient event, the continuous movement of the mechanical fuel pump rack is responsible for differentiations in the amount of injected fuel quantity per cylinder during the same transient cycle.

Figure 4.2. Photograph of a transient engine test bed

4.2.1 Dynamometers

Dynamometers used for transient measurements are either *engine*- or *chassis* units. The former type is used for engine-based transient measurements, where the engine is directly coupled to the dynamometer (Figures 4.1 and 4.2) and will be discussed in more detail in this section.

Chassis dynamometers, on the other hand, are applied when the whole (passenger or light-duty) vehicle performance is under study, for example during a Transient Cycle certification procedure.[1] In this case, the vehicle is usually placed on rollers and operated by a driver with the inertia of the vehicle during speed or load changes being simulated by a set of flywheels coupled to the rollers. The mechanical energy transmitted by the wheels to the rollers is absorbed by a brake, which is integrated with the rollers through a shaft; obviously, a chassis dynamometer facility is both expensive and highly sophisticated.

Engine dynamometers for transient measurements should, in general, meet the following criteria.

- Fast response and low inertia: typical response time lower than 1 s for torque reversal from –100% to 100% is desirable, depending on the dynamometer size, in order to be able to simulate real-life transients and vehicular engines behavior. This can be achieved, for example, by electric or eddie-current dynamometers or by hydrostatic or variable fill hydraulic brakes. On the other hand, the traditional sluice-gate water brake configuration, being characterized by relatively high moment of inertia and slow response, is not usually recommended for transient experimentation.
- Motoring capability: this is primarily important when Transient Cycles analysis is under request for heavy-duty engines, since part of the Transient Cycle certification procedure requires that the engine is motored (see also Appendix A - Section A.2.2.2); it can be established with a DC or AC electric brake only.
- Electronic, closed-loop control: the dynamometer should be capable, via specific software, of applying various loading profiles according to the transient schedule under study. The use of electronically controlled dynamometers ensures both command and reproducibility of the transient experiments.

Figure 4.3 illustrates a variable fill hydraulic dynamometer operating map; the full-load engine torque curve is also depicted in the same figure. The shape of the dynamometer torque–speed (T-N) characteristic influences the operation stability of the combined system. A series of rising dynamometer torque curves, such as the ones shown in Figure 4.3, should generally intersect the engine torque curves in such a way that dT/dN is greater for the dynamometer than for the engine; otherwise, a control system needs to be applied for establishing stable operation. Since the dynamometer is part of the system that undergoes transient process, its

---

[1] For heavy-duty diesel-engined vehicles, engine rather than chassis dynamometers are applied for Transient Cycles certification.

characteristics affect the speed response of the engine as it is dictated by the following equation describing Newton's second law of motion for a rotating system:

$$T_e - T_{dyn} = G_{e+dyn}\frac{d\omega}{dt} \tag{4.1}$$

where T_e and T_{dyn} are the instantaneous net engine and dynamometer torques, and G_{e+dyn} the total (engine plus brake) mass moment of inertia. For hydraulic dynamometers, the fluid inertia should also be accounted for in the total moment of inertia term in Equation 4.1. The steady-state dynamometer torque T_{dyn} needed in the above equation can be approximated by the following relation

$$T_{dyn} \approx c\ N^s \tag{4.2}$$

where s=1 for electric brakes (torque proportional to speed) and s=2 for hydraulic ones (torque proportional to the square of speed). For example, for the variable fill hydraulic brake of Figure 4.3, different values of constant 'c' correspond to different fill percentages, hence different steady-state torque curves with the maximum torque curve corresponding to 100% fill.

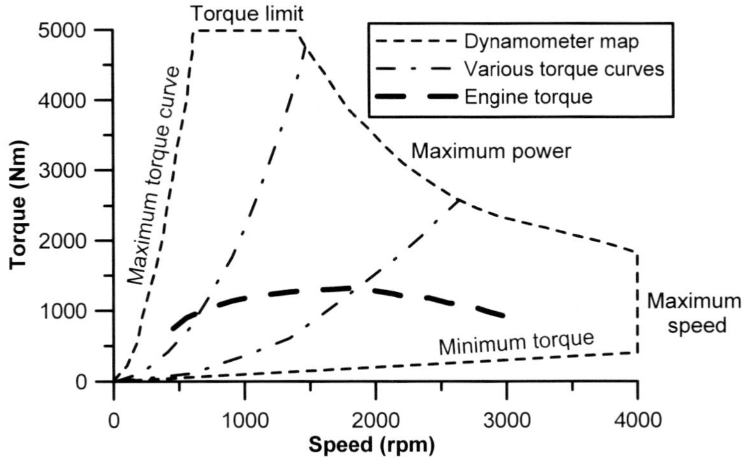

Figure 4.3. Typical torque/speed map of a variable fill hydraulic dynamometer

There are various dynamometer operating modes that are of interest during transient test bed experimentation; these can be classified as follows with reference also to Figure 4.4 [3].

1. Constant speed: if there is a small excursion from the speed set point, the dynamometer control system applies speed feedback control to make a

correction to the absorbed torque in order for the system to return to the demand speed.

2. Constant torque: here, the dynamometer controller uses torque feedback to hold torque constant at the prevailing demand setting.
3. Transient Cycle: in this case, which is the most complicated mode, the dynamometer torque is controlled according to the specific loading required by the simulated Transient Cycle under study.
4. Uncontrolled mode: the dynamometer setting (constant ‘c’ in Equation 4.2) remains unchanged (*e.g.*, constant field strength for an electric dynamometer or constant sluice gate position for a hydraulic one) and torque increases according to the square of speed for hydraulic brakes (propeller law) or linearly with speed for electric brakes following the engine acceleration. This operation is often met in marine engine test-beds as well as for simulation of vehicle aerodynamic resistance. However, when large speed changes (*e.g.*, more than 500 rpm) are encountered, such as during acceleration tests, deviations of the dynamometer torque from its steady-state curves are experienced, thus affecting accordingly the engine transient response profile.

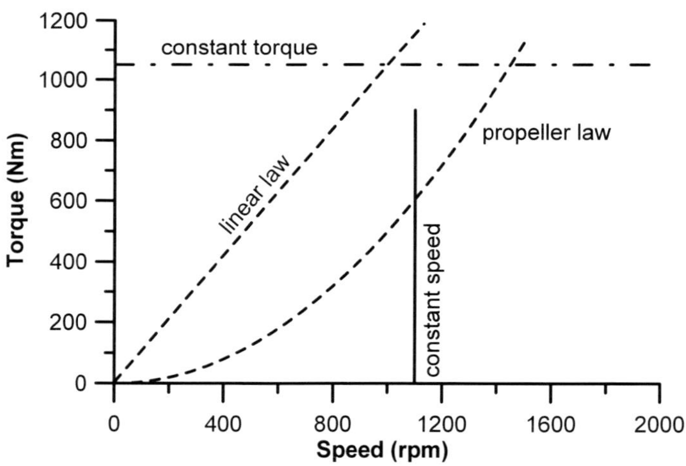

Figure 4.4. Hydraulic dynamometer modes for transient diesel engine experimentation

Perhaps, the most critical part of an engine–dynamometer setup is the integrated control algorithm. Figure 4.5 illustrates a typical dynamometer feedback control of a fast response, electronically controlled, variable fill hydraulic brake. The Control Module collects the various signals from the engine and dynamometer (speed, ‘throttle’ position, *etc.*) and processes them in order to calculate the loading of the dynamometer, according to the desired control mode. The controller is usually a computer-based digital system with the control algorithm applied through specific software; its characteristics can be varied so as to suit the specific engine under test.

The software is capable of identifying and applying different dynamometer loading/speed schedules, and may even simulate vehicle related resistances such as road grade, aerodynamic resistance, *etc.*, as well as the whole vehicle powertrain, mechanical or automatic gearbox, clutch or torque converter, driveshafts, differential, vehicle chassis, suspension and tires for Transient Cycles analysis; in the latter case, the controller also sends 'throttle' commands to the engine in order to achieve the desired engine speed as demanded by the specific Cycle. At any instant, the updated dynamometer load demand is directed, for a hydraulic dynamometer such as the one depicted in Figure 4.5, to a servo device controlling the position of the water outlet control valve. This can be achieved via an electric servo motor driving a pilot-operated back-pressure valve, or an electro-hydraulic system in which a hydraulic servo valve controls the magnitude and direction of flow to a rotary hydraulic actuator connected to a water outlet butterfly valve [4]. Steiber *et al.* [5] and Babbitt and Moskwa [6] describe such complex configurations capable of controlling engine performance during transients; the implementation makes it convenient to perform a number of studies, for example: 1. study new control algorithms for diesel (and gasoline) engines by developing advanced techniques to estimate engine torque; 2. perform engine mapping and calibration (optimization) using a combination of design of experiments (DOE), neural networks, physical-model based, fuzzy-logic, and non-constant variance based techniques; and 3. perform sensitivity analysis, durability studies, and accelerated testing to evaluate engine performance and engine control robustness.

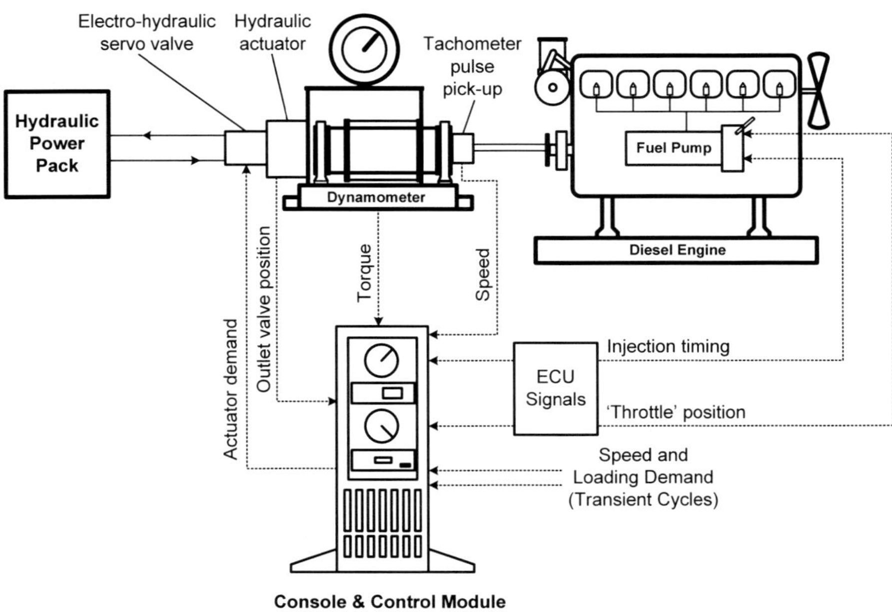

Figure 4.5. Electro-hydraulic servo-control system of a variable fill dynamometer showing engine, dynamometer and console-control module (adapted from Hodgson and Raine [4])

For modern engines equipped with electronic control modules, monitoring and control of the ECU data is also required, which can be accomplished via specific software that receives and translates the various ECU signals (*e.g.,* injection data, EGR position, VGT vanes position, fuel flow, air-mass flow-rate, *etc.*), and feeds them to the master controller and/or the data acquisition card for further processing together with the other measured variables.

4.2.2 Instantaneous Measurement of Exhaust Gas and Particulate Matter

Accurate, real-time measurement of exhaust gaseous and particulate emissions is of paramount importance, since it is these emissions that usually peak during transients way above their quasi-steady values. It is beyond the scope of this book to address in detail the operating principles, the merits and the drawbacks of the various exhaust gas emission techniques; only a brief description will be provided.

Conventional gas analyzers used for steady-state emission measurements are usually characterized by high response time of the order of a few seconds; hence, their dynamics cause distortion of the emission measurement requiring a comprehensive reconstruction technique (see for example [7–9] mainly in terms of NO_x emission analysis) to project real-time emissions. This kind of analysis was introduced by McClure [7] researching the exposure of miners to the exhaust emissions of diesel powered equipment. He measured the response of the gas analyzer describing it through a delay and empirically derived 'shape factors'; a transfer function was formulated, which was used to infer the variation of concentration in the exhaust gases. Beaumont *et al.* [8] designed a finite horizon reconstruction filter based on generalized predictive control theory, whereas Chan *et al.* [9] followed a physical rather than mathematical approach by characterizing the chemiluminescent analyzer using two phenomenological models, *i.e.,* diffusion and perfect gas mixing, in order to simulate the sampled gas emission state when passing through a series of tubes and surge volumes in the analyzer.

On the other hand, fast response CO, HC and NO_x analyzers are now commercially available with a very short response time (of the order of a few ms) that allows real-time emissions measurement. For NO_x, this can be achieved by locating the detectors in remote sample heads that are positioned very close to the sample point in the engine and conveying the sample gas to the detectors under the influence of a vacuum through narrow heated capillaries (Figure 4.6 [10]). An analyzer with 2 ms response time can capture 50 measurements during one cycle of a four-stroke engine running at 1200 rpm, thus being capable of successfully tracing the peaks in exhaust emissions.

For accurate gas sampling, it is important that no chemical reactions take place in the sampling line. This means that inert materials, *e.g.,* Teflon, have to be used for all parts being exposed to the gas. Moreover, filters may be installed in order to remove particulates, whereas the exhaust gas is usually diluted prior to measurement to avoid saturation of the volatile materials (see also Figure 4.12 later in the section). The exhaust gas line for HC and NO_x measurement is heated to prevent losses due to condensation in the sampling system.

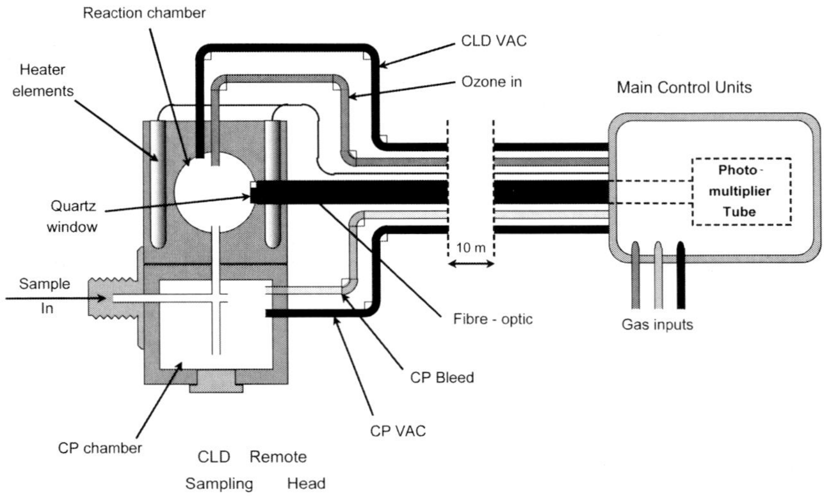

Figure 4.6. Schematic diagram of fast NO instrument operating principle (Copyright © Cambustion Ltd). The reaction between NO and ozone emits light; this reaction is the basis for the (very small) chemiluminescence detector (CLD) in which the photons produced are detected by a photo-multiplier tube. The reaction chamber is coupled to a constant pressure sampling system in order to avoid exhaust sample pressure interference on the signal. The sample enters the reaction chamber through this constant pressure chamber. The CLD output voltage is proportional to NO concentration. Demonstrated in this figure is the two-channel analyzer, with the reaction chambers located in remote sample heads on 10 m umbilicals

Things are more complex when instantaneous particulate matter (PM) measurements are involved. First, there are several properties of particulates that are of interest, *i.e.*, mass, size, number concentration and surface (discussed in more detail in Section 5.1.1); current legislation regulates PM mass, although recent medical reports have correlated greater health risk with smaller PM sizes. Second, the instantaneous rate of PM emissions consisting of elemental carbon, organic carbon, soluble organic fractions, metals and sulfates might be related to other emissions. For example, the elemental carbon might be related to the production of CO, whereas organic carbon might be related to instantaneous hydrocarbon values or lubricating oil effects. Further, PM are strongly dependent on time (and on the sampling system used), with dilution, condensation of volatile compounds, nucleation and coagulation, affecting the instantaneous particle mass and size distribution. Owing to this complexity, various techniques have been under study and development for instantaneous measurements of particulate mass, namely

1. inertia mass-based measurement such as the tapered element oscillating microbalance (TEOM) and the quartz crystal microbalance (QCM);
2. particle size distribution technique;

3. optical methods that primarily measure elemental carbon concentration (often considered as a surrogate to total PM mass emissions) and are based on
    - light scattering such as the nephelometer,
    - light absorption such as the aethalometer or the photoacoustic spectroscopy or the photo-electric aerosol sensor (PAS) or the laser induced incandescence (LII), and
    - light extinction such as the exhaust opacity meter;
4. surface measurement such as the epiphaniometer or the diffusion charging sensor.

Important criteria for evaluation of the above instruments/techniques are the degree of correlation of time-integrated results with conventional filter measurements, the reproducibility of the results, noise and interference from non-particulate mass sample properties, time resolution, dynamic measurement range (particularly important for testing of low emission vehicles equipped with particulate traps), simplicity of use and cost [1, 11–16]. A brief discussion of four of the most popular techniques will follow.

TEOM consists of a small filter cartridge placed on the end of a hollow tapered tube (Figure 4.7). The other end of the tube is fixed rigidly to a base. The tube with the filter on the free end is oscillated in a clamped-free mode at its resonant frequency. This frequency depends on the physical characteristics of the tube and the mass on its free end. A particle laden air stream is drawn through the filter, where the particles deposit and then pass through the hollow tube. A volumetric flow controller and a positive displacement pump maintain the flow-rate of the diluted exhaust gas constant.

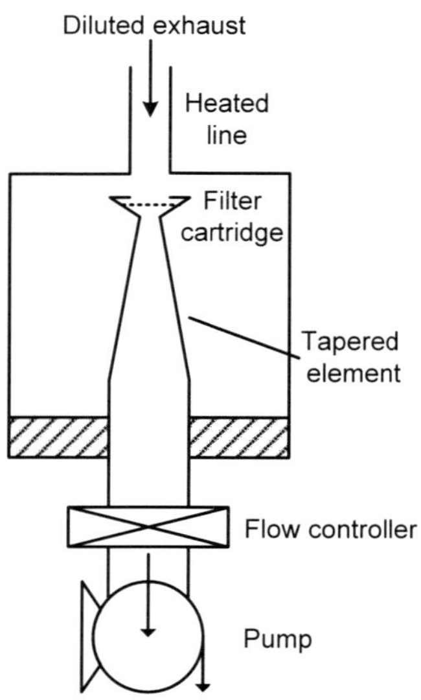

Figure 4.7. Schematic arrangement of tapered element oscillating microbalance device for instantaneous PM mass measurements

As particles deposit, the mass of the filter cartridge increases and the frequency of the system decreases. By accurately measuring the frequency change, the accumulated mass can be computed. Combining this accumulated mass with the volume of air drawn through the system during the same time period, yields the particle mass concentration. Mass change from the initial conditions is calculated by the following equation

$$\Delta m = k_o \left(\frac{1}{f_f^2} - \frac{1}{f_i^2} \right) \tag{4.3}$$

where k_o is the spring constant specific to the tapered element, f_i the initial oscillation frequency and f_f the oscillation frequency after the addition of the mass [17]. The system can be calibrated by placing a calibration mass on the filter cartridge and recording the frequency change due to this mass.

The main drawback of the tapered element oscillating microbalance technique is its susceptibility to moisture. Since the TEOM measures the mass of material retained on the filter attached to the oscillating element, the reported mass will vary with the filter moisture content. In general, the TEOM is used to sample from diluted exhaust gas; the humidity of the exhaust, however, is expected to increase with higher fueling and decrease at motoring conditions, hence water is deposited on the filter during accelerations and removed during decelerations. These water mass transfers cause the tapered element oscillating microbalance signal to indicate higher mass rates during accelerations or load increases and lower mass rates during decelerations; in fact, if the temperature or HC vapor concentration in the exhaust changes significantly from one instant to another, a negative mass concentration may be recorded. As a result, a comprehensive water correction approach needs to be applied, such as, for example, the one described in [14]. Moreover, the TEOM instrument is sensitive to vibrations typical in engine test cells; hence, its physical location requires isolation, careful installation and supervision.

Exhaust *opacity* is based on the fact that the mass of particulates is proportional to the total light extinction across the exhaust gas stream. Opacity meters (Figure 4.8) are operated without prior dilution; a volume or the total of the exhaust gas is *continuously* drawn and passed through a heated tube to prevent condensation. The opacity of the sample is determined from the measurement of the attenuation of visible light by the smoke in the air sample located between the light source (*e.g.*, an incandescent lamp or a green light emitting diode) and a light detector (a photocell or a photodiode). Attenuation of light is caused by scattering by the particles and absorption in the particles.

Opacity O (not to be confused with the very popular in terms of steady-state experimentation filter smoke number, which is, in general, based on the same principle but cannot provide continuous measurements) is then calculated by the following equation on a percentage basis, based on the Beer–Lambert law

$$O = 100\left(1 - \frac{I}{I_o}\right) = 100\left(1 - e^{-kL}\right) \tag{4.4}$$

where I (W/m^2) is the light intensity after passing though the particle cloud and I_o the incident light intensity; L is the path length of the light beam (m) and k is the extinction coefficient (m^{-1}). The latter is related to the number, size and properties of individual particles and can be defined as the inverse of the path length over which light intensity decreases by a factor of e (=2.718), as described by

$$k = -\frac{1}{L}\ln\left(\frac{I}{I_o}\right) \tag{4.5}$$

Since an opacity meter measures smoke, empirical correlations have been developed for estimation of particulate mass from opacity readings [*e.g.*, 18–20].

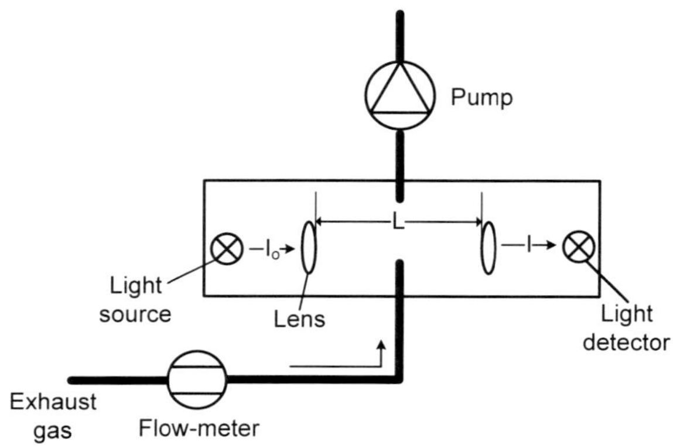

Figure 4.8. Schematic arrangement of exhaust opacity device

Opacity measurements have a very good time response and are relatively simple and inexpensive; the main drawbacks are the somewhat poor sensitivity to low soot mass concentrations (this is, in fact, very important since the detection limits of opacity meters are nowadays of the same magnitude as the low PM emissions produced by modern diesel-engined vehicles equipped with particulate filters), the insensitivity to small particle sizes (D_p<0.3 μm) as well as the inability to distinguish between black smoke (originating from particulates), blue smoke (droplets of unburned lubricating oil) and white smoke (mainly unburned hydrocarbons and water vapor experienced during cold starting).

Laser induced incandescence (LII) entails the heating of particulates using a high power laser pulse. Following the laser pulse the ensuing gray body radiation, when collected in a narrow band of wavelengths, yields information on local volume fraction. Figure 4.9 shows a schematic arrangement of a LII-based experimental setup developed by Gupta et *al.* [21] for transient PM measurements. This incorporates a 532 nm laser output shaped into a vertical sheet using a combination of cylindrical lenses, so that the focal plane lies at the center of a quartz sample cell. A fast response photocell placed downstream of the sample cell monitors the laser power to the required pulse energies. The LII signal was detected by a particulate mass detector placed at right angle to the incoming laser beam at 430±5 nm. The peaks of the LII signal were measured using a high-speed data acquisition card. The walls of the sample cell were maintained at a temperature slightly higher than the aerosol, to avoid particle deposition due to thermophoresis. The LII technique, which has also been applied to soot measurements in diesel flames, is characterized by fast response according to the laser repetition frequency (values as high as 50 Hz have been reported) and high sensitivity, but it is rather expensive due to the cost of the laser involved.

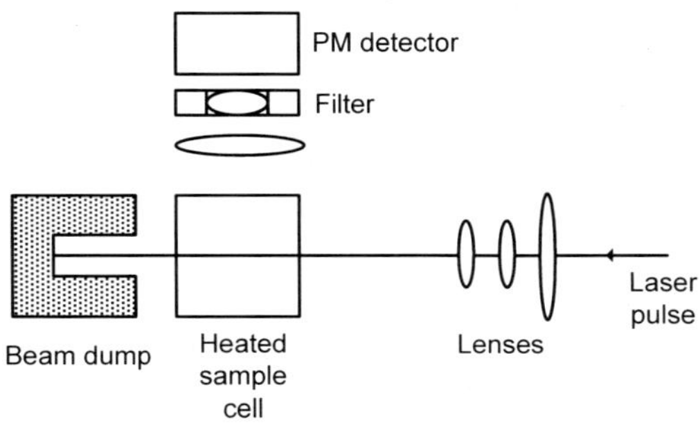

Figure 4.9. Schematic arrangement of laser induced incandescence device for instantaneous PM mass measurements (adapted from Gupta *et al.* [21])

The last case of instantaneous PM measurement discussed is the *particle size distribution* technique; this is based on a particle size spectrometer, with subsequent integration of the amount of detected particles/size for mass evaluation. PM number concentration and size distribution is typically accomplished through the following instruments that draw a continuous sample of the exhaust gas.

- The Scanning Mobility Particle Sizer (SMPS™): particles experience an electrostatic force and an opposing viscous drag; these serve to separate particles according to their electrical mobility, which is a function of diameter. The SMPS is able to measure submicron particles collectively within the 2.5 to 1000 nm range, with a relatively wide size resolution of up to 64 channels per decade. To obtain the size distribution, the Differential Mobility Analyzer (DMA) component of the SMPS is scanned through a range of voltages, which limits time resolution (scan time 30 to 120 s) and makes the device unattractive for transients testing. Particle number concentration is accomplished via the integrated Condensation Particle Counter (CPC).
- The Electrical Low-pressure Impactor (ELPI™): this separates particles into 12 impact stages (size classes) according to aerodynamic size ranging from 7 nm (with filter stage) up to 10 μm, using a low-pressure cascade impactor (Figure 4.10). The particles are first charged into a known charge level in the corona charger. Afterwards, they enter a cascade low-pressure

impactor with electrically insulated collection stages. The particles are then collected in the various impactor stages according to their aerodynamic diameter. Response time is reported to be less than 5 s.

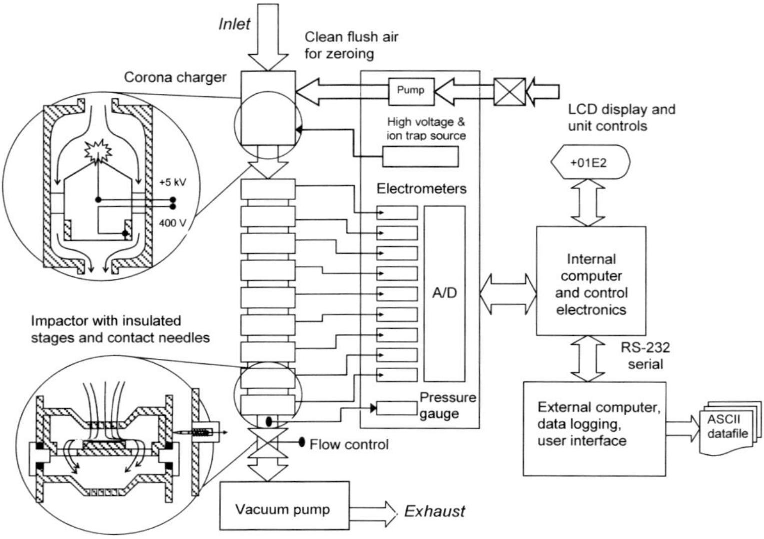

Figure 4.10. Operation block diagram of Electrical Low-pressure Impactor (copyright © Dekati Ltd)

- The Engine Exhaust Particle Sizer Spectrometer (EEPS™): measures distribution in the range from 5.6 to 560 nm in 32 channels (16 channels per decade), with 10 Hz data collection (scan time 0.1 s), which makes it ideal for transient experimentation; it determines particles diameter based on the measurement of charged particles mobility in an electrical field.
- The Differential Mobility Spectrometer (DMS™): based on the same operating principle as the DMA of the SMPS (Figure 4.11); it measures the number of particles and their spectral weighting in the 5 nm to 2.5 μm size range (16 channels per decade) with a scan time of 0.2 s.

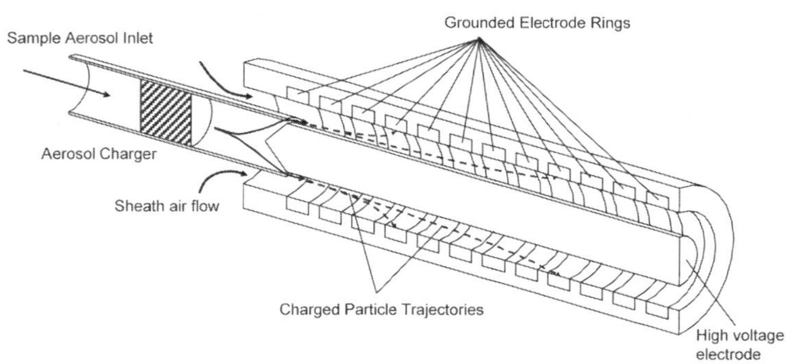

Figure 4.11. Operating principle of DMS500 fast particle size spectrometer (copyright © Cambustion Ltd)

- The (nano)-Micro Orifice Uniform Deposit Impactor (MOUDI™): this separates PM into 13 progressively smaller cut points (10 nm to 18 μm) based on the apparent aerodynamic diameter; this, too, is recommended for steady-state experimentation only.

Care has to be taken in the case of PM mass estimation from size distribution measurements, since agglomerates formed during diesel combustion are non-spherical and, therefore, their mass does not correlate with the cube of particle diameter. Further, there is no direct correlation between particle density and diameter. The use of particulate filters (traps) may induce an additional challenge for the accuracy of particle size measurement devices. To convert particle spectral density D_p to mass, Park *et al.* [22] suggested a general relationship of the following form

$$m = c\,(D_p)^b \tag{4.6}$$

where c and b are constants and b< 3. For example, Hagena *et al.* [23] assuming constant particle density for each resolution size ('bin') scanned by their differential mobility spectrometer, applied the following relation based on Equation 4.6:

$$m = 6.95 \times 10^{-3}\,(D_p)^{2.34}\,n_p \tag{4.7}$$

with n_p the number of particles for each bin scanned. In this equation, the non-spherical nature of particle shapes is accounted for by the diameter exponent that is less than 3. The leading coefficient acts as a 'pseudo-density' for the particles in that bin; its magnitude is impacted by particle constituents and the unit changes that occur with the non-integer particle diameter exponent. After the particle mass is calculated for each bin, the total mass is found by summing the masses in each bin and dividing by the number of bins per decade. This summation is used as an approximation that accounts for integration over a logarithmic scale.

$$\text{Total mass}\,(\text{kg}/\text{m}^3) = \frac{\sum_{\text{bins}} m}{\text{bins}/\text{decade}} \tag{4.8}$$

Unfortunately, none of the above mentioned methodologies has been found capable of accurately capturing both instantaneous peaks and total PM mass emissions, particularly during Transient Cycles; in fact, the results obtained are sometimes conflicting, depending strongly on the specific test bed and instrument, the calibration of the unit and the nature of the transient tests. For example, Black *et al.* [24], studying the performance of a modern, European, passenger vehicle diesel engine during the EUDC Transient Cycle, identified inability of the TEOM method to capture the fastest transient features in the emission behavior, whereas the opacity method was able to better pinpoint the various emission discrepancies encountered. Gupta *et al.* [21] found an LII-based instrument's signals to vary

±20% compared with SMPS measurements at steady-state conditions; on the other hand, the LII was found to have a considerably better time resolution (depending on the laser used) than the TEOM for step-change operations. Although Jarrett and Clark [14] applied a comprehensive water correction technique, negative PM values were still observed during decelerations when applying a TEOM procedure. It has been suggested that the aerodynamic diameter as measured by the ELPI may lead to an underestimation of the actual size of diesel particulates during transients, since the density of agglomerates is lower than unity. Likewise, the SMPS measuring the mobility diameter suffers from multiple charges on the agglomerates, thus, underestimating their actual size and mass [25]. An instructive comparative experimental study can be found in [26].

Owing to the uncertainties involved with the results obtained from the developed PM mass measurement methods, it is not surprising that none has been certified so far as a standard, routine experimental procedure.

4.2.2.1 Emission Measurements During Transient Cycles

A special case of exhaust gas measurement during transients is the constant volume sampling (CVS) technique illustrated in Figure 4.12. This has been established by the current legislation in many countries as the standard sampling methodology for the certification of light and heavy-duty engines/vehicles; details can be found, for example, in [11, 27]. In the case of passenger or light-duty vehicle certification, the CVS facilities are combined with a chassis dynamometer. CVS consists in the dilution of the total raw exhaust gas with clean air in a dilution tunnel,[2] by keeping the flow constant using a positive displacement pump or a vacuum pump and a critical flow venturi. The constant flow facilitates an easy and accurate calculation of mass-based emissions. A heat exchanger may be optionally installed for proportional particulate sampling and for flow determination. The (primary) tunnel is usually a polished and heated stainless steel tube and should be of sufficient length to allow for complete mixing of the exhaust gas with the dilution air. The tunnel diameter depends on the exhaust gas flow, being small enough to cause turbulent flow ($Re > 4000$); it is of the order of 250 to 300 mm for heavy-duty diesel engines. The amount of dilution air must be sufficient to cool the exhaust gas–air mixture at all engine operating conditions at a temperature of no more than 325 K.

Samples for continuous gas phase measurements may be drawn from the dilution tunnel at a point where the exhaust gas is thoroughly mixed with the dilution air (typically ≥ 10 tunnel diameters from the point where both streams are introduced into the tunnel). A small fraction of the CVS flow (*e.g.,* 1%) is sampled into a Teflon bag. Gas samples for bags are drawn from downstream of the heat exchanger or, in the case of gases that may suffer losses in the heat exchanger (HC, NO_x), heated bag sampling from upstream of the heat exchanger is the preferred practice. After the test, concentrations of the gases in the sample inside the bag are measured by means of a gas emissions analyzer. The highest risk of gaseous sample loss during CVS exists with hydrocarbons of high boiling point; HC

[2] Dilution ratio typically ranges from 10–30, favoring condensation and nucleation processes; this dilution ratio is way below that experienced in 'real-life' engine operation.

analysis, therefore, should be performed by a continuous, heated analyzer even if other gases are measured from bags by steady-state instruments.

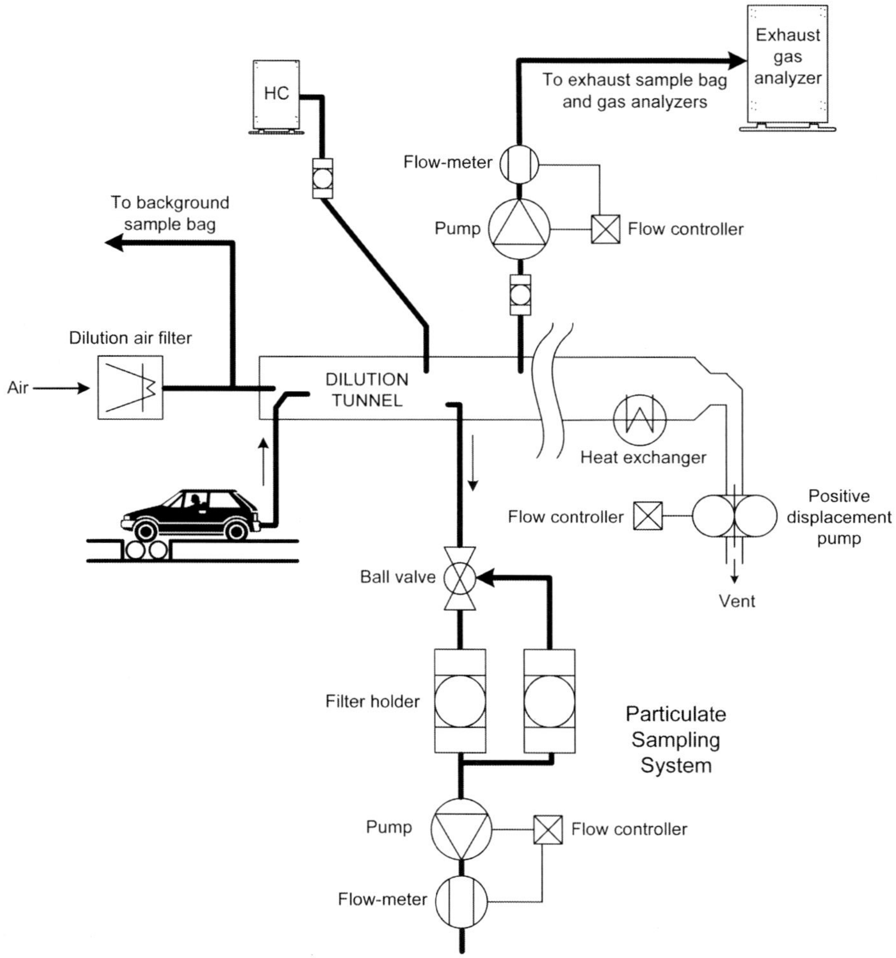

Figure 4.12. Typical constant volume sampling configuration with positive displacement pump for emission certification of diesel-engined vehicles

For the particle mass measurement, a sample of the total flow from the dilution tunnel is extracted and passed either directly ('single' dilution) or through secondary dilution ('double' dilution) to the particulate sampling system, where the particles are collected on Teflon coated, fiber-glass filters. To ensure reproducible measurement of volatile components absorbed on the filter, it is required that the filter equilibrates at a defined temperature and humidity level. Particle emission is determined from the total particle mass collected onto the filters during the Cycle. For comparison with the standard limits, an average over the Transient Cycle (g/test or g/km or g/kWh) is calculated for both gaseous and particulate emissions.

The use of the CVS system requires large space and cost; it is evident that with the constant volume sampling methodology, calculation of the *total* amount of exhaust emissions is accomplished without indication of either temporary overshoots or the time window in the test, where emissions are primarily produced.

In view of the high cost and large volume facilities of the CVS technique, the partial flow sampling system (PFSS) has been proposed. Here, a small fraction of the engine exhaust gas is sampled, while dilution air is transiently added so that the diluted exhaust flow-rate is constant through the filter. Since the flow of this sample must be proportional to the engine exhaust gas flow-rate, the latter parameter must be known and has to be measured. In the case of transient testing, a very fast response of the flow system is necessary to match the proportionality also during the transient phases (transient partial flow sampling system, TPFSS).

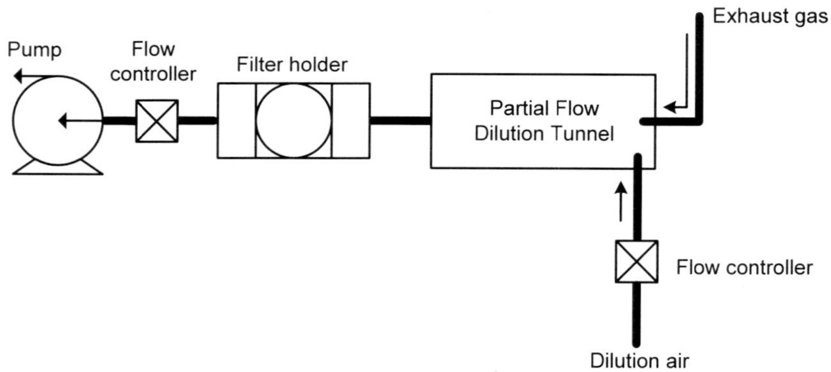

Figure 4.13. Typical arrangement of Partial Flow Sampling System

The small percentage of engine exhaust gas is diluted with air. The conditioning and mass flow measurement of the dilution air occurs within the PFSS (Figure 4.13). The dilution tunnel is mounted very close to the exhaust source to control the response time and temperature, and to minimize deposition. A second mass flow system within the PFSS measures the total flow through the filter, *i.e.*, the sum of the dilution air and sampled exhaust flow. The sampled exhaust flow is calculated from these two flows. As expected, the need to infer a small flow accurately using the difference between two large flows requires highly accurate flow measurement systems. As the engine intake air and exhaust gas flow-rates vary during transient testing, the TPFSS must react by sampling a proportionally consistent flow from the stack. To accurately conduct transient type tests, the system must have extremely fast response time, in general less than 300 ms. Therefore, such test cells must be equipped with appropriately fast and accurate intake air-flow measurement systems (response time lower than 100 ms to a 90% step change in air-flow). Where possible, it is best to include a real-time fuel flow measurement signal. Alternatively, the fuel rate signal from the engine ECU can be used. While TPFSS are substantially easier to operate and maintain than CVS tunnels, a solid understanding of air-flow measurement systems is necessary to ensure data quality. Particulate emissions calculations are then based on the cycle integrated exhaust mass-flow results, the integrated sample flow value from the TPFSS and the filter mass loading [11].

The post-test data processing formulae for cycle integrated PM calculations are documented in ISO 16183 [28]. A comparison between PFSS and CVS particulate mass measurements during steady-state and transient operation was conducted [29, 30] with promising results, but further investigation is still required.

4.2.3 Heat Release Analysis of Transient Pressure Data

After logging steady-state signals, a heat release analysis can follow in order to analyze the combustion characteristics and interpret cylinder pressure data; the latter have typically been treated with the Fast Fourier Transform in conjunction with a digital filter to remove the high frequency noise components. Moreover, the pressure data obtained for each cycle should be corrected for any drifting of the signals obtained from the charge amplifiers, by taking into account the mean (over each engine cycle) value of the pressure in the gas inlet period and 'setting' this equal to the experimentally obtained boost pressure value. Heat release analysis is usually based on single-zone modeling, assuming a uniform mixture of air and combustion products at each instant of time, with combustion assumed to take place incrementally in a homogeneous mode. A brief description of the procedure that is followed for naturally aspirated, steady-state pressure data analysis will be provided first.

The expression for the first law of thermodynamics during the closed part of the cycle, neglecting the sensible enthalpy of fuel, is given by the following equation (see also Section 9.3 for more details on the formulation of Equation 4.9 as well as the usual assumptions of single-zone modeling)

$$\frac{dQ_{net}}{dt} = \frac{1}{\gamma - 1}\left[p_g \frac{dV}{dt} + V \frac{dp_g}{dt} - p_g \frac{V}{m}\frac{dm}{dt} \right] + p_g \frac{dV}{dt} + h_{bl} \frac{dm_{bl}}{dt} \tag{4.9}$$

where p_g is the recorded gas pressure, V the instantaneous cylinder volume (Equation 9.30), m the instantaneous mass of the in-cylinder charge and index 'bl' corresponds to blow-by losses through the cylinder rings. The specific heat capacity ratio γ has values appropriate to air prior to combustion (≈ 1.35) and to burned gases at the overall equivalence ratio during the combustion process $(\approx 1.26 \text{ to } 1.3)$.

In order to calculate the corresponding uniform gas temperature from the perfect gas state equation ($p_g V = m R_s T_g$), the instantaneous in-cylinder mass must be known. For the gas exchange period, the mass flow to or from the engine cylinder can be calculated using Equations 9.74 accounting for one-dimensional, quasi-steady, compressible flow. An initial estimation of mass is made at the point of intake valve closure, which is later on checked for convergence.

During compression, combustion and expansion, the instantaneous mass can then be computed from the following relation

$$m(\varphi) = m(\varphi_{IVC}) - \int_{\varphi_{IVC}}^{\varphi} \frac{dm_{bl}}{dt} \frac{1}{6N} d\varphi \tag{4.10}$$

where φ_{ivc} is the crank angle at inlet valve closure, and dm_{bl} the mass lost from the cylinder due to blow-by in the step $d\varphi$, with the blow-by rate calculated, again, from Equations 9.74.

For the calculation of gross heat release rate, the instantaneous heat loss rate $dQ_L/d\varphi$ to the cylinder walls is needed as will be formulated in Section 9.3.2.2. Then, the gross heat release rate is

$$\frac{dQ_{gross}}{d\varphi} = \frac{dQ_{net}}{d\varphi} - \frac{dQ_L}{d\varphi} \tag{4.11}$$

Obviously, by knowing the lower heating value LHV of the fuel, the fuel burned mass rate can be computed from

$$\frac{dm_{fb}}{dt} = \frac{1}{LHV} \frac{dQ_{gross}}{dt} \tag{4.12}$$

If we now integrate Equations 4.11 and 4.12 over an engine cycle (corresponding, for example, to 4π rad for a four-stroke engine), we can obtain the amount of heat released by the total fuel and the total fuel itself, respectively.

During (turbocharged) engine transients, the above described heat release analysis procedure should proceed with caution owing to the following two issues that require attention.

- Cylinder gas pressure measurement via pressure transducers may be influenced by thermal shock owing to the high rate of engine load increase.
- The air-supply and fueling rate computation is highly uncertain from transient cycle to cycle, particularly for highly turbocharged diesel engines with increased valve overlap period.

The latter subject has been treated in more detail by Assanis *et al.* [31] and a few remarks from this work will be reproduced below. While data collected from one engine cycle during the transient event can be treated as quasi-steady and, thus, reduced using standard approaches, the characterization of the in-cylinder gas properties on a cycle-resolved basis presents special challenges. In particular, as was discussed at the beginning of this chapter, the capacitance of both the fuel and air systems make standard measurement techniques inappropriate for measuring air trapped and fuel injected in the cylinder for each cycle (*e.g.*, of the order of 0.1 s or even less). Knowledge of the latter quantities is crucial, though, in order to accurately determine the specific heat capacities ratio γ in Equation 4.9 that varies within an engine cycle and from cycle to cycle in a transient event. For example, the use of the following well known equation for the instantaneous injected mass flow-rate has been proposed assuming one-dimensional, quasi-steady, incompressible fuel flow through the orifice:

$$\dot{m}_{fi} = c_D A_N \sqrt{2\rho_f \, \Delta p_{inj}} \tag{4.13}$$

where c_D is the discharge coefficient, A_N the nozzle area, ρ_f the fuel density and Δp_{inj} the instantaneous difference between cylinder pressure and injection pressure, as recorded by the respective transducers and after having been correctly phased to the TDC marker signal. Particularly for the discharge coefficient, Assanis *et al.*, found that for their four-stroke, six-cylinder, DI engine equipped with unit injectors, this ranged between 0.68 and 0.72, remaining practically constant throughout most of the operating range, with somewhat increased values at extremely low loads.

The total mass of injected fuel per cycle M_{fi} can then be found by integrating Equation 4.13 over the injection period, from the start of injection (SoI) to the end of injection (EoI), *i.e.*,

$$M_{fi} = c_D A_N \int_{SoI}^{EoI} \sqrt{2\rho_f \, \Delta p_{inj}} \cdot d\varphi \tag{4.14}$$

Experimental determination of air-supply (term $m(\varphi_{IVC})$ in Equation 4.10) is much more complex, particularly in modern engines with EGR control. Assanis *et al.* attempted to evaluate the instantaneous air-mass flow-rate by splitting the induction process into two stages: 1. when the exhaust valves are closed, and 2. during the valve overlap period, and using the measured pressure and temperature in the intake manifold for calculation of the total delivered air-charge per cycle.

The response of the cylinder wall temperature during the transient event constitutes one more challenge for a successful heat release analysis. It might be assumed that knowledge of the exact wall temperature profile with respect to the engine cycle is required; this is not necessarily the case, since, as was discussed in Section 2.3.1, the wall temperature response is quite slow during the transient event. Hence, it is not actually a coarse approximation to assume constant cylinder wall temperature throughout the few seconds of the load or speed increase transient schedule.

Summarizing, heat release analysis during transients can proceed as follows:

- The signals for cylinder pressure and fuel injection of each transient cycle are initially processed to define indices for the start of fuel injection, end of fuel injection and start of combustion as, for example, described in [1,2].
- The heat release loop of the code begins at the point where combustion starts. Before performing the heat release analysis, cylinder composition must be determined. Knowledge of the fuel injection indices allows

integration of the pressure differential term in Equation 4.13, hence calculation of the amount of injected fuel per cycle.

- The instantaneous fuel flow-rate from Equation 4.13 can be evaluated based on the discharge coefficient c_D.
- The trapped mass of air is then calculated via the two-step induction process mentioned above. This allows determination of the overall fuel–air ratio, even though the exact composition inside the cylinder is not known.
- The mean gas temperature profile is generated using the perfect gas law equation.
- The global specific heat capacities ratio γ can then be calculated using a simplified equilibrium program. The specific heat capacities ratio treated in this way does not necessarily reflect the actual composition at any given location in the cylinder, but does reflect the variation of properties with global fuel–air ratio.
- Next, the cylinder pressure is filtered to remove high frequency noise using a standard Fast Fourier Transform and appropriate filter.
- The net heat release rate is computed from Equation 4.9, and after evaluation of heat loss to the cylinder wall (Section 9.3.2.2) the gross heat release is calculated from Equation 4.11. They can then be integrated to yield the overall heat release.

References

[1] Zhao H, Ladommatos N. Engine combustion instrumentation and diagnostics. Warrendale PA: Society of Automotive Engineers, 2001.

[2] Stone R. Introduction to internal combustion engines, 3rd edition. London: MacMillan, 1999.

[3] Hodgson PG, Raine JK. Computer simulation of a variable fill hydraulic dynamometer. Part 2: steady state and dynamic open-loop performance. Proc Inst Mech Eng 1992;206:49–56.

[4] Hodgson PG, Raine JK. Computer simulation of a variable fill hydraulic dynamometer. Part 3: closed-loop performance. Proc Inst Mech Eng 1992;206:327–36.

[5] Steiber J, Trader A, Treichel B, Tsujimoto D, Reese II R, Bedard M, Musial M. Development of an engine test cell for rapid evaluation of advanced powertrain technologies using model-controlled dynamometers. SAE Paper No. 2006-01-1409, 2006.

[6] Babbitt GR, Moskwa JJ. Implementation details and test results for a transient engine dynamometer and hardware in the loop vehicle model. International Symposium on 'Computer Aided Control System Design', Hawai, USA, August 22–27, Proc. IEEE, pp. 569–74, 1999.

[7] McClure BT. Characterization of the transient response of a diesel exhaust-gas measurement system. SAE Paper No. 881320, 1988.

[8] Beaumont AJ, Noble AD, Pilley AD. Signal reconstruction techniques for improved measurement of transient emissions. SAE Paper No. 900233, 1990.

[9] Chan SH, Chen XS, Arcoumanis C. Measurement and signal reconstruction of transient nitric oxide emissions in the exhaust of a turbocharged diesel engine. Trans ASME, J Dynam Syst Measurement Control 1997;119:620–30.

[10] Collier TR, Gregory D, Rushton M, Hands T. Investigation into the performance of an ultra-fast response NO analyzer equipped with a NO_2 to NO converter for gasoline and diesel exhaust NO_x measurements. SAE Paper No. 2000-01-2954, 2000.
[11] http://www.dieselnet.com, Ecopoint Inc.
[12] Kittelson DB, Arnold M, Watts WF. Review of diesel particulate matter sampling methods – final report. University of Minnesota, Dept of Mechanical Engineering, 1999.
[13] Burtscher H. Physical characterization of particulate emissions from diesel engines: a review. Aerosol Sci 2005;36:896–932.
[14] Jarrett RP, Clark NN. Evaluation of methods for determining continuous particulate matter from transient testing of heavy-duty diesel engines. SAE Paper No. 2001-01-3575, 2001.
[15] Hofeldt DL, Chen G. Transient particulate emissions from diesel buses during the central business district cycle. SAE Paper No. 960251, 1996.
[16] Moosmüller H, Arnott WP, Rogers CF, Bowen JL, Gillies JA, Pierson WR *et al.* Time resolved characterization of diesel particulate emissions. 1. Instruments for particle mass measurements. Environ Sci Technol 2001;35:781–7.
[17] Patashnick H, Meyer M, Rogers B. Tapered element oscillating microbalance technology. North American/9th US Mine Ventilation Symposium, Kingston, ON, Canada, June 2002, Proc. pp. 625–31.
[18] Christian R, Knopf F, Jaschek A, Schindler W. A new method for the filter smoke number measurement with improved sensitivity. MTZ 1993;54:16–22 (in German).
[19] Alkidas AC. Relationships between smoke measurements and particulate measurements. SAE Paper No. 840412, 1984.
[20] Lapuerta M, Martos FJ, Cárdenas MD. Determination of light extinction efficiency of diesel soot from smoke opacity measurements. Measurement Sci Technol 2005;16:2048–55.
[21] Gupta S, Shih J, Hilman G, Sekar R, Shimpi S, Martin W *et al.* Transient, real-time particulate measurements in diesel engines. Proc. 9th Diesel Engine Emission Reductions Conference, Newport, RI, August 24–28, 2003, pp. 1–6.
[22] Park K, Cao F, Kittelson DB, McMurry PH. Relationship between particle mass and mobility for diesel engine exhaust particles. Environ Sci Technol 2003;37:577–83.
[23] Hagena JR, Filipi ZS, Assanis DN. Transient diesel emissions: analysis of engine operation during a tip-in. SAE Paper No. 2006-01-1151, 2006.
[24] Black J, Eastwood PG, Tufail K, Winstanley T, Hardalupas Y, Taylor AMKP. Inter-correlations between smoke opacity, legal particulate sampling (LPS) and TEOM, during transient operation of a diesel engine. SAE Paper No. 2007-01-2060, 2007.
[25] Makkee M, van Gulijk C, Moulijn JA. The choice of instrument (ELPI and/or SMPS) for diesel soot particulate measurements. SAE Paper No. 2003-01-0784, 2003.
[26] Westlund A. Ångström H-K. Evaluation of techniques for transient PM-measurements. SAE Paper No. 2008-01-1680, 2008.
[27] Directive 1999/96/EC of the European Parliament and of the Council, 13 December 1999.
[28] ISO 16183:2002. Heavy duty engines – measurement of gaseous emissions from raw exhaust gas and of particulate emissions using partial flow dilution systems under transient test conditions.
[29] Schweitzer T, Stein HJ. A new approach to particulate measurement on transient test cycles: partial flow dilution as alternative to CVS full flow systems. SAE Paper No. 2000-01-1134, 2000.
[30] Khalek IA, Ullman TL, Shimpi SA, Jackson CC, Dharmawardhana B, Silvis WM, *et al.* Performance of partial flow sampling systems relative to full flow CVS for

determination of particulate emissions under steady-state and transient diesel engine operation. SAE Paper No. 2001-01-1718, 2001.

[31] Assanis DN, Filipi, ZS, Fiveland SB, Syrimis M. A methodology for cycle-by-cycle transient heat release analysis in a turbocharged direct injection diesel engine. SAE Paper No. 2000-01-1185, 2000.

5

Emissions

The deterioration of the heterogeneous combustion during turbocharged diesel engine transient operation discussed in Section 2.3.2, particularly during the early cycles, where the fuel–air equivalence ratio may assume higher than stoichiometric values, is accompanied by a temporary overshoot in gaseous, particulate and noise emissions. In the following paragraphs, the primary mechanisms that are responsible for transient exhaust emissions as well as the main differences in the emission process between transient turbocharged and steady-steady conditions will be described; the differences concerning naturally aspirated transient engine operation will also be highlighted at the end of each sub-section. Although the global concern for environmental pollution has directed great attention to the non-regulated exhausts too, such as benzene, low molecular carbonyl compounds or sulfur compounds, in the following paragraphs the analysis will primarily focus on the regulated exhaust constituents, *i.e.*, particulate matter (PM), nitrogen oxides (NO_x), hydrocarbons (HC) and carbon monoxide (CO). During steady-state operation, diesel engines produce high levels of NO_x owing to the increased gas temperatures, at high loads, and also increased amount of soot due to locally incomplete combustion at high fuel–air equivalence ratios; on the other hand, hydrocarbons and CO emissions are usually maintained at lower, acceptable levels. Detailed mechanisms of transient emissions will be discussed throughout this chapter; however, at this stage it is important to emphasize that the evolution of transient emissions after a speed or load increase differentiates (increases) considerably from the respective steady-state operation, it varies significantly according to the transient schedule, and it relies heavily on the specific engine and sub-systems configuration and control strategy.

5.1 Particulate Matter and Smoke

Diesel particulates consist mainly of combustion generated carbonaceous material (soot) on which some organic compounds (arisen mainly from unburned fuel and lubricating oil) have been absorbed. Particulate material is distributed over a wide size range typically from 20 nm to 10 μm, thus being respirable. PM is often separated into a soluble organic fraction (SOF) and an insoluble or dry fraction, which is often used as an estimation of soot. The soot percentage in particulate from diesel exhaust varies, but it is typically higher than 40–50%. Other particulate matter constituents are: un- or partially burned fuel/lubricant oil, bound water, wear metals and fuel-derived sulfate, as is also depicted in the upper sub-diagram of Figure 5.1 showing representative composition of diesel engine emitted PM for a 1990s technology engine tested during the US heavy-duty Transient Cycle. The sulfuric acid/sulfate fraction is roughly proportional to the fuel sulfur content. The fraction associated with unburned fuel and lube oil (*i.e.*, soluble organic fraction) varies with engine design and operating condition; it can range from less than 10% to more than 90% by mass.

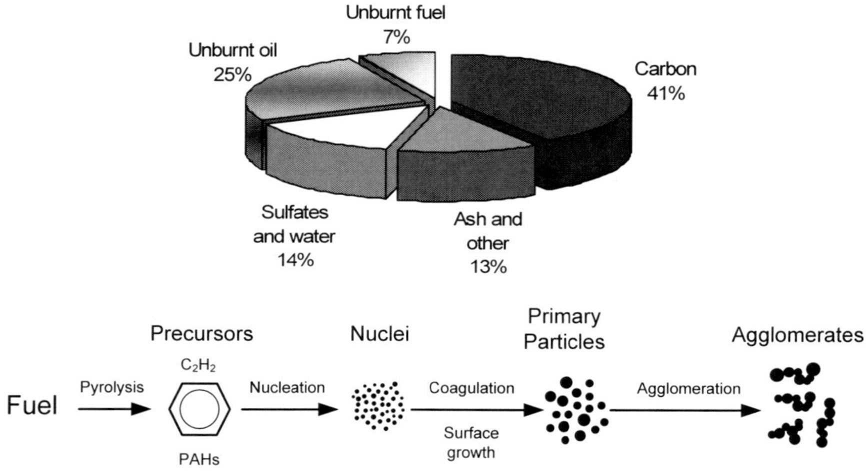

Figure 5.1. *Upper:* Typical particulate composition from a heavy-duty diesel engine during a Transient Cycle (reprinted from Kittelson [1], copyright (1998), with permission from Elsevier). *Lower:* Schematic diagram showing the steps of the soot formation process from gas phase to solid agglomerated particles (reprinted from Tree and Svensson [2], copyright (2007), with permission from Elsevier)

Soot is formed from unburned fuel that nucleates from the vapor phase to a solid phase in fuel-rich regions at elevated temperatures. Hydrocarbons or other available molecules may condense on, or be absorbed by soot depending on the surrounding conditions. The evolution of liquid- or vapor-phase hydrocarbons to solid soot particles, and possibly back to gas-phase products, involves six commonly identified processes, schematically depicted in the lower sub-diagram of Figure 5.1, namely [2]

- pyrolysis: endothermic process altering the molecular structure of fuel in the presence of high temperature and producing species such as polycyclic aromatic hydrocarbons (PAHs), considered as soot precursors;
- nucleation: spontaneous formation of a nucleus or nanoparticle from volatile materials such as hydrocarbons;

- surface growth: process of adding mass to the surface of a nucleated soot particle;
- coagulation: where roughly spherical particles collide and coalesce to form bigger spherical particles;
- agglomeration: where particles stick together to form large groups of primary particles in chain-like structure; and
- oxidation: occurs at each stage in the process, typically at temperatures higher than 1300 K, where soot is burned in the presence of oxidizing species to form gaseous products such as CO, CO_2 and H_2O.

The eventual emission of soot from the engine will depend on the balance between the process of formation and oxidation.

During steady-state engine operation, net soot production is mainly dependent on engine load. As the load increases, more fuel is injected into the cylinders, increasing the temperatures in the fuel-rich zones. Moreover, the duration of diffusion combustion is increased favoring soot formation, whereas the remaining time after combustion as well as the availability of oxygen – both of which enhance the soot oxidation process – decrease; thus, the production of soot is favored. The main mechanism of soot emissions during transients is largely based on this key behavior.

Figure 5.2 illustrates a complex transient event in terms of both load and speed changes of a two-stroke, turbocharged engined bus during the Central Business District Cycle[1] [3]; CO_2 emissions are also demonstrated in the same figure for comparison purposes. The latter reflect real-time fuel consumption and it is an important indicator of when we expect particulate concentrations to be large. Increases in CO_2 emissions occur during vehicle accelerations, while decreases to nearly zero concentration correspond to the braking periods during vehicle deceleration. Note that significant fluctuations in the fueling rate (*i.e.*, CO_2 concentration) are also evident at times just following discontinuities in the acceleration profile, and indicate that the driver is attempting to match the speed of the bus to the prescribed cycle profile. The decreases in CO_2 concentrations, which follow the local increases, may reflect either turbocharger lag intervals or driver accelerator changes.

The first thing to notice about the particulate matter concentration histories is that large increases in soot emissions occur during the same periods corresponding to vehicle acceleration. This was anticipated due to the higher in-cylinder fuel concentrations, which exist at higher loads. Soot emissions are, however, very non-linear with respect to fueling rates. In fact, the data on a coarse time scale suggest that about 80% of the particulate mass is emitted during the acceleration transients, where only about 45% of the fuel is consumed. Notice also that there is a strong correlation between sharp increases in CO_2 concentration and sharp increases in soot concentration. However, CO_2 concentration does not necessarily stay high for all the time soot concentration does. In some instances, high soot concentrations

[1] The Central Business District Cycle comprises 14 repeated modes, covering a total distance of 3.22 km, with an average speed of 22.3 km/h and a maximum speed of 32.18 km/h.

appear to lag instances of high fuel consumption by a small period of time, appearing slightly later when the fueling rate is actually decreasing.

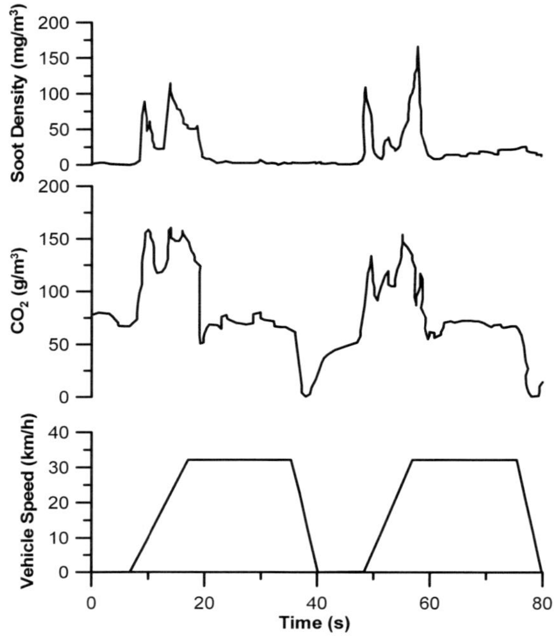

Figure 5.2. Soot and CO_2 emissions during part of an urban bus cycle (two-stroke, six-cylinder, 9 L displacement volume, turbocharged and aftercooled diesel engine – experimental results adapted from Hofeldt and Chen [3])

A question then arises as to the source of the particulates: are the PM emissions generated primarily during the steep transients in fueling rate (*i.e.*, CO_2 concentrations) and emitted during or shortly thereafter, or are they created somewhat more uniformly throughout the operating range conditions only to be re-entrained by turbulent fluctuations in the exhaust flow during transients? Knowledge of the source is of critical importance in determining how to reduce particulate matter. If the fueling rate transients are primarily responsible, then emissions could be reduced by controlling the rate of increase of injected fuel. In analyzing this question, note that PM emissions associated with idle to full-load acceleration are not consistently larger than subsequent emissions generated as a result of driver speed adjustments (times of the latter are indicated by the sudden increases in CO_2 concentrations during the cruise portion). In view of the above, although the magnitudes of the increases in fueling rate are generally not very large, the accompanying changes in PM emissions are often substantial. If re-entrainment of a substantial amount of particulates created outside the acceleration period was responsible for the PM emissions, then one would expect the soot concentrations recorded during the initial transient to be consistently the highest [3]. Since this is not the case demonstrated in Figure 5.2, the first important conclusion is that soot during transients is generated primarily during the fueling rate changes.

In the light of the previous finding, expansion of the PM production mechanism during transients can be accomplished if we isolate one particular transient event. Figure 5.3 compares quasi-steady and transient experimental particulate matter traces during an instantaneous 1–9 bar bmep step load increase of a turbocharged diesel engine at constant engine speed [4]. It is a good example to highlight the different evolution pattern of transient PM emissions compared with steady-state operation, from which useful conclusions can be drawn regarding the soot production mechanism during a load acceptance. As was discussed in Chapter 1, such transients are more often met in industrial or marine engines, which are characterized by higher ratings and greater load changes compared with their automotive counterparts. The primary observation lies in the large departure of experimental lines from the quasi-steady baseline during the first 2 s of the transient event; measurements indicated large and sharp transient spikes at the onset of a tip-in event. In contrast, the quasi-steady baseline appears to qualitatively follow the fueling history, thus underestimating peak particulate concentrations by at least one order of magnitude. Smaller differentiations between transient and quasi-steady emissions would be observed the slower the load application or the higher the engine (flywheel) mass moment of inertia, as in the latter cases the transient event would develop at a much slower rate, with significantly smoother turbocharger lag effects.

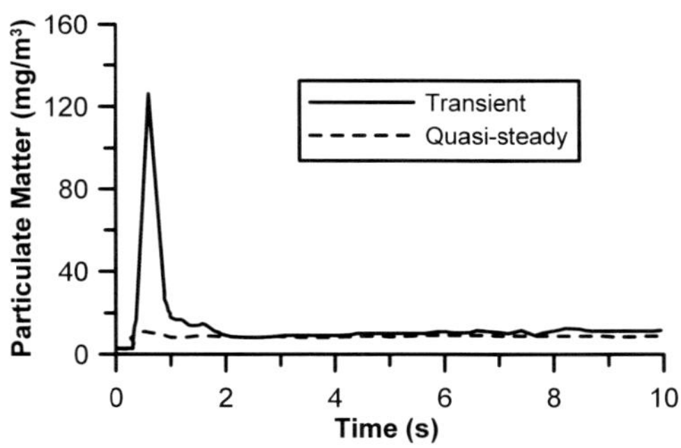

Figure 5.3. Comparison between transient and steady-state particulate matter emissions during turbocharged engine instantaneous 1–9 bar bmep step load transient at 2000 rpm (four-stroke, V-8, 6 L displacement volume, DI, VGT, heavy-duty diesel engine with electronic unit injectors – experimental results reprinted with permission from SAE Paper No. 2006-01-1151 [4], © 2006 SAE International)

As was discussed earlier in the chapter, the formation of soot is mainly dependent on engine load, with higher loads promoting higher temperature, longer duration of diffusion combustion and lower oxygen availability. The locally very high values of fuel–air ratio experienced during turbocharger lag enhance the above mechanism, which is more pronounced the higher the engine rating, *i.e.*, the higher the full-fueling to no-fueling difference. For the overshoot in particulate matter emissions observed in the first seconds of a transient load increase event therefore, the main cause is the instantaneous lack of air due to turbocharger lag, aided by the initial sharp increase in ignition delay during the early transient cycles. Rapid increases in fuel injection pressure upon the onset of the instantaneous load step cause the penetration of the liquid fuel jet within the combustion chamber to increase. Since the initial higher-pressure fuel jets are injected into an air environment that is practically unchanged from the previous steady-state condition, the higher-momentum fuel jet is not accompanied by equally enhanced gas motion. Liquid fuel impingement on the cool combustion walls increases, lowering the rate of mixture preparation and enhancing the heterogeneity of the mixture (see also the conceptual flow diagram of Figure 5.4). Moreover, the subsequent harder combustion course prolongs combustion and reduces the available time for soot oxidation [4,5]. Interestingly, the above behavior is enhanced in modern diesel engines by the amount of residual gas still present in the inlet manifold and cylinders, up to the point where the ECU shuts down the EGR valve. In fact, even after the EGR valve has closed, a small amount of residual gas is still present in the inlet manifold and piping, reducing accordingly the respective trapped air–fuel ratio.

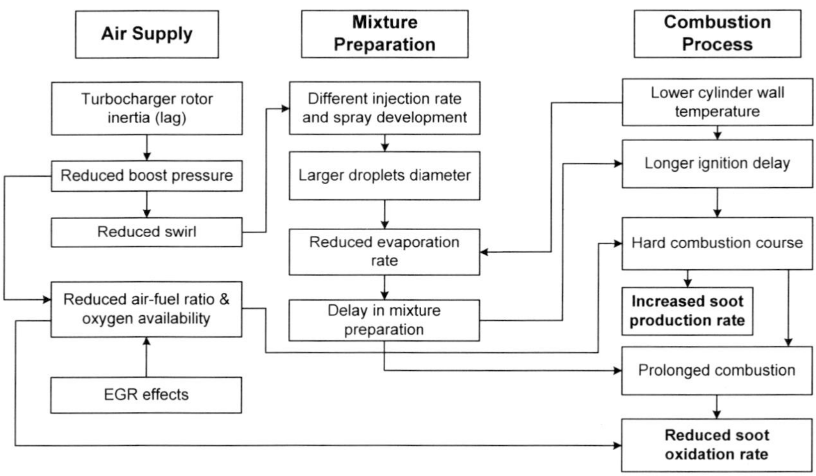

Figure 5.4. Causes of particulate matter increase during transients owing to air-supply, mixture preparation and combustion process discrepancies compared with steady-state operation (adapted from Harndorf and Kuhnt [5])

Summarizing, the main mechanisms for the soot emissions overshoot during a turbocharged diesel engine transient load increase (the same arguments hold for acceleration transients) are located in

- the low air-charge in the cylinder that cannot match the increased fueling, leading to high fuel–air equivalence ratio and temperature;
- EGR effects that further worsen the above behavior;
- discrepancies in the mixture preparation, which result in longer ignition delay period; and
- various other anomalies during the combustion process.

Further elaboration of soot development mechanism can be realized on a microscopic scale, *i.e.,* during an individual engine cycle. This can be accomplished with reference to Figure 5.5, which illustrates what happens during two intermediate engine cycles of a load acceptance transient event in terms of soot density development. Differences are actually noticed in the absolute values of soot density, while the development profile remains the same when comparing transient (during the turbocharger lag) with quasi-steady cycles. The intriguing finding from Figure 5.5 is that during turbocharger lag, soot density assumes lower maximum but higher exhaust values, compared with the respective steady-state conditions (same rotational speed and bmep as the transient cycles). This can be attributed to the slowing down of the whole combustion process under transients, *i.e.,* lower values of pressure, fuel mass burned at each time step and oxygen partial pressure in the fuel spray (the latter caused by poorer mixing). The result is that soot formation rate is actually lower during the turbocharger lag cycles, but this holds also true for the respective oxidation rate during the expansion stroke; the latter is outweighed by its formation counterpart, ultimately leading to higher 'exhaust' (at the point of EVO) soot values. The previously identified increase of the ignition delay period is further responsible for the slight delay in the beginning of soot formation compared with quasi-steady operation.

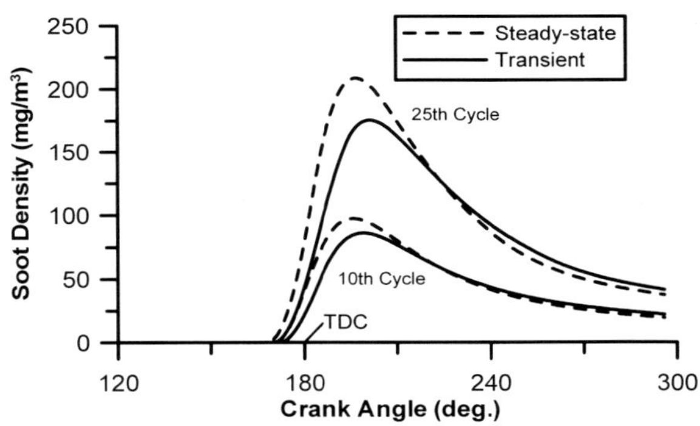

Figure 5.5. Development of in-cycle, cumulative soot emissions during the turbocharger lag period of a load acceptance event compared with quasi-steady conditions

A much harder case in terms of soot emissions is experienced in the everyday vehicle's up-gear change schedule. Particularly at the beginning of the acceleration, where the turbocharger speed is minimum and the available exhaust gas energy almost zero, the highest amount of particulates is experienced, owing to the increased turbocharger lag phenomena originating from the abrupt no-fueling to full-fueling transition; the higher the engine rating the more intense the above phenomenon. The latter behavior in terms of smoke emissions,[2] is demonstrated in Figure 5.6 for a six-cylinder, turbocharged, diesel-engined truck without fuel

[2] Particulate matter is measured on a mass basis, as was discussed in Chapter 4. On the other hand, smoke meters measure the relative quantity of light that passes through the exhaust or the relative reflectance of particulates collected on filter paper; they are used, therefore, as a surrogate for particulate mass emission levels.

limiting function. After the first gear has been engaged, the gradual increase in engine and turbocharger speed aids the vehicle's acceleration; however, temporary overshoot in smoke emissions is noticed practically during every gear change, unless some kind of strict fueling control is incorporated, which, however, affects speed response adversely as will be discussed in the next paragraph.

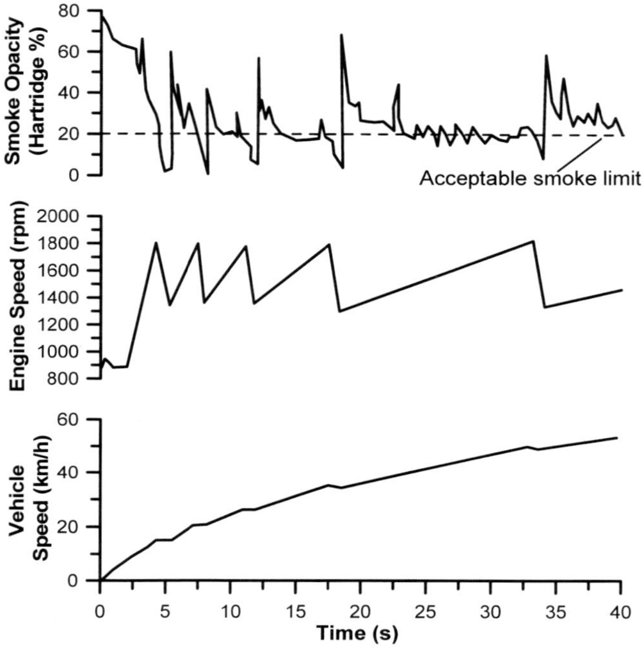

Figure 5.6. Smoke emissions during up-gear change of a turbocharged truck (four-stroke, six-cylinder, 11.32 L displacement volume, turbocharged diesel engine – experimental results reprinted with permission from SAE Paper No. 770122 [6], © 1977 SAE International)

Some of the most successful measures to control transient soot emissions are

1. shutting down the EGR valve as soon as the transient event commences (discussed in detail in the next section as this technique comes at a cost, namely increase of NO_x emissions);
2. application of fuel limiters (Section 2.2.2);
3. use of particulate filters or traps (nowadays, these can achieve at least 80% reduction in PM mass emissions under steady-state or transient conditions); and
4. use of oxygenated fuels such as biodiesel produced by transesterification of vegetable oils or animal fats (discussed later in the chapter, in Section 5.6), or adding synthetic oxygenates such as glycol ethers or maleates to the diesel fuel.

Perhaps the most popular of the above-mentioned measures for PM emissions control is the use of fuel limiters that adjust maximum fueling according to the current turbocharger boost pressure, as was discussed in Section 2.2.2. Figure 5.7, which is actually an enhancement of Figure 2.20, documents in the most explicit way the fuel limiter's influence on smoke emissions by comparing the acceleration of a turbocharged truck engine of modest rating (10 bar maximum bmep) with and without a system for controlling maximum fuel delivery as a function of boost pressure. Limiting maximum fuel delivery by 20% at zero boost, results in a substantial reduction of the total amount of emitted smoke owing to the fuel pump rack moving slowly towards full-fueling; however, the response time of the engine and turbocharger is doubled resulting in much poorer acceleration, hence vehicle driveability that may be unacceptable. The latter points out the necessity to cope with the core of the turbocharger lag problem, instead of just trying to damp out its consequences, as will be discussed in the next chapter.

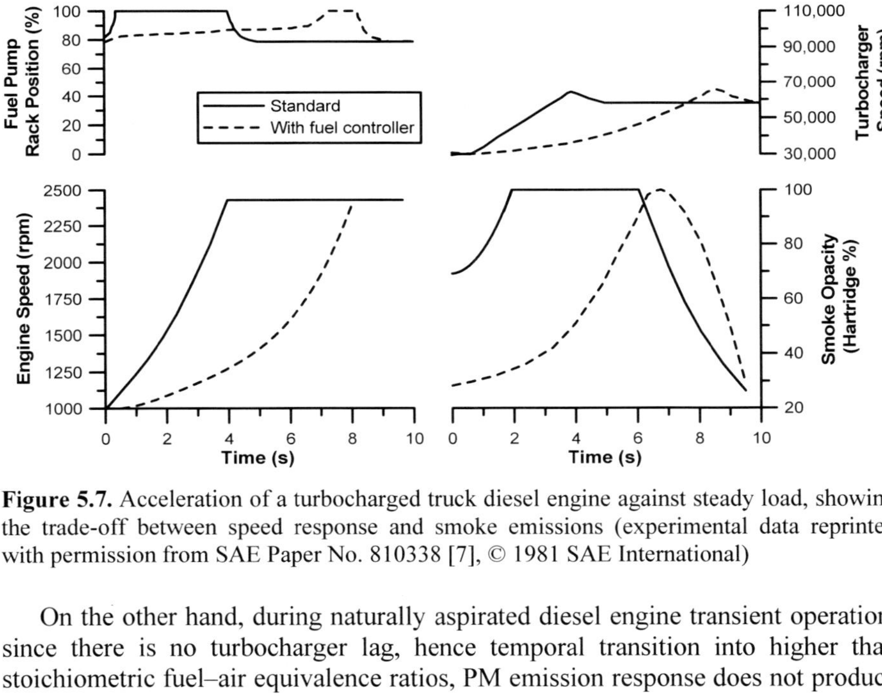

Figure 5.7. Acceleration of a turbocharged truck diesel engine against steady load, showing the trade-off between speed response and smoke emissions (experimental data reprinted with permission from SAE Paper No. 810338 [7], © 1981 SAE International)

On the other hand, during naturally aspirated diesel engine transient operation, since there is no turbocharger lag, hence temporal transition into higher than stoichiometric fuel–air equivalence ratios, PM emission response does not produce any notable initial overshoot and is characterized by minimum delay compared with the respective steady-state values. Soot emissions during naturally aspirated engine transient operation follow practically the steady-state pattern.

Particle Size Distribution

Currently, diesel particulate matter legislation is based on emitted particle mass (see also Tables in Appendix A). However, particle size distribution is gaining increasing attention in terms of air quality, as it is believed that the toxicity increases as the particle size decreases; in fact, a correlation between elevated ambient particulate matter concentration and hospital admissions has been suggested. In order to be able to proceed to accurate particle number concentration and size distribution measurements, real-time instruments that are capable of capturing fast changing particle size distributions over the entire size are required, as was discussed in Chapter 4.

Unlike mass, particle number is not conserved. Within the exhaust pipe of a diesel engine, the continuously changing conditions may lead to particulates nucleation (resulting in an increase in both particle number and mass concentration), condensation and coagulation (resulting in a reduction in the number concentration and increase in particulate size). The above phenomena are also noticed during dilution – the exact dilution ratio being of paramount importance in order to simulate real-life conditions – and sampling, making it very difficult, on the one hand, to obtain unambiguous results and, on the other, to design a standard.

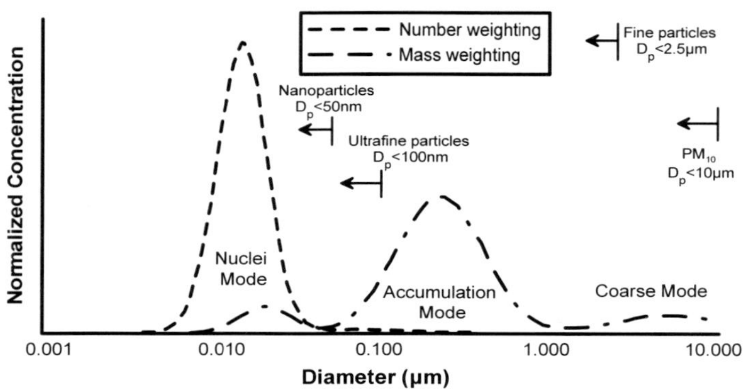

Figure 5.8. Typical engine exhaust particle size distribution (reprinted from Kittelson [1], Copyright (1998), with permission from Elsevier)

Figure 5.8, taken from Kittelson [1], shows idealized, trimodal, lognormal in form, diesel exhaust particle number and mass-weighted size distributions. The concentration of particles in any size range is proportional to the area under the corresponding curve in that range. Most of the particle mass exists in the so-called accumulation mode in the 0.1–0.3 μm diameter range. This is where the carbonaceous agglomerates and associated adsorbed materials reside. The nuclei mode typically consists of particles in the 0.005–0.05 μm diameter range. This mode usually consists of volatile organic hydrocarbon and sulfur compounds that form during exhaust dilution and cooling, and may also contain solid carbon and metal compounds. By number, more than 90% of the particles emitted by a diesel engine (the same is true for a DI spark ignition engine) fall into the nuclei mode, which, however, contains only 1–20% of the particle mass. The coarse mode contains 5–20% of the particle mass; it consists of accumulation mode particles that have been deposited on cylinder and exhaust system surfaces and later re-entrained. Also shown in Figure 5.8 are some definitions of size for atmospheric particles: PM_{10}, D_p (diameter) < 10 μm; fine particles, D_p < 2.5 μm; ultrafine particles, D_p < 0.10 μm; and nanoparticles D_p < 0.05 μm.

Particle size influences the environmental impacts of engine exhaust in several ways: it influences the atmospheric residence time of the particles, the optical properties of the particles, the particle surface area and ability to participate in atmospheric chemistry, and the health effects of the particles. The residence time of particles in the atmosphere is longest for particles in the 0.1–10 μm diameter range and is typically about one week. Larger particles are removed from the atmosphere quite quickly by settling, and smaller ones by diffusion and coagulation. A typical residence time for 10 nm particles is only about 15 min. The main mechanism for removal of these tiny particles is coagulation with particles in the accumulation mode. Thus, although these lose their identity as individual particles, they remain in the atmosphere for essentially the same time as the larger accumulation mode particles [1].

During steady-state engine operation, mostly accumulation mode particles are formed (80–110 nm), at least when high or medium speed automotive or truck diesel engines are involved. Early in the combustion phase, the majority of the particles is between 80–90 nm; later in the cycle, soot particles agglomerate and, therefore, particles of larger diameter are formed. During the soot oxidation period the number of particles is drastically reduced in the entire size spectrum. The trend is towards larger particles with increasing load, with nanoparticles favored mainly at idling conditions. As the load and so the equivalence ratio increases, more fuel is injected; this favors the formation of larger particles primarily owing to the longer diffusion combustion duration, to the higher combustion temperature, and to the reduced oxidation rate of the soot in the expansion stroke, since there is less time available after the end of the diffusion combustion and also lower oxygen availability. Number concentrations (particles per volume of gas), on the other hand, exhibit a decreasing trend with increasing fuel–air equivalence ratio through most of the operating range and may only increase at the highest loads.

Among the major factors that have been found to influence both particle number concentration and size distribution is the type of fuel used, with the ultra-low sulfur fuel being preferable for overall decreased PM emissions, although an increase in the nanoparticles concentration has been documented. The same holds true for higher fuel injection pressures or lower EGR rates. Nonetheless, it seems that a trade-off may exist between particle number distribution and mass, with lower aerodynamic diameter particles favored with decreasing PM mass emissions. Whereas diesel particulate filters can reduce particle number concentration by 1 to 2 orders of magnitude, mainly in the accumulation mode, concern has been expressed that the current DPF technology may be allowing nanoparticles to pass through, or even producing them at high engine loads during the fast oxidation process, *e.g.*, through oxidative fragmentation of soot aggregates.

As with almost all aspects of transient operation, particle concentration and size/mass distribution differentiates from the respective steady-state conditions, with changing load and engine speed affecting strongly both the number and size distribution of emitted particles. As is illustrated in Figure 5.9 for the American FTP Transient Cycle and Figure 5.10 for the Japanese JE-05 Light-duty Cycle, heavy accelerations (or load increases) are responsible for overall higher particle number concentrations.

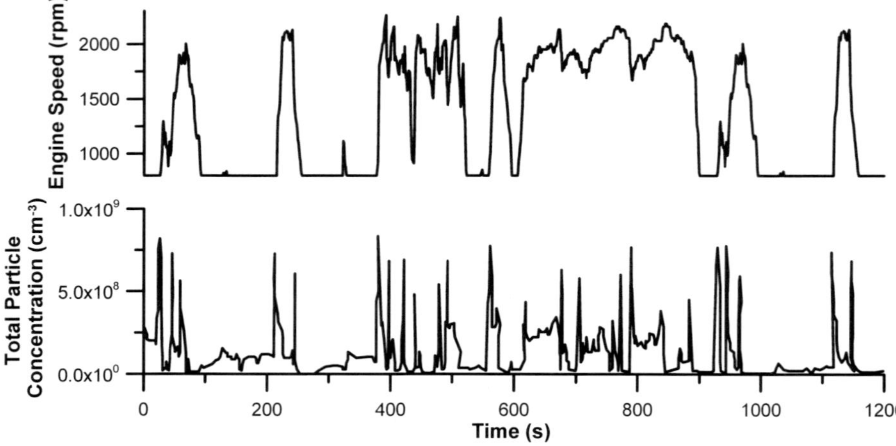

Figure 5.9. Total particle number concentration during the heavy-duty FTP Transient Cycle, measured via an Engine Exhaust Particle Sizer (5.9 L displacement volume, turbocharged diesel engine with EGR running on 350 ppm sulfur fuel – from Wang *et al.* [8], Aerosol Science & Technology. Copyright 1995. Mount Laurel, NJ. Reprinted with permission)

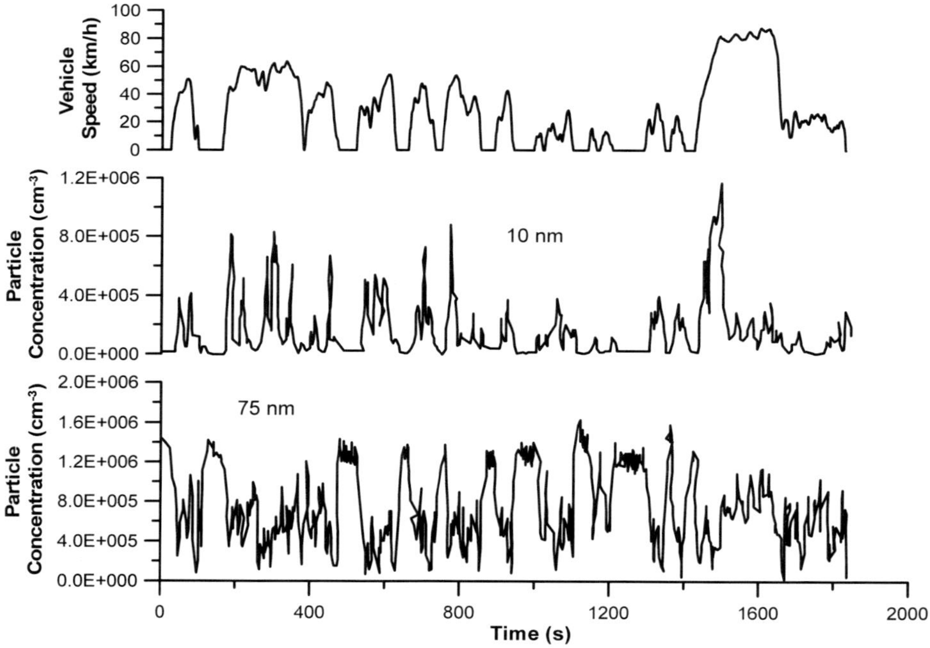

Figure 5.10. Particle number concentration for two distinct aerodynamic diameters during the Japanese JE-05 Cycle, measured with a DMA (four-cylinder diesel engine of 4.777 L displacement volume, rated at 96 kW at 3000 rpm – experimental results from Takeuchi *et al.* [9], reprinted with kind permission from Springer Science+Business Media)

The same holds true for higher (steady-state) loads, which explains why during cruising, higher number particle concentration is observed compared with idling (see also Figure 5.11). The latter remarks suggest that increase in the local fuel–air ratio is the dominant cause for the increase in the number of emitted particles, but other factors, such as the exhaust gas temperature or the use of after-treatment devices, may be involved.

Figure 5.11 quantifies the previous results by showing comparative average ultrafine ($D_p < 100$ nm) particle number concentrations for three different, diesel-engined buses over five operating schedules, namely idle, steady-state cruising at 55 mph (80 km/h), Central Business District (CBD) Transient Cycle, New York Bus (NYB) Cycle[3] and UDDS (*cf.* Section A.3.2.2). The important findings from the work by Holmén and Qu [10] is that over 90% of the total particle number concentrations over all five test cycles were measured in ELPI™ size corresponding to 29–164 nm, with ultrafine particles comprising more than 60% of the total particle number concentration. Moreover, for all driving cycles, the total average ultrafine particle number concentrations for the baseline diesel-engined bus equipped with oxidation catalyst were more than 100 times higher than those measured for the two alternative bus configurations; idle operation exhibited the lowest amount of ultrafine particles compared with the other tested Cycles.

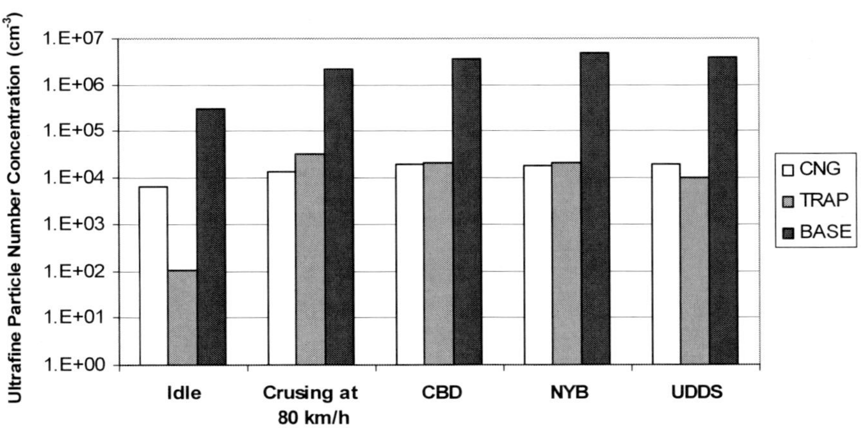

Figure 5.11. Average ultrafine particle number concentrations during five Cycles for three different bus configurations, measured with ELPI™ ('CNG' corresponds to compressed natural gas engined bus, 'TRAP' to diesel engine with oxidation catalyst and DPF, and 'BASE' to diesel-engined bus with oxidation catalyst; diesel engines are four-cylinder, 8.5 L displacement volume, running on ultra-low sulfur fuel – experimental results reprinted with permission from Holmén and Qu [10]. Copyright (2004) American Chemical Society)

Things are somewhat more complex in terms of particle size distribution [11]:

1. Nuclei mode. During transients, *e.g.*, during a Transient Cycle that comprises cold starting, and various speed and load changes at different gears, a

[3] The New York Bus Cycle has duration of 600 s for a driving distance of 0.99 km, with an average speed of 5.94 km/h and a maximum speed of 49.56 km/h with 11 stops.

large amount of nuclei mode particles is formed, primarily when the engine load and exhaust gas temperature are low (*i.e.*, during the cold start phase) or declining (*i.e.*, during load decreases). Most nuclei mode particles form during the dilution process or during expansion and cooling in the engine after combustion is complete; this also explains why a large amount of nuclei particles is formed when the load is being reduced and not just when the engine is idling. In addition to exhaust conditions, past research has positively correlated the formation of nuclei mode particles to unburned hydrocarbon emissions, which as will be discussed in Section 5.3 are formed at very low loads too.

2. Accumulation mode. During various transient events, a significant accumulation mode can be observed. This happens at high, medium, and increasing loads. As the engine load and exhaust gas temperature increase, volatile compounds have less opportunity to nucleate, and more accumulation mode particles are emitted due to higher fuel consumption inside the cylinder. This is especially true for increasing fueling conditions such as acceleration, which do not exist during steady-state operation. If over-fueling occurs during acceleration, *i.e.*, temporary increase of fuel–air equivalence ratio into higher than stoichiometric values, hydrocarbon emissions are increased promoting the above-mentioned mechanism. In contrast, in gasoline fueled engines, particle emissions are higher for all sizes during cold starting.

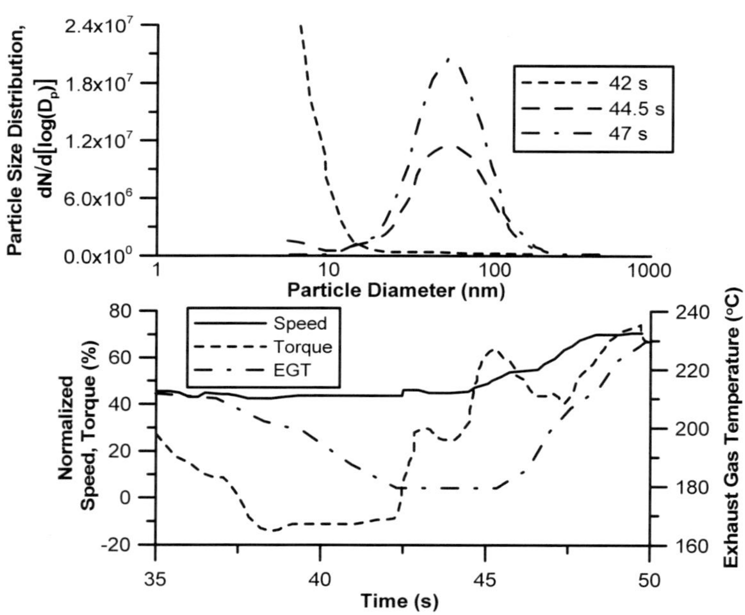

Figure 5.12. Particle size distribution during part of the FTP Transient Cycle (experimental results reprinted with permission from Liu *et al.* [11]. Copyright (2007) American Chemical Society)

The above remarks are documented in Figure 5.12 (upper sub-diagram) showing particle size distribution at three distinct phases during the heavy-duty

FTP Transient Cycle. The end of a load decrease occurred at 42 s into the Cycle, causing engine loads to be negative, speeds around 40% of maximum, and exhaust gas temperatures relatively low. The particle size distributions resulting from these engine conditions, displayed in Figure 5.12, begin with a large peak in the nuclei mode; several seconds further into the FTP cycle, a large peak of accumulation mode particles emerged as the engine began to accelerate, evidenced by the growth of the accumulation mode in the 44.5 and 47 s transient two size distributions. As the engine load increased, more fuel was injected, thus decreasing the air–fuel ratios, increasing soot emissions, and accelerating the rate of particulate agglomeration.

3. Bimodal distributions. Bimodal, *i.e.*, both nuclei and accumulation mode distributions are, usually, observed under quickly changing engine conditions. When the engine is accelerating, decelerating, or the load and exhaust gas temperature are changing from increasing to decreasing, or vice versa, conditions are often favorable for both types of particles to form. The relative magnitudes of these modes then depend on the precise load, speed, fuel consumption, and exhaust gas temperature at the time of particle formation. The results of these quickly changing conditions rarely can be seen during steady-state testing [8–11].

Some further aspects of particle mass and number concentration will be discussed in Section 7.1.4, with respect to engine cold starting operation.

5.2 Nitrogen Oxides

Besides particulate matter, nitrogen oxides are the most critical pollutants produced by diesel engines; they consist mostly of nitric oxide (or nitrogen monoxide) NO and nitrogen dioxide NO_2 (referred to, collectively, as NO_x) with traces of dinitrogen trioxide N_2O_3 and dinitrogen pentoxide N_2O_5. Nitrogen oxides are strongly dependent on temperature (primary dependence), local concentration of oxygen and duration of combustion. Other notable factors are injection timing, aspiration, charge-air cooling, swirl, fuel properties, *etc*. Recent research has indicated that NO_x is mainly formed during the diffusion rather than the premixed phase of combustion, on the weak side of the reaction zone in the post-flame gases. Nitrogen oxides are highly active ozone precursors playing an important role in the smog chemistry. The principal reactions governing NO formation from molecular nitrogen and oxygen during combustion of lean or near stoichiometric fuel–air mixtures are usually described by the extended Zel'dovic mechanism

$$\begin{aligned} &O + N_2 \rightleftarrows NO + N \\ &N + O_2 \rightleftarrows NO + O \\ &N + OH \rightleftarrows NO + H \end{aligned} \tag{5.1}$$

with the third equation being important only when the mixture is near stoichiometric. For NO_2, which is of interest at lower engine loads hence not that important during turbocharged diesel engine load increase or acceleration transients, the respective formation reaction is

$$NO + HO_2 \rightarrow NO_2 + OH \tag{5.2}$$

with a possible subsequent conversion of NO_2 to NO via

$$NO_2 + O \rightarrow NO + O_2 \tag{5.3}$$

Decomposition of NO_x during the expansion stroke is, unfortunately, very slow primarily owing to the lower gas temperatures involved. This means that the amount of NO_x produced during combustion practically 'freezes' at some point during expansion, *i.e.,* it is transferred to the exhaust.

The most effective method of reducing NO_x emissions at steady-state conditions is by lowering the peak cylinder temperature through exhaust gas recirculation or retarded injection timing. Unlike SI engines, exhaust gas recirculation in a diesel engine replaces oxygen, and so promotes a slight enrichment of the mixture. Consequently, as was discussed in Section 2.5, the usual ECU strategy during transients aims at shutting down the EGR valve in order to help build-up of air–fuel ratio and boost pressure, and limit intolerable smoke emissions. It is not surprising then that during this phase an overshoot of nitrogen oxides is noticed. The trade-off between NO_x and smoke emissions during turbocharger lag (experienced also at steady-state conditions) can be documented in an explicit way with reference to Figure 5.13, showing measurements after a 40–80% step increase in engine load for a heavy-duty turbocharged diesel engine [12]; three EGR rate patterns are compared, *i.e.,* 1. continuous, low EGR rate, 2. high EGR rate before transient to zero EGR rate during transient, and 3. continuous, high EGR rate. Case 3 involves an overall increased EGR rate during the whole transient event and, as expected, results in lower NO_x emissions compared with cases 1 and 2. It should be pointed out that for all three examined strategies, smoke was controlled to minimum levels at steady-state operation prior to load acceptance, however case 3 resulted in a notable 'smoke spike' occurring shortly after application of the new load. This was due to the considerable delay in the increase of inlet manifold pressure, resulting in a lower air–fuel ratio as was discussed in Section 2.1. Afterwards, when the inlet pressure and air–fuel ratio recover, smoke levels drop substantially. The magnitude of the smoke and NO_x spikes depends strongly on the EGR rate before and after load acceptance, but on an adverse manner. In case 1, the decrease in air–fuel ratio is small, and the smoke spike is rather low; however, in case 3 air–fuel ratio decreases considerably during the turbocharger lag phase and the resulting smoke spike assumes a large peak value. In case 2 on the other hand, which is the one practically followed by the engine ECU, smoke is maintained at acceptable levels but at the expense of NO_x emissions that peak for as long as the EGR valve remains closed.

The main mechanism of NO_x production during turbocharged diesel engine transient operation can be best explained with reference to Figure 5.14, which compares quasi-steady and instantaneous (transient) experimental NO_x traces during the 1–9 bar bmep step load increase transient at constant engine speed of the engine of Figure 5.3.

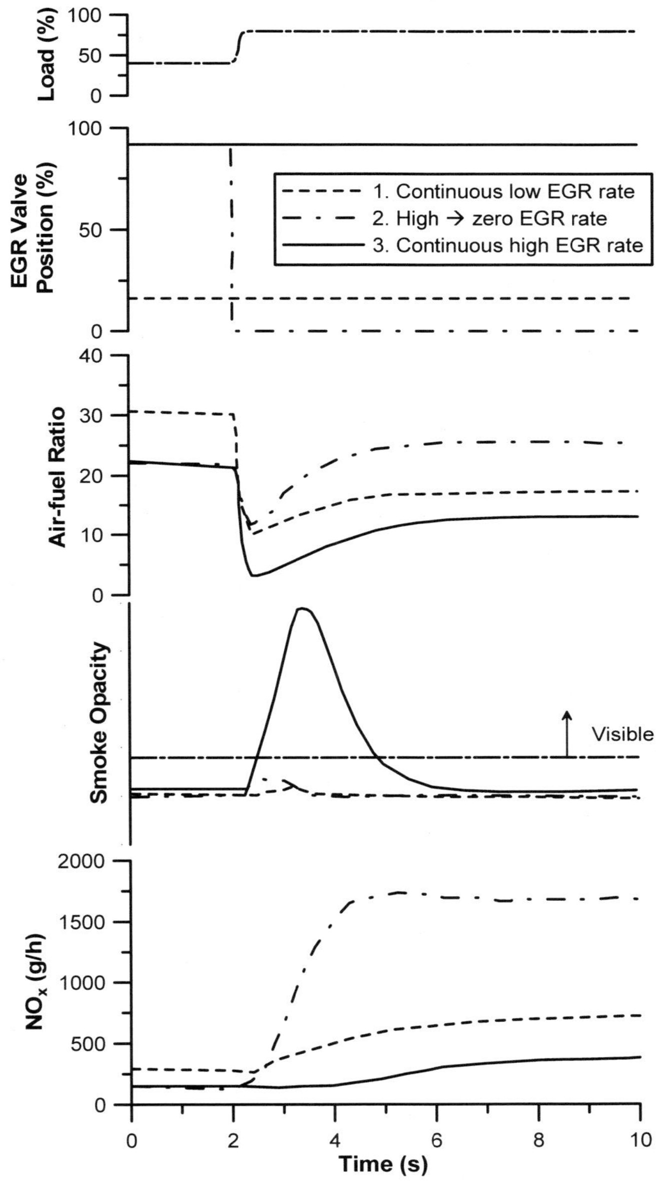

Figure 5.13. Trade-off between smoke and NO_x emissions after a step load increase for three different EGR strategies (four-stroke, six-cylinder, DI, turbocharged diesel engine of 13 L displacement volume, rated at 279 kW at 2200 rpm – experimental results reprinted with permission from SAE Paper No. 2004-01-0120 [12], © 2004 SAE International)

Noteworthy departures of the experimental transient lines from the quasi-steady baseline are observed during the first seconds of the transient event, with the NO_x spike being, however, of significantly lower magnitude than its PM counterpart of Figure 5.3, but extending over a wider time range. Since the quasi-steady baseline appears to qualitatively follow the fueling history, a clear underestimation of transient peak NO_x values by a factor of 1/3 is observed.

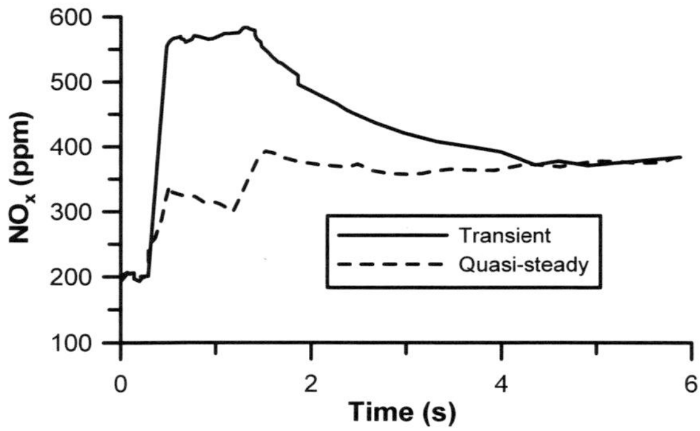

Figure 5.14. Comparison between transient and steady-state NO_x emissions during instantaneous 1–9 bar bmep step transient event at 2000 rpm (four-stroke, V-8, 6 L displacement volume, DI, VGT, heavy-duty diesel engine with electronic unit injectors – experimental results reprinted with permission from SAE Paper No. 2006-01-1151 [4], © 2006 SAE International)

In the case of NO_x emissions, it is, primarily, the lag between increased fueling and the response of the air-charging system, as well as the EGR starvation during the first cycles that is responsible for the increased emissions noticed in Figure 5.14 (*cf.* Figures 2.39 and 2.42 for EGR valve behavior during transients); the injection pressure and exhaust pressure history, on the other hand, play here a secondary but not negligible role. Since the main parameter affecting NO_x formation is the burned gas temperature, local high temperatures due to close to stoichiometric air–fuel mixtures are expected to increase NO_x emissions during the turbocharger lag cycles, as is obvious in Figure 5.14, for the first two seconds after the load change. At the same time, the cylinder wall temperature is still low, as the thermal transient develops much slower, a fact that leads to higher heat transfer rates to the cylinder walls, hence a slight reduction of the gas temperature. Further reduction of the cylinder gas temperatures is also favored by the injection timing retard that may be experienced during the load increase. The synergistic effect of the above factors produces the relatively small overshoot in NO_x emissions compared with their PM counterparts of Figure 5.3, with the EGR valve displacement playing an important role too. In general, the EGR valve closes immediately after load application, causing a sharp drop in the respective EGR rate to almost zero (tending to increase NO_x production), before gradually opening as the recovery period develops and the air–fuel ratio is built-up sufficiently (*cf.* Figure 2.39). The spike in NO_x emissions for the engine of Figure 5.14, exhibits peak values roughly 80% higher than the final steady-state level, with the load schedule profile being of significant importance (the more instant the load application the higher the NO_x peak) [4].

Figure 5.15 expands on the previous findings, focusing on the in-cycle development of NO concentration compared with their respective steady-state counterparts (*i.e.*, at the same load – defined by bmep – and engine speed) during an intermediate transient cycle of a similar load increase transient event as the one illustrated in Figure 5.14. Unsurprisingly, differences between steady-state and transient NO emission traces are observed concerning, mainly, their absolute values. On the other hand, the general development profile remains qualitatively unaltered for both operating modes. NO concentration, as depicted in Figure 5.15, assumes higher maximum and 'exhaust' values during the transient cycles compared with its steady-state counterpart. This is, primarily, attributed to the higher cylinder gas temperatures during the early transient cycles, resulting from higher values of fuel–air equivalence ratio; the latter being the direct consequence of turbocharger lag. Closer examination of the curves in Figure 5.15 reveals two other interesting facts. First, NO production practically lasts for a very small period, namely 20–25°CA. Second, NO during transients starts to evolve later in the cycle (a 'delay' of the order of 10°CA is observed compared with the respective steady-state cycles for the particular engine), implying that combustion begins later in the cycle, therefore the in-cylinder high temperatures occur later. This results from the combined effect of a variety of physical off-design phenomena occurring during transients, such as the later start of injection (originating in the lower values of the fuel delivery system residual pressure), the increased ignition delay period and the overall slowing down of the whole combustion process compared with steady-state operation.

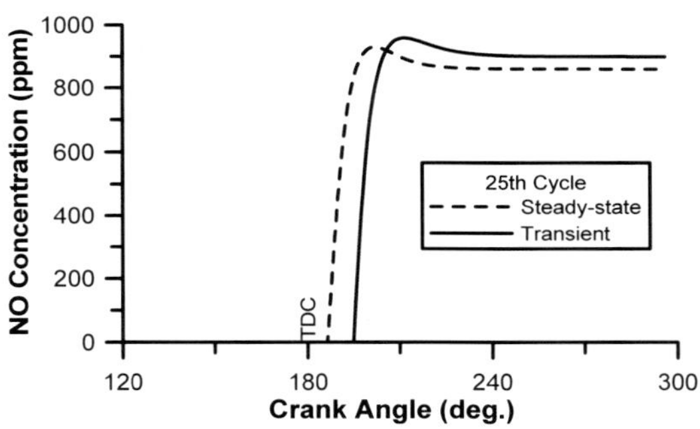

Figure 5.15. Development of transient NO concentration histories for an intermediate cycle during a 10–80% load increase compared with quasi-steady conditions

Similar arguments hold for the more complicated (acceleration) case during a Transient Cycle. The EGR valve closes while the engine is accelerating in order to reach the air-mass flow requirements and prevent increased smoke emissions. The latter, however, is accomplished at the expense of NO_x emissions that peak during the transient evolutions (Figure 5.16). On the other hand, when decelerating, very little amount of fuel is injected, resulting in a sharp decrease in NO_x emissions. During cruising, the EGR valve is open until its set point value and, consequently, NO_x emissions are successfully controlled.

As is the case for steady-state engine operation, every measure that can decrease (peak) cylinder temperatures, *e.g.*, charge-air cooling or retarded injection timing, can prove successful in limiting NO_x emissions during transients too; however, an adverse affect may be noticed concerning smoke (or HC) emissions as was discussed earlier in the section. In view of the above, a sophisticated, load-dependent, transient control strategy needs to be employed. This may be based on coordinated or model-based control theory, incorporating, for example, a fast-acting, electronically controlled EGR in order to bridge the differences in emitted smoke and NO_x while, at the same time, maintaining acceptable engine speed response/driveability [*e.g.*, 12–17]. Besides that, and in order to meet the forthcoming, more stringent emission legislation limits (*e.g.*, Euro VI), a urea-based SCR will be probably required for further management of nitrogen oxides even at steady-state conditions, as it is currently the case with large marine and industrial diesel engines.

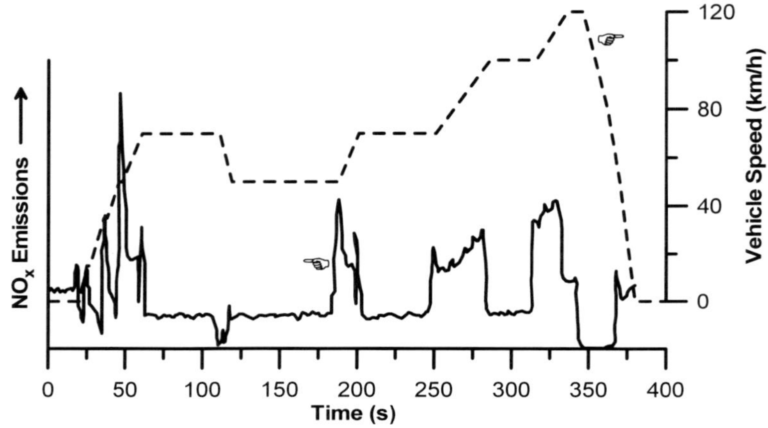

Figure 5.16. NO_x emissions during the EUDC Transient Cycle (four-stroke, four-cylinder, VGT, aftercooled, automotive, HSDI diesel engine with EGR – experimental results adapted from Black *et al.* [13])

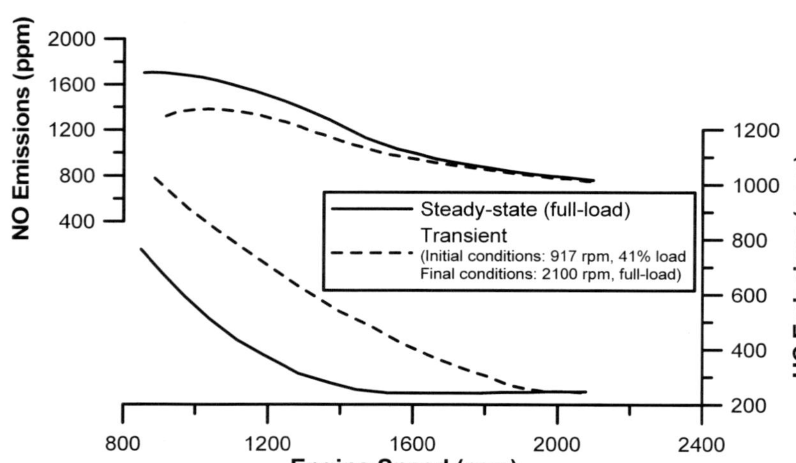

Figure 5.17. Comparison between transient and quasi-steady, naturally aspirated NO and HC emissions during acceleration (four-stroke, single-cylinder, water-cooled diesel engine of 1.455 L displacement volume – experimental results adapted from Meggyes [18])

NO_x trade-off between steady-state and transient operation evolves in a much different way in naturally aspirated diesel engines, as here the predominant delay of the turbocharger lag is absent. The main cause for the discrepancies observed in this case is the different cylinder wall temperatures involved, that ultimately affect nitric oxide production. Owing to the cylinder wall temperature being lower during most of the transient event, gas temperatures decrease too compared with the respective steady-state operation (recall that there is no turbocharger lag effects, hence excursion to high fuel–air equivalence ratios in naturally aspirated engine operation), ultimately leading to lower NO values, particularly during the early cycles of the transient event, as is demonstrated in Figure 5.17. Since the wall temperature is primarily determined by the engine load rather than speed, the differences observed are smaller the higher the initial load prior to acceleration [18, 19].

5.3 Hydrocarbons

Hydrocarbon emissions in diesel engines consist of decomposed fuel molecules or recombined intermediate compounds; another source is lubricating oil. During the ignition delay period, they form as the result of either very high fuel–air ratios or, mainly, undermixing of fuel that cannot ignite or support a flame. For the fuel injected after the ignition delay period, slow mixing of fuel and pyrolysis products with air, resulting in overrich mixture or quenching of the combustion products can result in incomplete combustion and unburned hydrocarbons. During idling and low-load operation, HC emissions are noteworthy; the same holds true for very rich mixtures (*e.g.*, during turbocharger lag). Primary sources of HC production are located around the perimeter of the reaction zone, where the mixture is too lean to burn, in the fuel retained in the injection nozzle sac (fuel leaving the injector nozzle at low velocity, late in the combustion process), in the spray cores and tail. In the atmosphere, hydrocarbons undergo photochemical reactions with NO_x leading to formation of smog and ground level ozone. In engine emission standards, hydrocarbons are commonly regulated as either total hydrocarbons (THC) or as non-methane hydrocarbons (NMHC), with the latter category excluding methane (CH_4). Hydrocarbon emissions are greatly influenced by load, ambient conditions, turbocharging and fueling system. It seems that a strong correlation exists between smoke and HC emissions (at least during load or speed increase transients), implying that information for the former emissions can be useful to the latter. However, contrary to soot, NO_x (or CO) that peak only during load increase or acceleration, turbocharged diesel engine hydrocarbon emissions are noticed at the onset of deceleration or load decrease too, since in the latter case, as was discussed at the end of Section 2.1, turbocharger lag effects lead to instantaneously very low fuel–air equivalence ratios.

The development of hydrocarbon emissions will be discussed with reference to Figure 5.18 illustrating what happens during the first 400 s of the New European Driving Cycle, which includes cold starting too. A typical, medium-size, European sedan was used for the analysis equipped with a 2.2 L HSDI diesel engine, EGR, common rail injection system, diesel oxidation catalyst (DOC) and diesel particulate filter (DPF).[4] Measurement of the real-time transient emissions from the engine was accomplished using fast-response, two-channel analyzers enabling engine-out and tailpipe emissions to be logged simultaneously. The engine-out instruments were sampled through a common heated sample line and filter to protect the instruments from the high levels of HC and particulates expected, primarily during the regeneration event. This arrangement was found to be effective at preventing span drift caused by the excessive amounts of engine-out soot and unburned fuel during the regeneration process. The tailpipe instruments were sampled through a shorter, unfiltered and unheated single sample line.

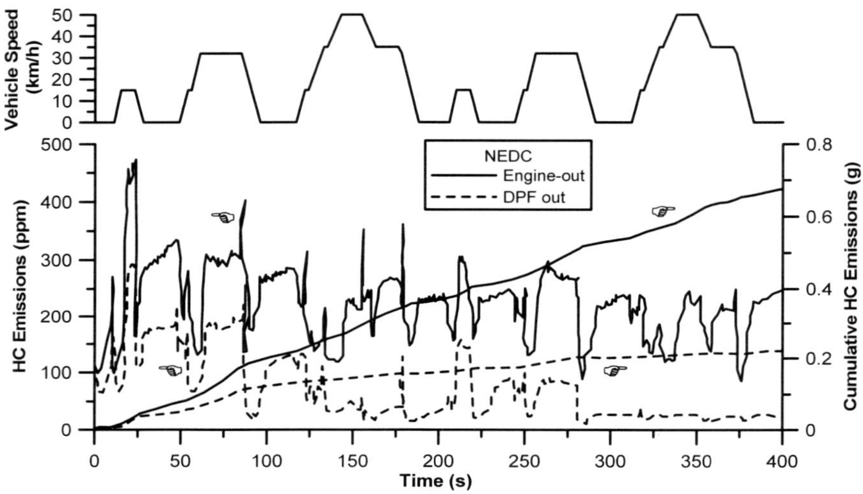

Figure 5.18. HC emissions during the first 400 s of the New European Driving Cycle before and after the diesel particulate filter (medium size, European sedan with four-cylinder, 2.2 L displacement volume, HSDI diesel engine – experimental results reprinted with permission from SAE Paper No. 2006-01-1079 [20], © 2006 SAE International)

Similar to NO_x or PM emissions, several spikes are observed during the various accelerations of the Cycle. As the engine load and fuel–air equivalence ratio increase, higher concentrations in the spray core are produced and more fuel is deposited on the walls. Particularly during the turbocharger lag phase, where Φ assumes higher than stoichiometric values, the shortage of oxygen favors

---

[4] The scheduled 'regeneration' of the DPF is of particular interest in a modern diesel engine, as it requires the engine to operate in quite different regimes compared with 'normal' operation, and with different control requirements. The regeneration event typically demands temperatures above 550°C; it is usually initiated under high speed driving conditions, where the gas temperatures are higher. Retardation of the main fuel injection and introduction of post-injection are two of the most popular means to achieve the high level of exhaust temperatures required for regeneration. The effect of the changes to the injection parameters is to increase DOC inlet temperatures directly and may cause a significant increase in HC, which reacts over the DOC to provide additional heat to the DPF.

production of the heavier unburned hydrocarbons. Tailpipe HC emissions are also provided in the same figure highlighting the DPF's ability for effectively coping with this kind of emissions; in fact, immediately from start, there is already an apparent 50% 'conversion'. This is caused by HC being adsorbed onto the after-treatment components, most probably the washcoat of the diesel oxidation catalyst. In this case, the light-off temperature of the DOC is sufficiently low (and the overall trapping capacity sufficiently high) for the trapped HC to be oxidized on the DOC before it is desorbed. From cold start, we can see that HC trapping 'efficiency' is approximately 50%. Once light-off occurs, the overall apparent conversion efficiency rises significantly.

During the early part of the Cycle where the engine is still cold, abrupt increases in feedgas HC are apparent during load transitions, these being particularly noticeable during transitions from cruise to deceleration. It is interesting that these only occur during the first 200 s, where the catalyst is not yet hot and therefore some of the HC emerges at the tailpipe (less that trapped on the DOC). Clearly, there would be opportunities for reducing this transient HC breakthrough, although it is not a dramatic proportion of the total HC emitted over this period. All in all, although short-duration spikes in HC emissions are noticed during load transients through the Cycle, it is the catalyst's light-off time that contributes more to the overall tailpipe emissions [20].

Summarizing the above findings, HC emissions during a Transient Cycle of a turbocharged diesel engine are enhanced

- during acceleration, load increases (and cold starting) owing to high local fuel–air equivalence ratios and low cylinder wall temperatures that ultimately lead to longer ignition delay periods (therefore limiting of maximum fueling or closing down of the EGR valve can prove beneficial, with the injection timing playing a secondary role); and
- at the beginning of deceleration, owing to the instantaneously very low mass of injected fuel that is excessively diluted into the still high air-charge, preventing the start or completion of combustion.

As regards naturally aspirated engine transients, THC emissions have been measured up to several times their final steady-state concentration right after the start of the acceleration, before decreasing to their final values after several engine cycles (Figure 5.17) that is, a completely different evolution pattern compared with their NO_x counterparts [18, 19, 21]. Again, THC emissions are mainly dependent on high fuel–air equivalence ratios being also unfavorably affected by the smaller wall temperatures during most of the transient event compared with the respective steady-state points; the differences show an increasing trend the higher the initial engine load as, in this case, more fuel is deposited on the cylinder wall enhancing the respective chemical reactions. For the same reason, the initial peak in naturally aspirated transient HC emissions is enhanced with fuels of higher viscosity or wider range of distillation temperature or lower cetane number (*i.e.*, longer ignition delay), as well as with pistons of smaller cavity, and with lower coolant temperature. The above behavior is mainly determined by the increase in the C_1 to C_8 HCs, as regards the low HC components, and only partly by the increase in ethylene C_2H_4, as regards the lower HC components, whereas unregulated toxic compounds such as benzene and 1,3 butadiene have a very small contribution [19].

5.4 Carbon Monoxide

Carbon monoxide is a colorless, odorless and tasteless gas that forms as an intermediate combustion product. Its emissions depend mainly on the fuel–air equivalence ratio, being usually of low significance during steady-state diesel engine operation. For fuel-rich mixtures, such as those experienced during the turbocharger lag phase, CO concentration in the exhaust increases steadily with increasing Φ. The latter is documented in Figure 5.19, where a comparison is made between quasi-steady and transient carbon monoxide emissions during part of the European ETC Transient Cycle (*cf.* Figure A.2) for a turbocharged diesel engine.

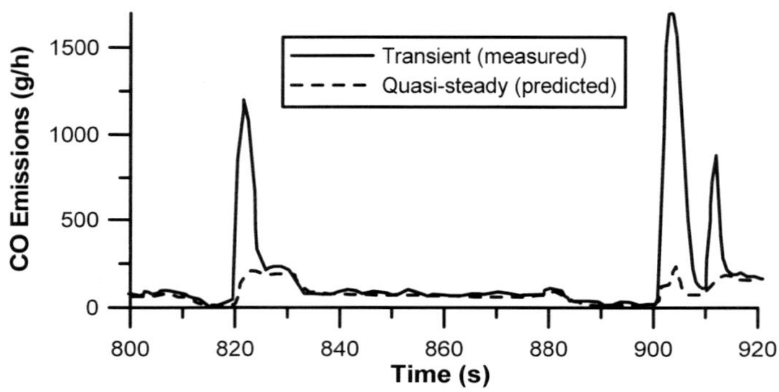

Figure 5.19. Comparison between quasi-steady and transient CO emissions during part of the ETC Cycle (four-stroke, 12 L displacement volume, turbocharged, heavy-duty diesel engine – experimental results reprinted with permission from SAE Paper No. 2005-01-3852 [22], © 2005 SAE International)

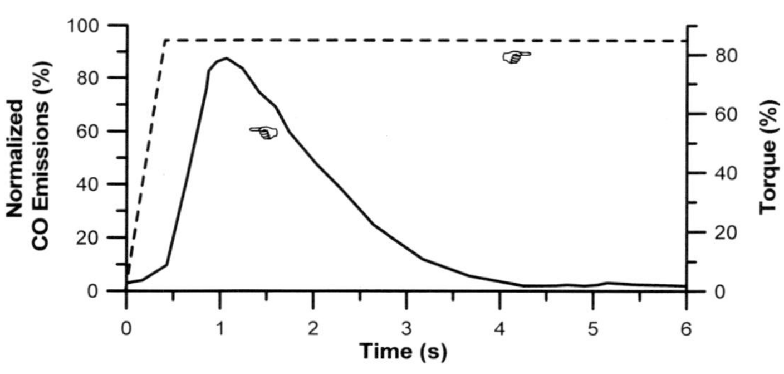

Figure 5.20. Development of normalized CO emissions after a 0–90% increase in engine load for a turbocharged diesel engine

Significant peaks are observed during acceleration (the same holds for load increases, as is demonstrated in Figure 5.20), where the local fuel–air equivalence ratio exceeds the stoichiometric value leading to incomplete combustion. Instantaneous, transient CO emissions were measured up to 8 times the respective quasi-steady values (at 905 s in the Cycle) highlighting once again the different evolution pattern and magnitude of exhaust emissions during transients compared with steady-state operation. On the other hand, when the engine is in cruise (*i.e.*, operates at quasi-steady conditions) or decelerates, fuel–air equivalence ratio assumes much lower values, hence CO emissions decrease substantially.

In accordance with the above remarks, Lindgren *et al.* [23] showed that during loading with a wheel loader (non-road mobile machinery), CO emissions increased more than 200% compared with steady-state conditions; the postulated reason being again the failure of the turbocharger to reach a fully spooled state at low engine speed in combination with high loading torque. During the same operation, fuel consumption, NO_x and HC emissions increased by 14, 16 and 60%, respectively.

On the other hand, as is the case with soot, CO emissions during naturally aspirated transient operation practically follow the steady-state pattern, with minimum overshoot observed.

5.5 Non-regulated Emissions and Odor

Non-regulated emissions from diesel engines correspond to organic and air-toxic pollutants such as

- aldehydes, *i.e.*, organic compounds containing a carbonyl group (carbon atom double bonded to oxygen) of the following general formula

  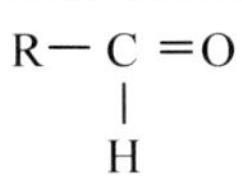

  where R is a radical; simplest aldehyde is formaldehyde HCHO;

- benzene, which is a colorless, flammable, carcinogenic aromatic hydrocarbon of the type C_6H_6 and of the formula

  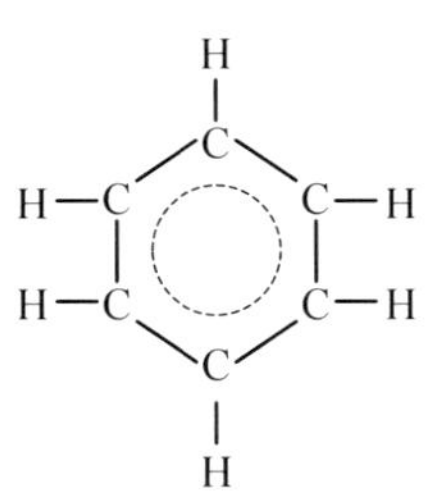

- alcohols, *i.e.*, organic compounds with a hydroxyl group (-OH) bound to a carbon atom of an alkyl or substituted alkyl group, having the general type $C_nH_{2n+1}OH$;
- naphthalene, which is a volatile, crystalline, aromatic, white, solid hydrocarbon of the type $C_{10}H_8$, comprising of two benzene rings sharing a pair of carbon atoms;

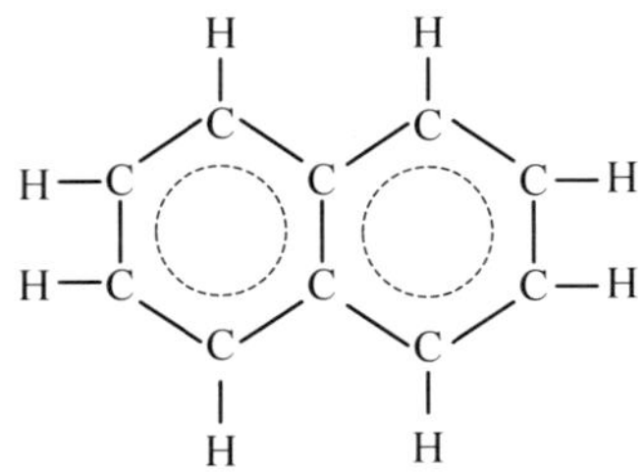

- ketones, *i.e.*, carbonyl group linked to two other carbon atoms or radicals

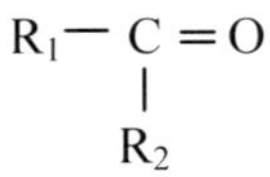

- polycyclic aromatic hydrocarbons, also known as PAHs; these are compounds similar to benzene but consisting of four, five or six rings joined together;
- nitrated polycyclic aromatic hydrocarbons, also known as NPAHs; these are similar to PAHs but with NO attached to one or more of the rings, *etc.*

Non-regulated emissions are, usually, more difficult to measure, owing to their very small concentration (of the order of a few ppm), whereas their detailed mechanism of production is not absolutely clear. Gullett *et al.* [24] reported on the real-time emission characterization of some of the above pollutants during cold starting, hot starting and load increase transients of a medium duty diesel engine. During discrete load increases, benzene and methylated naphthalenes were found to increase in concentration, mainly at high loads, owing to the higher fuel–air equivalence ratios experienced, whereas naphthalene showed little apparent change, contrary to benzene, whose development is illustrated in Figure 5.21.

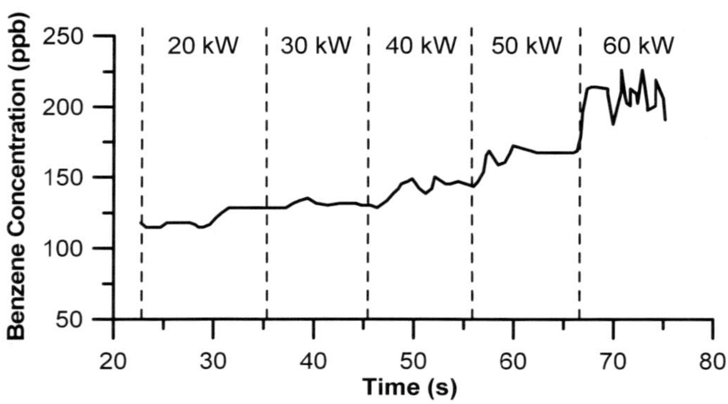

Figure 5.21. Benzene emissions after continuous load increases (four-stroke, six-cylinder, water-cooled, medium duty, turbocharged diesel engine – experimental results reprinted from Gullett *et al.* [24], copyright (2006), with permission from Elsevier)

Emission rates for selected toxic hydrocarbons and carbonyl compounds are presented in Table 5.1 for various diesel and SI engined MY 2000 vehicles [25].

Table 5.1. Emission rates of selected toxic hydrocarbons and carbonyl compounds (mg/km) during various US Transient Cycles, measured with the CVS methodology (missing values are below detection limits – experimental results reprinted from Graham [25], copyright (2005), with permission from Elsevier)

	FTP-75 Cold start phase	FTP-75 Transient phase	HWFET	NYCC	US06
Diesel vehicle 1 (four-cylinder, 1.7 L displacement volume, turbocharged diesel engine, rated at 66 kW at 4200 rpm, installed on a Mercedes-Benz A170)					
1,3-Butadiene	0.662	0.470	-	0.429	0.011
Benzene	0.873	0.844	0.038	1.67	0.094
Toluene	0.382	0.347	0.019	0.580	-
Formaldehyde	17.7	12.8	0.78	27.6	2.06
Acetaldehyde	9.85	5.96	0.41	12.6	1.23
Acetone	4.96	3.00	0.46	3.22	1.40
Diesel vehicle 2 (three-cylinder, 0.8 L displacement volume, turbocharged diesel engine, rated at 30 kW at 4200 rpm, installed on a SmartCar)					
1,3-Butadiene	0.879	-	-	0.470	0.066
Benzene	0.572	0.036	0.018	0.497	0.134
Toluene	0.267	0.007	0.035	0.131	0.096
Formaldehyde	15.7	0.75	0.33	12.5	4.30
Acetaldehyde	7.42	0.24	0.15	5.66	1.90
Acetone	3.67	0.38	0.11	3.75	0.94
Diesel vehicle 3 (three-cylinder, 1.2 L displacement volume, turbocharged diesel engine, rated at 45 kW at 4000 rpm, installed on a Volkswagen Lupo)					
1,3-Butadiene	0.344	-	-	0.228	-
Benzene	0.507	0.094	0.001	0.429	0.041
Toluene	-	0.123	-	0.013	0.111
Formaldehyde	5.10	0.97	0.01	7.93	0.64
Acetaldehyde	3.72	0.73	0.13	4.32	0.48
Acetone	1.84	1.26	-	3.99	0.44
Gasoline Vehicle (four-cylinder, 2.0 L displacement volume, SI engine, rated at 103 kW at 5500 rpm, installed on a Renault Megane)					
1,3-Butadiene	0.803	-	-	-	0.017
Benzene	5.20	0.045	0.003	0.071	0.260
Toluene	29.2	0.314	0.320	-	0.989
Formaldehyde	0.81	0.26	0.03	0.68	0.02
Acetaldehyde	0.72	0.10	0.01	0.65	-
Acetone	0.44	0.15	0.01	0.93	0.02
Gasoline HEV (four-cylinder, 1.5 L displacement volume, SI engine, rated at 52 kW at 4500 rpm, coupled to a 33 kW permanent magnet EM, installed on a Toyota Prius)					
1,3-Butadiene	-	-	-	-	-
Benzene	0.025	0.002	-	0.010	0.005
Toluene	0.205	0.019	0.020	0.026	0.058
Formaldehyde	0.12	0.07	0.07	0.39	0.08
Acetaldehyde	-	-	0.06	0.20	-
Acetone	-	-	-	-	-

The cold start produced the highest hydrocarbons (1,3 butadiene C_4H_6, benzene and toluene $C_6H_5CH_3$ in Table 5.1) emission rates of the FTP Test Cycle, though the increase compared with the hot-start (now shown) emissions was not that great. The high-speed driving of the HWFET and US06 Cycles produced lower emission rates due to the high temperature of the exhaust and the efficient operation of the oxidation catalyst. Only modest increases in the diesel vehicle emissions are seen in the US06 over the HWFET due to aggressive driving (*cf.* Table A.10). The emission rates for the NYCC are slightly higher than those observed over the FTP phases (other than cold start phase) for all vehicles. This is due in part to the short distance traveled during the test. For the gasoline HEV, the NYCC emissions are also very low, likely due to the hybrid design that allows the electrical drive system to move the vehicle at low speeds and loads while the engine, if still running, is operating in an efficient steady-state mode and is used to generate electricity to power the electric drive.

Past research has indicated that carbonyl compound (formaldehyde, acetaldehyde CH_3CHO and acetone CH_3COCH_3 in Table 5.1) emission rates for diesel-fueled vehicles are traditionally greater than for conventional gasoline vehicles; this trend is confirmed in the results given in Table 5.1. Carbonyl compound emissions from the diesel-engined vehicles are dominated by formaldehyde; acetaldehyde is the next most abundant carbonyl. Cold start emissions produce a larger number of detected carbonyl species. The US06 emission rates are greater than the HWFET emission rates due to the aggressive driving characteristics of the cycle. Carbonyl compounds appear in the exhaust due to incomplete combustion and at the low power demand and low exhaust temperatures of a Cycle. The overall higher emissions of carbonyl compounds of the diesel-engined vehicles were also responsible for a substantial higher (under some conditions up to one order of magnitude) ozone-forming potential compared with their gasoline counterparts [25]. Similar remarks were made by Kado *et al.* [26] concerning, mainly, diesel *vs.* CNG fueled bus engines during various American Transient Cycles.

An abrupt increase in engine load results also in increases of exhaust odor. Odor is not a property of substances, but the result of stimuli with subsequent signal processing in the brain. Perceiving an odor by human senses when a threshold concentration is exceeded is, therefore, a reflection of sensory impressions. The perception of odor constitutes quite an ecological issue today, as it is usually associated with toxicity although this may not always be the actual situation. It is hydrocarbon and aldehydes (in particular formaldehyde) in the exhaust gas that have been primarily found to correlate quite well with odor [27].

5.6 Biodiesel

Oxygenated fuels (alcohols such as bio-methanol or bio-ethanol) have been used as alternatives to gasoline fuel with success in reducing emissions from spark ignition petrol engines. Bio-ethanol and especially vegetable oils or their derived biodiesels (methyl or ethyl esters) are considered as promising fuels (oxygenated by nature) to supplant a fraction of petroleum distillates, having the added advantage of being derived from biological sources showing, thus, an *ad hoc* advantage in emitted CO_2 reduction [28]. Biodiesel is a renewable fuel that can be produced from a variety of vegetable oils such as rapeseed oil, soybean oil, sunflower oil and palm oil, or animal fats, in batch or continuous systems, mainly through the transesterification process, where one ester is converted into another. This reaction proceeds with catalyst (base or acid) or without catalyst by using primary or secondary monohydric aliphatic alcohols, where the glycerol–based triesters (or triacyl glycerides) that make up the fats and oils are converted into monoesters yielding free glycerol as a by product:

$$\text{Triglycerides} + \text{Monohydric alcohol} \rightarrow \text{Mono-alkyl esters} + \text{Glycerol}$$

It is evident from the above reaction that biodiesel fuels consist of a mixture of transesterified free fatty acids of high molecular weight. The more widely used biodiesels are rapeseed methyl ester (RME) in Europe and soybean methyl ester (SME) in the US, collectively known as fatty acid methyl esters (FAME).

The advantages of biodiesels as diesel fuel, apart from their renewability, are their minimal sulfur and aromatic content, higher flash point, higher lubricity, higher cetane number and higher bio-degradability and non-toxicity. On the other hand, their disadvantages include the higher viscosity and pour point, lower calorific value and lower volatility (Table 5.2).

Table 5.2. Properties of typical diesel fuel, various biodiesels and vegetable oils

Fuel Properties	**Density at 15°C (kg/m^3)**	**Kinematic viscosity at 40°C (mm^2/s)**	**Lower heating value (kJ/kg)**	**Cetane number (–)**
Diesel fuel	837	3	42,700	50
Cottonseed oil methyl ester	885	4	37,500	52
Soybean oil methyl ester	885	4.1	37,300	51
Sunflower oil methyl ester	880	4.4	37,500	50
Rapeseed oil methyl ester	885	4.7	37,300	53
Palm oil methyl ester	870	4.5	37,200	50
Cottonseed oil	910	34	36,800	38
Soybean oil	925	33	37,000	38
Sunflower oil	920	34	36,500	37
Corn oil	915	35	36,300	38
Olive kernel oil	925	32	37,000	39

Furthermore, their oxidation stability is lower, they are hygroscopic, and as solvents may cause corrosion of components, attacking some plastic materials used for seals, hoses, paints and coatings. They show increased dilution and polymerization of engine sump oil, thus requiring more frequent oil changes. For all the above reasons, it is generally accepted that blends of standard diesel fuel with up to 20% (by volume) vegetable oils or biodiesels can be used in existing diesel engines without any modifications, but there are concerns about the use of higher percentage blends that can limit the durability of various components, leading to engine malfunctioning. Thus, neat (*i.e.*, 100%) vegetable oils or biodiesels are not viable options at present, but their addition to diesel fuel at low concentrations can be considered as equivalent to other oxygenated fuel additives, of course with the added advantage of renewability and emitted CO_2 reduction.

Biodiesel fuels have been found capable of substantially decreasing particulate matters, hydrocarbon and CO (although a slightly adverse impact on NO_x emissions has been reported by some researchers[5]) during steady-state [30, 31] and transient conditions [32–36] of both light-duty and heavy-duty diesel engines.

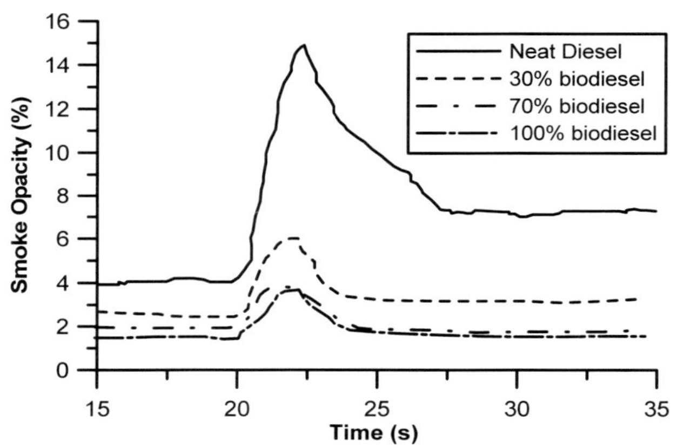

Figure 5.22. Smoke opacity development during a 26–90 Nm load increase transient event at 1661 rpm for different diesel–sunflower biodiesel blends (four-stroke, four-cylinder, turbocharged and aftercooled, DI diesel engine of 2.2 L displacement volume, rated at 85 kW at 4000 rpm – experimental results from Armas *et al.* [35], copyright (2006), with permission from Elsevier)

Figure 5.22 illustrates typical sunflower biodiesel effects on smoke opacity development during a load increase transient event at constant engine speed. Clearly, smoke opacity was found to decline the higher the biodiesel content in the fuel blend with remarkable decrease actually observed when comparing (the rather extreme cases of) 70% or 100% biodiesel blends to neat diesel fuel. The latter effect has been attributed primarily to the following factors:

- increased biodiesel oxygen concentration, which aids the soot oxidation process. Soot formation, caused by high temperature decomposition,

[5] There is also at least one reference [29] that connects the decrease in particulate matter emissions from biodiesel use with increase in the number of (the more toxic) nanoparticles.

mainly takes place in the fuel-rich zone at high temperatures and pressures, specifically within the core region of each fuel spray. If the fuel is partially oxygenated, it could reduce locally fuel-rich regions and limit soot formation, thus reducing PM emissions. The same remark holds true for the case of diesel blends with (at least) up to 10% v/v bioethanol [37]);

- combustion is mixing controlled, which means that there is more time available after diffusion combustion for soot oxidation;
- longer carbon chains;
- absence of aromatic and sulfur compounds that are generally considered to act as soot precursors (see also Figure 5.1).

Figure 5.23 expands on the latter arguments by showing cumulative PM emissions from a heavy-duty diesel engine during the US EPA Transient Cycle (*cf.* Figure A.9) with respect to the biodiesel percentage fuel blend. Reduction up to 66% (hot start) or 63% (cold start) was measured when running the engine with 100% biodiesel (derived from transesterification of soybean oil) compared with conventional diesel fuel, without sacrifice in engine efficiency. At the same time, HC and CO emissions were reduced up to 45% with the respective NO_x emissions increasing up to 11%.

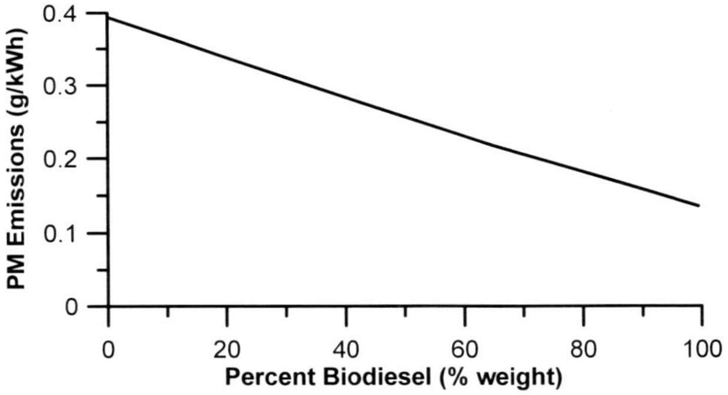

Figure 5.23. Effect of soybean biodiesel content on averaged cumulative particulate matter emissions during the EPA Transient Cycle of heavy-duty diesel engines (experimental data adapted from Graboski *et al.* [32])

Some further aspects of biodiesel combustion explaining the behavior demonstrated in Figures 5.22 and 5.23 will be discussed below. When using small blend ratios, *i.e.,* 10% and 20% by volume, which is the most usual case, the injection rate or the macroscopic behavior of the spray is almost identical for the neat diesel fuel and the biodiesel blended fuels used for the same engine operating conditions of injection timing, speed and load (brake mean effective pressure). The increased values of compressibility of the (liquid) biodiesel compared with the diesel fuel, which causes an earlier injection of fuel into the engine cylinder, is not expected to play an important role, as this injection advance difference is at most of 1°CA (crank angle) even for the neat biodiesel case.

The same remark is anticipated to hold true for the influence of the cetane number, which is a little higher for the biodiesel than for the diesel fuel. This influences the premixed combustion, by decreasing its duration (decreased ignition delay) the higher the cetane number. The only marked difference occurs in the atomization process, given that the mean droplet size is larger when biodiesel is used against the diesel fuel case. This is mainly due to the biodiesel having a higher kinematic viscosity than that of the neat diesel fuel. This fact and the different distillation curves, *i.e.*, higher distillation curve for biodiesel than that of the neat diesel fuel, indicate that the evaporation process will be slower for the biodiesel and thus could affect the combustion process.

As discussed above, the macroscopic behavior of the spray is almost identical for the neat diesel fuel and any other blended fuel used, so that the amount of instantaneous entrained and mixed air is also the same. Taking into account that the stoichiometric air–fuel ratio is of the order of 12.5 with the neat biodiesel and 15 with the neat diesel fuel, this means that the air–biodiesel oil mixtures can reach the stoichiometric conditions nearly 15% faster than the air–diesel fuel mixture. Therefore, this difference in the air–fuel ratio, caused by the presence of bound oxygen in the biodiesel, must influence combustion. More specifically, if the combustion process is mixing controlled, the use of biodiesel injection has a beneficial effect, while if the combustion process is evaporation controlled, the effect is adverse. For turbocharged engines, where high temperatures and pressures exist during mixture preparation in the cylinder, the combustion process both for the diesel fuel and the biodiesel blends would tend to be rather mixing controlled.

Prevailing theories on soot formation state that the inception of soot particles starts with polymerization of species containing a double carbon bond. Likewise, the absence of sulfur from the biodiesel should lead to lower particulate matter (PM) emissions. The somewhat higher cetane number of the biodiesel, compared with the neat diesel fuel case, may decrease NO_x emissions due to the relatively lower ignition delay and thus shorter premixed combustion during which NO_x is mainly formed; to this end also contributes the absence of aromatic compounds from the biodiesel. However, all these may be obscured by the delicate distribution of the fuel–air 'packets' inside the sprays, given that NO_x production is favored by high local temperatures and near to stoichiometric (and slightly to the 'lean') local conditions.

As was the case with the particulate matters in Figure 5.23 and CO emissions, the use of biodiesel blends of at least 20% by volume, can help reducing substantially non-regulated emissions too such as PAHs, NPAHs (recall that biodiesel does not contain any aromatics or PAH compounds), aldehydes (up to 50%), sulfates and ketones. This was the result reached by Sharp *et al.* [36] for three turbocharged (bus and truck) diesel engines equipped with diesel oxidation catalysts during the 20 minute US heavy-duty Transient Cycle.

Synthetic oxygenates such as glycol ethers or maleates can also prove quite efficient in reducing PM emissions during transients with limited (if any) NO_x penalty, when added to the diesel fuel in small percentages (*e.g.*, 5% v/v). Kozak *et al.* [38] studied the performance of a typical EURO IV, European diesel-engined passenger vehicle during the New European Driving Cycle and the American FTP-75 Cycle and found that the maleated fuels were responsible for up to 25% PM emission reduction during the NEDC but, surprisingly, did not affect the respective emissions for the (much harder in terms of acceleration) FTP-75

Transient Cycle. Although NO_x and CO_2 emissions as well as fuel consumption were more or less kept at the same level as for the neat diesel fuel operation, a considerable increase in CO and HC emissions was reported.

5.7 Combustion Noise

The final class of emissions discussed in this chapter is engine noise. The three primary sources of noise generation in a diesel engine are (see also Figure 5.24)

1. Gas-flow;
2. Mechanical processes; and
3. Combustion.

From an acoustic point of view, the diesel engine is a very complex system comprising various dynamic forces acting on an equally complex structure of varying stiffness, damping and response characteristics. Gas-flow noise, usually low frequency controlled, is associated with the intake and exhaust processes, including turbocharging, and the cooling fan. Mechanical noise originates from inertia forces causing piston slap, from gears, tappets, valve trains, timing drives, fuel injection equipment and bearings.

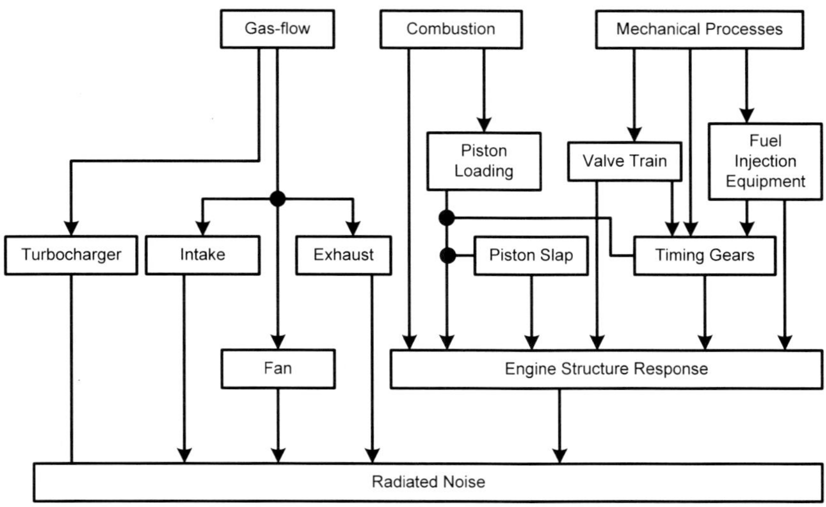

Figure 5.24. Engine noise generation (from Lilly [39], copyright Elsevier)

The high rate of cylinder pressure rise, mainly after the ignition delay period causes vibration of the engine block, resulting ultimately in combustion noise radiation;[6] the latter manifests itself as the characteristic diesel combustion knock.

[6] Both combustion and inertia forces cause vibration of the engine structure resulting in noise emissions. Since gas forces in a diesel engine are higher than their inertia counterparts (*cf.* Figure 3.2), combustion noise prevails over other, mechanically originated, noise radiation.

Injection parameters, *e.g.*, timing and amount of fuel injected during pre- and main injections define the rate of heat release playing, thus, a principal role. In order to analyze combustion noise, the cylinder pressure signal is examined on the frequency spectrum [39, 40]; Russell and Haworth [41] and Pischinger *et al.* [42] discuss various methods for measuring and analyzing combustion noise.

Overall engine noise measured 1 m away from the engine surface typically ranges from 80–110 dBA depending, mainly, on engine size, speed and injection system. Particularly for diesel-engined vehicles, the unpleasant combustion knock is also a matter of passengers and pedestrians discomfort; it is not surprising then that the European Union (directives 70/157/EEC and 96/20/EC [43]), Japan and the US have imposed stringent regulations concerning noise emissions from vehicles, railroads and airplanes.

Anderton and Baker [44] discuss the effect of various parameters on noise generated from diesel engines during steady-state conditions such as two *vs.* four-stroke operation, naturally aspirated *vs.* turbocharged, cylinder configuration *etc.* The fundamental conclusion reached was that turbocharging the diesel engine produces lower high-frequency (this ultimately results in a slightly lower level of combustion noise) and higher low-frequency excitations; the same results are observed when the engine operates a two-stroke rather than a four-stroke cycle.

Unsurprisingly, (combustion) noise development during a speed or load increase transient differs substantially from the respective steady-state operation [41, 45–47]. Head and Wake [46] were among the first to study the influence of accelerating rate, intake temperature, cylinder wall temperature and fuel injection rate on the noise emitted during accelerating operation of a diesel engine. The main finding was that combustion noise is generally higher during transients, typically of the order of 4–7 dBA, compared with the respective steady-state operating points. As will be analyzed later in the section, this was mainly attributed to the lower cylinder wall temperature during the first cycles of the transient event. An increase of piston slap generated noise during transients was also reported.

Figure 5.25 taken from the work by Dhaenens *et al.* [48] illustrates experimental results for a 2.5 s, 1000–2200 rpm acceleration of a turbocharged, truck diesel engine. The experiments were performed in a semi-anechoic test cell according to the vehicle accelerated drive-by test (ISO R362); intake and exhaust orifice noise were eliminated and the gearbox was acoustically encapsulated in order to focus on pure engine noise.

The analysis revealed that transient *overall* engine noise exceeded steady-state levels by 5 dBA maximum (measured at 1 m distance from the engine surface), while it was also characterized by a broadband level noise increase combined with amplified resonance effects. The structure vibration revealed higher acceleration levels especially at the engine liner, oil pan flange and oil pan. Intake and exhaust side noise was found 5 dBA higher during transients, whereas combustion noise (typically between 1 and 2 kHz octave frequency range, which are primarily sensed by human ear) increased even more, by 7 dbA, compared with steady-state operation. The latter is demonstrated in Figure 5.26, which shows noise excitation at the 1200 rpm operating condition of the transient test, based on third octave spectra. A similar analysis for the 1700 rpm intermediate operating point, however, revealed much lower differences, probably due to the already higher cylinder wall temperatures. Similar results were reached by Rust and Thien [47] for naturally aspirated diesel engines, who also extended the analysis to load acceptance transients.

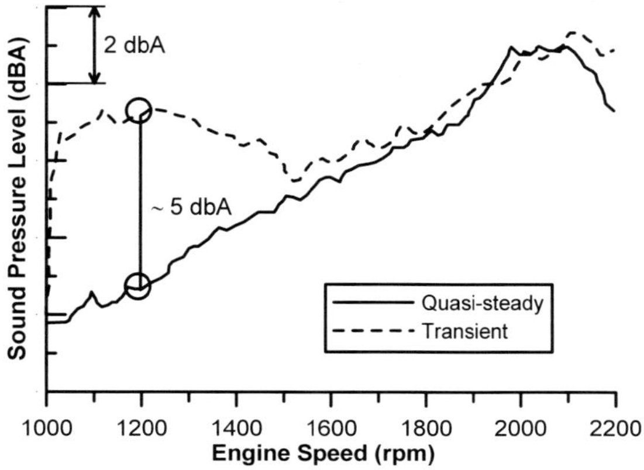

Figure 5.25. Transient *vs.* quasi-steady overall engine noise emissions during acceleration (EURO III, six-cylinder, DI, turbocharged diesel engine of 12.6 L displacement volume, rated at 355 kW at 2000 rpm – experimental results reprinted with permission from SAE Paper No. 2001-01-1566 [48], © 2001 SAE International)

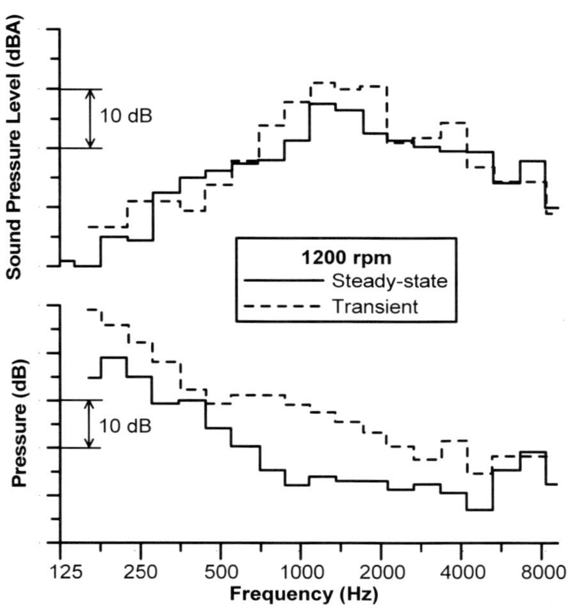

Figure 5.26. Quasi-steady *vs.* transient combustion noise excitation at 1200 rpm conditions of a turbocharged diesel engine acceleration event (experimental results reprinted with permission from SAE Paper No. 2001-01-1566 [48], © 2001 SAE International)

The main mechanism behind the increase in combustion noise radiation during transients lies in the operating principles of a transient event. In the first cycles, after a load or speed increase, the injected fuel quantity has already increased substantially cooling down the charge-air temperature; however, the cylinder wall temperature is still low (up to 100°C lower than the corresponding steady-state conditions) as the thermal transient proceeds at a much slower rate (*cf.* Figures 2.22 and 2.23) due to the cylinder wall thermal inertia; this is highlighted in Figure 5.27. As was discussed in Section 2.3.2, the combination of increased fueling with the still low cylinder wall temperatures and 'cooled' charge-air temperatures results in longer ignition delay, hence more intense premixed combustion periods leading to steeper cylinder pressure gradients $dp/d\varphi$ and, consequently, higher combustion noise levels (see also the conceptual diagram of Figure 5.28). The latter is enhanced by the lower charge-air temperatures, advance of dynamic injection timing and lower turbocharger compressor boost pressure due to turbocharger lag.

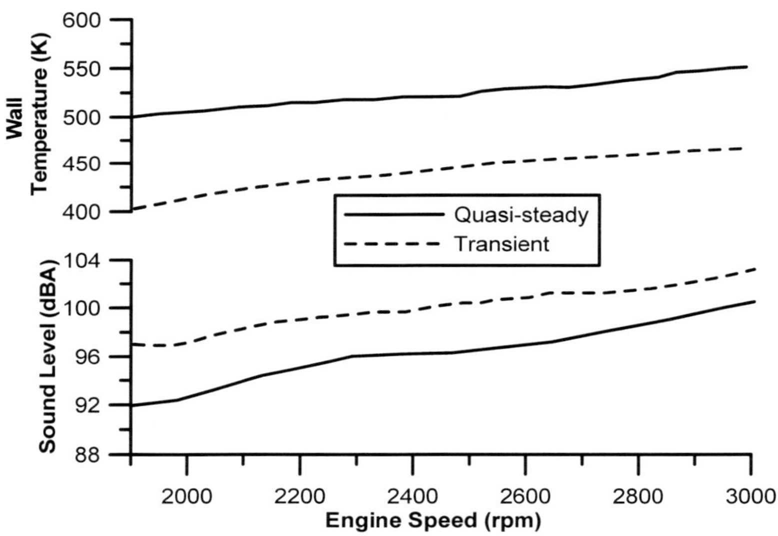

Figure 5.27. Sound level and cylinder wall temperature development during steady-state and transient accelerating conditions (two-cylinder, naturally aspirated diesel engine of 1.7 L displacement volume, rated at 27 kW at 3000 rpm – experimental results adapted from Rust and Thien [47])

The above phenomena are even more prominent during cold starting, where the much lower cylinder wall temperatures lead to even longer ignition delay periods, hence, harsher combustion and higher noise emissions. Modern, electronically controlled, common rail injection systems with pre-injections can help by facilitating chemical and physical preparation of the air–fuel mixture, thus, reducing premixed combustion and limiting overall noise radiation during cold start; however, noise increase during transients compared with steady-state operation is still apparent [49].

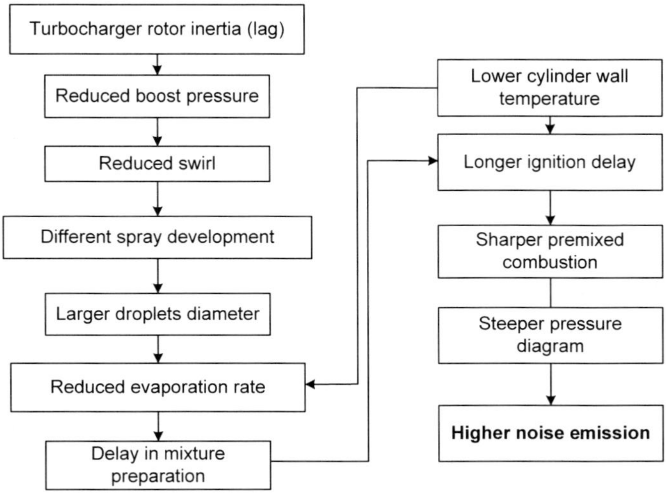

Figure 5.28. Mechanism of combustion noise radiation increase during transients

Since the primary source of increased transient noise emissions is the longer ignition delay period during the turbocharger lag cycles, improvement measures should aim in reducing ignition delay, for example, by [48]

1. optimization or reduction of start of injection shift during transients;
2. optimization of injection rate, *e.g.*, by pilot injection, in order to control the amount of premixed fuel as it is depicted in Figure 5.29;[7]

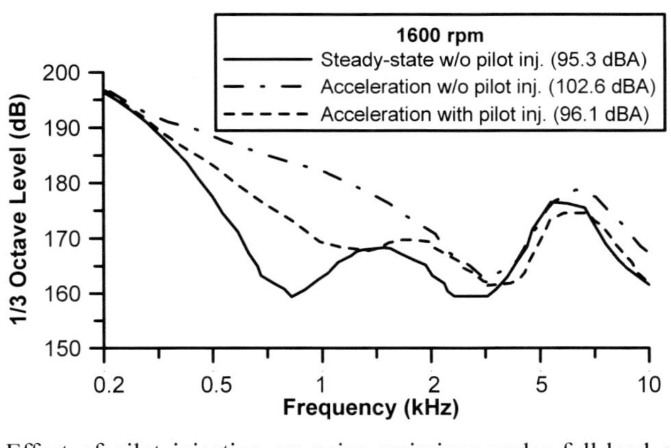

Figure 5.29. Effect of pilot injection on noise emissions under full-load acceleration at 1600 rpm (six-cylinder, naturally aspirated diesel engine of 11.4 L displacement volume, rated at 177 kW at 2200 rpm – experimental results reprinted with permission from SAE Paper No. 870989 [47], © 1987 SAE International)

---

[7] It should be noted that pilot injection is considered a successful measure to reduce the overall combustion noise emission level rather than the increase during transients.

3. faster increase of turbocharger compressor boost pressure as will be analyzed in Chapter 6;
4. combustion chamber temperature rise, *e.g.*, through load regulated EGR (depicted in Figure 5.30 – *cf.* conflict between EGR opening and soot emissions during transients (Figure 5.13)) or LHR operation;
5. optimization of structure attenuation behavior with respect to combustion excitation; and
6. effective encapsulation of those engine components that favor transient noise radiation.

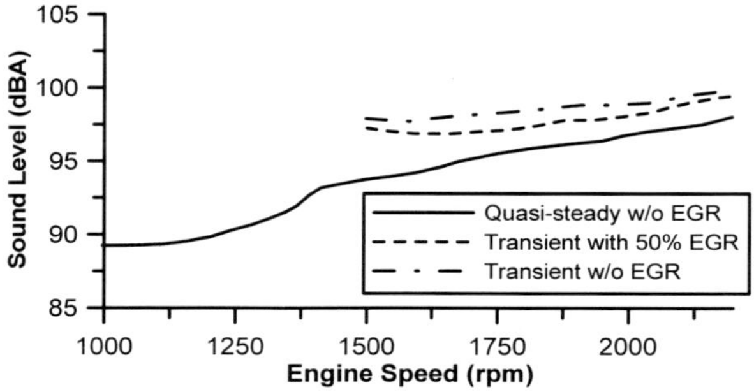

Figure 5.30. Effect of EGR on noise emissions under full-load acceleration (six-cylinder, naturally aspirated diesel engine of 11.4 L displacement volume, rated at 177 kW at 2200 rpm – experimental results reprinted with permission from SAE Paper No. 870989 [47], © 1987 SAE International)

References

[1] Kittelson DB. Engines and nanoparticles: a review. J Aerosol Sci 1998;29:575–88.
[2] Tree DR, Svensson KI. Soot processes in compression ignition engines. Prog Energy Combust Sci 2007;33:272–309.
[3] Hofeldt DL, Chen G. Transient particulate emissions from diesel buses during the central business district cycle. SAE Paper No. 960251, 1996.
[4] Hagena JR, Filipi ZS, Assanis DN. Transient diesel emissions: analysis of engine operation during a tip-in. SAE Paper No. 2006-01-1151, 2006.
[5] Harndorf H, Kuhnt H-W. Improvement of transient behavior of turbocharged diesel engines through additional air injection in the turbocharger. MTZ 1995;56:20–8 (in German).
[6] Winterbone DE, Benson RS, Mortimer AG, Kenyon P, Stotter A. Transient response of turbocharged diesel engines. SAE Paper No. 770122, 1977.
[7] Watson N. Transient performance simulation and analysis of turbocharged diesel engines. SAE Paper No. 810338, 1981.
[8] Wang J, Storey J, Domingo N, Huff S, Thomas J, West B. Studies of diesel engine particle emissions during transient operations using an engine exhaust particle sizer. Aerosol Sci Technol 2006;40:1002–15.
[9] Takeuchi K, Yabumoto J, Okada Y, Kawai T, Montajir RM, Goto Y. A new dual-type DMA for measuring nanoparticles emitted from combustion engines. J Nanoparticle Res 2005;7:287–93.

[10] Holmén BA, Qu Y. Uncertainty in particle number modal analysis during transient operation of compressed natural gas, diesel and trap-equipped diesel transit buses. Environ Sci Technol 2004;38:2413–23.
[11] Liu ZG, Ford DC, Vasys VN, Chen D-R, Johnson TR. Influence of engine operating conditions on diesel particulate matter emissions in relation to transient and steady-state conditions. Environ Sci Technol 2007;41:4593–9.
[12] Yokomura H, Kouketsu S, Kotooka S, Akao Y. Transient EGR control for a turbocharged heavy duty diesel engine. SAE Paper No. 2004-01-0120, 2004.
[13] Black J, Eastwood PG, Tufail K, Winstanley T, Hardalupas Y, Taylor AMKP. Diesel engine transient control and emissions response during a European extra-urban drive cycle (EUDC). SAE Paper No. 2007-01-1938, 2007.
[14] Narusawa K, Odaka M, Koike N, Tsukamoto Y, Yoshida K. An EGR control method for heavy-duty diesel engines under transient operations. SAE Paper No. 900444, 1990.
[15] Duffy KP, Miller RL, Bowyer RO, Bromnick PA, Mason JD, Stobart RK *et al.* Model-based EGR control development for an HSDI engine. ASME ICE Fall Conference, Ann Arbor MI, 16–20 Oct 1999, Proc. ASME, pp. 1–12.
[16] Wijetunge RS, Hawley JG, Vaughan ND. Application of alternative EGR and VGT strategies to a diesel engine. SAE Paper No. 2004-01-0899, 2004.
[17] Van Nieuwstadt M, Kolmanovsky I, Moraal P, Jankovic M, Stefanopoulou A. EGR-VGT control schemes: experimental comparison for a high speed diesel engine. IEEE Control Systems Magazine 2000;20:63–79.
[18] Meggyes A. Contribution to the emission characteristics of diesel engines at transient operation. MTZ 1975;36:111–5 (in German).
[19] Miyamoto N, Ogawa H, Shibuya M, Fuwa N. Time series analysis of diesel exhaust gas emissions under transient operation. SAE Paper No. 930976, 1993.
[20] Campbell B, Peckham M, Symonds J, Parkinson J, Finch A. Transient gaseous and particulate emissions measurements on a diesel passenger car including a DPF regeneration event. SAE Paper No. 2006-01-1079, 2006.
[21] Raihan KA, Takimoto F, Ogawa H, Miyamoto N. Time-resolved behavior of unburned hydrocarbon components in diesel exhaust under transient operations. SAE Paper No. 2001-01-1259, 2001.
[22] Ericson C, Westerberg B, Egnell R. Transient emission predictions with quasi stationary models. SAE Paper No. 2005-01-3852, 2005.
[23] Lindgren M, Pettersson O, Hansson P-A, Norén O. Engine load pattern and engine exhaust gas emissions from off-road vehicles and methods to reduce fuel consumption and engine exhaust gas emissions. Report Agriculture and Industry 308, Swedish Institute of Agricultural and Environmental Engineering, Uppsala, 2003.
[24] Gullett BK, Touati A, Oudejans L, Ryan SP. Real-time emission characterization of organic air toxic pollutants during steady state and transient operation of a medium duty diesel engine. Atmos Environ 2006;40:4037–47.
[25] Graham L. Chemical characterization of emissions from advanced technology light-duty vehicles. Atmos Environ 2005;39:2385–98.
[26] Kado NY, Okamoto RA, Kuzmicky PA, Kobayashi R, Ayala A, Gebel ME *et al.* Emissions of toxic pollutants from compressed natural gas and low sulfur diesel-fueled heavy-duty transit buses tested over multiple driving cycles. Environ Sci Technol 2005;39:7638–49.
[27] Hamm E, Hohenberg G, Standt U-D, Zelenka P. The effect of fuel specifications and different aftertreatment systems on exhaust gas odour and non-regulated emissions at steady state and dynamic operation of DI-diesel engines. SAE Paper No. 1999-01-3559, 1999.
[28] Directive 2003/30/EC of the European Parliament and of the Council of 8/5/2003 on the promotion of the use of biofuels or other renewable fuels for transport.

[29] Tsolakis A. Effects on particle size distribution from the diesel engine operating on RME-biodiesel with EGR. Energy Fuels 2006;20:1418–24.
[30] Rakopoulos CD, Antonopoulos KA, Rakopoulos DC, Hountalas DT, Giakoumis EG. Comparative performance and emissions study of a direct injection diesel engine using blends of diesel fuel with vegetable oils or bio-diesels of various origins. Energy Convers Manage 2006;47:3272–87.
[31] Rakopoulos CD, Rakopoulos DC, Hountalas DT, Giakoumis EG, Andritsakis EC. Performance and emissions of bus engine using blends of diesel fuel with bio-diesel of sunflower or cottonseed oils derived from Greek feedstock. Fuel 2008;87:147–57.
[32] Graboski MS, Ross JD, McCormick RL. Transient emissions from No. 2 diesel and biodiesel blends in a DDC Series 60 engine. SAE Paper No. 961166, 1996.
[33] Starr ME. Influence on transient emissions at various injection timings, using cetane improvers, bio-diesel, and low aromatic fuels. SAE Paper No. 972904, 1997.
[34] Wang WG, Lyons DW, Clark NN, Gautam M, Norton PM. Emissions from nine heavy trucks fueled by diesel and biodiesel blend without engine modification. Environ Sci Technol 2000;34:933–9.
[35] Armas O, Hernández JJ, Cárdenas MD. Reduction of diesel smoke opacity from vegetable oil methyl esters during transient operation. Fuel 2006;85:2427–38.
[36] Sharp CA, Howell SA, Jobe J. The effect of biodiesel fuels on transient emissions from modern diesel engines, part II unregulated emissions and chemical characterization. SAE Paper No. 2000-01-1968, 2000.
[37] Armas O, Cárdenas MD, Mata C. Smoke opacity and NO_x emissions from a bioethanol-diesel blend during engine transient operation. SAE Paper No. 2007-24-0131, 2007.
[38] Kozak M, Merkisz J, Bielaczyc P, Szczotka A. The influence of synthetic oxygenates on EURO IV diesel passenger car exhaust emissions – Part 2. SAE Paper No. 2008-01-1813, 2008.
[39] Lilly LRC. Diesel engine reference book. London: Butterworths, 1984.
[40] Austen AEW, Priede T. Origins of diesel engine noise. Institution of Mechanical Engineers, Symposium on 'Engine Noise and Noise Suppression', pp. 19–32, 1958.
[41] Russell MF, Haworth R. Combustion noise from high speed direct injection diesel engines. SAE Paper No. 850973, 1985.
[42] Pischinger FF, Schmillen KP, Leipold FW. A new measuring method for the direct determination of diesel engine combustion noise. SAE Paper No. 790267, 1979.
[43] Commission Directive 96/20/EC of 27 March 1996 adapting to technical progress Council Directive 70/157/EEC relating to the permissible sound level and the exhaust system of motor vehicles.
[44] Anderton D, Baker J. Influence of operating cycle on noise of diesel engines. SAE Paper No. 730241, 1973.
[45] Watanabe Y, Fujisaki H, Tsuda T. DI Diesel engine becomes noisier at acceleration – the transient noise characteristic of diesel engine. SAE Paper No. 790269, 1979.
[46] Head HE and Wake JD. Noise of diesel engines under transient conditions. SAE Paper No. 800404, 1980.
[47] Rust A, Thien GE. Effect of transient conditions on combustion noise of NA-DI diesel engines. SAE Paper No. 870989, 1987.
[48] Dhaenens M, van der Linden G, Nehl J, Thiele R. Analysis of transient noise behavior of a truck diesel engine. SAE Paper No. 2001-01-1566, 2001.
[49] Alt N, Sonntag H-D, Heuer S, Thiele R. Diesel engine cold start noise improvement. SAE Paper No. 2005-01-2490, 2005.

6

Methods of Improving Transient Response

6.1 Introduction

Turbocharger lag is the most notable feature of diesel engine transient operation that drastically differentiates the torque pattern from the respective steady-steady conditions. Turbocharger lag is caused because, although the fuel pump responds rapidly to the increased fueling demand after a load or speed increase, the engine air-supply cannot match this higher fuel flow instantly but only after a number of engine cycles owing to the inertia of the whole system; the above phenomenon is enhanced by the unfavorable turbocharger compressor characteristics at low-loads and speeds. As a result of this slow reaction, the relative air–fuel ratio during the early cycles of a transient event assumes very low values (even lower than stoichiometric), deteriorating combustion and leading to slow engine (torque and speed) response and overshoot in particulate, gaseous and noise emissions. There are three approaches to cope with this problematic behavior, namely

1. control of the fuel flow;
2. speed-up of the system (engine, manifolds and turbocharger) response; and
3. increase of the number of transmission gears for vehicular applications.

In Sections 2.2 and 5.1, boost sensitive fuel limiters were identified as one successful and popular means of fueling control and consequently smoke emissions reduction, though at the expense of slower engine response and poorer (vehicle) driveability. This is, however, a 'passive'[1] method (and so is the increase of the number of transmission gears, which actually worsens driveability owing to the much more frequent gear-changes required), since it tries to cure one symptom, *i.e.*, black smoke coming out of the exhaust pipe, rather than the cause, which is the

---

[1] The use of terms 'active' and 'passive', here, is different from the terminology proposed in [1]. Winterbone used the term active to characterize those methods of improving response that require an external energy source, while the passive methods were those that relied on the engine configuration or control system alone.

turbocharger delay. In this chapter, we will focus on the 'active' methods for improving the transient response of the engine, manifolds and turbocharger. In general, turbocharger lag and the other associated system delays, hence the torque and speed response of a diesel engine, is a complex phenomenon that is influenced by a variety of dynamic, thermodynamic and design parameters, *e.g.*, engine dynamics (moment of inertia, manifolds volume), governor characteristics, turbocharger dynamics, match, configuration and turbine nozzle area, engine fueling and valve timing; the most important of them will be analyzed in terms of load acceptance or acceleration transients and the major conclusions derived will be summarized in Table 6.1 at the end of the section.

For a successful improvement in turbocharged diesel engine transient response, we need to look back into the turbocharger compressor characteristics (Figure 2.5). Recall that these are characterized by low delivery pressure and mass flow at low rotational speeds, as well as by relatively narrow flow range. The points that need to be addressed for better engine response are located exactly in this behavior (see also Figure 6.1 for a general layout). Improvement can be achieved, when focusing on the engine exhaust side, by somehow accelerating the turbine. Consequently, an increase in the power delivered to the compressor is realized, hence boosting its operating point. Alternatively, we can focus on the inlet side by directly increasing the air-supply delivered to the cylinders. At the same time, we should try to increase the delivery pressure overall (this will facilitate higher exhaust back-pressure, faster turbocharger response, hence shorter fuel limitation and better driveability, as well as higher EGR rates) and widen the compressor mass flow range. For the latter, primarily desirable for automotive engines, the compressor surge limit poses a significant restriction.

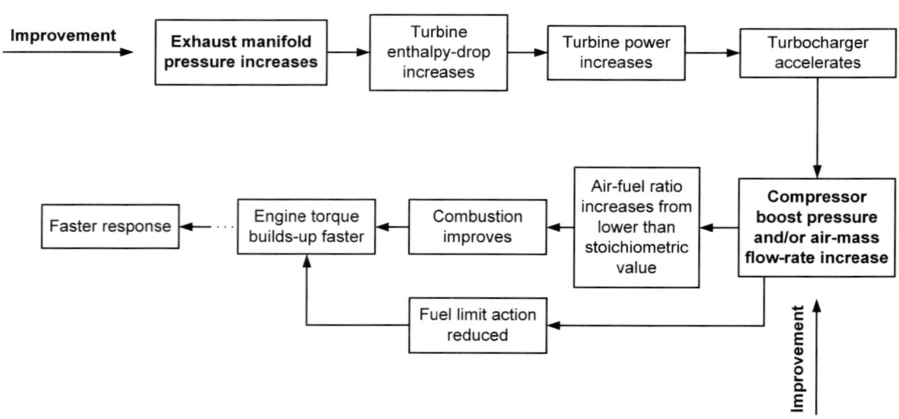

Figure 6.1. Main mechanism of transient response improvement

The various techniques that have been studied or developed so far for improving transient response can be classified as follows:

1. According to the system whose response is improved, such as

- methods that focus on the turbocharger, *e.g.,* variable geometry turbine (VGT), combined supercharging, two-stage turbocharging, low inertia turbocharger *etc.;* and
- methods that deal with the engine, such as fueling or valve timing control, low-heat rejection cylinder, manifolds volume and configuration, *etc.*

2. According to the specific approach through which the improvement is established; in this case, reference is made to (see also Figure 6.2)

- methods that deal with the *dynamics* of the system, namely the engine's or the turbocharger's mass moment of inertia, such as smaller turbocharger size or (to a certain degree) two-stage or sequential turbocharging, since it is well acknowledged that the turbine's acceleration poses the most significant obstacle to the engine transient response. In the same context, manifolds volume and configuration influence the system's (flow) inertia;
- methods through which more *favorable air-supply characteristics* are established. These may focus on
  - the compressor side, *e.g.,* the combined supercharging or the two-stage turbocharging configuration,
  - the engine side, *e.g.,* retarded injection timing or early opening of the exhaust valve,
  - the turbine side, with emphasis on the turbine nozzle area, *e.g.,* variable geometry turbine or sequential turbocharging configuration; two-stage turbocharging is applicable here too,
  - a completely different supercharging scheme, *e.g.,* mechanical supercharging or pressure wave supercharging;
- methods that utilize an external energy source such as electrical or hydraulic assistance, *e.g.,* electrically assisted turbocharging or hybrid-electric powertrain or air-injection onto the compressor impeller.

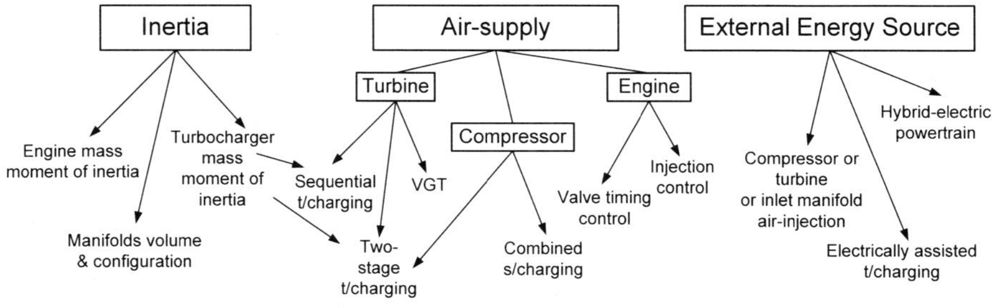

Figure 6.2. Classification of various methods of improving transient response

It should be pointed out that the need for fast speed response and low exhaust emissions during transients may pose conflicting requirements on the engine calibration (typically optimized for low fuel consumption and exhaust emissions at steady-state conditions), and this constitutes a significant challenge for the engine designer endeavoring to achieve best overall performance; at the moment, most engine control systems actually make interpolations between the steady-state maps in order to cope with the peculiarities during transients. The majority of the measures that have proven successful so far require re-matching of the engine with the turbocharger and an additional control system. This ranges from a simple mechanical actuator (waste-gate valve configuration in smaller-frame, low-inertia turbocharger units) to sophisticated and expensive electronic control units. The latter monitor via sensors various engine and turbocharger operating parameters, process them on a high-speed processor unit and decide on the optimum operation. The control system communicates with the engine management unit and may even be integrated in it.

Most of the cases/results that will be discussed in the following sections stem from experimentally validated simulation codes and have also been confirmed in practice; however, since the majority of the transient models follow the single-zone, filling and emptying approach, the availability of exhaust emission results is rather limited. Nonetheless, it is relatively safe to assume that minimization of turbocharger lag not only leads to faster engine response, but also produces lower amounts of exhaust and particulate emissions.

Before proceeding to the various methods for improving turbocharged diesel engine transient response, a brief description of the impact of some important dynamic parameters on transient operation will be addressed in the following section, with special reference to the magnitude, type and duration of the applied load change. The insight gained by the analysis of Section 2.1 has provided a starting point for analyzing and understanding these parameters effect. For all cases examined in the next section, results from a moderately turbocharged and aftercooled, six-cylinder diesel engine are used for the analysis.

6.2 Type and Features of Applied Load, and the Effect of Various Dynamic and Thermodynamic Parameters

The *magnitude* of the applied load plays a primary role in engine response, mainly as regards maximum and final speed droop (Figure 6.3). At the initial steady-state conditions, the engine and load torques are equal and the air–fuel ratio high due to the low-loading. As soon as the new, higher load is employed, a significant deficit is observed in the net (engine minus resistance) torque, since the engine torque cannot instantly match its increased load counterpart, so that engine speed drops (*cf.* block diagram of Figure 1.7).

Obviously, the higher the applied load, the higher this torque deficit during the early cycles of the transient event. This, in turn leads to a 'harder' turbocharger lag period, lower air–fuel ratio, higher crankshaft deceleration and lower engine speed, thus initiating larger governor displacement, greater fuel pump rack position, and higher values of engine torque, cylinder peak pressure and turbocharger speed. Turbocharger lag effects are more pronounced the higher the applied load, a fact leading also to higher soot emissions. As is revealed by Figure 6.3, the cycle where the engine speed assumes its minimum value seems to be advanced the higher the final load, whereas the recovery period is less affected by the intensity of the applied load change (although this strongly depends on the specific governor characteristics). Higher applied loads put the whole engine under severe test, proving very demanding in terms of acceptable engine response and low emissions. Depending on the specific load change magnitude, even engine stall might occur, particularly if a 0–100% step load is applied without temporary over-fueling or if the engine operates under fuel-limiting schedule. The key parameters in Figure 6.3 that affect the whole engine behavior during the transient event are the specific governor characteristics (the tighter the governor curves the lower the maximum speed droop) as well as the total mass moment of inertia (the lower the moment of inertia the larger the speed deviation, *cf.* Figure 6.7).

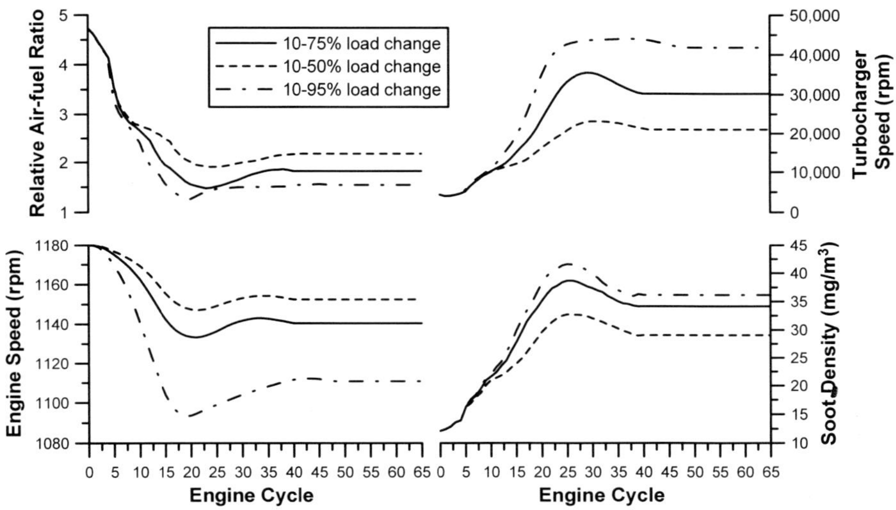

Figure 6.3. Effect of load change magnitude on engine transient response

Figure 6.4 illustrates the impact of the *type* of the loading connected to the engine. It is recalled that the resistance torque term τ_L in the crankshaft torque balance Equations 3.15 and 3.19 can be approximated by the following relation

$$\tau_L(\omega_L) = C_1 + C_2\omega_L^{C_3} \tag{6.1}$$

where

1. for a linear load-type (electric brake, generator) $C_3=1$;
2. for a quadratic load-type (hydraulic brake, fixed pitch propeller, vehicle aerodynamic resistance) $C_3=2$; and
3. C_1 is the speed-independent load term (*e.g.*, road slope).

In Figure 6.4, all three loading types (quadratic, rigid and linear) are investigated for the same step load increase. Obviously, the stronger the dependence of the load torque on speed, in other words the higher the C_3 value in Equation 6.1, the smaller the observed speed droop, as now the load torque 'follows' closely the engine speed response, and the faster the final equilibrium is achieved. On the other hand, larger speed droops (originating from smaller values of exponent 'C_3') lead to more abrupt governor displacement and, consequently, higher fueling rate and cylinder pressure/gas torque; thus, greater crankshaft deformation values are experienced, together with longer duration of the transient event. The rigid loading (C_3=0) is the most difficult case, since the response of the system depends on the engine alone. Nonetheless, since the particular engine of Figure 6.4 possesses a relatively high moment of inertia and the applied load change of 10–75% is definitely within the engine's 'capability', the differences observed between the three examined resistance types are not that great. For this reason, a fourth case is also illustrated in Figure 6.4. Here, a 10–100% load change is examined for the rigid loading case; the fuel pump rack reaches its 100% position after almost 17 cycles (roughly 1.7 s) and remains there unchanged, but it cannot prevent the engine from stalling since the load torque is higher than the maximum torque the engine can produce at these speeds unless some kind of over-fueling is permitted.

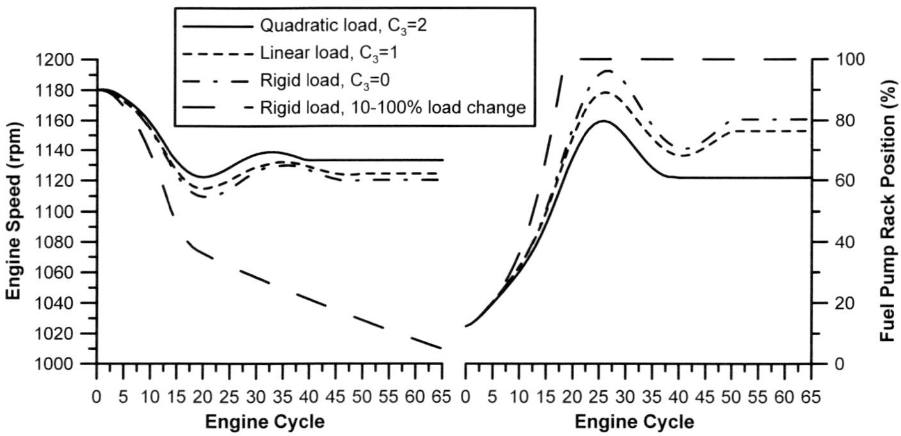

Figure 6.4. Effect of load type on engine transient response after a 10–75% step increase in engine load

Figure 6.5 focuses on the effect of the *duration* of the employed load change. Fast load application increases considerably the initial crankshaft angular deceleration, thus affecting the whole speed profile and the cycle at which the minimum speed is observed (10th cycle for the instantaneous load application compared with the 33rd cycle for the '3 s' case). Consistent with engineering intuition, the final equilibrium conditions are practically the same for all cases examined, since both the magnitude and the type of loading are the same. The most demanding case is the one with instant load application. Here, the load torque increases abruptly reaching its final value during the first cycle, while its engine counterpart responds with a delay, following the governor response to the speed drop as determined by the individual governor characteristics. Consequently, there is a considerable torque deficit during the first cycles of the transient event that leads to earlier, and considerably higher crankshaft deformations and friction overshoot compared with the other cases. The harder turbocharger lag effects are reflected into higher peak and cumulative (up to the cycle where the engine speed equilibrates) smoke emissions in Figure 6.5. An unacceptable oscillatory form of engine recovery may also be observed in this case depending on the specific governor characteristics. On the other hand, a long load change duration causes smoother development of all engine and turbocharger properties as the whole transient event develops at a much slower, hence, 'safer' rate. Summarizing, although the examined load–time schedules in Figure 6.5 cover quite a large spectrum, the final engine speed droop remains practically unaffected, whereas the maximum speed droop and soot emissions are seriously influenced.

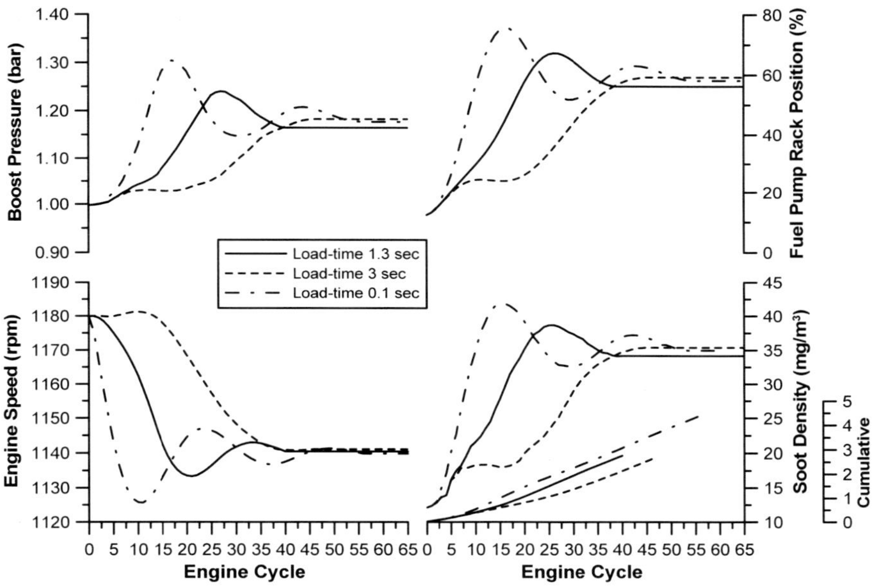

Figure 6.5. Effect of load change duration on engine transient response after a step increase in engine load

It is a matter of the final applied load and type of resistance involved for an engine with one problematic cylinder to successfully cope (though with greater speed droop) with a specific load change. This interesting case is illustrated in Figure 6.6, where the static injection timing is delayed for cylinder No. 3 of the six-cylinder, turbocharged engine due to some malfunctioning of the fuel pump. Two cases are investigated as follows: in the first one, static injection timing is 0°CA after TDC and in the second case 20°CA after TDC (in the nominal case the static injection timing is 26°CA before TDC). Delayed injection reduces considerably the efficiency of the third cylinder; in fact, as it is revealed from the bsfc sub-diagram, in the early cycles of the 20°CA after TDC case the third cylinder is motored by the other five cylinders, thus retarding the whole speed response of the engine by reducing the available total torque. Consequently, the fuel pump rack is forced to reach more extreme positions drifting the boost pressure too (not shown). Even though only one of the six cylinders suffered a delayed injection, this was sufficient for a notable increase (12%) in the engine speed droop compared with the nominal case.

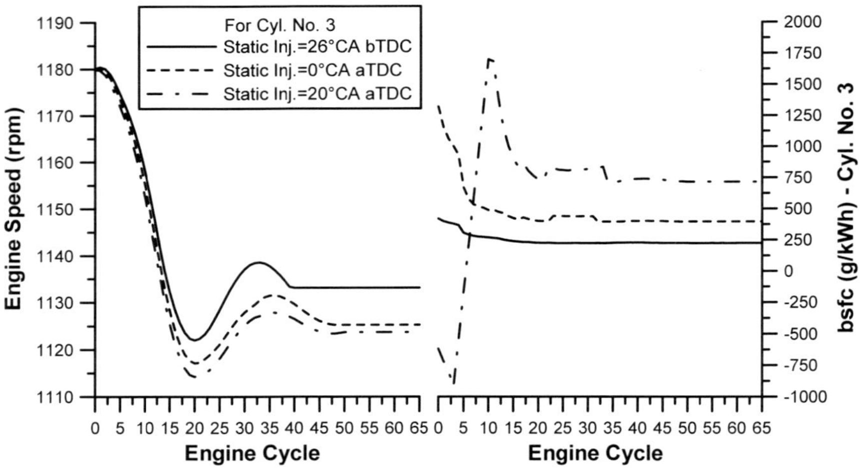

Figure 6.6. Effect of retarded injection timing for one cylinder of a six-cylinder diesel engine after a step increase in engine load

The results from several other thermodynamic and dynamic parameters on engine transient response are summarized below.

- A high engine mass moment of inertia (Figure 6.7) slows down the transient response, by lowering the maximum speed droop at the expense of increased recovery period. The obvious advantage here is the ability of the engine to cope successfully with higher and faster applied load changes, a fact that is important in industrial applications. Since the whole transient event develops more smoothly, with lower crankshaft decelerations and slower governor clutch movement, turbocharger lag effects are suppressed with positive effects on exhaust emissions. On the other hand, when automotive acceleration transients are involved, the increase in the recovery period imposed by a high moment of inertia manifests itself as poor vehicle driveability.
- The compression ratio plays a secondary role during transients, affecting thermodynamic (*e.g.*, peak cylinder pressures and temperatures, NO_x emissions) but not dynamic properties.
- Variation of the cylinder wall temperature during transients has a minimal effect on engine response and final conditions. However, the initial wall temperature slightly affects response, with the hot walls (of either cylinder or exhaust manifold) being more favorable, as now smaller amounts of exhaust gas energy are lost during the load or speed increase aiding faster turbine acceleration; a detailed analysis will follow in Section 7.3, in terms of low-heat rejection engine operation.

- Decreased ambient temperature slightly improves boost pressure build-up and speed response owing to the increased inlet air density.
- Engine response at high altitude worsens owing to the decrease in both inlet air density and partial pressure of oxygen.
- Reduced fuel cetane number is mainly responsible for prolonged ignition delay, affecting only marginally the engine acceleration.
- As was discussed in Section 3.2.2, geometrical and technical characteristics of the governor affect the engine response seriously in terms of speed droop and recovery period, as well as the whole speed response profile. The impact of each term is sometimes conflicting when comparing speed droop and recovery period.

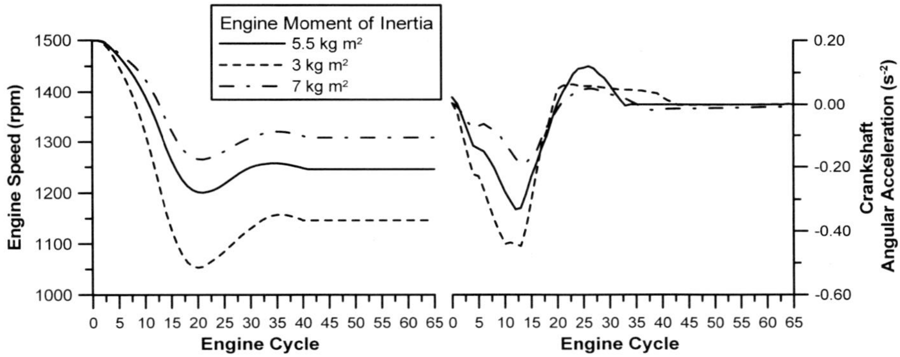

Figure 6.7. Effect of engine moment of inertia on engine transient response after a step increase in engine load

6.3 Air-injection

During the critical early cycles of a load or speed increase event, turbocharger delay is reflected into slow build-up of compressor boost pressure and shortage of combustion air, resulting in fuel–air ratio overshoot and combustion deterioration. One obvious reaction here is to aid engine response by supplying additional amounts of air, *e.g.*, by injecting pressurized air stored in a bottle or tank into the compressor or inlet manifold or the turbine wheel. Injecting air onto the compressor wheel is, generally, more successful than air-injection onto the turbine blades, since it is more efficient to drive the compressor directly instead of increasing its boost pressure mechanically from the turbine side; moreover, the air that is injected in the compressor side is instantly available to the engine for combustion. Consequently, direct increase of air-supply is achieved and the duration of combustion discrepancies is reduced since the high air-flow into the engine cylinders can match the increased fuel quantity in just a few cycles, with the resulting high exhaust gas energy aiding faster turbocharger and engine response.

The method of air-injection into the compressor has proven successful in terms of speed recovery (the success is not so pronounced at cold starting transients), but this depends strongly on the duration and onset of injection. However, it falls short in as much as it requires engine (space) modifications, *e.g.*, nozzles on the turbocharger housing, see also Figure 6.8 (care has to be taken in order to avoid blade resonance or unbalanced forces on the compressor rotor), an air tank, larger than normal air-brake (for truck engines) or air-starting (for industrial or marine engines) compressor capacity as well as integration of a control system (Figure 6.9). Incorporation of an air-injection system into the compressor operation results also in slight modification of the compressor characteristics, namely an increase in the delivery pressure ratio for the same speed and mass flow, and reduction in the required torque to drive the compressor; effects on the surge line have been reported to be negligible, at least as regards four-stroke engine operation.

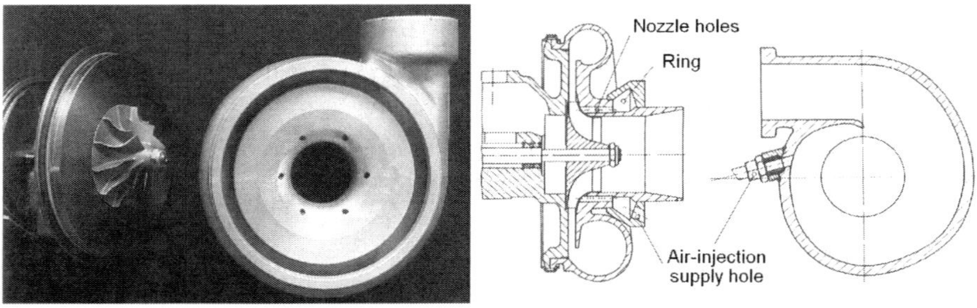

Figure 6.8. View and design of a modified turbocharger for air-injection (reprinted with permission from Harndorf and Kuhnt [2])

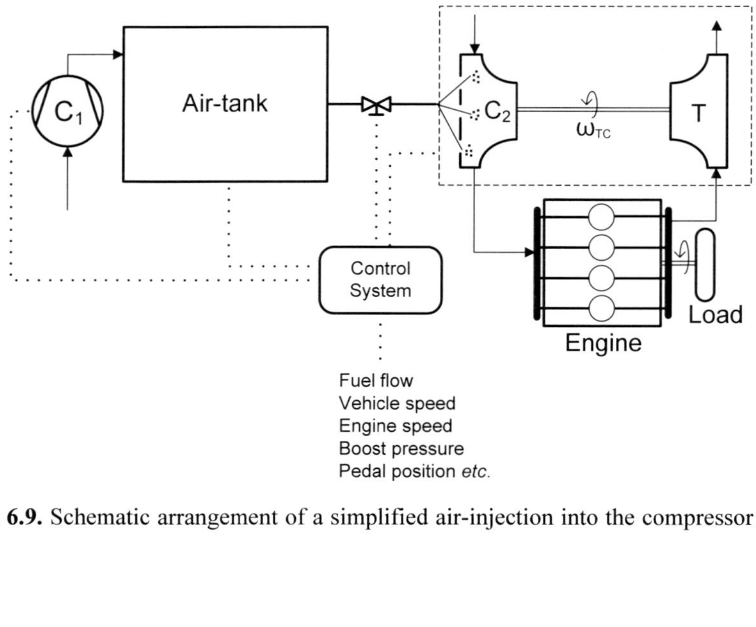

Figure 6.9. Schematic arrangement of a simplified air-injection into the compressor control system

Ledger *et al.* [3, 4] and Winterbone *et al.* [5] proposed a method of air-injection into the centrifugal compressor with a control system interfaced to the four-stroke engine in order to support the modifications required. This consisted of a closed-loop controlling the air-injection via the air–fuel ratio. The latter was measured as the quotient of air manifold pressure and fuel pump rack position. When the value was lower than a specified threshold (being a function of engine speed), air-injection was initiated. Typical results are reproduced in Figure 6.10 for a 0–60% load acceptance aided by a 2 s air-injection duration right after the onset of the transient event. Following the assisted compressor operation, the air–fuel equivalence ratio was kept at adequately high values during the early cycles of the transient event, preventing combustion deterioration and establishing faster turbocharger response. As a result, the engine response time was approximately halved and the transient speed droop was significantly reduced. Further tests were performed with 100% step load applications; the standard engine was not capable of accepting this load, whereas when air-injection was applied (again for 2 s) satisfactory load acceptance was achieved [1].

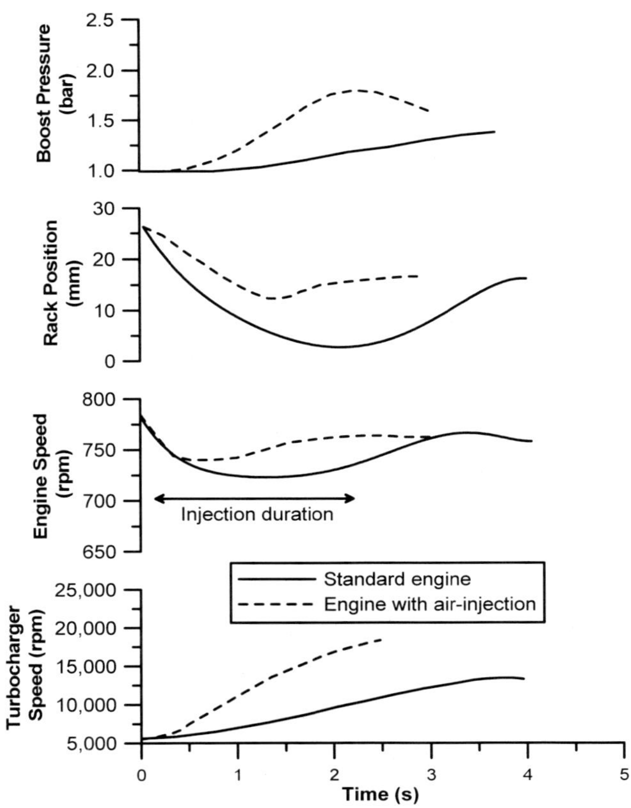

Figure 6.10. Effect of air-injection on transient response after a load step of 60% full-load (medium-speed, turbocharged diesel engine rated at 1410 PS at 750 rpm – simulation results from Winterbone [1], reprinted by permission of Oxford University Press)

Further application of the latter air-injection assist on a six-cylinder, light-duty turbocharged diesel engine of 11.32 L displacement volume revealed a reduction in the time period with peak smoke from 3 to 0.5 s as well as improvement in the speed response from 4 to 1.5 s for a severe load acceptance transient at 1000 rpm. This was accomplished by applying 1 s air-injection duration, which provided an air consumption of about 0.1 kg. Similarly encouraging results were reported for acceleration transients. Harndorf and Kuhnt [2] quantified the transient soot emissions improvement for a similar engine configuration at approximately 50% the amount of emissions under operation without air-injection; when combined compressor and turbine air-injection was applied, the improvement was even higher.

Of particular importance in air-injection configurations, is the exact injection schedule, *i.e.*, onset and duration of injection as well as the pressure of the injected air. Figure 6.11 illustrates engine response after a severe load increase with and without air-injection for various air-assist schedules. Injecting air for 1 s at 3.45 bar was proven insufficient to prevent stalling as was also the case with the baseline, unassisted engine; however, if the injection duration lasts for 2 s, even lower injection pressures can prove beneficial for the successful engine recovery. Of course, the higher the injection pressure, the lower the maximum speed droop and the shorter the recovery period up to the final equilibrium. Although it was argued that higher injection pressures for a shorter duration were more effective, this is actually a subject of optimization for the engine under study, with the optimum injection pressure being a matter of careful choice in order to avoid the possibility of the compressor moving onto surge.

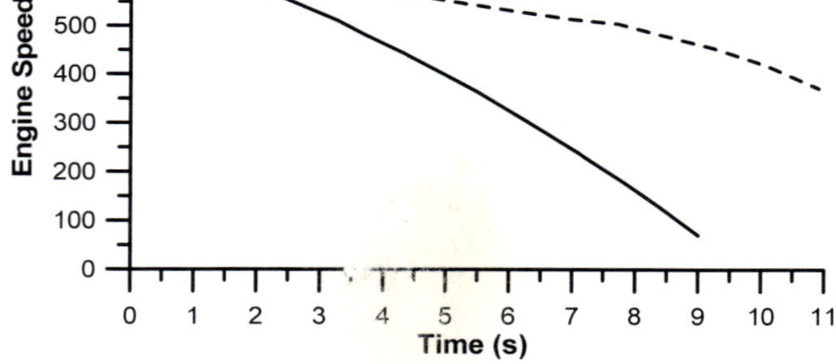

Figure 6.11. Comparison of transient response after a 0–16.88 bar bmep load increase with various air-injection schedules (four-stroke, six-cylinder, turbocharged diesel engine of 68 L displacement volume, rated at 746 kW – simulation results adapted from Ledger *et al.* [4])

Another important parameter of the air-injection assist scheme is the quantity of injected air, as this will determine the size of the air-reservoir necessary. For the medium-speed engine of Figure 6.11, a reservoir of 0.45 m^3 containing air at 6.89 bar would supply sufficient air for a 3 s injection at 3.45 bar via a suitable reducing valve. A similar air-injection for a truck engine of 7 L displacement volume and a maximum rated speed of 2200 rpm, would require a reservoir of approximately 0.15 m^3 mounted under the chassis and pressurized from the air-braking system [4].

Air-injection through the inlet valve directly into the engine cylinders has been proposed as an even more effective technique because, in this case, the increased engine torque was instantly available. This was revealed by the work of Boy [6], who employed an air-tank in parallel to the turbocharger compressor with an air pressure of 30 bar, for a medium-speed marine diesel engine acceleration.

6.4 Turbocharger Configuration

Turbocharger properties and configuration play the most significant role in engine transient response, as it is the turbine's rotor acceleration that contributes mostly to the phenomenon of turbocharger lag. In Section 2.1, the (flow inertia) impact of the specific turbocharging system, *i.e.,* constant pressure *vs.* pulse turbocharging was discussed; in this section, we will focus on other important turbocharger properties, namely

1. moment of inertia and
2. configuration *viz.,*
   - combined turbocharging;
   - two-stage turbocharging;
   - variable geometry turbine;
   - electrically assisted turbocharging; and
   - sequential turbocharging.

During transient operation, the turbocharger shaft torque balance is described by the following equation based on Newton's second law of motion for rotational systems

$$\eta_{mTC}\tau_T - |\tau_C| = G_{TC}\frac{d\omega_{TC}}{dt} \tag{6.2}$$

In Equation 6.2, τ_T and τ_C represent the turbine and compressor torque, respectively, and η_{mTC} is the mechanical efficiency of the turbocharger shaft. Improvements in response can therefore be achieved by increasing the turbine torque or by reducing the moment of inertia or by reducing the turbocharger shaft mechanical losses.

It should be noted that, in general, faster turbocharger response means faster engine response and reduced smoke emissions. On the other hand, slow turbocharger response results in the engine behaving as naturally aspirated for a considerable portion of the transient event resulting in long fuel limiting action (in order to avoid excessive smoke emissions), hence slow engine response and poor driveability, with a possibility of stall when large load increases are employed in industrial applications. From the above-mentioned methods, the reduction in the turbocharger moment of inertia adopting lighter materials is the most uncomplicated to apply since it does not modify the steady-state characteristics of the turbo-machines. Hence, the matching of the engine with the turbocharger remains unaltered as opposed to the similar philosophy method of choosing a smaller turbocharger frame. On the other hand, combined or sequential- or two-stage turbocharging as well as variable geometry turbines affect strongly the steady-state performance of the engine, requiring a much more complex matching procedure during the design process as well as complicated and expensive control systems for their successful implementation.

6.4.1 Turbocharger Mass Moment of Inertia

Perhaps the most obvious technique for faster turbocharger, hence engine response, is associated with the former's rotational moment of inertia. This subject has been treated in detail by Watson [7]. Following Equation 6.2, changes in turbocharger rotating inertia G_{TC} have been found to have the strongest influence on the latter's rotational speed response; it is not surprising then that turbocharger manufacturers try to design for the lowest possible inertia. This can be achieved

- By adapting lighter materials, *e.g.,* titanium aluminide TiAl, or ceramic ones such as silicon nitride, in order to decrease the turbine rotor mass. This is the most straightforward technique since it does not influence the turbine nozzle area, hence steady-state operation remains unaltered.
- For multi-cylinder ('V'-type) engines, a reduction in inertia can be achieved by replacing a single unit with two smaller ones operating in parallel (Figure 6.12 – this kind of configuration is sometimes referred to as 'bi-turbo'); in this case, the nozzle area is altered (decreased) but so does the mass flow through each turbine.

Figure 6.12. V-20, marine diesel engine producing 9,000 kW at 1000 rpm, with two turbochargers in parallel each feeding one bank of cylinders (courtesy of MAN Diesel)

- If the flow requirements for a specific engine application lie in the mid-range of one turbocharger frame, the same requirements might be met by another unit at the top of a smaller size provided, of course, that the efficiency is maintained at sufficiently high levels. In this case, however, the lower turbine nozzle area of the smaller turbocharger frame alters the whole (steady-state) matching between engine and turbocharger.

Most of the turbocharger inertia (70% or even more) is typically associated with the turbine wheel owing to the high density material with respect to that of the compressor impeller, and large diameter compared with the shaft. The inertia of the turbocharger disc is proportional to the fourth power of diameter, but since tip width and rotor length tend to increase with diameter, turbine inertia is more closely related to the fifth power of diameter, *i.e.*, $G_{TC} \propto d^5$. This suggests that reducing turbine (and compressor) rotor diameter will strongly influence turbocharger and, consequently, engine speed response, albeit at the expense of turbocharger efficiency, mainly owing to the higher rotational speeds encountered (see also Figure 2.5 – the higher turbocharger speeds are usually associated with lower isentropic efficiencies). Hopefully, as diameter is reduced, the penalty in efficiency changes much more slowly than inertia.

Acceleration of the turbocharger to a desired speed implies that a certain kinetic energy has been imparted to the rotor. Assuming zero initial speed for simplicity, the kinetic energy change of the rotor is

$$E_{kin} = \frac{1}{2} G_{TC} \omega_{TC}^2 \tag{6.3}$$

Hence, the energy input to accelerate the turbocharger is proportional to inertia and the square of rotational speed. Thus,

$$E_{kin} \propto d^5 \left(\frac{1}{d} \right)^2 = d^3 \tag{6.4}$$

This is a strong reason to reduce diameter, at the expense of increased speed, a technique that has the additional advantage of reducing turbocharger size for a specified mass flow-rate [7].

By reducing the turbo-machines diameter, hence turbocharger inertia, faster acceleration of the turbocharger shaft is established; the higher achieved rotational speed moves the compressor operating point towards higher boost pressure and air-mass flow-rate values. The increased air-supply to the engine cylinders minimizes the fuel limiter's function, thus allowing higher amounts of fuel to be injected without fear of smoke emissions. That, in return, increases the exhaust gas energy, hence turbine power, and the transient event develops much faster and with lower amounts of emissions. The above arguments are highlighted in an explicit way in Figure 6.13 for an acceleration event of a highly-rated turbocharged diesel engine.

Initially, the engine is running at low-speed and low-load conditions, hence turbocharger speed is low and boost pressure is almost equal to atmospheric pressure. Afterwards, the speed demand lever is rapidly moved to its maximum position, with the load being applied according to the propeller law. Initial engine acceleration from 1000–1400 rpm seems not to be influenced by turbocharger inertia, since the engine has no boost during the first 0.5 s, even with the lightest turbocharger. Thus, the engine accelerates in response to the no-boost fuel controller limit on fuel delivery. In any case, engine response in Figure 6.13 is rapid, since the applied load is relatively low.

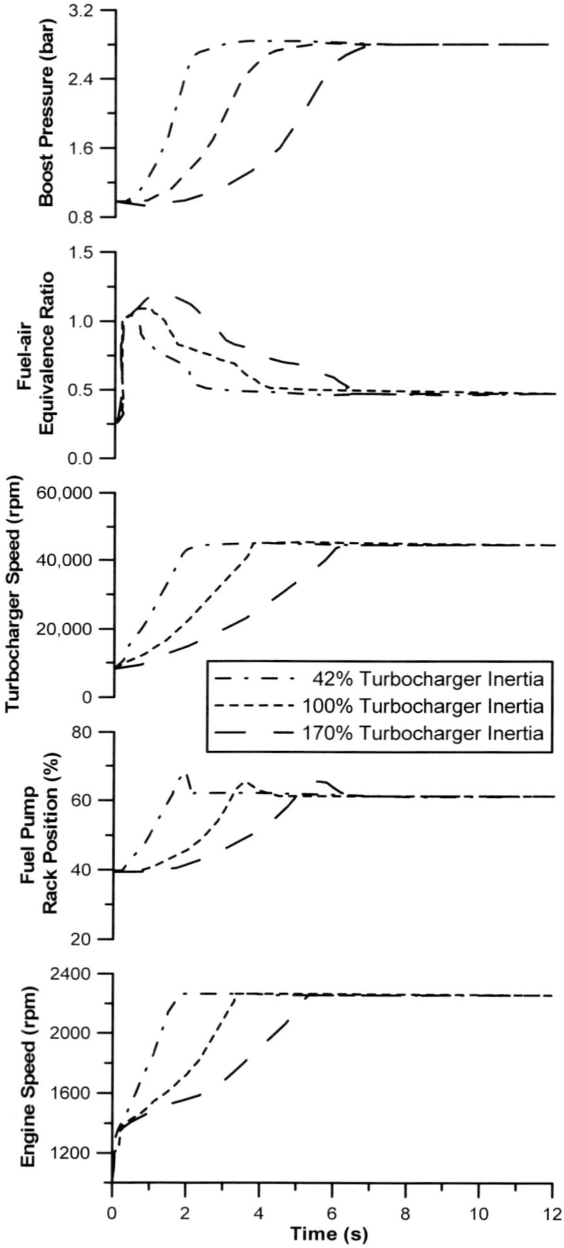

Figure 6.13. Effect of turbocharger moment of inertia (same frame) on propeller law engine acceleration of a highly-rated turbocharged diesel engine (simulation results reprinted with permission from SAE Paper No. 810338 [7], © 1981 SAE International)

Turbocharger acceleration is significantly influenced by changes in inertia. This affects the subsequent engine response via the fuel controller. Thus at the 1 s point, the lowest inertia turbocharger has developed a pressure ratio of 1.4, allowing the fuel controller to inject more fuel. This enables the engine to continue accelerating rapidly. In contrast, the highest inertia turbocharger takes 2 s to deliver any boost, and three times as long to reach a pressure ratio of 2; fuel delivery is limited by the fuel controller, hence engine response is very slow [7].

Similar results hold for load acceptance transients, such as the one illustrated in Figure 6.14. Smaller turbocharger inertia (frame) reduces maximum and final engine speed droop by facilitating a steeper turbine response ('circled' in Figure 6.14). Combustion deterioration cannot be totally prevented, but the much smaller duration of the turbocharger lag ensures a smoother response and lowers soot emissions during the early cycles of the transient event. It should be noted, however, that a very low turbocharger inertia may, under some circumstances (*e.g.*, depending on the specific governor characteristics), lead to an oscillatory engine recovery, which is unacceptable both in terms of recovery time and crankshaft stress.

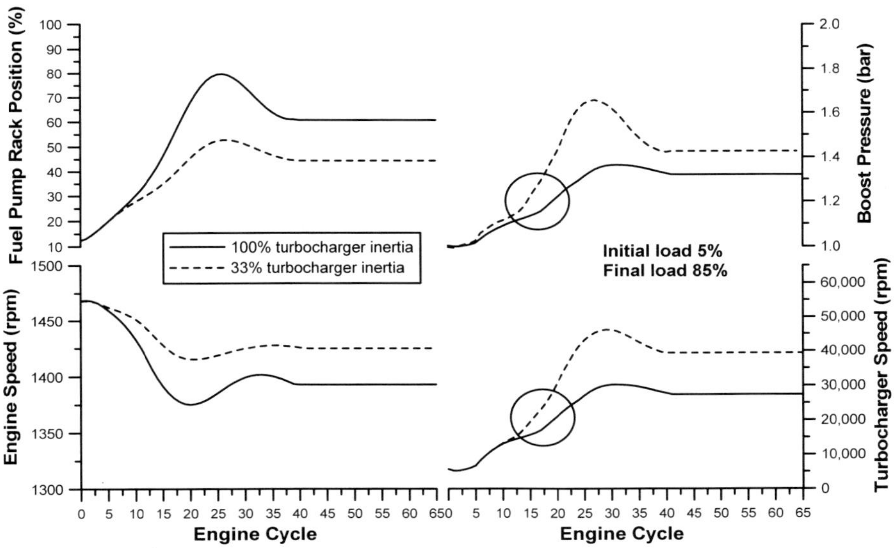

Figure 6.14. Effect of turbocharger mass moment of inertia on engine transient response (six-cylinder, turbocharged and aftercooled, DI diesel engine)

A particular aspect of low-inertia turbochargers, originating from application of smaller turbocharger frame, is the fact that at high engine speeds and loads, over-boosting of the engine might be experienced owing to the very high-pressure ratios induced at the high turbocharger speeds for the small turbine nozzle area; this may prove extremely dangerous for safe engine and turbocharger operation. The usual approach here is to install an exhaust waste-gate valve (Figure 6.15); the latter is internally built into the turbine casing (although external waste-gate valves do exist), and consists of a spring loaded valve, which gradually opens in response to the inlet manifold pressure allowing by-pass of an amount of exhaust gas directly to the atmosphere. At the same time, the effective flow area out of the exhaust manifold is increased. This reduces the exhaust gas pressure, with obvious benefits in terms of reduced pumping work, and better engine efficiency.

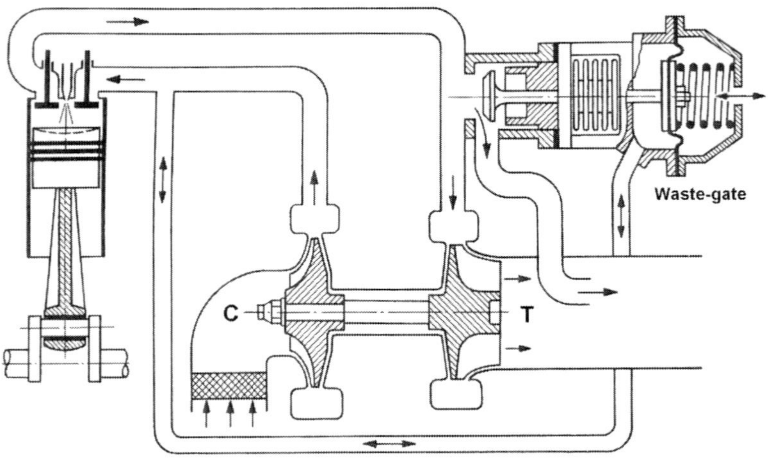

Figure 6.15. *Upper:* Schematic arrangement of waste-gate controlled exhaust gas turbocharging system (from Zinner [8], reprinted with kind permission from Springer Science+Business Media). *Lower:* Modern automotive turbocharger with waste-gate valve (courtesy of BorgWarner Turbo Systems)

Following Equation 6.2, a reduction in the turbocharger shaft frictional losses results in higher turbocharger efficiency η_{mTC} and, therefore, a higher amount of mechanical power is delivered to the compressor, hence quicker compressor boost pressure and air-mass flow build-up is accomplished. However, the improvement is very modest compared with the inertia effects.

6.4.2 Combined Supercharging

In Section 2.1, the merits of mechanically assisted supercharging were introduced in terms of transient engine operation. Two important parameters that favor engine transient response with such a supercharging scheme were identified. Specifically, compressor speed that is directly related to engine speed, whereas high boost pressure can be achieved even at low (engine) speeds and air-mass flow-rates owing to the specific flow characteristics of displacement type compressors. Nonetheless, application of this supercharging configuration is limited owing to the higher fuel consumption due to the power drawn from the engine crankshaft to drive the mechanical compressor, and the design considerations originating in the mechanical connection between crankshaft and compressor. The combined or 'hybrid' supercharging scheme (Figure 6.16) attempts to merge the advantages of mechanically assisted supercharging and exhaust gas turbocharging in one configuration.[2] A positive displacement compressor – typically 'Roots' blower or screw or Wankel type – usually of small size (C_1) is directly coupled to the engine crankshaft via a (fixed ratio) electro-mechanical clutch, feeding the turbocharger aerodynamic type compressor (C_2).[3]

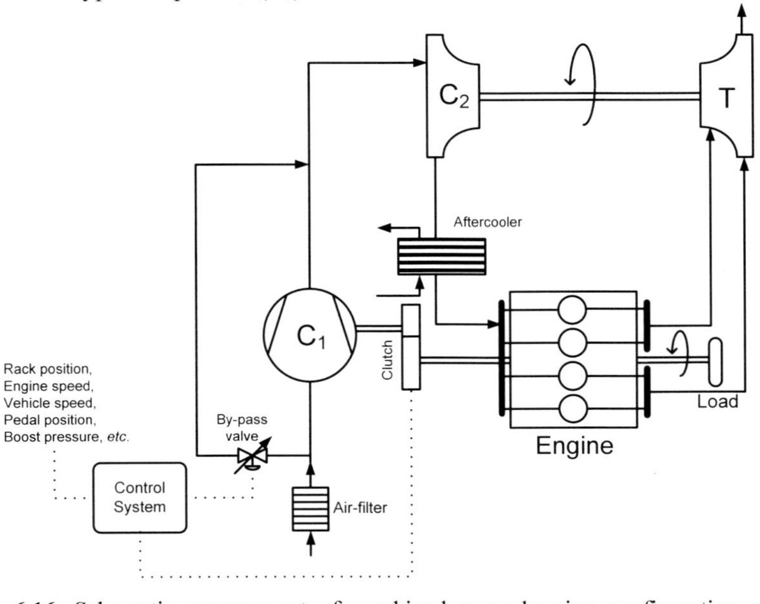

Figure 6.16. Schematic arrangement of combined supercharging configuration with its simplified control system, incorporating a twin-entry turbine

---

[2] In the literature, the use of more than one turbo- or superchargers is sometimes referred to as 'sequential' or 'twin' or 'two-stage' turbocharging. In this book, the term 'combined supercharging' will be used for the series arrangement of a mechanical supercharger and an exhaust gas turbocharger, the term 'two-stage turbocharging' (Section 6.4.3) will be used to describe two exhaust gas turbochargers installed in series, and the term 'sequential turbocharging' (Section 6.4.6) will be used to describe two or more exhaust gas turbochargers operating in parallel.

[3] Such a supercharging scheme is also met in two-stroke engines, with the displacement type compressor assisting the scavenge process at low engine loads and speeds.

This kind of arrangement is illustrated in Figure 6.17 for a spark ignition engined vehicle.

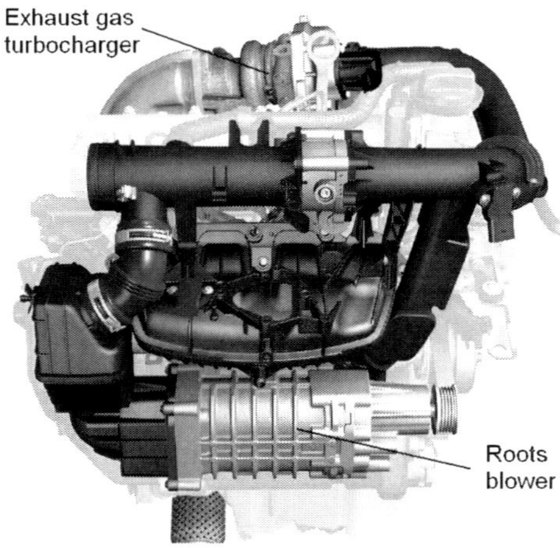

Figure 6.17. Combined supercharging configuration incorporating a Roots mechanical blower and an exhaust gas turbocharger connected in series (courtesy of Volkswagen Group)

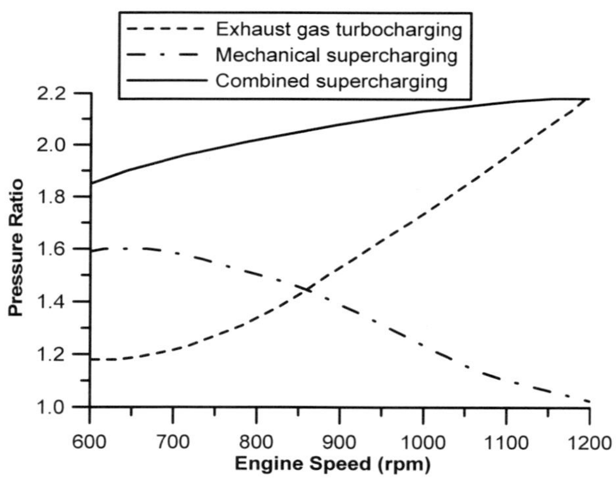

Figure 6.18. Comparison of boost pressure between mechanical supercharging, exhaust gas turbocharging and combined supercharging (reprinted with permission from Schmitz *et al.* [9])

Obviously, the engine benefits from this setup at steady-state conditions too, with the air-supply being higher than that of exhaust gas turbocharging alone since the two delivery pressure ratios are multiplied. Figure 6.18 describes in an explicit way the advantages of combined supercharging compared with exhaust gas turbocharging or mechanical supercharging in terms of achieved boost pressure. At low engine speeds and loads, where the exhaust gas energy is not adequate to provide enough boost to compressor C_2, compressor C_1 is active; likewise, at the onset of an acceleration transient event in the low and medium speed range, compressor C_1 aids faster engine response since it follows directly the engine acceleration. By so doing, the unfavorable flow characteristics of the exhaust gas turbocharger compressor at low engine speeds and loads are remarkably improved since the mechanical supercharger provides boost pressure irrespective of engine load. Moreover, since the delivery pressures from the two compressors are multiplied, the use of a smaller turbocharger frame is enabled with obvious benefits in terms of faster turbocharger response. For steady-state engine operation in the middle speed range as well as for steady-state and transient operation at high engine speeds, the air-supply by-passes the mechanical compressor and its turbocharger counterpart is the sole booster device. The most significant drawbacks of combined supercharging are the mechanical connection between engine and positive displacement compressor as well as the need for a complex control system that takes into account various engine and supercharger variables such as the engine speed and load (pedal position), vehicle speed, boost pressure, exhaust back-pressure, *etc.* (Figure 6.16).

By closer examination of the curves in Figure 6.18 the following merits of combined supercharging can be identified

- relatively flat torque curve;
- considerable increase of engine torque in the low-speed range compared with conventional exhaust gas turbocharging, which results in
- narrower range of fuel limiting control; and
- primary use of the exhaust gas turbocharger in the medium and high engine speed range, where its characteristics are better suited to the reciprocator.

Application of the above combined supercharging configuration on a vehicle yields the acceleration response demonstrated in Figure 6.19. When using conventional exhaust gas turbocharging, the recovery period up to the demand speed is 19 s; with combined supercharging, on the other hand – applying a Wankel type mechanical compressor – a remarkable 30% reduction is accomplished without overshoot in smoke emissions. The main mechanism behind this considerably faster acceleration is the fact that sufficiently high amounts of engine torque are available still from the early cycles of the transient event, owing to the additional boost offered by the mechanical compressor (recall that a displacement compressor (Figure 2.7) is characterized by steep pressure ratio curves). Consequently, the engine runs in smoke-limited operation for much smaller time as the turbocharger lag is minimized compared with the conventionally turbocharged operation, and acceleration proceeds much faster.

Notice the change in the slope of the engine speed curve for the exhaust gas turbocharged vehicle at approximately t = 13 s, which is attributed to the increase in boost pressure a few seconds earlier. In fact, for the first 8 s of the transient event, the conventional turbocharger's boost pressure has not increased at all (the engine operates in naturally aspirated, hence fuel limiting mode during all this time), whereas for the combined supercharged vehicle, an immediate rise in boost pressure is accomplished. However, right after the mechanical unit was disengaged the charge pressure dropped to almost 70% of the maximum before recovering relative quickly. Moreover, the combined supercharged vehicle accelerated from a higher initial boost pressure (hence turbocharger operating point and rotational speed), a fact that further contributed to the faster response.

Improvement in engine starting behavior through combined supercharging has been also reported [10]; this was accomplished with a slightly different configuration, where the mechanical compressor was fitted downstream of the aerodynamic one and operated via the crankshaft (preferred) or the starter motor of the vehicle.

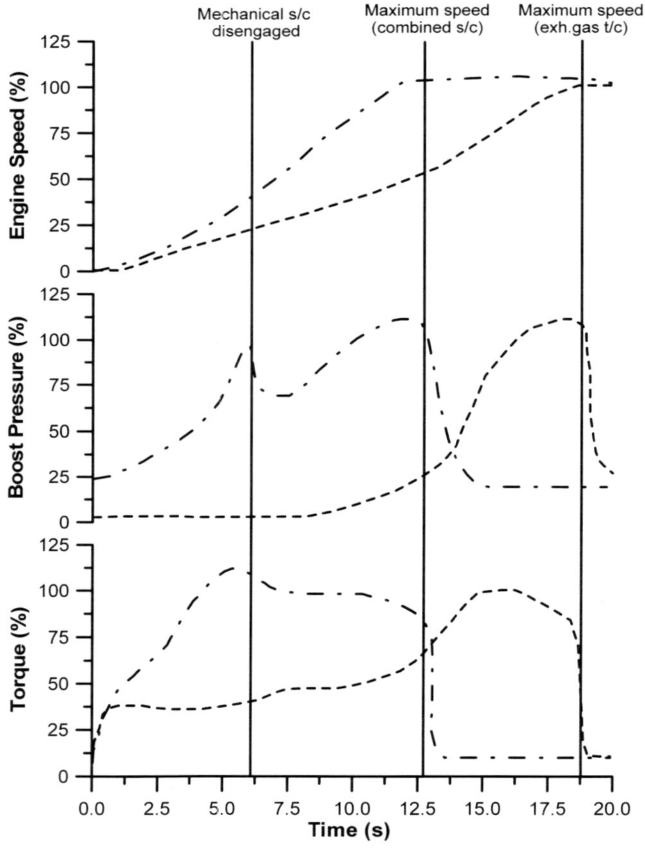

Figure 6.19. Improvement in engine transient response using a combined supercharging configuration for a no-load 500–1900 rpm acceleration (40 t truck operating in tenth gear; dashed line corresponds to exhaust gas turbocharging and dash-dotted line to combined supercharging – simulation results reprinted with permission from Schmitz *et al.* [9])

6.4.3 Two-stage Turbocharging

Two-stage turbocharging primarily addresses the need for increased bmep, typically above 25 bar often required in large trucks or industrial-marine applications. In order to achieve such a specific power production, an inlet pressure of at least 3.5–4 bar is needed, which is not recommended with single-stage units owing to the low efficiency and narrow flow range of aerodynamic compressors at high-pressure ratios. Instead, two-stage turbocharging is applied (Figures 1.1 and 6.20), where the two turbochargers are connected in series. The first is the low-pressure (LP) unit fitted upstream of the (smaller) high-pressure (HP) one, with an aftercooler and, possibly, an intercooler between the two compressors. Depending on the specific application (automotive, marine) the following may be part of a more sophisticated two-stage configuration: a valve downstream of the LP compressor for by-passing the HP unit in order to avoid over-speeding, one or two waste-gate valves, twin-entry and/or variable geometry turbine arrangement.

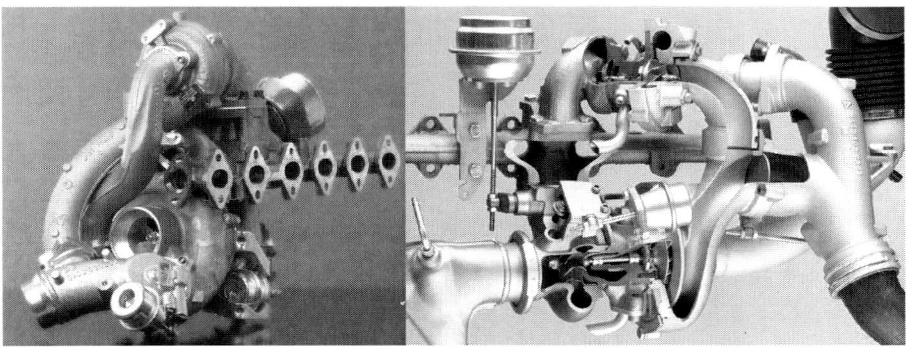

Figure 6.20. Regulated two-stage turbocharging unit for automotive use (left photo courtesy of BorgWarner Turbo Systems; right photo courtesy of BMW AG)

Two-stage turbocharging achieves higher boost pressures, owing to the multiplication of the delivery pressure ratios, hence the air-supply to the engine cylinders is improved in the low-speed region increasing torque and facilitating the use of higher EGR rates without smoke penalty. When considering the operation of two-stage units, the most important issues that have to be addressed is the control system (Figure 6.21) as well as the matching between engine and turbochargers, which is clearly a more difficult task (size relation between the two units, effective area of the turbines, flow range, surge avoidance, altitude operation, *etc.*) compared with single-stage turbocharging, as the second unit induces one more degree of complexity. In the case of automotive or truck engines, the matching issues are enhanced by the additional requirements for wide speed range and good torque back-up. There are particular aspects of two-stage turbocharging that favor transient operation; these originate in both flow and dynamic grounds, namely

- the reduced speed change of the most influential in terms of response HP unit (this may be of variable geometry arrangement), which also operates, generally, at higher rotational speeds; and
- the reduced moment of inertia of the individual stages compared with the single unit.

The HP turbine can benefit from the LP one due to the fact that the kinetic energy needed to accelerate the two sets is lower than that for a (larger) single unit (recall that turbocharger inertia is related to the fifth power of diameter). The use of two (smaller) units results in much lower total inertia.

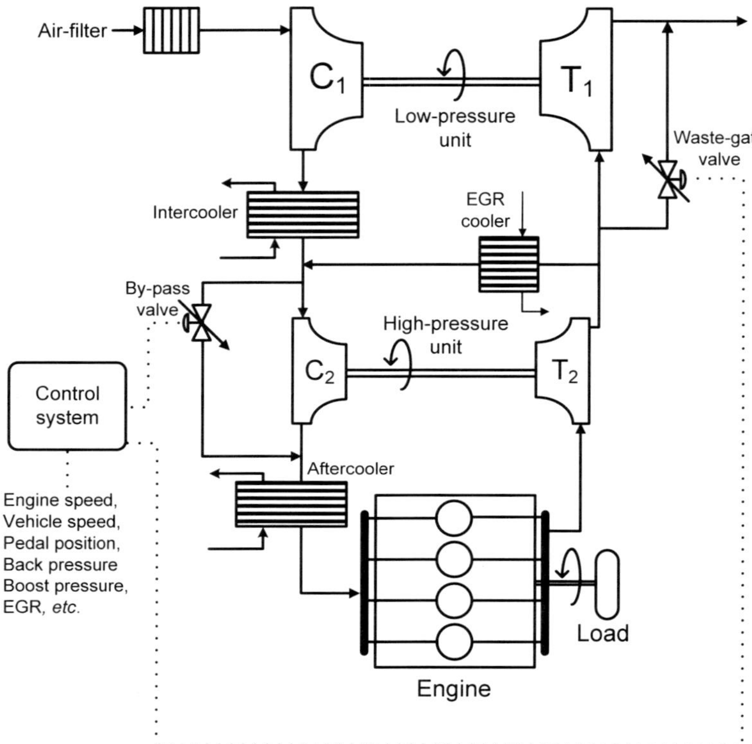

Figure 6.21. Schematic arrangement of an automotive two-stage turbocharging configuration with its simplified control system

As is revealed in Figure 6.22, the lower the HP unit's inertia, the faster turbine acceleration is achieved, hence, air-supply build-up and engine response. Whereas the single-stage unit needed almost 3.5 s for the boost pressure to achieve its final steady-state value, the low-inertia two-stage unit required only 2 s, that is a remarkable 40% reduction. However, a very small HP unit comes at a cost, namely engine efficiency may worsen at steady-state conditions requiring a comprehensive control algorithm for optimum performance.

In fact, Figure 6.23 reveals that the two-stage unit with the small HP turbine exhibits worse fuel consumption than the single-stage turbocharger. Nonetheless, at low engine speeds the compressor of the single-stage system was running in the surge area during a large part of the cycle, whereas at high engine speeds its speed exceeded the safe running limits. This fact, which is attributed to the small operating range of a centrifugal compressor at high rotational speeds, was counterbalanced by applying the two-stage system that was capable of providing excess air at medium to high engine speeds; the latter also incorporated a waste-gate valve for safer HP unit operation at high speed and load conditions.

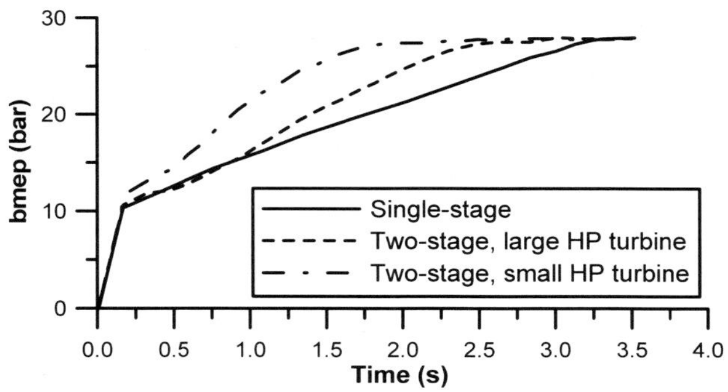

Figure 6.22. Improvement in engine transient response using a two-stage turbocharging configuration (four-stroke, EURO III engine of 12.6 L displacement volume rated at 450 kW at 1900 rpm – simulation results from Sturm and Kruithof [11], reprinted with permission from the Aachener Kolloquium)

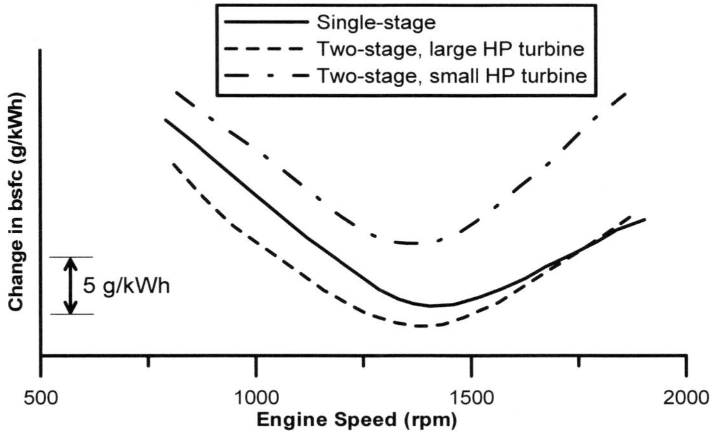

Figure 6.23. Comparison of steady-state fuel consumption with single- and two-stage turbocharging configurations (simulation results from Sturm and Kruithof [11], reprinted with permission from the Aachener Kolloquium)

Further results concerning an extreme case experienced by an industrial engine during severe load acceptance are illustrated in Figure 6.24. During the load increase, the margin of torque developed instantaneously by the single-staged turbocharged engine over that applied by the load is not sufficient to prevent stall. Since the trapped equivalence ratio, without fuel control, reaches 1.3 (the minimum operational fuel–air ratio for the engine was set at 18:1), there is a lack of sufficient combustion air. On the other hand, with two-stage turbocharging, the HP unit initially accelerates at almost the same rate as the single unit but the work done by the LP unit helps to raise inlet manifold pressure more rapidly, thus preventing stall

[12, 13]. Notice the higher initial rotational speed of the HP unit compared with its LP counterpart.

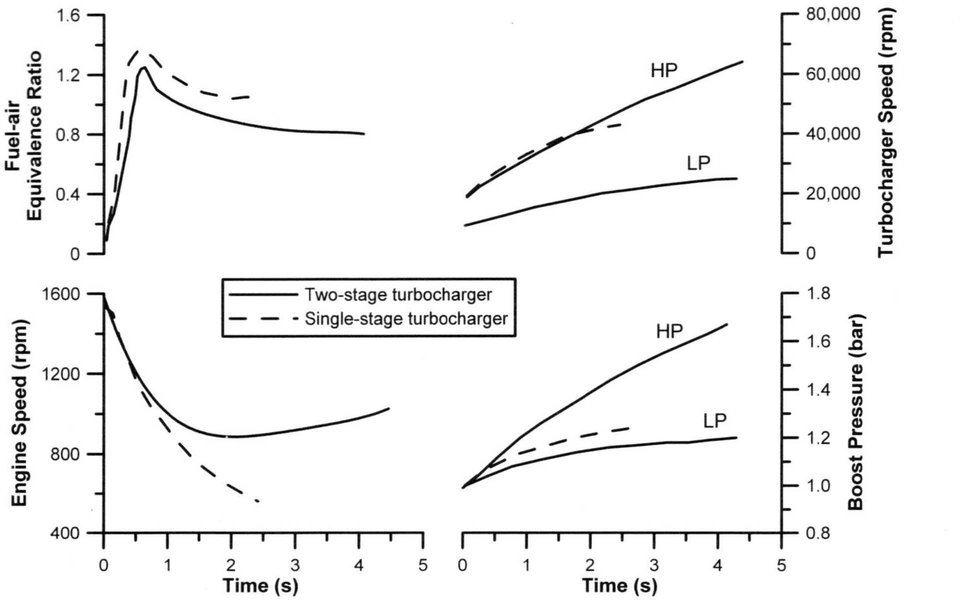

Figure 6.24. Comparison of load acceptance with single- and two-stage turbocharging for 0–15 bar bmep load change in 0.25 s (two-stage turbocharged engine without fuel limit control – simulation results from Watson *et al.* [12], reprinted with permission from the American Society of Mechanical Engineers)

6.4.4 Variable Geometry Turbine

The basic aspects of VGT operation were discussed in Section 2.4, with Figure 2.38 describing the main VGT control strategy during transients and Figure 2.39 illustrating a typical VGT engine response. In this section, we will focus on the driveability effects of variable geometry turbocharger configurations compared with fixed geometry, conventional units. During transient operation (either load or speed increase) the main strategy of the VGT is the reduction of the nozzle area by closing down the vanes so as to increase the back-pressure and enthalpy drop across the turbine, thereby boosting the compressor operating point. By so doing, faster build-up of the engine air-supply is established, minimizing fuel limiting function and improving driveability and emissions. At the same time, the EGR valve is closed to support filling of the cylinders with fresh air. As soon as the air-supply to the engine has been built-up, gradual opening of the VGT vanes follows in order to prevent over-boosting. The positive effects of VGT during transients come therefore as the result of improvement in flow characteristics and air-supply, whereas the dynamics (inertia) of the turbocharger are not altered.

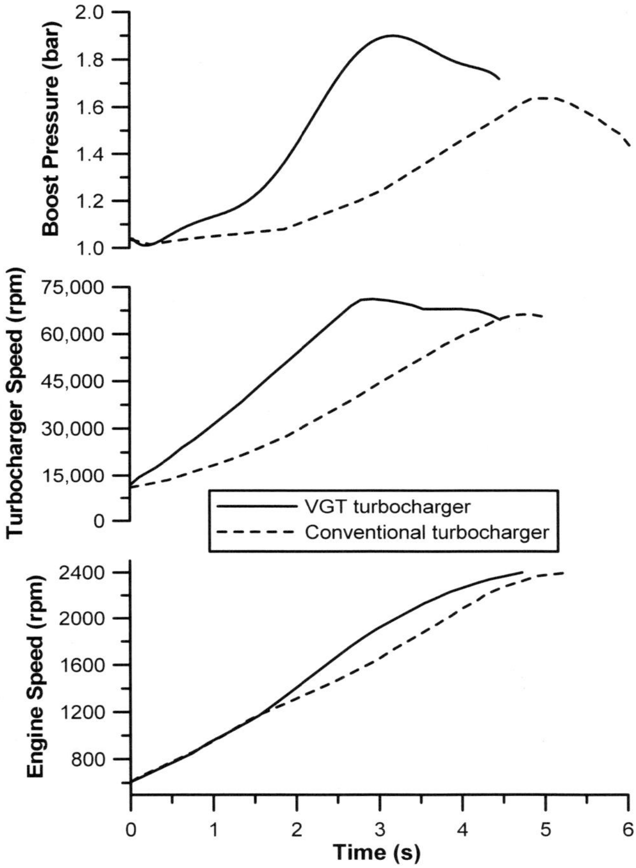

Figure 6.25. Effect of VGT on engine speed development during second gear acceleration from 600 rpm (six-cylinder, turbocharged diesel engine of 11 L displacement volume, installed on a 20 t truck – experimental results adapted from Matsura *et al.* [10])

The advantages of VGT engine transient response compared with its conventional turbocharged counterpart are illustrated and quantified in Figures 6.25 and 6.26 for two typical automotive transient tests, namely engine acceleration and performance over a Transient Cycle, respectively. Figure 6.25 focuses on a low-load acceleration of a 20 t vehicle powered by a four-stroke, six-cylinder diesel engine. For the VGT engine a speed- and load-dependent VGT control strategy is adopted as follows: in the extremely low-load and low-speed range, the VGT vanes are scheduled to minimize nozzle area in order to improve response by maintaining adequately high boost pressure. In the middle-load range, the vanes maximize nozzle area in order to improve fuel consumption by reducing the pumping loss caused by the positive difference between exhaust- and inlet manifold pressure. For the high-load operating range, nozzle vanes occupy three positions (minimum, middle and maximum) according to the increase in engine speed.

Closing down of the VGT vanes right after the onset of the transient event results in much faster turbocharger acceleration and boost pressure build-up compared with the conventionally turbocharged engine, hence speed recovery is faster. For example, time up to the intermediate speed of 2100 rpm is reduced by 0.6 s (or 15%) for the VGT compared with the conventional vehicle. As soon as the boost pressure has assumed an adequate value, VGT vanes gradually open to avoid over-boosting, that is why a reduction in the delivery pressure at approximately t = 3 s is observed.

Figure 6.26 expands on the previous results by highlighting a much more difficult and demanding case in terms of engine and VGT response during the US Highway Driving Cycle (*cf.* Figure A.5 in Appendix A). For the results given in this figure, an emissions oriented VGT control strategy was adopted, *i.e.,* exhaust manifold pressure being always 0.5 bar higher than its inlet manifold counterpart, in order to maintain flow through the EGR. A P-I controller was implemented to adjust the blade settings based on the target pressure difference. The intention was to examine the repercussions of a control strategy that utilizes an adjustable EGR valve, and prefers the authority of the EGR valve over residual flow-rates to that of the VGT nozzle setting. By so doing, the emphasis is on the assessment of VGT to provide desirable conditions in the intake air/exhaust sub-systems during transients and on engine/turbocharger matching and response issues [14].

Sudden and very rapid changes of engine speed and fueling cause dramatic fluctuations of exhaust manifold pressure, thus triggering equally dynamic changes of VGT settings as the P-I controller struggles to provide sufficient correction. Closing down of the VGT vanes during the accelerations of the Transient Cycle resulted in significant boost pressure increase, still from the onset of the test. Overall, the VGT blade position (not shown) varied between approximately zero (almost fully closed) and 0.4, *i.e.,* the flow through the turbine was restricted during much of the Cycle in order to provide the required back-pressure for EGR flow. As a result, the fuel–air equivalence ratio was much lower, minimizing fuel limiting function and aiding faster engine torque response. The observed reduction of fuel–air ratio illustrates the potential of the VGT to minimize the side effects of turbocharger lag in terms of exhaust gaseous and particulate emissions described in Chapter 5. Except for the near idle operation at the beginning of the Cycle, the VGT controller was able to provide the target pressure difference across the engine throughout the transient test. However, the rather aggressive emissions oriented strategy caused, on the one hand, the turbocharger to operate at much higher speeds throughout the whole Cycle, and, on the other hand, the compressor operating points moved closer to the surge line, a fact that underlines the importance of a careful full-range matching procedure. A more conservative pressure difference between exhaust and inlet manifold would have to be adopted, particularly at high engine speeds in order to compensate for the above inconvenience. This is actually the case illustrated on the upper sub-diagram of Figure 6.26; an alternative, performance oriented, VGT strategy for high-load conditions (VGT blade position adjusted so as to achieve desired boost pressure depending on engine speed) is employed here. This managed to attain an almost constant turbocharger speed throughout the whole Transient Cycle at the expense of lower boost pressure and higher fuel–air ratios (not shown in Figure 6.26), compared with the emissions oriented strategy. Consequently, the variations of the pressure in the intake manifold with engine load were kept at very low level too.

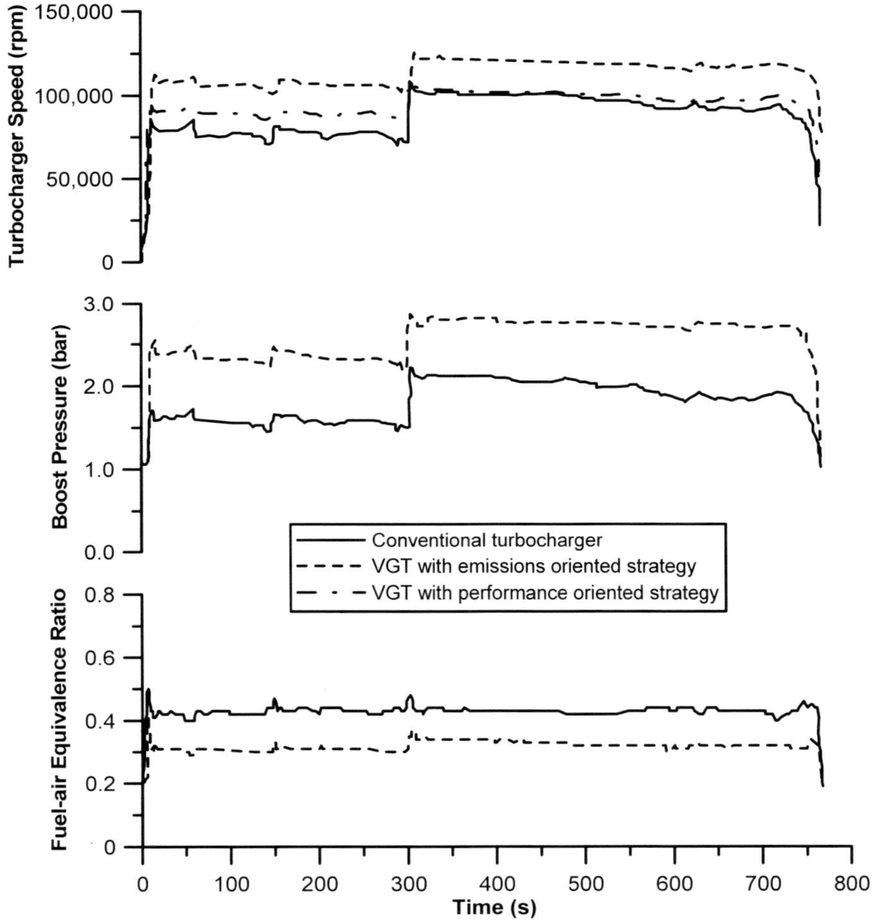

Figure 6.26. Effect of VGT strategy on engine response over the US Highway Driving Cycle (V-8, turbocharged and aftercooled diesel engine of 7.3 L displacement volume, rated at 210 PS at 2400 rpm – simulation results reprinted with permission from SAE Paper No. 2001-01-1247 [14], © 2001 SAE International)

6.4.5 Electrically Assisted Turbocharging

The ability of a (fixed geometry) turbocharger turbine to extract power from the exhaust gases depends on the latter's mass-flow and temperature. Consequently, at low engine speeds and loads the energy potential of the turbine is limited, producing low boost pressure and delaying engine torque response. A possible solution would be an electrically assisted turbocharging configuration; this makes use of an asynchronous (induction) electric motor or a synchronous type (*e.g.*, reluctance motor or permanent magnet) mounted on the turbocharger shaft (Figure 6.27 upper sub-diagram).

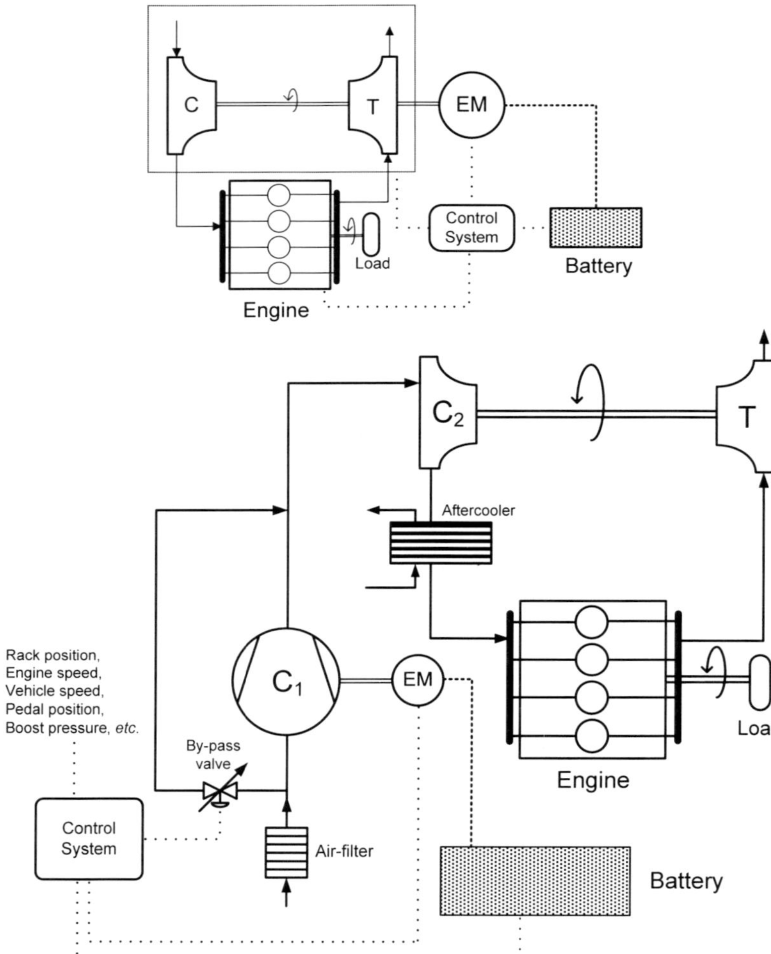

Figure 6.27. *Upper:* Schematic arrangement and simplified control diagram of diesel engine with turbocharger electric motor (EM) assist. *Lower:* Alternative combined supercharging configuration with electrically driven compressor

The electric motor is called to offer additional power to the compressor when a transient or starting event is detected, with its primary application being in automotive engines. The power to drive the electric motor is typically drawn from the battery of the vehicle, a fact that is expected to affect efficiency, particularly at low engine speeds. A comprehensive control algorithm determining the onset, magnitude and duration of electrical power assist must be also integrated in the engine control unit, taking into account various operating parameters of the engine, battery and turbocharger. Cost of the system is also a parameter that has to be taken seriously under consideration. An alternative configuration makes use of an electric motor driving a separate compressor (Figure 6.28); the latter is installed upstream or downstream of its turbocharger counterpart in an arrangement that resembles a combined supercharging scheme (Figure 6.27 lower sub-diagram). The analysis that follows will focus on the electrically assisted turbocharging demonstrated in the upper sub-diagram of Figure 6.27. A comprehensive analysis of a combined supercharging configuration with electrically driven compressor can be found in [15], whereas in [16] a comparison between the two electrical assisting systems is performed.

Figure 6.28. Generation 2 VTES electric supercharger (courtesy of Controlled Power Technologies Ltd)

Equation 6.2 described the normal, unassisted turbocharger shaft torque equilibrium; during the period when the electrical turbocharger assistance is provided, the respective balance is given by

$$\eta_{mTC}\tau_T + \tau_{EM} - |\tau_C| = G_{TC+EM}\frac{d\omega_{TC}}{dt} \tag{6.5}$$

with τ_{EM} the electric motor torque (the respective maximum power is of the order of a few kW for a typical automotive diesel engine of 2 L displacement volume) and G_{TC+EM} the combined turbocharger and EM mass moment of inertia. For a successful integration of the electric motor, the established turbocharger shaft acceleration must fulfill the following relation

$$\frac{\eta_{mTC}\tau_T + \tau_{EM} - |\tau_C|}{G_{TC+EM}} > \frac{\eta_{mTC}\tau_T - |\tau_C|}{G_{TC}} \tag{6.6}$$

Zellbeck *et al.* [17] studied vehicle performance with an electrically assisted turbocharger in comparison to its ‘nominal’ counterpart fitted with a waste-gate valve. The former was able to improve vehicle full-load acceleration from 40–80 km/h in sixth gear up to 20% compared with the nominal case. The results are depicted in Figure 6.29, showing mean effective pressure, boost pressure and vehicle speed for a four-stroke, four-cylinder diesel engine.

Notice the rather slow build-up of vehicle speed owing to the high gear engaged, which results in a high total moment of inertia (discussed in Section 3.5.1). Following Equation 6.5, the electric motor's assist is best highlighted in the initial abrupt increase of boost pressure right after the transient event commences ('circled' in Figure 6.29), with obvious benefits in terms of smoke emissions. In fact, the brake mean effective pressure of the electrically assisted turbocharged engine managed to achieve steady-state values a few seconds after initiation of acceleration, much sooner compared with the 'normal' engine operation. After the initial sharp increase in the boost pressure, however, the air-supply build-up of the electrically turbocharged engine proceeded at a similar rate compared with its conventional counterpart, since the electrical assistance was discontinued.

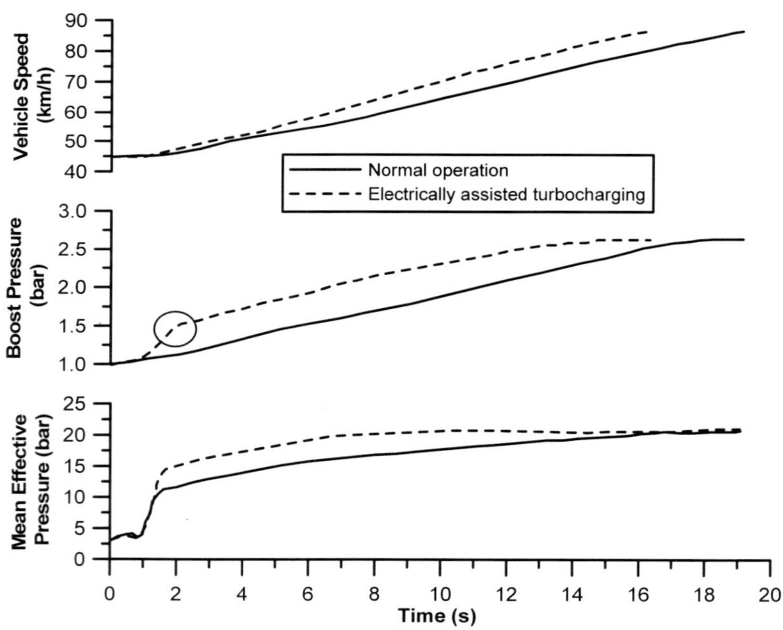

Figure 6.29. Acceleration performance of a diesel-engined vehicle with electrically assisted turbocharging (simulation results reprinted with permission from Zellbeck *et al.* [17])

Further results, this time for a severe load acceptance case are illustrated in Figure 6.30. The baseline engine as well as the electrically assisted engine with a rather small asynchronous electric motor (0.2 Nm maximum EM torque at 40,000 rpm and 6×10^{-5} kgm^2 inertia) could not accept the new load, leading to stall; successful transient performance was achieved when applying an electric motor-generator mounted on the turbocharger shaft. The latter was characterized by 0.57 Nm maximum torque at 20,000 rpm and 2.87×10^{-5} kgm^2 inertia, consuming approximately 1 Ah for a 30 s transient assistance, and producing an angular acceleration of the order of 19,860 r/s^2 compared with 3,333 r/s^2 of its EM counterpart; this resulted in better turbocharger assistance according to Equation 6.6, hence air-supply and engine torque were built-up effectively

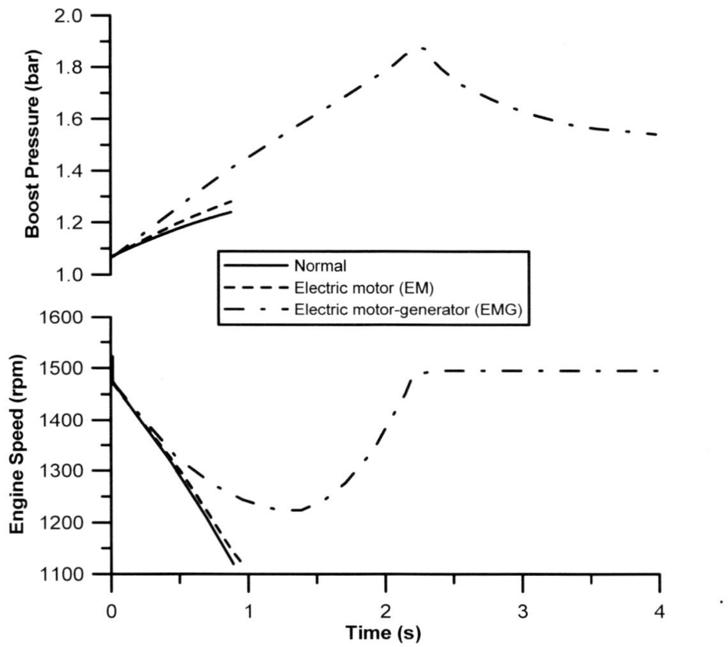

Figure 6.30. Effect of turbocharger electrical assistance on engine transient response after 0–97 kW load increase (six-cylinder, turbocharged and aftercooled diesel engine, rated at 162 kW at 2400 rpm – simulation results reprinted from Katrašnik *et al.* [18], copyright (2005), with permission from Elsevier)

A particular aspect in electrically assisted turbocharging units that needs attention is the fact that incorporation of an electric motor on the turbocharger shaft might require a renewed matching for optimum performance and surge avoidance, and a re-assessed balancing of the shaft owing to the instability induced by the increased length and the introduction of electromagnetic forces. Cooling of the EM is also an important aspect of its development. Moreover, a higher capacity battery may be needed in vehicular applications in order to assist the electric motor's frequent function in the continuously changing driving conditions. There are, however, two additional benefits from an electrically assisted turbocharging configuration that are worth mentioning.

1. The ability to operate the engine at *overall* higher boost pressure levels, hence torque, even at steady-state conditions with obvious advantages in terms of the fuel-limited low-speed region, while maintaining the same turbocharger size. In this case, however, since the electric motor will be called upon to operate more

frequently, a higher capacity battery is required (or better still a motor-generator configuration – see point 2. below) as well as care to avoid surge, since the operating range of the compressor is widened at the critical low-speed regime.

2. Integration of an electric motor-generator system. At low engine speeds, when the turbocharger needs supplementary power to overcome the (steady-state) torque limitations originating in the smoke-limited operation, as well as during transients, electrical power is supplied to the turbocharger shaft aiding its fast acceleration. At high engine speeds, the available exhaust gas energy will more than offset the required compressor power; instead of rejecting part of the exhaust gas through a waste-gate valve, work recovery can be established in the electric motor-generator that may subsequently be used in other electrical appliances of the engine powerplant, or even in a turbo-compound configuration. As a result, a motor-generator system is able to address three important aspects of engine driveability, namely, high torque at low engine speed, fast transient response, as well as good fuel economy, of course at the expense of a much more complex control system and higher cost. However, in order for a motor-generator turbocharging scheme to be beneficial, the driving cycle of the vehicle must favor regeneration periods that produce excess of energy to be stored in the batteries. The latter can be achieved, for example, during highway driving, whereas exclusive urban driving schedules are unlikely to produce any noteworthy benefit; such a configuration is described by Algrain [19] with the primary aim to improve the overall fuel efficiency of the engine. The system consists of a turbocharger with an electric generator integrated into the turbocharger shaft, and an electric motor integrated into the engine crankshaft. The generator extracts surplus power from the turbocharger turbine, and the electricity it produces is used to run the motor mounted on the engine crankshaft, recovering otherwise wasted energy in exhaust gases. Three control strategies were investigated, namely control of the engine exhaust temperature, control of the turbocharger speed or control of the boost pressure, which was found superior. An improvement in bsfc of the order of 5–10% was claimed without, however, any remarks on engine driveability and exhaust emissions during transients. Engine driveability data for both acceleration and Transient Cycle performance were provided by Millo *et al.* [20], who simulated the operation of a motor-generator electrically assisted turbocharging system (incorporating an asynchronous EM, DC-DC converter, super capacitors and control system) for a heavy-duty diesel engine. The electrically assisted turbocharged vehicle was capable of reducing the time up to the desired boost pressure and turbocharger speed by 25% compared with the VGT engine for a typical acceleration event.

One case in point: since the introduction of 42 V electrical systems in vehicular engines has been delayed, particular requirements are imposed on the classic 12 V system to support the above described motor-generator role.

An alternative turbine assist method has been discussed by Winterbone *et al.* [5], who studied the injection of high-pressure oil onto a Pelton wheel fitted to the turbine rotor. This method, too, resulted in a decrease of the required torque from the turbine to drive the compressor but necessitated a lot of modifications on the existing engine setup. The results, although considered as successful, were not sufficiently promising to justify mass application.

6.4.6 Sequential Turbocharging

The sequential turbocharging technique, *i.e.,* sequential use of two or more turbochargers installed in *parallel* (in contrast to the *series* arrangement of two-stage turbocharging units), is mostly met in marine engines (Figures 1.1 and 6.31 left) in order to improve part load and transient performance of highly-rated units.

The primary objective of sequential parallel turbocharging is to widen the flow range for a specific boost pressure, a fact that can prove quite beneficial for automotive applications too (Figure 6.31 right), which typically operate at wide air-mass flow range. Exactly here lies its main advantage compared with two-stage units (Section 6.4.3), which are characterized by narrower flow range since the entire air-supply passes through both compressors.

Figure 6.31. *Left:* Sequential turbocharging arrangement of a marine diesel engine (notice the large volume exhaust gas receiver, typical in constant pressure turbocharging), incorporating a large turbocharger (right), a smaller unit (middle) operating at medium-high loads and three control valves (courtesy of Mitsubishi Heavy Industries). *Right:* Sequential turbocharger for automotive use (courtesy of PSA Peugeot Citroën, Direction de la Communication – S. Meyer)

Figure 6.32 illustrates the schematic arrangement of the modern, automotive sequential turbocharging unit pictured in Figure 6.31. At part load, only the first turbocharger is active and the control valves are closed (during this period, control valve 2, located upstream or downstream of turbine T_2, is in fact slightly open, and so is the recirculation valve for the second turbocharger to work at a very low operating point in order to avoid thermal stress when this is suddenly engaged). The first turbocharger runs throughout the whole engine operating range, but is smaller and lighter than a single-stage unit; this permits faster acceleration at low and medium loads owing to the reduced inertia. The second unit, which is applicable only in the upper load range, may be even lighter. At high loads, the control valves 1 and 2 are opened and the second turbocharger is fully engaged; the engine practically operates in a regulated 'bi-turbo' configuration, adjusting the effective turbine total nozzle area, while at the same time reducing the total moment of inertia of the active units compared with a single-stage configuration.

Owing to surge risk, it may not be recommended to operate both turbochargers at low-speed/high-load conditions; likewise, use of only the first turbocharger is not recommended at high speed/high loads due to risk of over-speeding.

During the transition from one to two turbochargers, exhaust gas is progressively directed to the second turbine through control valve 2 until compressor C_2 has developed adequate boost pressure. At this point, control valve 1 opens and compressor C_2 is fully active, while the recirculation valve is closed. It should be pointed out that the control valves are not opened simultaneously in order to avoid compressor surge, since during the transition period, boost pressure and air-mass flow-rate for compressor C_1 slightly reduce. For the opposite case, *i.e.*, transition from two to one turbocharger, the second unit's flow is progressively minimized and compressor C_2 is disconnected through gradual closing of control valve 1. A comprehensive control algorithm is also required in order to take full advantage of the system's capabilities (of particular importance here is the accurate activation of the control valves), as well as a complex matching process, which is further complicated by incorporation of aftercooler, exhaust gas recirculation and even variable geometry turbines [21–23]. Similar schedule is followed for marine engines incorporating three or even four turbocharger units.

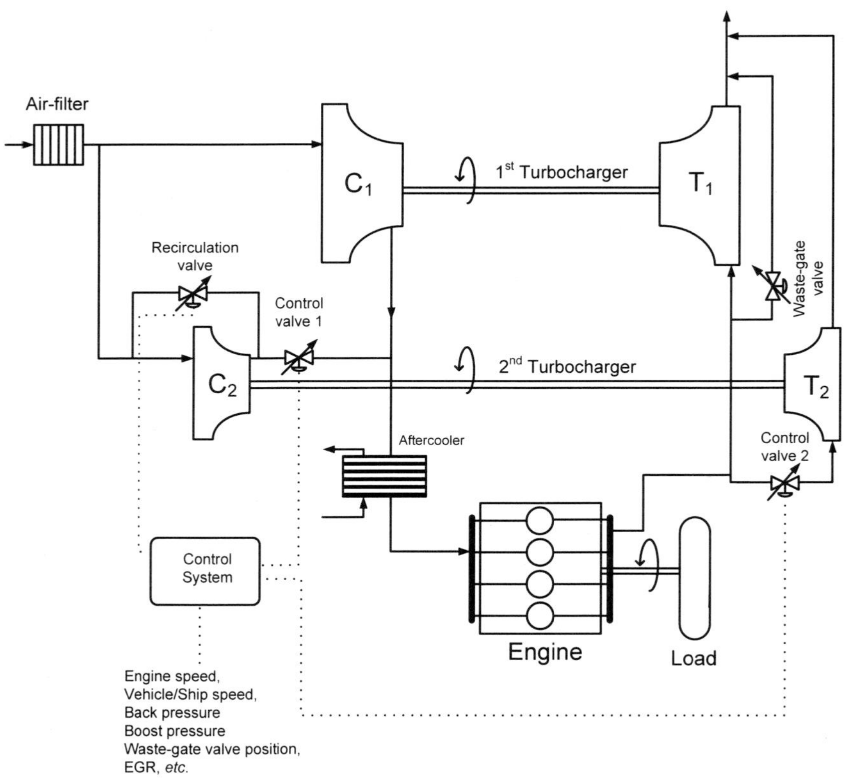

Figure 6.32. Schematic arrangement and simplified control diagram of sequentially turbocharged automotive diesel engine with two turbochargers

The potential of sequentially turbocharging in terms of improved transient operation is demonstrated in Figure 6.33. With appropriate control, the 2.2 L sequentially turbocharged engine bettered the response of a similar displacement volume VGT engine and in fact achieved the same torque response as a 2.7 L engine equipped with VGT; further results revealed that acceleration from 30 to 60 km/h was improved by almost 20% as regards recovery period.

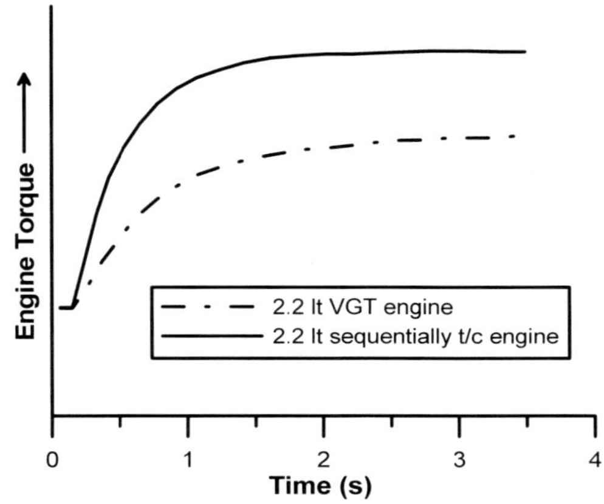

Figure 6.33. Comparison of load transient at 1500 rpm for VGT and sequentially turbocharged engine (16 valve, four-cylinder, 2.2 L displacement volume, European passenger car diesel engine – experimental results adapted from Galindo *et al.* [23])

Summarizing, the sequential turbocharging concept is actually based on the variable geometry mechanism, *i.e.,* adjustment of turbine total nozzle area according to the operating conditions for maximization of turbine expansion work. This is accomplished by using more than one turbocharger unit, but it has the added advantage of reduced inertia, hence faster acceleration of the unit operating at the low-load region compared with the VGT's single-stage; on the other hand, the system is realized only at the expense of a complex configuration in terms of matching, space requirements, control system and cost.

6.5 Engine Configuration

It has been documented in the previous chapters that accelerating the turbocharger from a very low-speed operating point poses significant difficulties and worsens turbocharger lag and smoke emissions owing to the unfavorable flow characteristics of aerodynamic type compressors at low rotational speeds. One of the most obvious ways of addressing this discrepancy is by trying to maintain the compressor operating point at higher levels, by 'somehow' increasing the amount of available energy to the turbine. This increase can be realized, among other measures that have already been discussed – such as the VGT or the combined supercharging – also from the engine point of view, through appropriate fueling or valve timing control. An alternative engine-based approach for transient response improvement involves external (electric) power supply in the form of a hybrid-electric powertrain; this will be discussed at the end of this section.

6.5.1 Fuel Injection Control

The most popular fueling control applied during both steady-state and transient turbocharged diesel engine operation is the boost-sensitive fuel limiter that was discussed in Section 2.2.2. It was further demonstrated in Figure 5.7 that (fueling) control of smoke emissions during acceleration or load increase is combined with poorer driveability and slower speed response, hence a compromise needs to be made for optimum performance. In this section we will focus on other fueling strategies that are based on injection timing control.

Injection timing influences engine performance and emissions in many ways; it affects the ignition delay, hence heat release profile, by determining the mass burned in premixed or diffusion combustion, peak cylinder pressure *etc.* Advancing the injection timing increases the ignition delay as the fuel is injected in a less favorable air environment. This results in a higher percentage of fuel burned in premixed combustion, producing higher peak cylinder pressures and combustion noise. In contrast, retarding injection has, in general, the opposite effects with the lower level of pressures and temperatures reducing the amount of NO_x emissions but adversely affecting engine startability, smoke emissions and power output. Deviation from the optimum timing in either direction affects negatively engine efficiency. The choice of the optimum injection timing as a function of engine speed and load is a difficult and laborious task that is typically established by the lowest bsfc; it is usually determined for steady-state operation and expressed in the form of a 3-D map. The need for good transient response, however, may impose different (conflicting) requirements and lead to differentiated timing values compared with steady-state operation. In recent years, the use of electronically controlled injection systems, particularly for automotive engines, has given the opportunity to apply various injection control strategies according to the engine's operating conditions.

An intriguing fueling control method from the transient point of view, consists in the adoption of a retarded injection strategy (beyond TDC), mainly at low-speed/low-load conditions, where the turbocharger normally produces very low boost pressure and operates at a low rotational speed. By so doing, heat release in the cylinder is altered as combustion is forwarded later in the expansion stroke reducing peak cylinder pressures. The obvious advantage here is the higher exhaust gas pressure and temperature at the point of exhaust valve opening, which is transformed into higher exhaust manifold pressure, and therefore higher available turbine energy favoring higher turbocharger rotational speeds. Clearly, the retarded injection timing alters the balance of fuel energy conversion with a significant part transferred from the piston to the turbine. However, important issues need to be addressed, namely fuel mixture flammability, smoke and HC emissions as well as the reduced piston work, hence engine efficiency, which necessitates a comprehensive control system and rather limits application of this technique to vehicular engines that experience frequent transient events.

Figure 6.34 is based on the work of Watson [24] reported in Section 2.1 concerning the comparative performance of the engines of Table 2.1 and Figure 2.11. In this case, the static injection timing for the turbocharged and aftercooled, six-cylinder engine is retarded to 25°CA after TDC at low-speeds and loads (aftercooler needs to be by-passed in this case in order to achieve sufficiently high gas temperatures for combustion initiation). During acceleration transients, injection timing is advanced to the normal 15°CA before TDC in order to maximize engine power. The advantages over the nominal engine operation are noteworthy. In fact, acceleration time up to 2300 rpm is reduced by 40%, from 4.1 to 2.5 s, since the turbocharger accelerated from a substantially higher operating condition (60,000 rpm and 2.1 bar compared with 10,000 rpm and 1.0 bar of the nominal un-retarded operation). This significantly minimized the turbocharger lag; however, the naturally aspirated engine's performance still could not be outperformed.

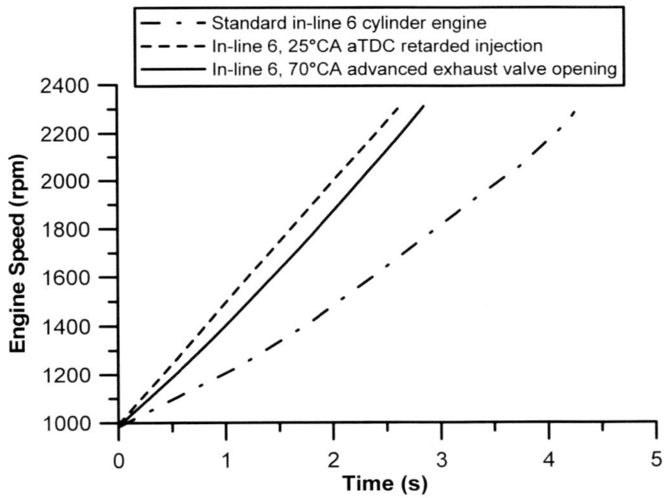

Figure 6.34. Effect of retarded injection timing and early opening of exhaust valve on engine acceleration (engine data available in Table 2.1 – simulation results adapted from Watson [24])

An alternative (extreme) case of fueling strategy has been the Hyperbar system [25]. Following the same general context as the retarded injection method, it consists of fuel injection and subsequent combustion in the exhaust manifold. Consequently, the exhaust gas enthalpy prior to the turbine is maintained at high levels even at no-load conditions, favoring turbine acceleration. This method, has not proven, however, very attractive owing to increased cost and complexity, reduced reliability and poor low-load performance.

Arcoumanis *et al.* [26] and Bazari [27] investigated additional injection strategies that focused on improved engine indicated efficiency or reduced emissions during transients. It was revealed that although the engine response was indeed affected by injection timing control, the gain was not that significant (particularly at low loads) to justify the use of electronic control for the sake of improved driveability only. At elevated loading, there is a higher gain potential probably owing to the fact that the increased fueling results in more fuel burning later in the cycle, hence a higher share of energy goes to the exhaust system. It was proposed that it may be better to optimize injection timing for lowest exhaust emissions rather than just speed response, a fact that has gained significant attention in recent years.

6.5.2 Valve Configuration

In general, there are two possible ways to improve transient performance through valve configuration, namely by manipulating

- valve timing, *e.g.,* advanced exhaust valve opening (EVO) or retarded intake valve opening (IVO) or better still, variable valve timing; and
- valve or cam profile.

Valve timing plays an important role from the engine configuration point of view, as it determines the pressure and temperature of the exhaust gas leaving the cylinder. The higher this level the higher the available energy at the turbine inlet, reducing accordingly the turbocharger lag and aiding faster engine speed response and lower smoke emissions. Historically, early opening of the exhaust valve was proposed as a viable means for improving transient performance, however at the expense of lower engine efficiency (since only part of the potential expansion work is actually obtained) and increases of engine-out soot emissions (since the soot oxidation process inside the cylinder is incomplete) (Figure 6.35).[4] By advancing the opening of the exhaust valve, the effective expansion stroke is reduced and the balance of fuel energy conversion is altered with a significant part transferred from the piston to the turbine.

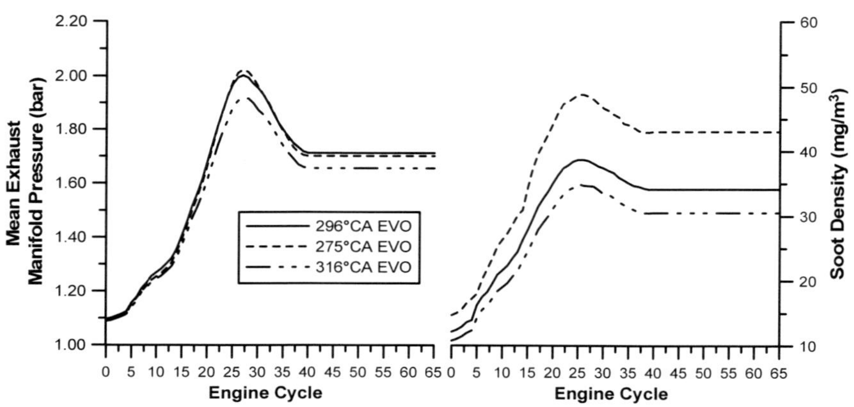

Figure 6.35. Effect of EVO timing on (mean) exhaust manifold pressure and soot emissions after a load increase transient event [28]

[4] The high exhaust temperature that is experienced through advanced exhaust valve opening can, however, prove beneficial for diesel particulate trap regeneration even at low loads, that may ultimately reduce vehicle-out soot emissions.

In the context of retarded fuel injection timing, the early opening of the exhaust valve is realized at low engine speeds and loads in order to boost the turbocharger operating point. During the transient event, valve timing would revert to normal, with the engine benefiting from the higher initial compressor boost pressure. The research of Watson [24] on the same engines as the ones discussed in Section 6.5.1, showed that advancing the complete exhaust event of the six-cylinder engine for 70°CA at the 1000 rpm, no-load starting point of the transient event resulted in turbocharger speed raise from 10,000 to 43,000 rpm and boost pressure to 1.5 bar, which ultimately led to a 33% response improvement over the base-line operation of Figure 2.11. Even earlier openings would be required to match the retarded injection effects as is revealed in Figure 6.34. However, the added bonus here was the much lower fuel consumption penalty compared with the retarded injection timing method.

An additional unfavorable issue during turbocharger lag that is associated with the valve timing schedule, is the back-flow of exhaust gas into the cylinder and/or inlet manifold during the early transient cycles. In Section 2.1 (remarks concerning Figure 2.8), this was attributed to the much slower increase of boost pressure compared with its exhaust manifold counterpart, negatively affecting the engine response. The above phenomenon is enhanced the larger the valve overlap period, *i.e.,* with higher engine ratings (typically more than 100°CA valve overlap duration), and is also noticed during low-load, steady-state engine operation, where the boost pressure is very low. Modern automotive engines equipped with variable geometry turbines suffer from this phenomenon too, as was discussed in Section 6.4.4. The obvious reaction here is to reduce the valve overlap period during the transient event, *e.g.,* by delaying inlet valve opening until close to TDC; ideally, a variable valve timing schedule should be adapted. Marzouk and Watson [29] investigated theoretically this case; during the transient event, the inlet valve opening was delayed by 40°CA reducing accordingly the valve overlap period. This control strategy ultimately resulted in a considerable reduction in engine speed droop of the order of 35% and in the recovery period of 27% after a load increase. Since the amount of exhaust gas reversing back into the cylinder was reduced, the trapped fuel–air equivalence ratio was lower too, favoring smaller duration of fuel limiting action and faster engine torque build-up and turbocharger acceleration with the added bonus of lower smoke emissions.

Similar results were reached by Boy [6], as illustrated in Figure 6.36 for a marine engine acceleration. The nominal 110°CA overlap in Figure 6.36 is actually very close to the optimum. Increasing the valve overlap leads to serious back-flow of exhaust gas mainly during the early cycles of the transient event; a reduction of the overlap however to 70°CA ensures that no back-flow takes place, hence the air–fuel ratio builds-up faster and so does the torque and the speed response of the engine.

Apart from manipulating valve timing, an additional possibility for increasing the energy available at the turbine inlet would be the reduction of the exhaust gas energy loss during the exhaust stroke. Most of this energy is now wasted during the opening and closing down of the exhaust valves, where the gas flow reaches sonic velocity owing to the high-pressure drop across the valves and the small gas flow sections. Increasing the exhaust valve opening and closing speed, therefore, produces a high oscillatory flow level in the exhaust line, which proves beneficial for the engine torque (and speed) response.

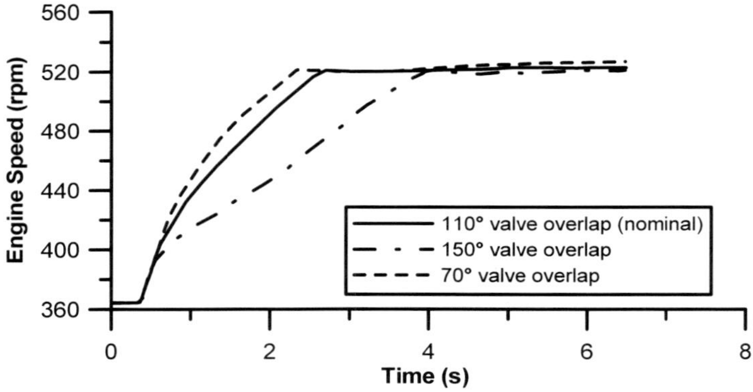

Figure 6.36. Effect of valve overlap on the acceleration performance of a medium speed, V-18, marine diesel engine (simulation results adapted from Boy [6])

Figure 6.37 expands on the previous arguments by studying the synergistic effect of valve timing and lift profile on the performance and smoke emissions response, during a load acceptance transient event at 1000 rpm. Three cases are examined, namely, fixed geometry twin-entry turbine (FGT) with EURO 3 emission limits against two single-entry, variable geometry turbine engines operating on the same VGT control schedule (desired boost pressure level as a function of speed and fueling) but with different cam lobe profiles; one of them with a short effective valve overlap (Cam No. 2) and the second with a longer overlap (Cam No. 1 – also applied on the engine with the fixed geometry turbine) [30]. In general, a smaller amount of back-flow is expected from the exhaust to the intake manifold with camshaft No. 2.

From the engine torque development illustrated in Figure 6.37 as well as the air-mass flow (now shown), it was concluded that with camshaft No. 1 the response is actually faster with the FGT compared with the same camshaft VGT engine. This behavior is attributed to the high positive value difference between exhaust pressure and intake pressure for the VGT engine owing to the closing down of the VGT vanes, leading to significant back-flow during the valve overlap period, which dramatically reduces the engine volumetric efficiency. On the other hand, boost pressure is always higher than exhaust pressure for the FGT configuration aiding engine response. Another relevant factor that influences this behavior is that the FGT has a twin-entry geometry, avoiding the interference between exhaust pulses from consecutive cylinders. In contrast, the VGT is a single entry unit and suffers from high exhaust interference of consecutive firing cylinders. Nonetheless, when camshaft No. 2 (less overlap) is combined with the VGT, the high exhaust back-pressure has no detrimental influence on the boost pressure, because with camshaft No. 2 there is no intake back-flow during the early transient cycles. As a consequence, a higher volumetric efficiency is established allowing higher fuel mass flow to be injected, without fear of excessive smoke emissions aiding faster engine response. In fact, the VGT engine with camshaft No. 2 managed to reduce the recovery period by almost 29% compared with the camshaft No. 1 operation. The previous remarks would not significantly change if a twin-entry VGT were used, since the main cause of volumetric efficiency reduction is the exhaust gas back-flow during valve overlap.

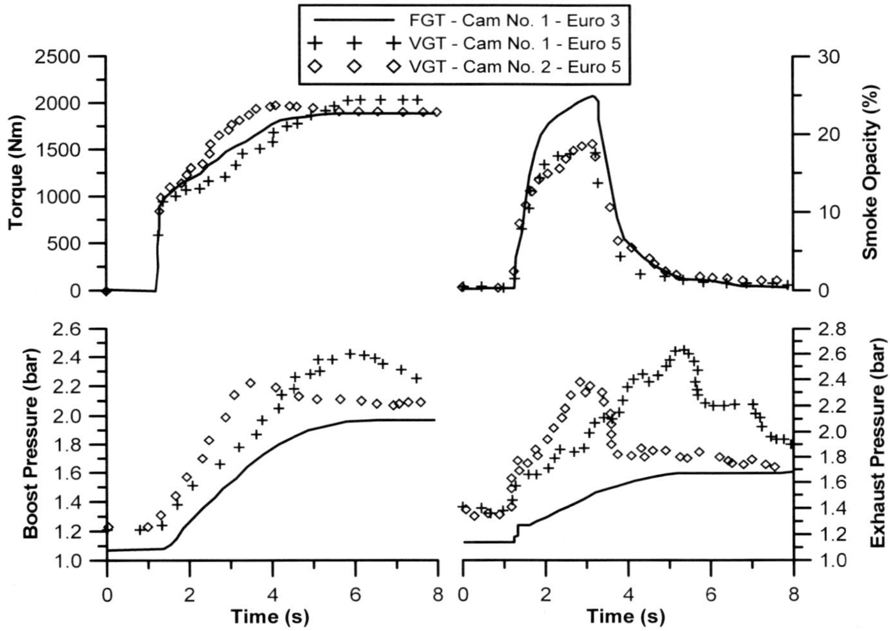

Figure 6.37. Comparison of transient evolution at 1000 rpm between fixed geometry turbine and two engines with VGT and different valve schedule (four-stroke, six cylinder, turbocharged diesel engine with 1400 bar common rail injection system, rated at 300 kW at 1800 rpm – experimental results from Galindo *et al.* [30], reprinted with permission from Inderscience Publishers)

6.5.3 Manifolds Configuration

The exhaust manifold volume and configuration can markedly affect the engine and turbocharger response after a ramp increase in load (or speed), leading even to unsteady and oscillatory operation if a very small manifold volume is chosen [31, 32]. A large manifold volume (as in constant pressure turbocharging (Figure 6.31)) increases the flow inertia of the system and slows down the response, since the amount of gas, which has to be 'accelerated', is much larger now (Figure 2.2). Smaller manifold volumes, on the other hand, can aid faster turbine response, since one of the system decelerators, *i.e.,* filling of the exhaust manifold volume with higher pressure gas, is accomplished much faster. For a specific engine, reducing manifold volume increases turbine power and compressor boost, but since the turbine area has been initially selected to provide a desired boost pressure at steady-state conditions, this will result in over-boosting the engine at high rotational speeds; hence, a turbine rematch might be needed. Further reduction of the exhaust manifold volume from the optimum brings no beneficial effects as regards response. Figure 6.38 demonstrates these remarks for a moderately turbocharged diesel engine operating on a 10–85% load increase transient event. The smaller the exhaust manifold volume, the higher the achieved exhaust manifold and boost pressure facilitating lower fueling and smaller speed droop.

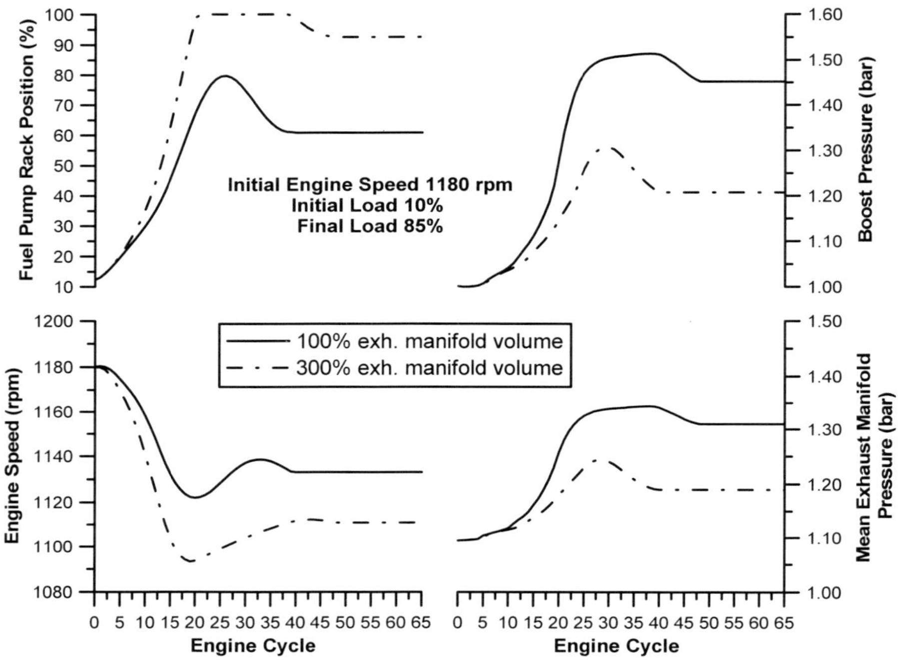

Figure 6.38. The effect of exhaust manifold volume on the engine transient response after a 10–85% load change (four-stroke, six-cylinder, turbocharged and aftercooled diesel engine)

Likewise, Figure 6.39 illustrates the impact of the exhaust manifold configuration for the same turbocharged engine operation. The cases examined are those of a single- and a twin-entry turbine; in the latter the exhaust manifold consists of two parts, the first connected to cylinders 1, 2 and 3 and the second connected to cylinders 4, 5 and 6 (ignition sequence of the six-cylinder engine under study is 1-5-3-6-2-4). As was anticipated, the twin-entry turbine case leads to faster boost pressure response due to the smaller volume of each one of the two 'exhaust manifolds', which is therefore filled sooner with the exhaust gas and accelerates the turbine faster. Moreover, there is no pressure wave interference between the cylinders, whereas the lower speed droop leads to slightly smaller full pump rack positions and fuel–air ratios.

Contrary to exhaust manifold volume, changes in inlet manifold volume do not result in significant engine response effects. Inlet manifold tuning, however, can under circumstances prove useful. Banisoleiman *et al.* [33] studied the transient improvement through inlet manifold tuning, using either ram pipe or Cser-

Helmholtz resonators. Appropriate grouping of the inlet manifold pipes could have quite promising results, although the positive aspects can not apply to all engine operating conditions and transient schedules.

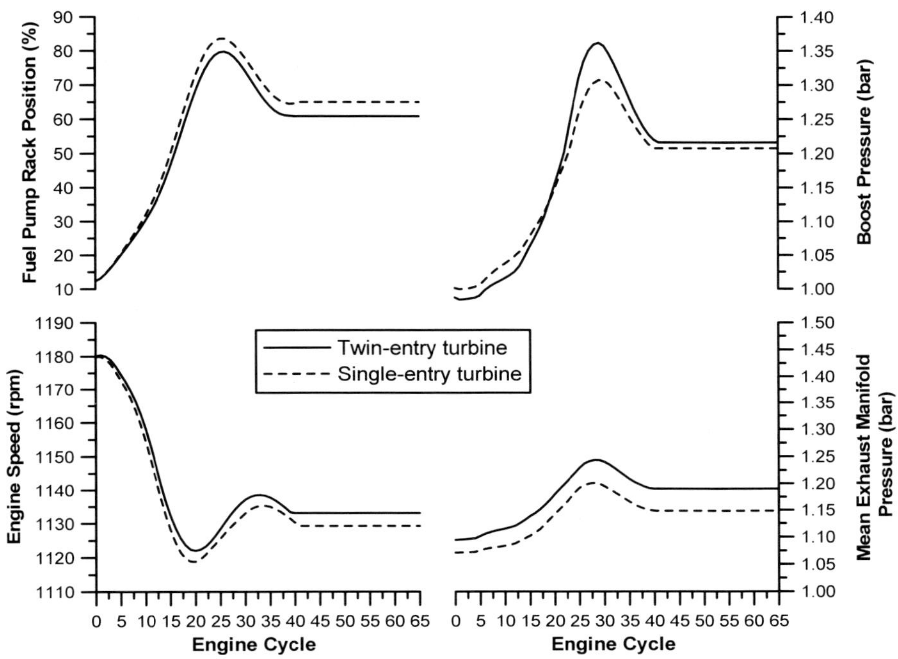

Figure 6.39. The effect of exhaust manifold configuration and turbine entries on the engine transient response after a step increase in engine load (four-stroke, six-cylinder, moderately turbocharged diesel engine)

6.5.4 Hybrid-electric Engine and Vehicle Operation

Global concern regarding decreasing oil supplies, environmental pollution and greenhouse gas effect has initiated the research/development of alternative vehicle powertrains. Such an alternative propulsion system is the hybrid-electric vehicle (HEV), which is widely considered as an interim stage in the path towards the zero emission vehicle (ZEV); the latter will be, most probably, in the form of fuel cell or purely battery propulsion. The conventional powertrains described in Section 3.5 can be combined with electric motors/generators/inverters/batteries to become part of a hybrid powertrain. This combination offers the potential to decouple the energy conversion process from the duty cycle of the vehicle. Consequently, the (diesel) engine can run at optimum efficiency for a higher proportion of its operation and, also, to be subject to less external disturbance resulting in significant fuel consumption reductions of at least 15–20% compared with a conventional vehicle. Two major challenges have to be addressed, namely, vehicle cost, which is typically 40–50% higher than its conventional counterpart, and system weight, integration and control. The various components in the powertrain must be carefully coordinated to maximize the benefits; advanced optimization is required both in system and control strategy design. The emergence of 42 V vehicle electrics makes the integration of electric motor/generators, turbochargers and engine ancillaries an intriguing proposition for future advanced powertrains [34–40].

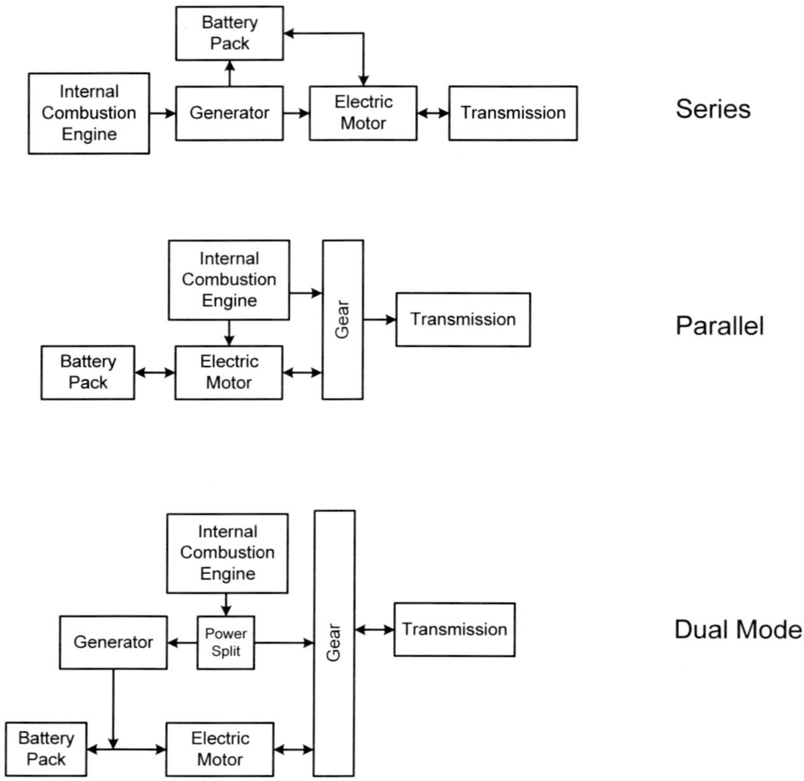

Figure 6.40. Schematic arrangement of the various hybrid powertrain configurations

Hybrid powertrains can be classified into three main types (see also Figure 6.40):

- series;
- parallel; and
- dual mode, which is actually a combination of series and parallel powertrain.

In a *series* hybrid powertrain the sole prime-mover is an electric motor, which receives electric power either by a set of batteries or by an internal combustion engine powered generator. The engine is typically smaller in a series drivetrain because it only has to meet average driving power demands; since it is not coupled directly to the transmission it runs at a specified speed/load operating envelope, where the efficiency is high, or it may be temporarily switched off. The battery pack is generally powerful in order to provide the remaining peak driving power needs, adding significantly to the weight and cost of the vehicle.

In a *parallel* hybrid configuration, both the engine and the electric motor generate the power for propulsion being permanently and independently coupled to the transmission. Since, in this setup, the engine is connected directly to the wheels, it eliminates the inefficiency of converting mechanical power to electricity, a fact making this kind of hybrid efficient for highway driving.

A *series-parallel* (dual mode) setup merges the advantages and complications of both parallel and series drivetrains. Here, the engine can both drive the wheels directly and be effectively disconnected from them so that only the electric motor powers the wheels. This system results in higher cost than a pure parallel hybrid because it needs a generator, a larger battery pack, and a more complex and sophisticated control system. However, the series-parallel drivetrain has the potential to perform better than either of the systems alone.

In general, the advantages of hybridization are the consequence of

- Smaller size (downsizing of) internal combustion engine (ICE);
- ICE temporary switch-off;
- ICE running at a specified speed/load operating envelope of relatively high efficiency; and
- Regenerative braking.

In series HEV configuration, the engine is practically ‘protected’ from transient operation, as it is not directly coupled to the wheels. Consequently, it need not follow the duty cycle of the vehicle; the control strategy being mainly determined based on the engine efficiency and exhaust emissions steady-state maps. From the diesel engine’s transient operation point of view, however, the most interesting configuration is the parallel hybrid, since in this set-up the diesel engine retains the prime mover role with the electric motor/battery packs assisting when needed. In Figure 6.41 the extension of the conventional powertrain system of Figure 3.30 is illustrated, configured now as a parallel HEV.

The coupling at the (fixed ratio) gear engages or disengages the electric motor depending on the operation mode of the hybrid. Hence, the diesel engine and/or electric motor are linked to the propeller shaft that drives the differential and (two) driveshafts. The EM charges the battery pack when working as a generator (*e.g.*, during regenerative braking, or when the batteries state of charge is low) and discharges the batteries while assisting the diesel engine in powering the vehicle (*e.g.*, during acceleration) when working as a motor. Overall, there are five driving modes possible in such a configuration, namely: the zero emission vehicle mode (engine shut off, primarily employed during city driving), the engine mode (as in a conventional vehicle – mainly during highway driving), the engine and EM hybrid mode, the EM generating hybrid mode (engine providing both the propulsion power and the power for charging the batteries) and the regeneration mode (typically during braking).

Figure 6.42 demonstrates two successful examples of hybrid-electric vehicles available in the Japanese and European markets. Four factors primarily influence the performance of a hybrid powertrain, *i.e.,* the scaling of the engine, the scaling of the electric motor, the number of battery packs, and the control strategy adopted in order to co-ordinate the various energy sources.

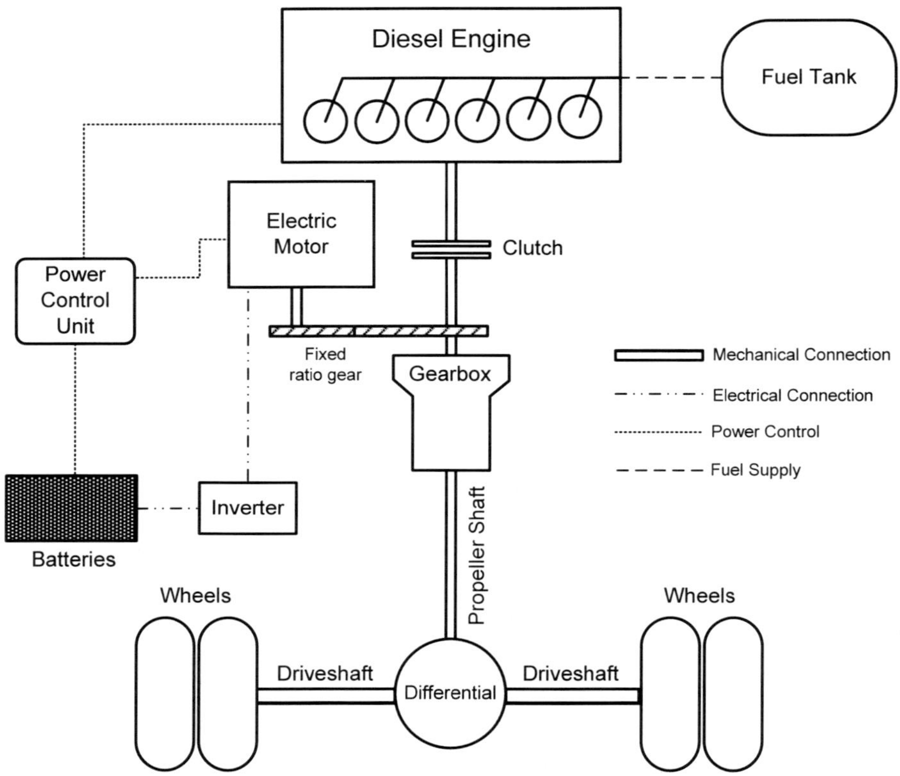

Figure 6.41. Schematic diagram of parallel HEV powertrain

As expected, increasing the power output of the powertrain components or the number of batteries also increases the vehicle weight, a fact that influences fuel economy. In order to get the best performance of a hybrid vehicle, the various energy sources have to be optimized (applying fuzzy logic, neural network techniques, genetic algorithms, *etc.*) according to the vehicle's primary usage. For example, to obtain best performance at continuous start-stop driving situations, an almost 50–50 split of power between diesel engine and electric motor is required; for highway driving this ratio would shift towards the engine, a situation termed 'mild' hybrid configuration (*i.e.,* hybridization ratio, defined as the ratio of electric power to total power, being lower than 30%). Particularly for parallel HEVs, a popular power distribution algorithm aims to keep diesel engine fuel consumption minimal by running the engine away from low-efficiency operating points, a methodology known as load leveling; however, a successful control strategy should always take into account the batteries' state of charge.

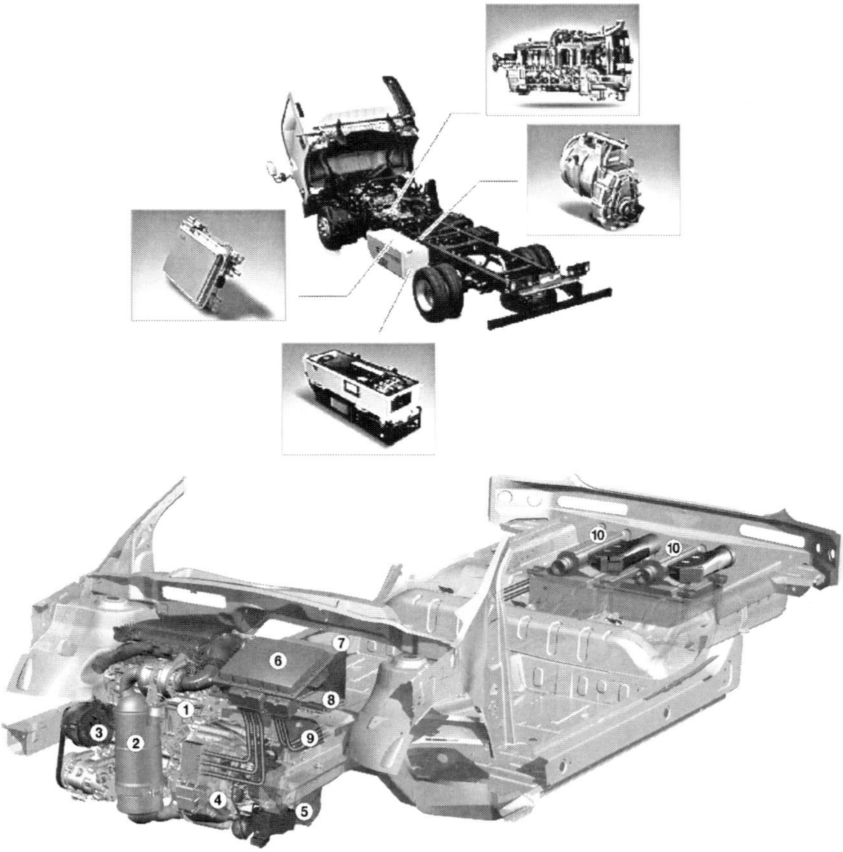

Figure 6.42. *Upper:* Parallel hybrid-electric powertrain installed on an Isuzu Elf truck: clockwise from top; transmission system, permanent magnet synchronous EM-generator rated at 25.5 kW at 1100 rpm, 5.5 Ah lithium-ion battery, inverter (courtesy of Isuzu Motors Ltd). *Lower:* Peugeot 307 'Hybride HDi' hybrid-electric passenger car powertrain: 1: diesel engine, 2: DPF, 3: Start and stop system, 4: EM (16 kW), 5: Automated manual transmission, 6: Inverter, 7: 12 V battery, 8: Control unit, 9: High-voltage cables, 10: 288 V NiMH battery (courtesy of PSA Peugeot Citroën)

An alternative, simpler, (mild) hybrid powertrain makes use of an integrated starter-generator system (ISG, Figure 6.43), which combines the automotive starter and alternator into a single machine, possibly operating on 42 V. An ISG can be mounted on the engine flywheel or between the engine and transmission or even externally to the engine and connected to the crankshaft via a drive-belt. Its primary aim is reduction of fuel consumption and exhaust emissions by turning off the vehicle engine at rest using a start-stop function and starting it again when the accelerator pedal is pressed (particularly useful for urban vehicular use); regenerative braking is usually facilitated too. From the transient operation point of view the advantage of such systems is assist of the crankshaft acceleration by supplying additional torque from the alternator, typically for a few seconds. The starter is typically a low speed, high current DC machine, while the alternator can be a variable speed, three-phase AC machine. System auxiliaries are, usually, an AC/DC converter to rectify the generator output voltage, a DC/DC converter to supply the vehicle's electrical power system and a comprehensive control system. In the most usual configuration, since the ISG is mounted on the engine flywheel, its speed is determined by the engine speed via a constant ratio; consequently, the additional power available from the ISG can be regulated by adjusting its torque. The latter can be positive or negative depending on the specific operating mode. In the motor mode, the ISG contributes power to the driveline by drawing electrical energy from the battery. In the generator mode, it absorbs power from the driveline and charges the battery.

Figure 6.43. Integrated starter-generator system (source: Mercedes-Benz)

One of the main advantages of HEV over conventional vehicle, apart from exhaust emissions and fuel consumption reduction, is that it offers scope for improving the engine transient response. Unlike all the previously discussed methods, however, where the improvement was located in the turbocharger operation, assisting faster boost pressure and air-supply build-up and restricting the fuel limiting action, HEV operation increases the crankshaft torque, through the electric motor contribution, whereas the turbocharger response and flow characteristics are not directly affected. Consequently, a HEV or an ISG configuration is not expected to increase part load air-supply as was the case with VGT or combined supercharging arrangements.

A typical comparison between conventional and parallel hybrid vehicle speed profiles, for 0–60 mph acceleration, is illustrated in Figure 6.44. The data concerns a 7,258 kg, 4×2 hybrid truck. This figure demonstrates in an explicit way the advantages of an HEV from the transient operation point of view. The hybrid achieved the desired speed of 60 mph slightly earlier than the conventional truck, primarily due to better performance immediately after launch. Recall that the torque production of a turbocharged engine is fuel limited, particularly, at low engine speeds; this results in significantly poorer acceleration. Unlike turbocharged diesel engines, electric motors can produce a significant amount of torque irrespective of engine speed (provided, of course, that the batteries are sufficiently charged). Thus, the HEV benefited from the electric motor's immediate torque contribution and successfully compensated for the slower response of the diesel engine due to turbocharger lag. A significant decrease in fuel consumption was also reported.

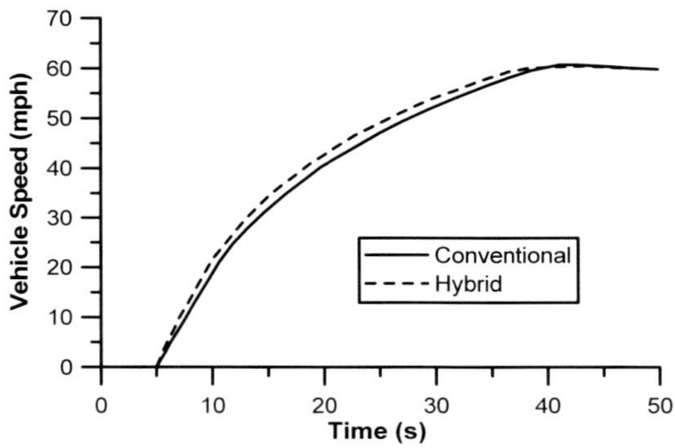

Figure 6.44. Speed profile comparison between conventional and hybrid truck during 0–60 mph acceleration (V-6, 5.5 L displacement volume, turbocharged diesel engine, rated at 157 PS at 2400 rpm, coupled to a 49 kW electric motor compared with a V-8, 7.3 L, turbocharged diesel engine rated at 210 PS at 2400 rpm – simulation results reprinted with permission from SAE Paper No. 2001-01-1334 [37], © 2001 SAE International)

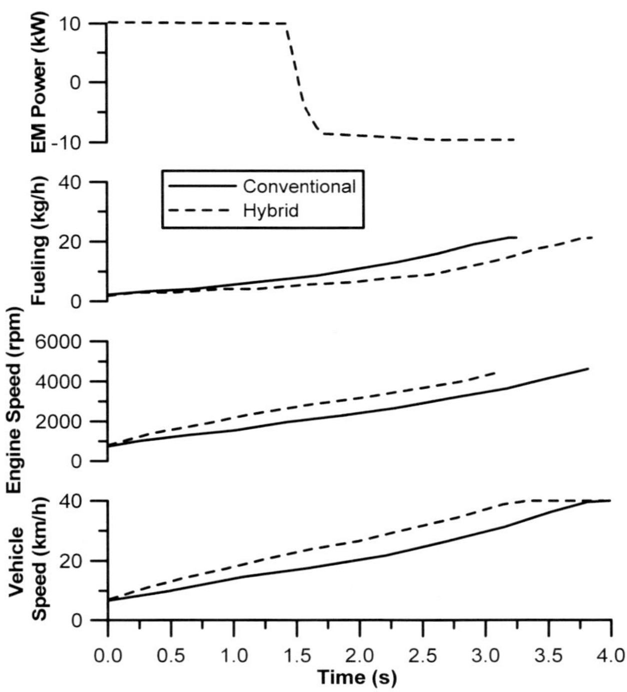

Figure 6.45. Comparison between conventional and parallel HEV acceleration in first gear (turbocharged diesel engine with 10.3 kW electric motor – simulation results reprinted with permission from SAE Paper No. 2000-01-0519 [40], © 2000 SAE International)

Similar results were obtained by Kolmanovsky and Stefanopoulou [40]. A comparison between conventional and hybrid vehicle taken from their study is reproduced in Figure 6.45. Initially, the engine is running at neutral idle conditions (800 rpm) and an acceleration at first gear is requested up to 40 km/h. The optimization procedure for the EM assist was based on minimization of acceleration time to a specified final velocity, for a fixed gear ratio. Again, the contribution of the EM during the first, critical, seconds of the transient event compensated for the turbocharged diesel engine's slow response and resulted in faster acceleration of the engine, turbocharger (not shown in Figure 6.45), and of the whole vehicle. At about t = 1.5 s, the EM assist was no longer needed, since the vehicle had successfully coped with the speed demand; consequently, the EM was gradually shut down and the vehicle kept running in conventional form. Fuel consumption during the acceleration event was found to be lower than with the conventional powertrain.

Obviously, the above-mentioned advantages of parallel HEV operation can be expanded in the case of performance during a Transient Cycle. Figure 6.46 illustrates the corresponding HEV speed profile during the US Federal Urban Driving Cycle (also known as FTP-72 – see also Section A.3.2.1) [37]. The vehicle launches from stop using only the electric motor to avoid inefficient engine operation under low power demands. However, the diesel engine is 'turned on' very quickly, since the power demand requires the output from both the engine and the electric motor. From 450 to 477 s, the power required to cruise at the speed of 35 mph is less than the 'engine on' power level, hence the engine is disengaged and the electric motor supplies all the torque required at the wheels (ZEV operation). When the truck decelerates, the regenerative braking is employed; hence, the motor operates as a generator to recover the energy that would otherwise be dissipated in brakes (477 to 490 s, and 530 to 537 s interval). It should be noted that when the battery state of charge (SOC) hits the lower bound (55%), at 523 s into the schedule (upper sub-diagram of Figure 6.46), the engine is immediately turned on to power the truck, as well as to recharge the battery. Hence, the electric motor is switched to the generator mode and its torque becomes negative as it charges the batteries.

Two control algorithms were investigated; the first, termed rule based algorithm, applies a simple analysis of component efficiency maps and the second is a dynamic programming based algorithm, which aims at minimizing fuel consumption rate by exploring the efficiency of the whole HEV powertrain and not just of the diesel engine. A more advanced version of the latter control strategy would have included exhaust emissions minimization too. A significant reduction in fuel consumption compared with the conventional truck during the Transient Cycle was observed, ranging from 22% for the rule based algorithm up to 33% for the dynamic programming one. For the former, the improvement originated, mainly, from regenerative braking, while for the latter, a better co-ordination of engine–transmission–electric motor-battery setup was the key for even higher fuel economy. Similar gains in fuel consumption have been reported for spark ignition engined hybrid-electric vehicles, highlighting the promising aspects of hybridization in the daily vehicle operation.

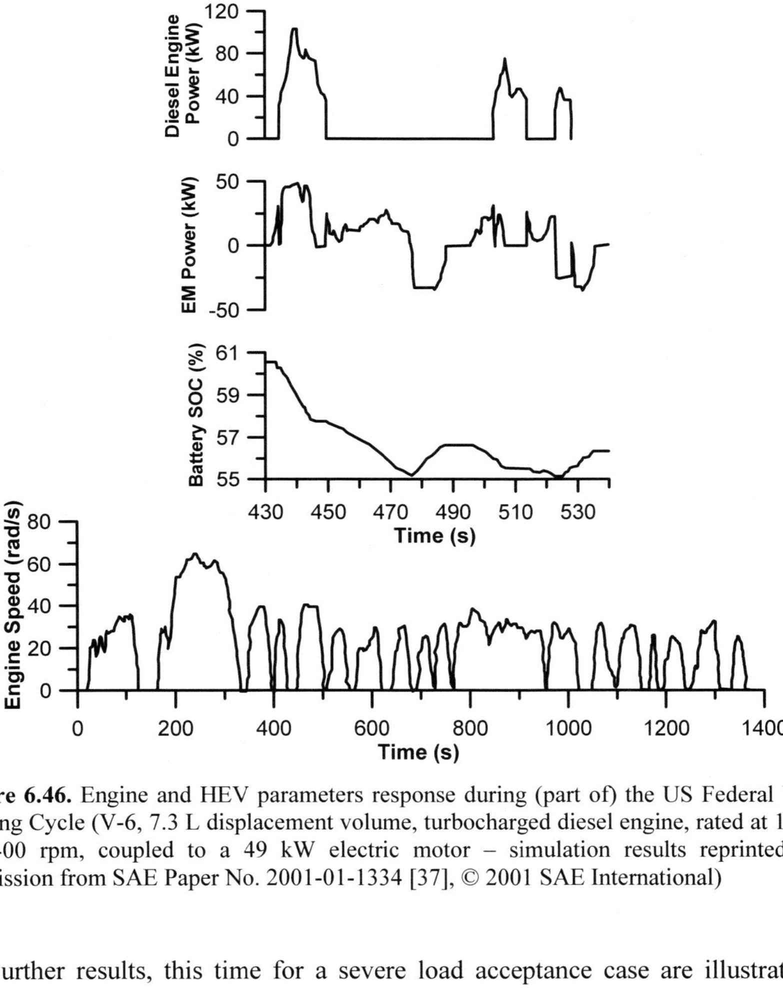

Figure 6.46. Engine and HEV parameters response during (part of) the US Federal Urban Driving Cycle (V-6, 7.3 L displacement volume, turbocharged diesel engine, rated at 157 PS at 2400 rpm, coupled to a 49 kW electric motor – simulation results reprinted with permission from SAE Paper No. 2001-01-1334 [37], © 2001 SAE International)

Further results, this time for a severe load acceptance case are illustrated in Figure 6.47 for an engine equipped with an integrated starter-generator system. Whereas the baseline engine could not accept the new load leading to stall, successful transient performance was achieved when applying an integrated starter-generator mounted on the engine flywheel (the ISG consumed up to 2.2 Ah at 42 V for a 30 s acceleration). By so doing, engine crankshaft response was improved for a few seconds, without affecting turbocharger dynamics. Consequently, boost pressure and turbocharger acceleration (not shown) were only marginally influenced. Even better response was achieved by a combination of the ISG and an electric motor-generator mounted on the turbocharger shaft. The latter was characterized by 0.57 Nm maximum torque at 20,000 rpm and 2.87×10^{-5} kgm^2 inertia assisting turbocharger acceleration, hence boost pressure and air-supply build-up and limiting fuel restriction during the first critical seconds of the transient load acceptance.

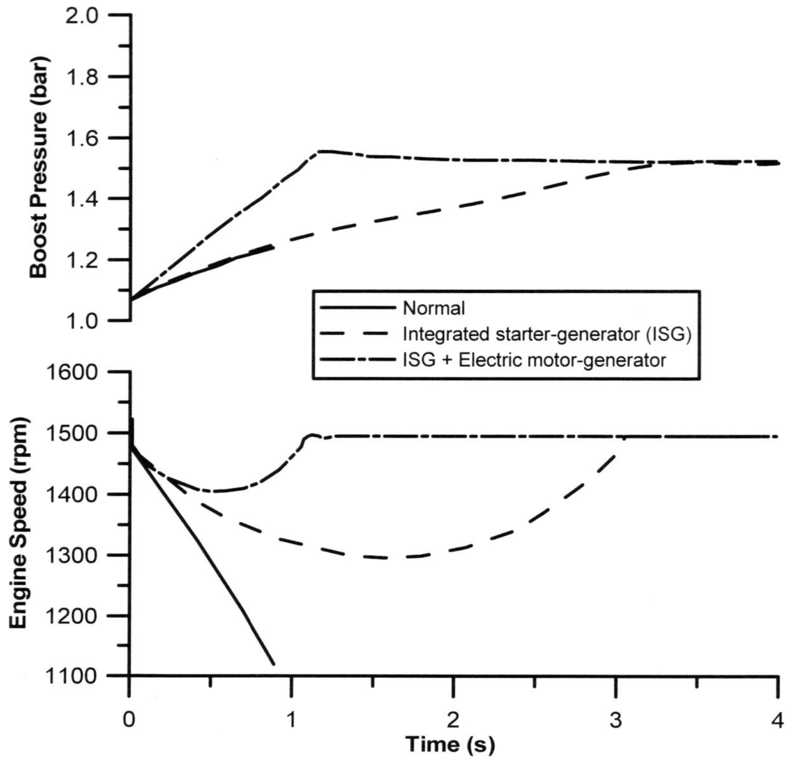

Figure 6.47. Effect of electrical assistance on engine transient response after 0–97 kW load increase (six-cylinder, turbocharged and aftercooled diesel engine, rated at 162 kW at 2400 rpm – simulation results reprinted from Katrašnik *et al.* [18], copyright (2005), with permission from Elsevier)

Comparing the integrated starter-generator with the electrically assisted turbocharging (Figure 6.30), it was argued that the ISG was a more successful electrical solution in case of a large load change, facilitating smaller speed droops, whereas it did not outperform the electrically assisted turbocharging in terms of electric energy consumption. For vehicular applications, however, the electrically assisted turbocharging configuration was considered superior to the integrated starter-generator, since the typical operational times of electric assisting systems last for more than a few seconds thus offering a better ratio of engine dynamics improvement to electric energy consumption; the higher the rating of the engine, the greater the benefit of the electrically assisted turbocharging compared with ISG [41].

Finally, Table 6.1 summarizes the main methods for improving transient response discussed in the previous sections providing a brief note on the main area of improvement achieved, the degree of improvement in terms of driveability, as well as the major system disadvantages.

Table 6.1. Classification of various methods for improving turbocharged diesel engine transient response with emphasis on the basic area of engine or turbocharger improvement, degree of driveability enhancement and system disadvantages

	Method	Improvement	Degree of improvement with successful control	Major drawbacks
1	Compressor/inlet manifold air-injection	Air-supply increase	Considerable	Space, cost, system modifications
2	Manifolds' tuning	Flow optimization	Small	-
3	Low-heat rejection cylinder	Decrease of heat transfer delays	Small	NO_x emissions
4	Late injection timing	Higher turbine back-pressure	Moderate	Higher bsfc
5	Early exhaust valve opening	Higher turbine back-pressure	Moderate	Higher bsfc
6	Hyperbar	Higher turbine back-pressure	Moderate	Cost, higher bsfc, complexity
7	Hybrid-electric powertrain	Engine torque increase	Moderate-considerable	Control, space, cost, weight
8	Integrated starter-generator	Engine torque increase	Moderate	Control, cost, weight,efficiency
9	Exhaust manifold insulation	Decrease of heat transfer delay	Insignificant-Small	-
10	Manifolds' volume (pulse *vs.* constant pressure)	Minimization of flow inertia	Considerable	Smaller t/c efficiency
11	Exhaust manifold configuration (multiple turbine entries)	Reduction of flow inertia	Moderate	Smaller t/c efficiency
12	Lighter turbocharger (same frame)	Smaller t/c moment of inertia	Considerable	-
13	Smaller turbocharger frame	Smaller t/c moment of inertia	Considerable	Over-boosting/speeding
14	Combined supercharging	Better low-end t/c characteristics, smaller t/c inertia	Considerable	Cost, space
15	Variable geometry turbine	Better low-end t/c characteristics, faster air-supply build-up	Considerable	Control system, matching, bsfc
16	Two-stage turbocharging	Better low-end t/c characteristics, smaller moment of inertia	Moderate-significant	Control system, space, cost
17	Sequential turbocharging	Wider flow range, smaller moment of inertia	Moderate-significant	Control system, matching, space, cost, surge
18	Electrically assisted turbocharging	Faster t/c acceleration	Considerable	Control system, cost
19	Turbine air-injection	Faster t/c acceleration	Moderate	Space, cost, system modifications
20	Pelton-wheel	Faster t/c acceleration	Moderate	Cost, space, system modifications
21	Turbocharger friction	Lower turbocharger losses	Insignificant	-

References

[1] Winterbone DE. Transient Performance. In: Horlock JH, Winterbone DE (eds). The thermodynamics and gas dynamics of internal combustion engines, Vol. II. Oxford: Clarendon Press, 1986; 1148–212.

[2] Harndorf H, Kuhnt H-W. Improvement of transient behavior of turbocharged diesel engines through additional air injection in the turbocharger. MTZ 1995;56:20–8 (in German).

[3] Ledger JD, Benson RS, Furukawa H. Performance characteristics of a centrifugal compressor with air injection. Proc Inst Mech Eng 1973;187:425–34.

[4] Ledger JD, Benson RS, Furukawa H. Improvement in transient performance of a turbocharged diesel engine by air injection into the compressor. SAE Paper No. 730665, 1973.

[5] Winterbone DE, Benson RS, Mortimer AG, Kenyon P, Stotter A. Transient response of turbocharged diesel engines. SAE Paper No. 770122, 1977.

[6] Boy P. Investigation of several influences on the dynamic performance of medium-speed diesel engines, Pt. 2: influences of different factors on the dynamic performance. MTZ 1980;41:491–6 (in German).

[7] Watson N. Transient performance simulation and analysis of turbocharged diesel engines. SAE Paper No. 810338, 1981.

[8] Zinner K. Supercharging of internal combustion engines. Berlin Heidelberg New York: Springer-Verlag, 1978.

[9] Schmitz T, Holloh K-D, Juergens R. Potential of additional mechanical supercharging for commercial vehicle engines. MTZ 1994;55:308–13 (in German).

[10] Matsura Y, Nakazawa N, Kobayashi Y, Ogita H, Kawatani T. Effects of various methods for improving vehicle startability and transient response of turbocharged diesel trucks. SAE Paper No. 920044, 1992.

[11] Sturm WL, Kruithof J. Development and testing of a HD diesel engine with two-stage turbocharging. 9. Aachener Kolloquium Fahrzeug- and Motorentechnik 2000.

[12] Watson N, Marzouk M, Baazaari Z. An evaluation of two stage turbocharging for efficient high-output diesel engines. ASME Energy Technology Conference & Exhibition, Houston, TX, Paper No. 78-DGP-2, 1978.

[13] Watson N, Janota MS. Turbocharging the internal combustion engine. London: MacMillan, 1982.

[14] Filipi Z, Wang Y, Assanis D. Effect of variable geometry turbine (VGT) on diesel engine and vehicle system transient response. SAE Paper No. 2001-01-1247, 2001.

[15] Münz S, Schier M, Schmalzl HP, Bertolini T. eBooster: design and performance of a innovative electrically driven charging system. BorgWarner Turbo Systems Academy.

[16] Fieweger K, Paffrath H, Schorn N. Drivability assessment of an HSDI diesel engine with electrically assisted boosting systems. Institution of Mechanical Engineers, 7th International Conference on 'Turbochargers and Turbocharging', Paper C602/009/2002, London 2002, pp. 283–93.

[17] Zellbeck H, Friedrich J, Berger C. Electrically assisted turbocharging as a new boosting concept. MTZ 1999;60:386–91 (in German).

[18] Katrašnik T, Medica V, Trenc F. Analysis of the dynamic response improvement of a turbocharged diesel engine driven alternating current generating set. Energy Convers Manage 2005;46:2838–55.

[19] Algrain M. Controlling an electric turbo compound system for exhaust gas energy recovery in a diesel engine. IEEE International Conference on 'Electro-Information Technology', 22–25 May 2005.

[20] Millo F, Mallamo F, Pautasso E, Mego GG. The potential of electric exhaust gas turbocharging for HD diesel engines. SAE Paper No. 2006-01-0437, 2006.

[21] Borila YG. A sequential turbocharging method for highly-rated truck diesel engines. SAE Paper No. 860074, 1986.
[22] Benvenuto G, Campora U. Dynamic simulation of a high-performance sequentially turbocharged marine diesel engine. Int J Engine Res 2002;3:115–25.
[23] Galindo J, Luján JM, Climent H, Guardiola C. Turbocharging system design of a sequentially turbocharged diesel engine by means of a wave action model. SAE Paper No. 2007-01-1564, 2007.
[24] Watson N. Eliminating rating effects on turbocharged diesel response. SAE Paper No. 840134, 1984.
[25] Andre-Talamon T. A new Hyperbar engine for ship propulsion. SAE Paper No. 830506, 1983.
[26] Arcoumanis C, Bazari Z, Chan SH. The transient response of a turbocharger vehicle diesel engine with injection timing control. Proceedings of the Institution of Mechanical Engineers, 4th International Conference on 'Turbocharging and Turbochargers', Paper C405/022, 1990, pp. 83–97.
[27] Bazari Z. The transient performance analysis of a turbocharged vehicle diesel engine with electronic fuelling control. SAE Paper No. 900236, 1990.
[28] Giakoumis EG, Dimaratos AM, Rakopoulos DC. A parametric analysis of exhaust emissions prediction during transient turbocharged diesel engine operation. 21st International ECOS Conference, Krakow-Gliwice, Poland, June 24–27, 2008, Proc. Vol. I, pp. 507–14.
[29] Marzouk M, Watson N. Load acceptance of turbocharged diesel engines. Institution of Mechanical Engineers, Conference on 'Turbocharging and Turbochargers', Paper C54/78, London, 1978, pp. 45–57.
[30] Galindo J, Serrano JR, Vera F, Cervelló C, Lejeune M. Relevance of valve overlap for meeting Euro 5 soot emissions requirements during load transient process in heavy duty diesel engines. Int J Vehicle Design 2006;41:343–67.
[31] Rakopoulos CD, Giakoumis EG, Hountalas DT, Rakopoulos DC. The effect of various dynamic, thermodynamic and design parameters on the performance of a turbocharged diesel engine operating under transient load conditions. SAE Paper No. 2004-01-0926, 2004.
[32] Rakopoulos CD, Giakoumis EG. Review of thermodynamic diesel engine simulations under transient operating conditions. SAE Paper No. 2006-01-0884, SAE Trans, J Engines 2006;115:467–505.
[33] Banisoleiman K, Smith LA, French BA. The interaction of diesel engine turbocharging and tuned inlet manifold systems under steady state and transient operation. Proc Inst Mech Eng, Part A, J Power Energy 1991;205:269–81.
[34] Brace CJ, Deacon M, Vaughan ND, Horrocks RW, Burrows CR. An operating point optimizer for the design and calibration of an integrated diesel/continuously variable transmission powertrain. Proc Inst Mech Eng, Part D, J Automobile Eng 1999;213:215–26.
[35] Baumann BM, Washington G, Glenn BC, Rizzoni G. Mechatronics design and control of hybrid electric vehicles. IEEE/ASME Trans, Mechatronics 2000;5:58–72.
[36] Johnson VH, Wipke KB, Rausen DJ. HEV control strategy for real-time optimization of fuel economy and emissions. SAE Paper No. 2000-01-1543, 2000.
[37] Lin C-C, Filipi Z, Wang Y, Louca L, Peng H, Assanis D, Stein J. Integrated, feed-forward hybrid electric vehicle simulation in SIMULINK and its use for power management studies. SAE Paper No. 2001-01-1334, 2001.
[38] Markel T, Brooker A, Hendricks T, Johnson V, Kelly K, Kramer B. *et al.* ADVISOR: a system analysis tool for advanced vehicle modeling. J Power Sources 2002;110:255–66.

[39] Liao GY, Weber TR, Pfaff DP. Modelling and analysis of powertrain hybridization on all-wheel-drive sport utility vehicles. Proc Inst Mech Eng, Part D, J Automobile Eng 2004;218:1125–34.
[40] Kolmanovsky I, Stefanopoulou AG. Evaluation of turbocharger power assist system using optimal control techniques. SAE Paper No. 2000-01-0519, 2000.
[41] Katrašnik T, Trenc F, Medica V, Markič S. An analysis of turbocharged diesel engine dynamic response improvement by electric assisting systems. ASME Trans, J Eng Gas Turbines Power 2005;127:918–26.

7

Special Cases of Transient Operation

7.1 Cold Starting

7.1.1 Introduction

Unlike spark ignition engines, where combustion is initiated by the spark and is aided by the high gasoline volatility, in diesel engines, the auto-ignition process may prove unreliable during cold starting. One of the most influential engine parameters in this case is the compression ratio; actually, it is usually the starting requirements that determine the lower limit of compression ratio in diesel engines. On the other hand, the increasing degree of turbocharging in recent years has led to a gradual decrease of compression ratio in order to avoid excessive mechanical stress of the cylinder components (see also Table 2.1), thus putting the cold start behavior of the engine under test.

When studying vehicle motors starting, apart from the engine characteristics, the transient event is influenced by the starter motor or starter system behavior. Figure 7.1 shows typical response of a single-cylinder, naturally aspirated diesel engine during starting by an electric starter. During the first four cycles, the engine accelerates rapidly with the assistance of the electric starter. Afterwards, up to cycle No. 18, engine speed continues to increase without the need for external assistance, until the point where stabilization to the idling speed is achieved. An increased amount of hydrocarbons and CO is expected during the cold start phase, particularly if misfiring occurs, which is likely the colder the ambient conditions. Misfiring is the most critical problem encountered during cold starting, which diversifies completely the response pattern from load acceptance or acceleration transients, ultimately leading to combustion instability or even complete failure. Under misfiring conditions, which are influenced by a variety of factors, combustion cannot supply enough power to drive the engine and overcome the increased friction losses; the latter emanate from the high lubricant viscosity at the low temperature. The intriguing fact is that, unlike acceleration or load acceptance, during cold starting, naturally aspirated engines suffer equally to their turbocharged counterparts.

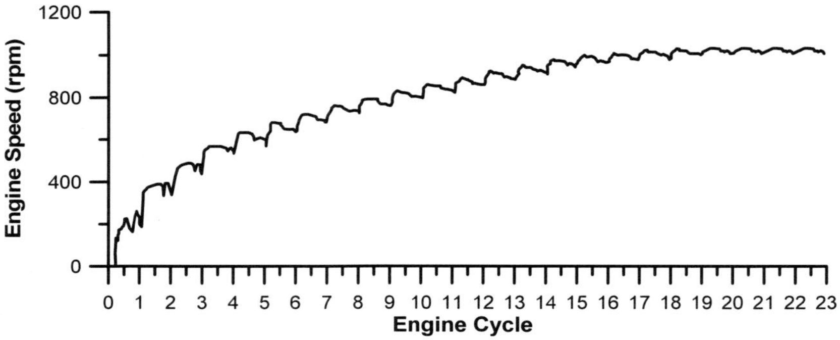

Figure 7.1. Engine speed development during starting (four-stroke, single-cylinder, naturally aspirated diesel engine)

Diesel engines driving emergency electrical generators experience the most challenging starting event. Here, as soon as the engine is started and assumes its nominal rotational speed, (severe) load acceptance takes place. Such a complex transient event is illustrated in Figure 7.2 for a turbocharged, industrial diesel engine.

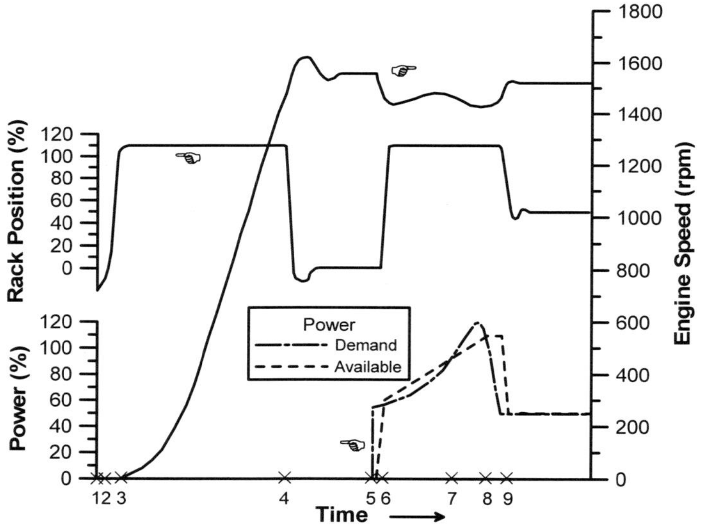

Figure 7.2. Starting of a diesel engine driving an emergency generator with a subsequent 50% load acceptance (experimental data reprinted with permission from Schulmeister [1])

The procedure is as follows

- the operator gives the order for engine start-up (point 1);
- the governor moves the fuel pump rack towards increased fueling (2);
- the engine speed begins to rise (3);

- the nominal engine speed of 1500 rpm is achieved and the governor returns to the no-load position (4);
- the new load is imposed and, consequently, the governor moves again the fuel pump rack towards increased fueling. The demand power exceeds its available counterpart at this point; as a result, engine speed falls (5);
- the available power exceeds demand power; the surplus torque leads to an increase of engine speed (6);
- torque deficit owing to the abrupt increase in demand power leads again to engine speed fall (7);
- torque surplus leads to reduction of fueling and increase of engine speed (8); and
- demand power stabilizes at the requested value, and the engine speed is regulated at its nominal value (9).

Although the imposed load was only 50% of the maximum, this was enough to cause, on the one hand, considerable deviations of the engine speed around its nominal value and, on the other hand, a relatively long recovery period [1].

7.1.2 Combustion Instability

Combustion development during diesel engine cold starting constitutes a very challenging operating condition, mainly in terms of successful process completion. There are three particular aspects that influence strongly (turbocharged or naturally aspirated) diesel engine cold starting performance, namely

- low coolant temperature, hence cylinder wall temperature that increases heat transfer rate to the cylinder walls (Equation 2.3) and prolongs ignition delay;
- low oil temperature, which increases significantly the lubricant's viscosity, hence frictional losses (analyzed in more detail in Section 7.1.3); and
- low engine rotational speed that allows more time for the above mentioned losses to develop.

As an opening statement, due to the low wall temperature during cold starting, the air-charge in the cylinder may not reach temperatures capable of vaporizing the injected fuel. Consequently, formation of a combustible air–fuel mixture may be prohibited, ultimately leading to complete combustion failure, but more likely to combustion instability with the engine compression ratio and starting aid playing a primary role. A key parameter is also the low injection pressure encountered at the low cranking speed that leads to poor spray penetration, atomization and fuel evaporation.

During the compression stroke and after fuel injection, heat is transferred from the air-charge to the liquid fuel spray, supplying its sensible heat and latent heat of evaporation. This process causes a drop in the local air temperature further enhanced by heat losses to the cold cylinder walls and by the blow-by losses. The resulting mixture of air and fuel vapor undergoes then pre-ignition chemical reactions. If these processes occur during the compression stroke when heat is added to the system, their rate is favored. On the other hand, if some of these processes occur during expansion, where energy is taken away from the mixture, then their rate decreases. In fact, Meyer and DeCarolis [2] argued that the criterion for ignition to occur in a diesel engine is that compression temperature and pressure should be high enough to reduce ignition delay to the time available between injection and some point prior to TDC. Even if (part of) the mixture is finally vaporized, the auto-ignition process is still subject to failure as will be discussed in the next paragraph. In any case, during cold starting, ignition delay is prolonged, leading to harder premixed combustion phase and higher combustion noise. One case in point: the increased ignition delay duration in ms does not necessarily correspond to longer ignition delay in degrees crank angle (compared with normal operation), owing to the very low rotational speeds encountered during cold starting. Hence, it has been argued that it may be advantageous to retard (single-stage) fuel injection during starting compared with warmed-up operation, in order to promote mixture preparation in a more favorable air environment.

Together with the physical preparation mechanism, the chemical auto-ignition process develops; auto-ignition in a diesel engine develops in two stages; first, slow reactions occur that form intermediate radical compounds, such as peroxides and aldehydes, and second, once a critical concentration of the intermediate compounds has been reached, very fast chain reactions occur leading to the formation of auto-ignition nuclei. If the critical concentration is not reached, or, as it is most usually the case during cold starting, since the reactions are decelerated owing to the low charge temperature, auto-ignition, hence combustion may fail. Combustion failure can be caused by the inability of the auto-ignition sources to burn the surrounding mixture, by the dilution of the charge with the residuals from the previous firing cycle, or if combustion is unable to supply enough power to drive the engine and overcome the increased frictional losses. The important finding from the analysis of cold starting pressure diagrams is that combustion failure is intermittent, with the specific in-cylinder conditions during one cycle affecting (positively or negatively) the next one, and leading to a series of firing and misfiring cycles as illustrated in Figure 7.3.

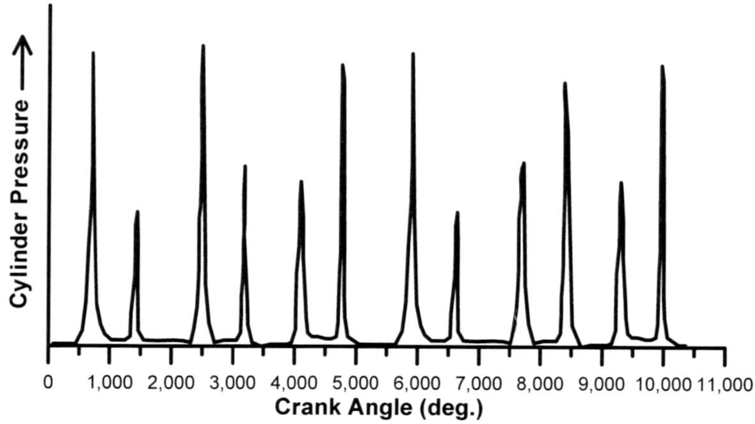

Figure 7.3. Typical diesel engine cold starting combustion instability

This behavior is actually the result of an imbalance between engine dynamics and combustion kinetics. Namely, the high rate of acceleration after a firing cycle reduces the time available for the physical and chemical processes to be completed prior to TDC. Likewise, after a misfiring cycle, engine rotational speed decreases allowing more time for pre-ignition chemical reactions around TDC, thus favoring ignition in the next cycles. At the same time, the gradual increase in cylinder wall and lubricant temperature decrease the amount of heat and frictional losses. Moreover, any unburned fuel quantity deposited in the cylinder after a misfiring cycle increases both the fuel–air ratio and the compression ratio of the succeeding cycle and promotes cold flame reactions that may favor successful firing in the next cycle. In support of the last argument, analysis of the relevant cylinder pressure diagrams as the ones depicted in Figure 7.3 has revealed that during the cold starting firing cycles, ignition delay was actually shorter, combustion proceeded more rapidly and peak cylinder pressures were higher than 'normal' combustion under fully warmed-up conditions [3–9].

Particularly for turbocharged diesel engines, turbocharger lag effects worsen the above-mentioned combustion instability situation; during cold starting, the engine speed is much lower than the required governing speed, hence the fuel pump rack is in the maximum fueling position (see also Figure 3.14). Since there is a lack of sufficient air-flow due to the low engine and turbocharger rotational speeds, locally high fuel–air ratios are experienced (possibly aggravated by EGR effects), leading to flame quenching and deterioration of combustion.

Further to the previous arguments, Figure 7.4 is provided that shows the rate of heat release (RoHR) in cycle 23 (upper sub-diagram), where combustion was successful, and cycle 24 (lower sub-diagram), where combustion failed, for a naturally aspirated DI diesel engine's cold starting event [6]. The rate of heat release is given in terms of the equivalent mass of fuel burned in mg per degree crank angle.

In firing cycle 23, injection started at 345°CA; at 352°CA, the sum of the energy production rates from combustion and the compression work was equal to the sum of the rates of all other losses. After 352°CA, the rate of energy released by combustion exceeded that of the other losses. At 373°CA, the pressure rise due to combustion was detected. During cycle No. 24 (lower sub-diagram of Figure 7.4), fuel injection started again at 345°CA (point (a)); point (b) shows the point where the rate of energy released by premixed combustion was equal to the rate of losses. The period between the start of fuel injection and point (b) was referred to as ID_h. After point (b), the energy produced from the premixed combustion started to exceed the other losses. After point (c), however, the combustion process failed to produce energy in excess of the energy consumed. During the latter phase, wall impingement phenomena were characterized by low impact velocities, whereas the bouncing of the still liquid drops worsened the process of air–fuel mixture formation. While ignition took place in cycle 24, the expansion work and the other associated losses (heat transfer to the cylinder wall, friction, blow-by) practically froze the reactions and combustion failed.

Figure 7.5 expands on the previous results by illustrating the ignition delay period ID_h for cycles 13–35 of the same cold starting event. The cycles with successful combustion are designated with the letter F (firing). It is observed that the ignition delay ID_h for the misfiring cycles was always shorter than for the firing ones. This indicates that longer ID_h ignition delay is actually the result of more fuel evaporation and prolonged mixing.

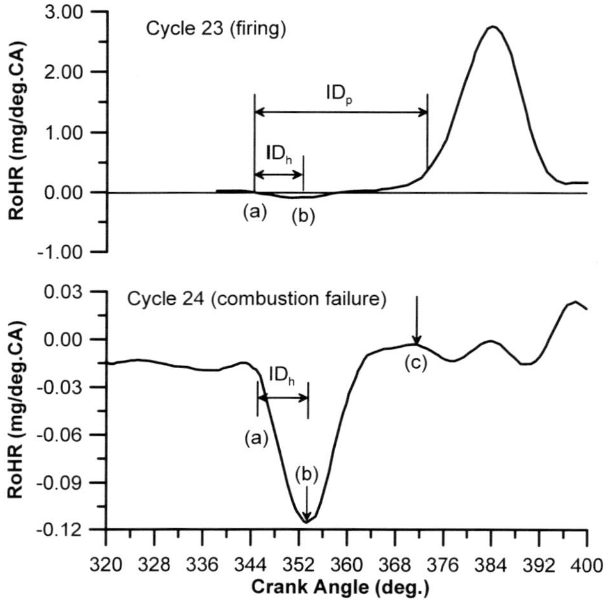

Figure 7.4. Rate of heat release (RoHR) for cycles 23 and 24 of a cold starting event (four-stroke, single-cylinder, naturally aspirated, DI diesel engine – experimental results reprinted with permission from SAE Paper No. 920005 [6], © 1992 SAE International)

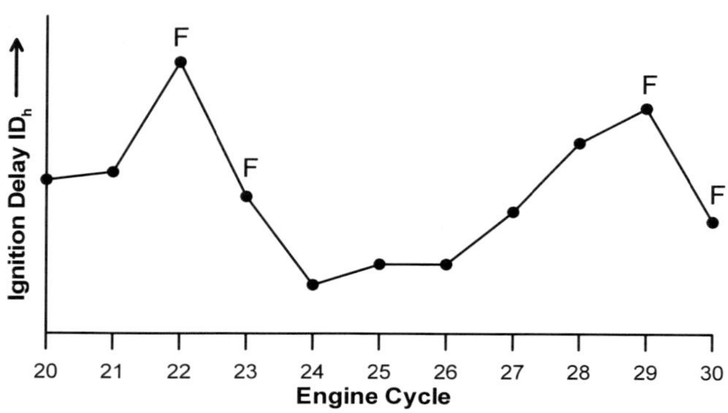

Figure 7.5. Ignition delay ID_h (defined in Figure 7.4) for cycles No. 20–30 of a cold starting event; 'F' denotes firing cycle (four-stroke, single-cylinder, naturally aspirated, DI diesel engine – experimental results adapted from Henein *et al.* [6])

Summarizing the previous discussion, Figure 7.6 illustrates the primary mechanisms that are responsible for combustion failure during cold starting of a diesel engine.

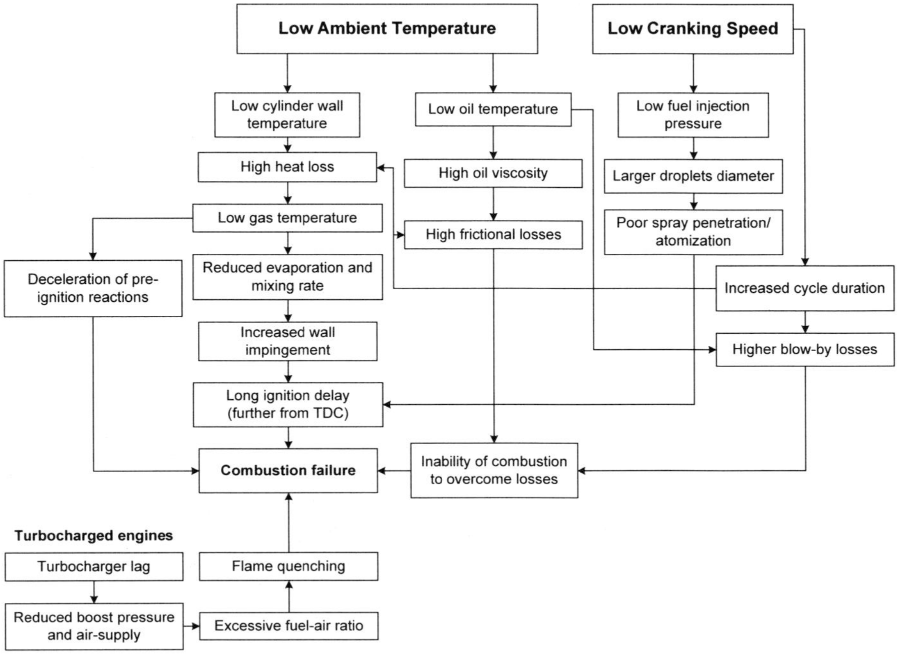

Figure 7.6. Combustion failure mechanisms during diesel engine cold starting

As was documented in the previous paragraphs, diesel engine cold starting behavior is characterized by a high degree of cycle-to cycle variation in terms of cylinder pressure (as was depicted in Figure 7.3) and crankshaft angular velocity; Henein *et al.* [6] argued that this variation is not erratic, but rather follows a defined mode. A summarization of the influencing parameters follows.

1. Ambient temperature: the lower the ambient temperature the lower the cylinder wall and oil temperatures, retarding the physical and chemical preparation mechanisms of the air–fuel mixture, prolonging ignition delay, increasing combustion noise and frictional losses, and favoring cycle-to-cycle variation. Typical results are demonstrated in Figure 7.7 for a naturally aspirated diesel engine in terms of torque percentage (with respect to fully warmed-up conditions) for acceleration right after cold starting. At –20°C, the torque ratio reached only 53% after 5 s and 68% after 30 s. Even when the temperature increased to 0°C, torque ratio was not more than 71% after 5 s acceleration.
2. Fuel properties: higher cetane numbers (Figure 7.7), are slightly favored, as well as higher fuel volatility, both affecting optimum injection timing. Increase in fuel cetane number (for example using a fuel improver) results in a slight decrease of the minimum starting ambient temperature of the order of a few degrees Celsius.
3. Compression ratio: it determines the charge-air pressure and temperature at the point of fuel injection, affecting strongly the ignition delay duration.

Reduction in the minimum starting temperature can be achieved for engines with increased compression ratios.

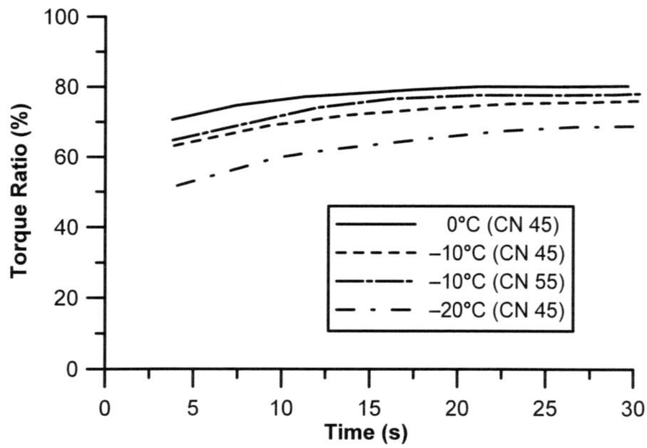

Figure 7.7. Effect of ambient air temperature and fuel cetane number (CN) on engine cold starting (four-stroke, four-cylinder, swirl chamber, water-cooled, naturally aspirated diesel engine – experimental results reprinted with permission from SAE Paper No. 850113 [4], © 1985 SAE International)

4. Blow-by losses: they affect the amount of trapped air. Owing to the low lubricant temperatures during cold starting, expansion and contraction of different engine components take place, thus the clearance between piston rings/liner and cylinder are increased favoring blow-by losses.
5. Cranking speed: the lower the cranking speed the more the available time for heat and blow-by losses, hence at extremely low engine speeds the compression temperature is lower than that at higher speeds prolonging ignition delay. Experiments have shown that a minimum cranking speed exists, below which starting is unviable. The above remarks confirm the importance of maintaining high cranking speed under cold starting conditions (even though this results in shorter time available for pre-ignition reactions around TDC) as an effective means of achieving relatively high compression pressure and temperature. As is the case with compression ratio, reduction in the minimum starting temperature can be achieved for engines with increased cranking period (most probably requiring a higher capacity battery). In contrast, whenever conditions of starting are favorable, increase in the cranking speed does not have much effect.
6. Fuel injection profile: cold starting has been found to improve when retarded injection is applied. This is attributed to better spray development, since early fuel injection in an unfavorable air environment promotes wall impingement. However, different theories have been also reported, where actually an advance in injection timing was found preferable, as it allowed the long ignition delay period to complete before TDC. Pre-injection may prove beneficial as regards prevention of misfiring and lowering of the

respective white smoke emissions, with the exact timing and quantity of pre-injected fuel being a matter of optimization procedure. The latter behavior is documented in Figure 7.8 for a heavy-duty, turbocharged diesel engine; application of a pilot-main injection strategy resulted in a significant decrease of ignition delay that ultimately led to a 68% improvement in cold starting duration (from 25 s to 8 s at –15°C), while at the same time reducing considerably white smoke emissions and combustion noise.

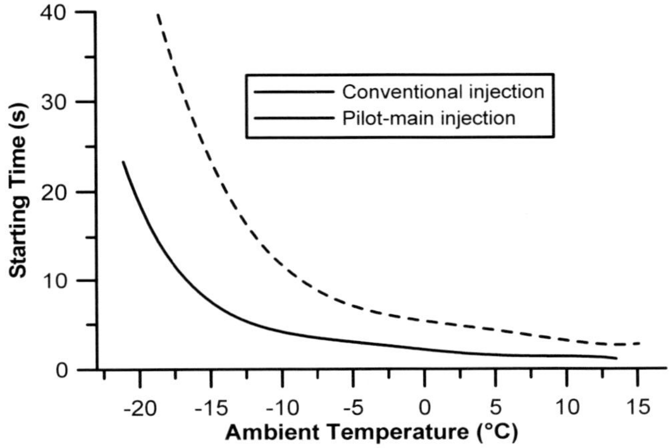

Figure 7.8. Effect of injection strategy on cold startability (six-cylinder, heavy-duty, DI diesel engine – experimental results reprinted with permission from SAE Paper No. 940586 [9], © 1994 SAE International)

7. Residual gas: exhaust gases recirculating from one cycle to the next affect the composition and temperature during the next compression stroke.
8. Fuel accumulation in the cylinder: any fuel remaining from the previous cycle affects the fuel–air ratio and raises the compression ratio of the following cycles, causing cold flame reactions that may facilitate ignition in the next cycle. This amount of remaining fuel may reach up to 30–40% of the injected fuel quantity.
9. Combustion chamber design: shape of piston bowl, swirl, squish, fuel injector location, dead volume, DI or IDI type. In general, the longer the distance between fuel injector and cylinder wall, the better the possibility of the fuel to evaporate before the spray impinges on the wall. Experiments with different combustion chamber geometries led to the conclusion that it is desirable if non-uniform mixing with a wide range of air–fuel ratios can be achieved, whereas combustion chambers that promote swirl (as for example prechamber IDI engines) and squish do not favor startability.
10. Starting-aid devices: glow plugs, coolant heaters, as well as cranking motor characteristics, such as cranking duration and motor torque. Of primary importance is the cycle where the starter is disengaged, as this point is likely to produce deceleration in the next cycles.

7.1.3 Dynamics and Friction Development

For the special case of automotive engines starting, Equation 3.19 that described the crankshaft torque balance becomes

$$\tau_{start}(\varphi)+\tau_e(\varphi)-\tau_{fr}(\varphi)-\tau_L(\varphi)=G_{tot}\frac{d\omega}{dt} \qquad (7.1)$$

where, for an automotive electrical starter, the respective motor torque is given by the following equation, as is illustrated in Figure 7.9

$$\tau_{start}=d_1 e^{-d_2 N} \qquad (7.2)$$

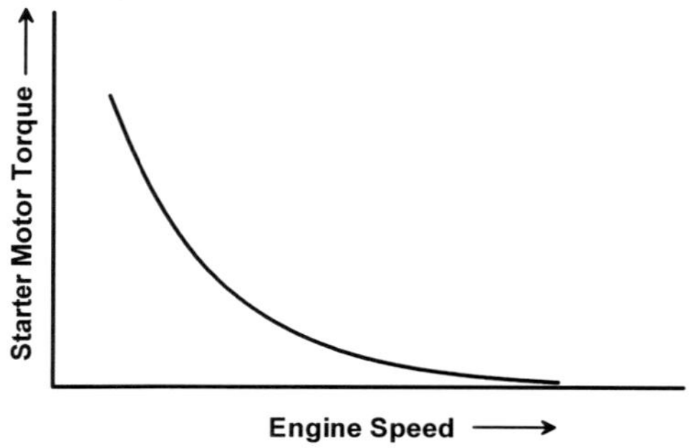

Figure 7.9. Typical starter motor torque *vs.* speed curve

Owing to the significant influence of oil temperature on friction torque, it is generally expected that friction terms during cold starting will be more pronounced (actually they have been estimated at four- or even five-fold their fully warmed-up counterparts [10]) and, thus, of particular scientific importance.

During warmed-up conditions a logarithmic relationship between oil kinematic viscosity and temperature exists as described by the following Vogel equation (C_{oil}, Θ_1 and Θ_2 are constants depending on oil type, and Θ_{oil} is the mean, over an engine cycle, oil temperature in °C):

$$\mu_{oil}=C_{oil}\cdot e^{\left(\frac{\Theta_1}{\Theta_{oil}+\Theta_2}\right)} \qquad (7.3)$$

During the first seconds after cold starting, however, the above equation is not valid, resulting in much higher oil viscosities and frictional losses. At low temperatures and low motoring speeds, piston rings assembly and the connecting rod big end bearing contribute most to the total friction. The crankshaft assembly (main bearings, thrust bearings, and oil seals) is generally the second largest contributor. Both the piston rings assembly and the crankshaft assembly exhibit similar friction characteristics to those of the complete engine, with an initial transient period during which friction levels fall eventually onto the logarithmic variation with bulk oil viscosity expressed by Equation 7.3. During the early seconds of engine operation, frictional dissipation at the rubbing surfaces raises the temperature of the oil films and the adjacent parts of the engine structure. The increase in film temperature lowers oil viscosity, therefore reduces friction at the hydrodynamically lubricated surfaces. Hence, friction behavior and changes in local thermal conditions are strongly coupled. In addition, the high oil shear rates can give rise to a temporary shear viscosity loss, although the effect of shear rate is less significant than the film temperature. Shayler *et al.* [11] carried out motoring tests on an unloaded crankshaft, in order to examine friction levels and the influence of local thermal conditions in and around the main bearings at cranking speed range and from initial ambient temperatures down to –20°C. They concluded that bearing friction force during this period was proportional to $N^{0.6}\mu^{0.8}/c$, with μ the dynamic viscosity, N the relative bearing surface speed and c the bearing clearance. Changes in viscosity produced by changes in film temperature were the most significant influence on bearing friction during cold running at a constant speed.

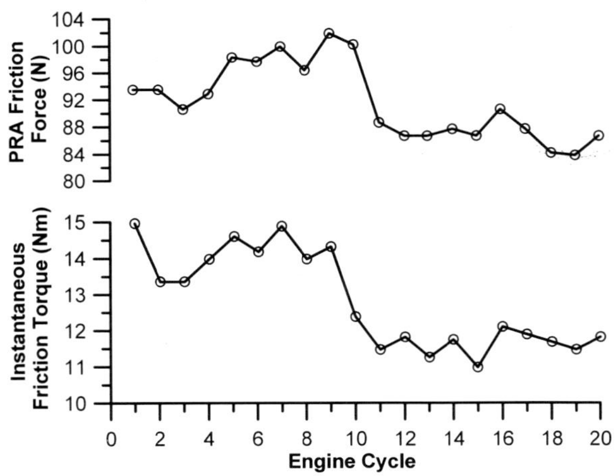

Figure 7.10. Cycle-averaged instantaneous friction torque and piston rings assembly (PRA) friction force during cold starting, with electric starter disengaged after cycle No. 9 (single-cylinder, naturally aspirated, diesel engine – experimental results adapted from Marek and Henein [12])

Marek and Henein [12] measured instantaneous friction torque (IFT) and piston rings assembly (PRA) friction force during cranking and starting of a single-cylinder, naturally aspirated, air-cooled diesel engine. For the friction torque measurement, the method of Equation 3.27 was applied, whereas for the piston rings assembly friction, the axial force acting on the connecting rod was measured by means of a resistance type strain-gauge. The main results from their analysis are reproduced in Figure 7.10. In general, both IFT and PRA friction force were reduced as the starting process progressed, with the highest values occurring at the first cycle. Drop in oil velocity caused by warming-up, development of oil film between the different sliding surfaces and drop in the elasto-hydrodynamic friction forces, particularly in the valve train, were identified as the main reasons for this reduction. A small increase observed between cycles 3 and 7 was attributed to the increase in combustion pressure and angular velocity. Friction during the inlet stroke was higher than during exhaust; as expected, the highest values were noticed during expansion.

7.1.4 Exhaust Emissions

Exhaust gases during cold starting have recently gained increased attention owing to their significant contribution to total emissions from diesel-engined vehicles (see for example, the New European Driving Cycle, Section A.2.2.1 or the US FTP-75, Section A3.2.1, where the emissions are sampled with the engine cold started). Cold starting exhausts are strongly influenced by the low cylinder wall and coolant temperatures that prolong ignition delay and lead to incomplete combustion aided, in the case of turbocharged engines, by the high fuel–air equivalence ratios following turbocharger lag; they consist of several types of gaseous or particulate components with different chemical compounds and optical properties, namely,

- black particles that are, mainly, agglomerated carbon crystals;
- white smoke particles defined as minute condensed particles of unburned fuel (hydrocarbons) and water vapors;
- fuel in vapor form;
- carbon monoxide; and
- sulfurous and other organic compounds.

During turbocharged diesel engine cold starting, engine speed is much lower than the required governing speed; hence, the fuel pump rack is in the maximum fueling position (in modern engines, the fuel rack position is temperature sensitive, adjusting fuel starting quantity according to ambient conditions) and the fuel–air ratio high owing to the mismatch between fueling and air-supply. As it is demonstrated in Figure 7.11, this behavior results in overshoot of smoke emissions (this is not the case when naturally aspirated engines are involved); for the same reason, unburned HC emissions (a constituent of white smoke) are expected to peak as well, way above their steady-state values. As soon as combustion becomes stable and the engine speed reaches or exceeds the governing self-sustained speed, the rack moves towards the direction of reduced fueling, with a subsequent decrease of soot emissions; the latter are strongly influenced by the coolant temperature. The lower the ambient/coolant temperature, the higher the heat loss to the cylinder walls resulting in lower charge temperature, longer ignition delay and more prolonged period of increased smoke emissions until the engine is warmed-up (10°C coolant temperature case in Figure 7.11).

Adjustment of the maximum fueling position can prove beneficial for the above-mentioned emissions control; for example, maximum HC emissions during cold starting of a light-duty turbocharged diesel engine were reduced by 50% by cutting off 20% of the maximum fueling [13]; however, the engine startability may be adversely affected. Another influential engine parameter on cold start emissions is injection timing. The later the timing, the more white smoke emitted owing to misfiring. It has been argued [14] that opacity during cold start, particularly during the initial phase, could be further reduced by advancing the exhaust valve closing timing. The latter results in higher residual gas fractions, shorter ignition delay and increased combustion stability facilitating lower smoke emissions, however at the expense of higher NO_x emissions. Another option is oxygen enriched intake air-supply [15].

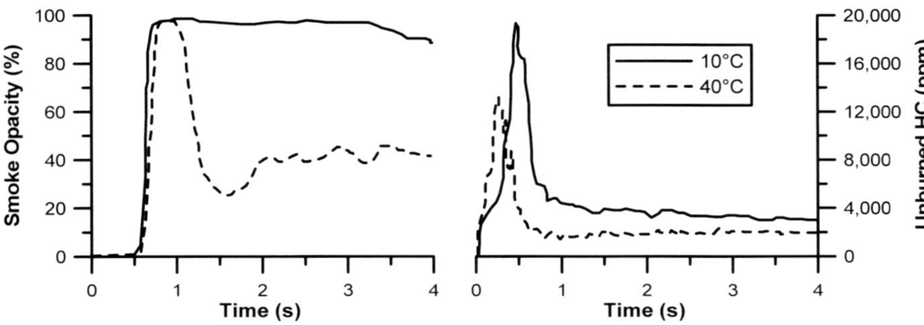

Figure 7.11. Smoke opacity and unburned HC emissions during engine cold starting for two different coolant temperatures (four-stroke, six-cylinder, 8.2 L displacement volume, turbocharged and aftercooled, DI diesel engine – experimental results adapted from Arcoumanis and Yao [13])

In general, every measure that can help avoid misfire, as for example more volatile fuels with higher cetane number or pre-injection or starting aid, is also likely to produce lower amounts of soot (turbocharged engines) and white smoke emissions.

Interestingly, another parameter that was documented in Section 5.6 to improve (transient) soot, namely biodiesel, exhibits an adverse behavior during cold starting; this is attributed to the higher initial boiling point of biodiesel with respect to conventional diesel fuel, which leads to more difficult fuel evaporation at low ambient temperatures, and the higher viscosity of biodiesel, which reduces the rate of spray atomization. Both phenomena, which are enhanced with very high biodiesel blends, lead to worse fuel–air mixing and, thus, more intense soot formation at low temperatures; the latter increase was measured up to 80% for an automotive, HSDI diesel engine, when comparing neat biodiesel blend with neat diesel operation. Later in the warm-up phase, when the engine assumes its normal operating temperature, the advantages of biodiesel combustion prevail, and soot emissions decrease compared with neat diesel fuel [16].

The predominant misfiring phenomena at low ambient temperatures have been also found to increase the total number of emitted particles, as is demonstrated in Figure 7.12, as well as the particles active surface area. In the same figure, particle number emissions from typical, port-injected SI engine (PISI) equipped with three-way catalyst are also provided for comparison purposes. Particularly for the DPF equipped diesel engine, the few minutes after cold start are responsible for more than 70% of the total amount of emitted particles (similar trend is valid for the SI engine), a contribution which decreases the higher the ambient temperature. The latter behavior may be partly due to the fact that, as past research has indicated, during cold starting, nucleation mode particles are favored, which may escape DPF treatment. Total particle number for the DPF engine was measured lower by a factor of 100,000 compared with the untreated diesel engine (right-hand sub-diagram of Figure 7.12) and lower up to a factor of 100 by mass [17]. Similar results were reached for three small automotive diesel-engined vehicles equipped with oxidation catalysts during the US FTP-75 Transient Cycle ; the amount of HC and CO produced in the cold start phase was twice that emitted during the similar hot phase of the Cycle [18].

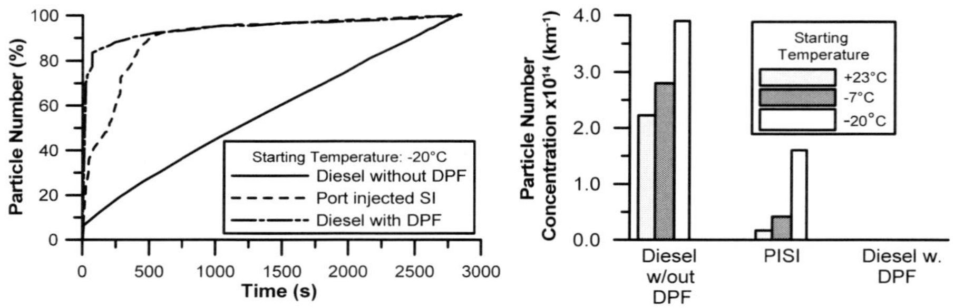

Figure 7.12. Particle number emissions during the French IUFC15 Transient Cycle (duration 2835 s, length 15 km, maximum speed approx. 44 km/h) including cold starting, measured with a CPC (four-stroke, four-cylinder, high-speed, diesel engines with or without DPF, running on 18 ppm sulfur diesel fuel and SAE 15W-40 oil – experimental data reprinted from Mathis *et al.* [17], copyright (2005), with permission from Elsevier)

During diesel engine cold starting, a significant overshoot in CO emissions is noticed too; typical results are reproduced in Figure 7.13. In fact, CO emissions for the HSDI engine of Figure 7.13 exceeded the 1,000 ppm threshold for almost 2 minutes after cold starting, before decreasing significantly as the warm-up period developed. At the same time, peak cold start CO was almost six-fold its hot start counterpart; again, the critical parameter is the coolant temperature, as it is observed when comparing cold to hot start emissions.

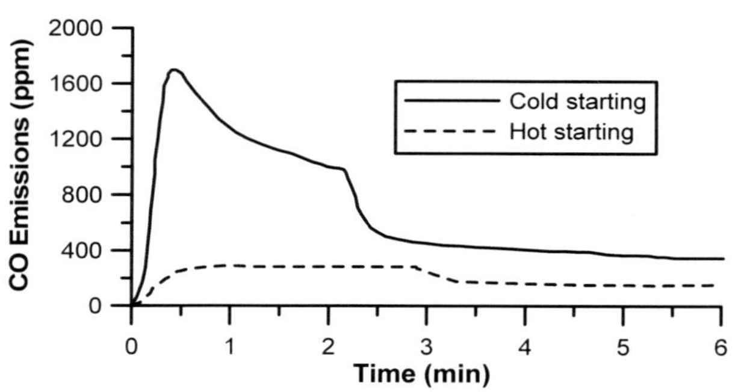

Figure 7.13. CO emissions during cold and warm starting and idle run (four-stroke, four-cylinder, high-speed, turbocharged diesel engine of 2 L displacement volume – experimental data reprinted with permission from SAE Paper No. 2001-01-1260 [19], © 2001 SAE International)

The main mechanism behind the cold start CO emissions development is incomplete combustion due to turbocharger lag during the early cycles as well as misfiring owing to the low ambient/coolant temperatures. Later in the warm-up phase as well as during hot starting, the higher wall temperatures promote smaller ignition delay periods and more complete combustion, hence CO emissions decrease considerably; at the same time, NO_x emissions, which are always of much lesser importance owing to the relatively low gas temperatures involved, are expected to gradually increase.

Gullett *et al.* [20] reached some very interesting findings concerning non-regulated exhaust emissions during starting, illustrated in Figure 7.14. After cold starting, an initial sharp emission peak for benzene and other BTEX (benzene, toluene, ethylbenzene, o-,m-,p-xylenes) compounds, as well as for all gas-phase PAHs and methyl-substituted PAHs was noticed. The cold start benzene peak intensity (≈15 ppm), which represents the maximum benzene concentration measured, was found to be about 100 times higher than the final steady-state emission levels (≈150 ppb, 1 min average) recorded after 5 minutes of a medium duty engine run time. Similarly, the cold start naphthalene concentration (≈1.5 ppm) was 14 times higher than its final steady-state value (≈110 ppb, 1 min average).

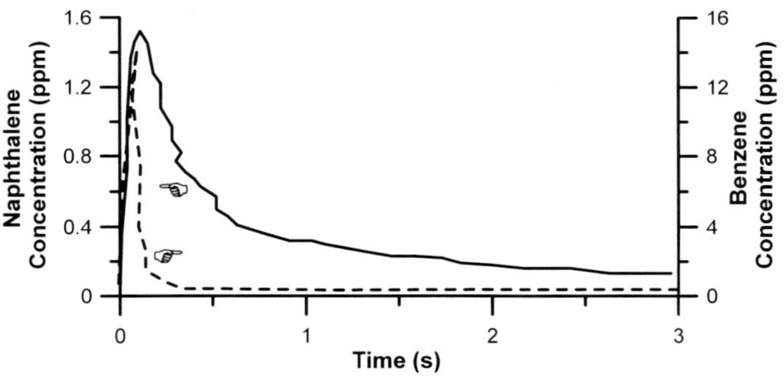

Figure 7.14. Naphthalene and benzene emissions after diesel engine cold starting (four-stroke, six-cylinder, water-cooled, medium duty, turbocharged diesel engine – experimental results reprinted from Gullett *et al.* [20], Copyright (2006), with permission from Elsevier)

Naphthalene concentration behavior is consistent with the presence of unburned fuel during the initial fuel-rich combustion. The time duration from the peak concentration to the steady-state concentration varied from seconds to several minutes; it was argued that this was a function of the specific compound and perhaps, partly, an artifact of the transfer time to the detector. Specifically for pollutants whose concentration is mainly dependent on formation reactions inside the cylinder, transient concentration discrepancies lasted only a few seconds. This was the case with benzene, whereas naphthalene took several minutes to stabilize at the final steady-state value. The transient benzene emissions after cold starting were about 20 times those during the period of steady-state operation, meaning that about 3.5 minutes of steady-state benzene emissions were equivalent to those emitted during the few seconds of cold starting. For naphthalene, a longer transient period after cold starting of about 6 min resulted in a four-fold increase over the same period steady-state emissions. Similar transient concentration phenomena were observed for hot starting events; however, peak concentrations were lower, while the duration up to final steady-state concentrations were generally shorter compared with the respective cold starting transients.

7.2 Compressor Surge

7.2.1 Surge Fundamentals

Successful matching between engine and turbocharger is of paramount importance for optimum steady-state and transient performance of a turbocharged diesel engine. Propulsion applications, particularly those encountering high rotational speeds and wide flow range such as automotive engines, require that the whole compressor operating range is exploited. As a result, under steady-state conditions, the maximum engine torque operating line on the compressor map is often located close to the compressor surge line. Although this surge margin is usually adequate under steady-state conditions, during dynamic engine operation there are certain cases where the turbocharger compressor may be driven beyond its stable operation leading to compressor surge [21–24], as a result of

- abrupt and fluctuating load changes in the vicinity of full-load conditions;
- turbocharger wear, *e.g.*, in the nozzle ring or turbine blades;
- turbocharger clogging, *e.g.*, in the intake filter/silencer, nozzle ring, turbine blades or scavenge air cooler;
- operation at high altitude;
- wear of fuel injection valves;
- switch from 1 to 2 or from 2 to 3 turbochargers in sequential turbocharging configurations (Section 6.4.6); and
- emergency shut-down.

During surge, the compressor operating point is characterized by high pressure and low air-mass flow-rate values. Moreover, the compressor flow exhibits large amplitude oscillations, resulting in severe fluctuation of the pressure downstream of the compressor, which can adversely affect the turbocharger and piping due to the vibrations generated. The almost instantaneous and reversing changes of the compressor torque introduce severe torsional loading to the turbocharger shaft.

Some of the above arguments are documented in Figure 7.15 that illustrates instantaneous compressor flow coefficient, tip Mach number and isentropic head coefficient for a deep surge cycle in a radial impeller-vaneless diffuser free-spool system that resembles a centrifugal turbocharger compressor. Surge was initiated in this case by a reduction in the area of a throttle valve located downstream of the compressor. A large plenum exists between compressor outlet and throttle valve, as is the case with inlet manifold in internal combustion engines. Three different time (frequency) scales can be identified in Figure 7.15; a Helmholz oscillation of approximately 7 Hz, a longer one (roughly 0.65 s) associated with the plenum blow-down and recovery (deep surge), and a still longer time scale associated with the rotor speed fluctuations. Four regimes exist in this surge cycle. The first is a quiescent period, with the tip Mach number increasing and the flow coefficient decreasing; during this period the compressor is moving to the left of its unstalled characteristics. In the second phase, fluctuations grow as the compressor operates in mild surge. In the third phase, the mass flow reverses and the plenum pressure decreases during blowdown. A high flow-rate is thus registered during the back-flow phase. In the fourth phase the blow-down finishes and the flow recovers rapidly to the forward direction and then gradually decreases to the flow-rate corresponding to the quiescent flow [25].

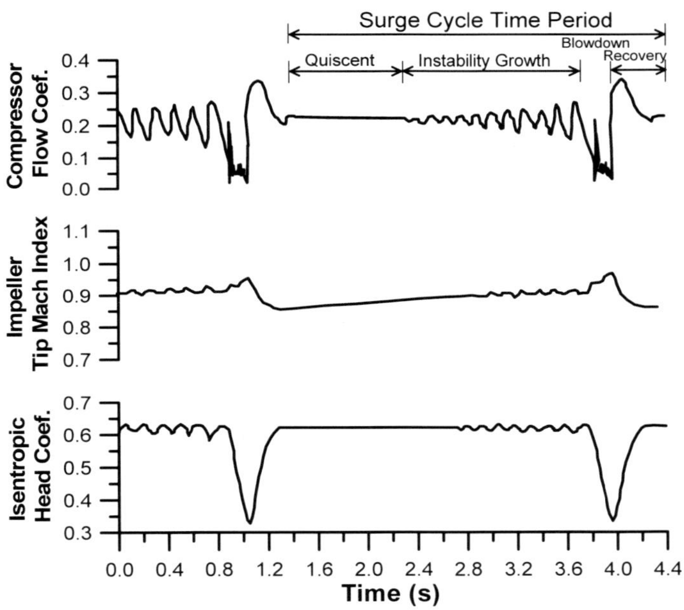

Figure 7.15. Various compressor parameters response during deep surge in a radial impeller-vaneless diffuser free-spool system (experimental results reprinted from Fink *et al.* [25], with permission from the American Society of Mechanical Engineers)

7.2.2 Compressor Surge during Diesel Engine Transient Operation

A typical, extreme engine operating case, which caused compressor surging is illustrated in Figure 7.16. In order to explain in more detail what is happening during the surge cycles, Figure 7.17 is also provided showing the oscillating loop of the compressor operating point.

Initially, the engine was running at 1200 rpm – near full-load conditions. Then, the load was rapidly reduced to 9%. Following the initial load decrease, the engine speed increased owing to the surplus of crankshaft angular momentum. Thus, the engine air-flow increased too, as this is mainly determined by engine speed (for four-stroke engines), and the compressor operating point moved towards the choke region (from 'Start' to point A in Figure 7.17). Very soon the engine load increased sharply to 87%, causing now a rapid drop of the engine speed; this was sensed by the governor that responded with increased injected fuel quantity. However, the relative delay of the governor response and, mainly, the inertia of the turbocharger rotor resulted in a still decreasing engine speed, which at about 0.6 s was lower than 800 rpm. At this moment, the compressor operating point moved from point A to point B (Figure 7.17) and surge commenced, since the still low boost pressure could not match the decreased air-mass flow-rate moving the compressor operating point to the left of the surge line into the unstable region. From this point on, the oscillating surge loops were initiated, which were consecutively evolved along the route B-C-D-E-B leading also to flow reversals. Finally, after almost 0.3 s from surge initiation, while the governor still increased the injected fuel mass in order to meet the final demand speed of 1200 rpm at 87% of full-load, the engine torque, and thus speed and air-flow, increased, resulting in the compressor operating point moving back into the stable region ('End' in Figure 7.17).

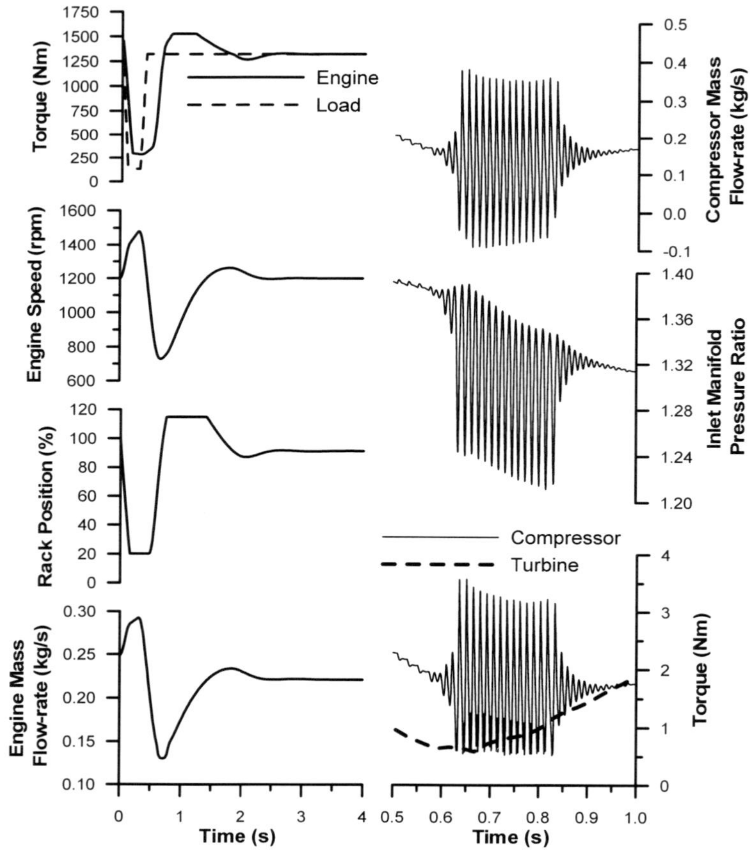

Figure 7.16. Prediction of engine transient behavior after a 96%–9%–87% load change schedule with turbocharger compressor experiencing surge (four-stroke, six-cylinder, turbocharged and aftercooled marine diesel engine) [24]

Analyzing more deeply the surge cycle, one can divide it into four different parts: E-B, B-C, C-D and D-E in Figure 7.17. Parts E-B (inlet manifold pressure build-up) and C-D (inlet manifold discharging) last longer than the other two parts, determining the filling and emptying time, respectively, of the inlet manifold and defining in this way the surge cycle period. During these two parts, the compressor air-flow changes gradually and the operating point moves in the stable region of the compressor map. The other two parts of the surge cycle, B-C and D-E, are much shorter; here, the air-supply changes very rapidly and the operating point moves mainly along unsteady flow regions of the compressor map. As shown in Figure 7.16, during compressor surging the torque absorbed by the compressor impeller varies considerably, mainly because of the air-flow reversals. In this way, severe torsional loadings were introduced to the turbocharger shaft. It is evident that such air-flow, inlet manifold pressure and turbocharger torque oscillations can harm both the engine and turbocharger operation if surging continuously occurs.

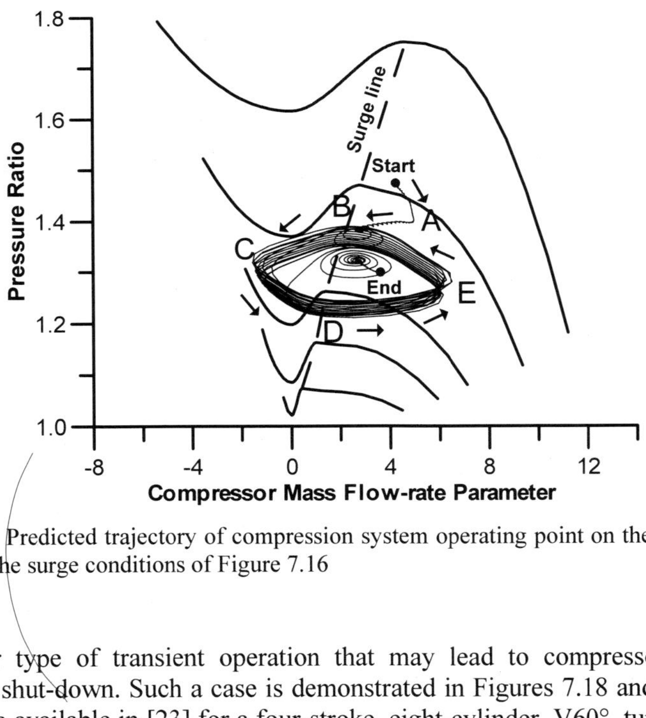

Figure 7.17. Predicted trajectory of compression system operating point on the compressor map during the surge conditions of Figure 7.16

Another type of transient operation that may lead to compressor surge is emergency shut-down. Such a case is demonstrated in Figures 7.18 and 7.19; they refer to data available in [23] for a four-stroke, eight-cylinder, V60°, turbocharged, marine diesel engine with constant pressure turbocharging, having 34.5 L displacement volume and 15.2 bar maximum bmep. For the particular engine, the emergency shut-down valve is located between the compressor outlet and the engine air-cooler, and it closes instantaneously in any case an emergency alarm is activated. Thus, the air-supply to the engine cylinders is immediately cut-off and the engine shuts down due to the lack of air to maintain combustion in the cylinders.

Initially, the engine was operating at 1500 rpm and 100% load. The emergency shut-down procedure was initiated at t = 0.9 s (point 'Start' in Figure 7.18 upper sub-diagram). At this point, the emergency shut-down valve started closing and it fully closed at 1 s. During the closing of the emergency shut-down valve the pressure after the compressor (Figure 7.19) was gradually increased, from 2.85 bar at 0.94 s to almost 3 bar at about 0.98 s. Therefore, the compressor operating point was driven into the unstable region, and the compressor flow-rate became almost instantly negative at about 0.98 s. However, the surge cycle was not completely followed since, after the negative to positive flow reversal, the shut-down valve had been fully closed, so the compressor flow became almost zero.

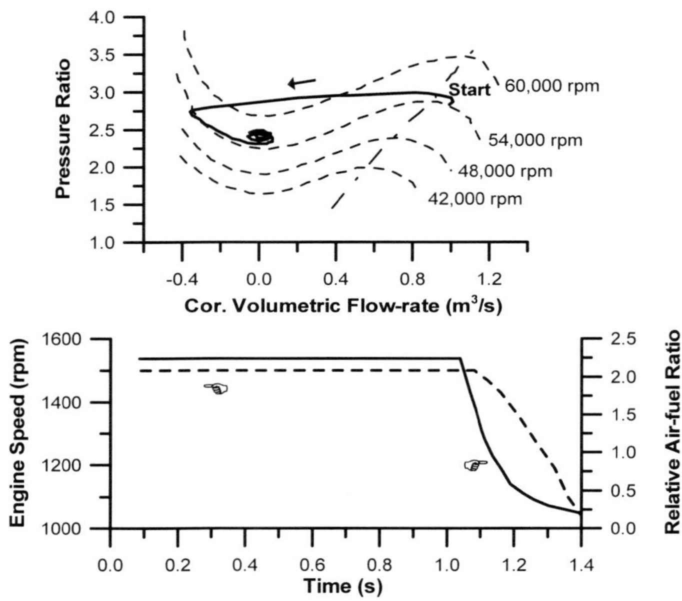

Figure 7.18. Predicted results for engine transient operation during emergency shut-down (four-stroke, eight-cylinder, V60°, turbocharged, marine diesel engine with constant pressure turbocharging – simulation results reprinted from Theotokatos and Kyrtatos [23], with permission from the American Society of Mechanical Engineers)

Owing to flow reversal, the pressure downstream the compressor was reduced to 2.3 bar at 1 s, whereas turbocharger speed increased from 55,000 rpm at 0.98 s to 57,600 rpm at about 1.1 s, because the compressor impeller absorbed torque was reduced due to the reduction of the compressor mass flow. It was only after 1.1 s that the turbocharger speed was reduced due to the reduction of the exhaust gas energy, which resulted in the reduction of the turbine torque. After the closing of the shut-down valve, and due to the turbocharger still rotating, the pressure downstream of the compressor followed the changes of the turbocharger shaft speed. The closing of the emergency shut-down valve resulted in lack of combustion air; the relative air–fuel ratio dropped below 0.6, which is the limit for combustion. Thus, the engine torque was reduced to almost zero within 0.2 s and, since the load torque was constant, the engine rapidly decelerated to stall.

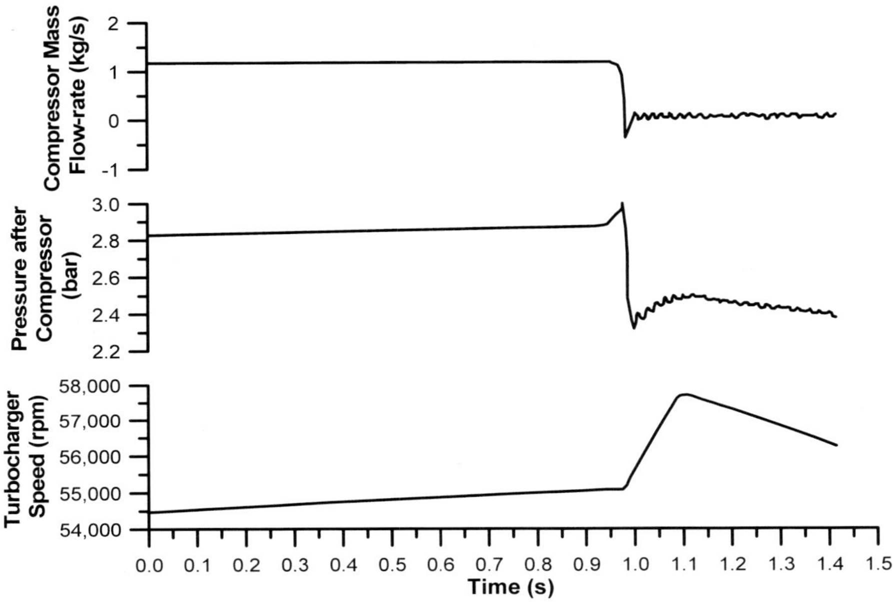

Figure 7.19. Predicted turbocharger parameters response for engine transient operation during the emergency shut-down of Figure 7.18 (simulation results reprinted from Theotokatos and Kyrtatos [23], with permission from the American Society of Mechanical Engineers)

As with almost all aspects of transient operation, compressor surge is governed by a variety of engine and turbocharger parameters; a few remarks are provided below.

- For the engine to avoid operation with compressor surge, everything that can limit speed drop is helpful, *i.e.*, high engine moment of inertia (this has a negative impact on fast speed response as was shown in Chapter 6), or low turbocharger moment of inertia, or long load change schedule (duration). The influence of the turbocharger mass moment of inertia on the compressor surge is illustrated in Figure 7.20 (*left*) for the same engine of Figure 7.16. For a clear representation three cases are examined, *viz.* 50%, 75% and 200% of the nominal value of the turbocharger inertia. Clearly, the smaller the turbocharger inertia, the greater the possibility for the compressor to avoid surge. On the contrary, as turbocharger inertia increases, its transient response becomes slower. Because of this lag, engine speed drops further, and so does the compressor air-mass flow-rate. Only when the inertia reduces to 50% of its nominal value, does the turbocharger of the present engine set-up manage to respond quickly enough, thus avoiding surge initiation for the particular load change schedule.

- Inlet manifold volume does not influence the amplitude of the compressor mass flow and pressure oscillations; however, it defines the frequency of the surge cycles. An increase in the inlet manifold volume results in reduction of the surge loops frequency.

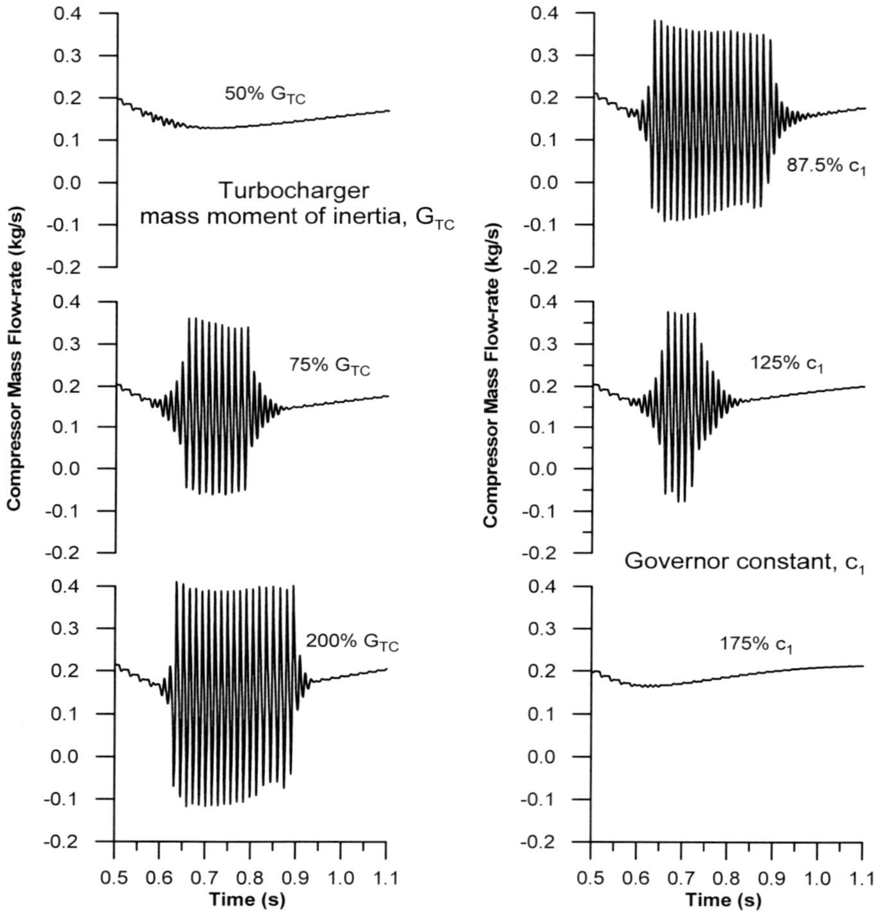

Figure 7.20. Effect of turbocharger mass moment of inertia G_{TC} (*left*) and speed governor 'c_1'-constant (defined in Equation 9.107b) (*right*) on compressor surge (simulated results for a four-stroke, six cylinder, turbocharged and aftercooled marine diesel engine, operating on severe load changes commencing from conditions close to the surge line)

- The compressor duct length has been shown to have no traceable effects either on the amplitude of the oscillations or the surge cycle period, particularly so for changes within reasonable limits.

- The impact of the governor P-constant 'c_l' on the compressor behavior during surge is demonstrated in Figure 7.20 (right). This constant is also known as 'the governor gain', and appears in the numerator of the second term on the right-hand side of Equation 9.107b; it describes how quickly the governor responds to the changes in engine rotational speed. The difference in the surge pattern, generated by the governor gain variation, is depicted for three values, *i.e.*, 88.5%, 125% and 175% of the nominal one. It is obvious that an increase in the governor gain leads to surge elimination, owing to the smaller engine speed droop. On the other hand, as its value decreases, surge intensity is amplified, rendering the system more unstable. In particular, reduction of the governor gain at half its nominal value results in almost double surge duration. Besides, excessive decrease in the governor gain value results in unacceptable oscillating engine response, together with a prolonged recovery period. Values of coefficient c_l higher than 150% of the nominal one keep the compressor operating point away from the surge area, for the particular engine-load set-up.

7.3 Low-heat Rejection Operation

The importance of heat transfer to the combustion chamber walls of reciprocating engines has been recognized from the early stages of their development. In recent years, the interest in the heat transfer phenomena in internal combustion engines (both spark and compression ignition) has been greatly intensified because of their major importance, among other things, on the thermal loading at critical places in the combustion chamber components. Heat transfer in compression and spark ignition engines is extremely complex, since the relevant phenomena are of transient nature even under steady-state engine operation, three-dimensional, and subject to rapid variations in cylinder gas pressures and temperatures during an engine cycle.

Moreover, during the last decades, there has been an increasing interest in the low-heat rejection (LHR, or sometimes loosely termed 'adiabatic') diesel engine. The objective of a low-heat rejection cylinder is to minimize heat loss to the walls, ultimately eliminating the need for a cooling system. This is achieved through the increased level of (wall) temperatures resulting from insulation of the cylinder walls, piston crown, cylinder head, or valves. By so doing, a reduction can be established in ignition delay and thus combustion noise, and also in hydrocarbon and particulate matter emissions. At the same time, an increase in exhaust gas energy is noticed; the latter can be recovered using some kind of bottoming device, *e.g.*, Rankine cycle or a turbo-compound configuration. A major issue here is the decrease in the volumetric efficiency, which affects adversely the power output, and the increase in NO_x emissions. Worsening of combustion as well as of the lubricating oil properties have also been reported because of the increased levels of insulation [26–31].

7.3.1 Load or Speed Increase Transients

The effects of low-heat rejection operation on the engine transient response will be discussed through the following representative case study, which concerns a four-stroke, six-cylinder, turbocharged and aftercooled diesel engine operating on load acceptance transients [32]. The baseline, non-insulated, case configuration is that corresponding to the engine in hand, possessing a cast iron wall thickness of 10 mm (conductivity 54 W/(m K), thermal difussivity 14×10^{-6} m^2/s). Three, quite different insulation schemes are examined, namely

1. 4.0 mm thick silicon nitride SN (conductivity 10 W/(m K), thermal diffusivity 2.80×10^{-6} m^2/s) slice sandwiched on the gas side of a 6 mm thickness cast iron wall;
2. 1.0 mm thick partially stabilized (PSZ) plasma spray coating (conductivity 1.0 W/(m K), thermal diffusivity 0.90×10^{-6} m^2/s) applied on the gas side of a 9 mm thickness cast iron wall; and
3. 1.5 mm thick partially stabilized plasma spray coating applied on the gas side of a 8.5 mm thickness cast iron wall.

A case with aluminum wall of 10 mm thickness (conductivity 180 W/(m K), thermal diffusivity 78×10^{-6} m^2/s) will be also included in the analysis for comparison purposes.

In Figure 7.21 the response of six important engine values is illustrated, *i.e.*, engine speed, fuel pump rack position, NO_x emissions, boost pressure, turbocharger speed and volumetric efficiency. Clearly, the engine speed as well as the other variables of the engine and turbocharger are only slightly affected by the level of temperatures inside the cylinder, with the highly insulated cases leading to slightly smaller speed droops. Similar speed response improvements were reached by Watson [33] and Schorn *et al.* [34] for acceleration transients (see also Figure 7.24 later in this section).

As was discussed in Section 2.1, the most notable feature of turbocharged diesel engine transient operation is the turbocharger lag. Turbocharger lag is caused by the lack of mechanical connection between turbocharger compressor and engine crankshaft. Consequently, the power delivered to the turbine must first accelerate the turbocharger shaft in order for the compressor to be able to produce the increased boost pressure. Other important delays concern the use of fuel limiters, heat losses to the cylinder and exhaust manifold walls, and the acceleration of the rotating masses. Therefore, a higher wall temperature is generally expected to improve turbocharger lag and thus speed response, as the effect of one of the above-mentioned 'decelerators', namely heat loss to the walls (owing to the increased fueling after the new loading has been applied) decreases considerably. Similar effects have been reported for the case where the exhaust manifold wall is insulated.

As a result, higher turbocharger speeds are noticed at the end of the transient event, although the overall improvement in turbocharger lag is, in general, not so pronounced, particularly so for the engine–load configuration of Figure 7.21, which is further characterized by tight governing, moderate turbocharging and high total mass moment of inertia that slows down the whole transient response. Things are quite different, however, as regards NO_x emissions development, as these depend strongly on gas temperatures; hence, they increase accordingly for the highly insulated engine configurations.

Closer examination of the curves in Figure 7.21 reveals that, apart from NO_x emissions, the volumetric efficiency (based on atmospheric conditions) exhibits a different profile; namely, a clearly decreasing trend is observed with higher degree of insulation schemes. The increased level of cylinder wall temperatures during the induction process transfers heat to the incoming charge, thus reducing its density and, hence, volumetric efficiency throughout the transient event. This behavior goes along with the findings reached for steady-state operation.

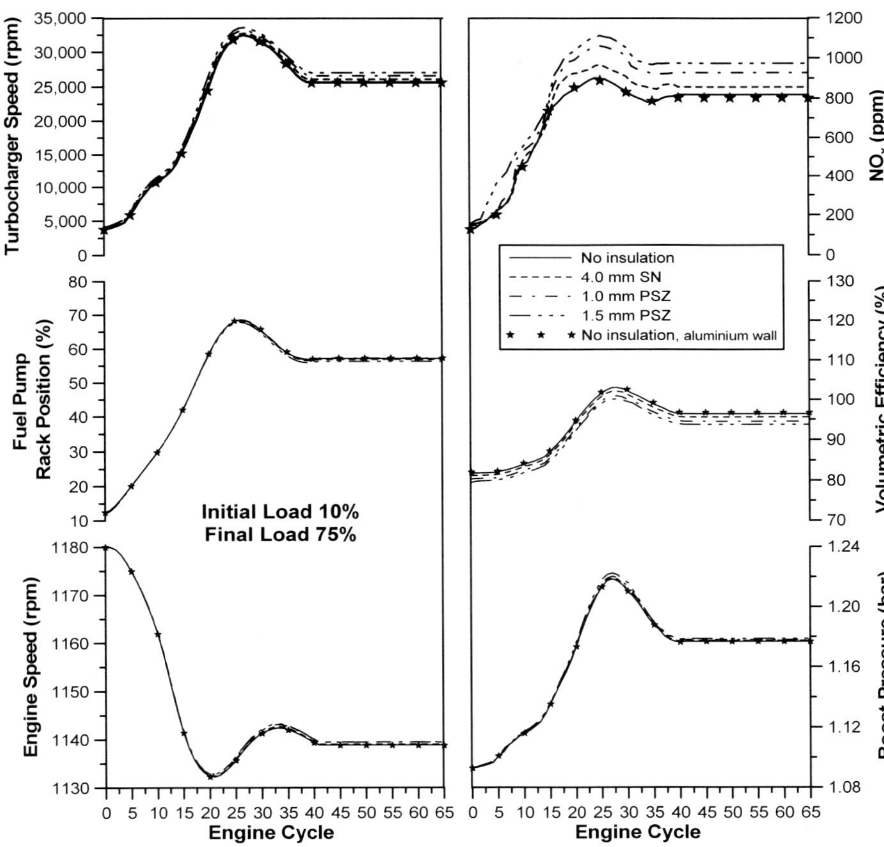

Figure 7.21. Response of various LHR engine parameters to a 10–75% increase in load (four-stroke, six-cylinder, moderately turbocharged and aftercooled diesel engine – unless otherwise noted, the cylinder wall is cast iron) [32]

Figure 7.22 extends the results of Figure 7.21 by showing the response of the transient indicated specific fuel consumption (isfc) of the engine, highlighting the benefit, gained from the insulation. The increased level of temperatures during each cycle has led to 'fuller' pressures diagrams due to increased expansion work and slightly reduced pumping losses, thus the efficiency has increased up to 3.65% for the 1.5 mm PSZ case at the 20th cycle of the transient event. It is obvious that the engine handles now the transient test in a slightly more efficient manner. However, the increase in isfc is much smaller than would be expected judging from the corresponding increase in gas temperatures (Figure 7.23) (*cf.* the second-law analysis in Section 8.5). Various theories have been reported on the subject, with the most profound being perhaps the one proposed by Alkidas [29]. He argued that an increased level of insulation leads to some kind of 'deterioration' of the combustion process, by shifting a larger part of fuel burning into the diffusion phase, *i.e.,* later in the cycle. Schwartz *et al.* [35] extending the results of Alkidas, concluded that an improved time phasing of premixed and diffusion burning must be accomplished if a desirable heat release is to be derived in a LHR engine, as the heat release pattern exhibits a great sensitivity to even small changes.

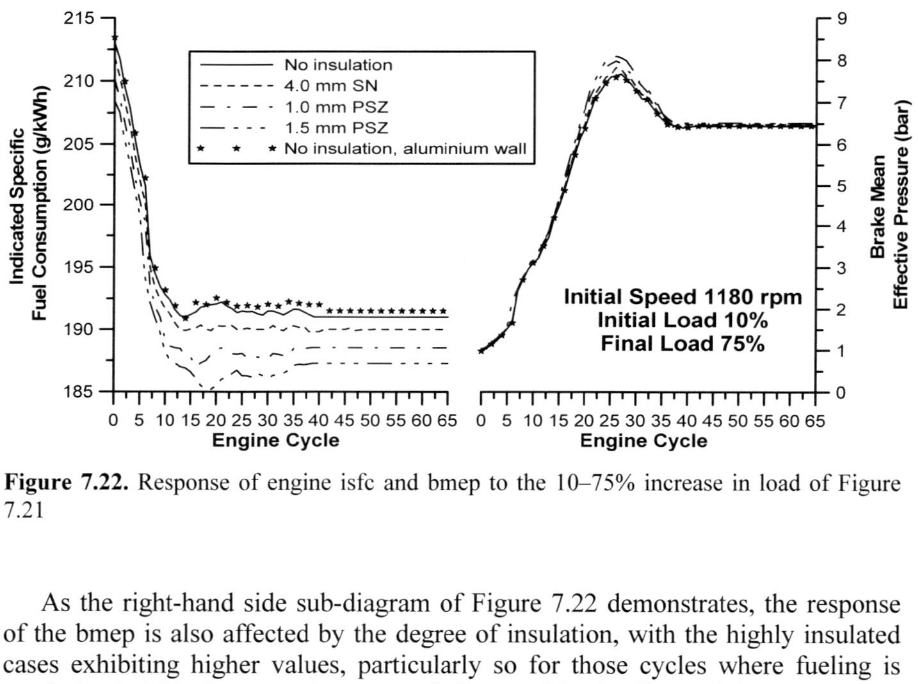

Figure 7.22. Response of engine isfc and bmep to the 10–75% increase in load of Figure 7.21

As the right-hand side sub-diagram of Figure 7.22 demonstrates, the response of the bmep is also affected by the degree of insulation, with the highly insulated cases exhibiting higher values, particularly so for those cycles where fueling is maximized. This happens because of the combined effect of improved isfc, decrease in the volumetric efficiency and, mainly, change in the amount of injected fuel due to the transient operation of the fuel pump.

In Figures 7.21 and 7.22 the difference observed between aluminum and cast iron non-insulated walls is modest, suggesting the small effect of the cylinder wall material on the cycle transient response (although the respective profiles of the wall temperatures and heat flux to the cylinder walls in Figure 7.23 are slightly differentiated). The same negligible difference is observed if similarly insulated cases are examined (*e.g.,* cast iron or aluminum walls both coated with 1.0 mm PSZ). Clearly, it is the insulator's, and not the wall's, thermal properties that are primarily responsible for the decrease in the total thermal conductivity of the insulated cylinder wall.

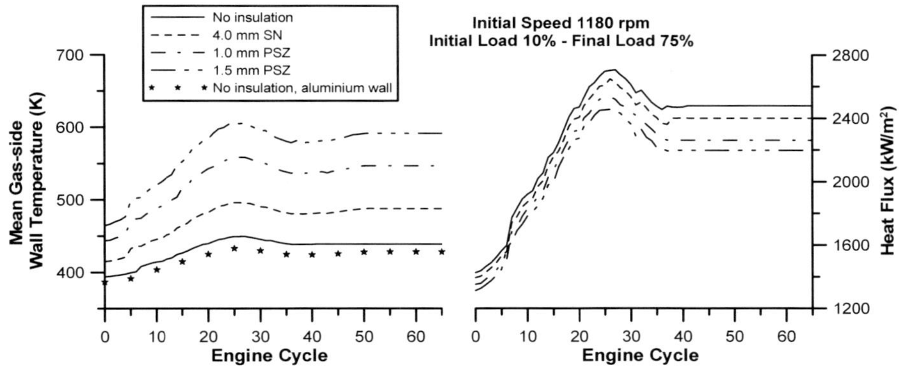

Figure 7.23. Response of mean gas-side wall temperatures and respective heat flux for the load increase of Figure 7.21

In general, similar trends hold for acceleration transients. Figure 7.24 illustrates the relatively modest (although not negligible) improvement obtained through insulation with Al_2TiO_5 of the exhaust manifold walls, on the full-load (high gear) acceleration of a turbocharged, passenger vehicle engine commencing from 2000 rpm.

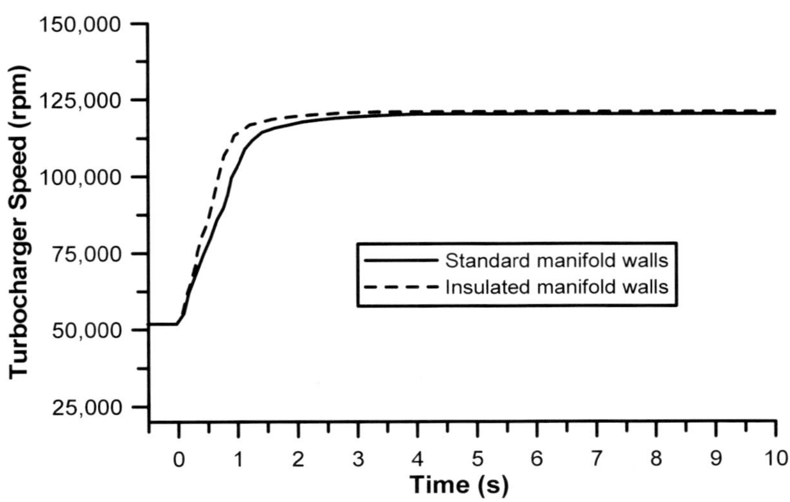

Figure 7.24. Influence of insulated exhaust manifold walls on the turbocharger acceleration of a four-cylinder engine on a 1500 kg passenger vehicle (initial conditions: engine speed 2000 rpm, vehicle speed 37.8 km/h, fuel pump rack position at full-load, without gear shift change – simulation results adapted from Schorn *et al.* [34])

Overall, it can be concluded that

- an increased degree of insulation during a transient load or speed increase results in increased temperatures and heat transfer coefficients;
- the main engine and turbocharger properties are moderately affected;
- speed response and recovery period are slightly improved owing to the reduced amount of heat lost to the cylinder walls during the period of increased fueling. The specific engine characteristics, *e.g.*, governor curves or engine moment of inertia play an important role on the exact degree of improvement; for example, the tighter the engine governing or the higher the moment of inertia, the lower the established gain from insulation. Moreover, the gains from insulation compared with uninsulated transient operation are expected to fade away the higher the engine speeds, owing to the smaller time available for heat transfer;
- the transient response of the engine parameters exhibits a behavior that goes along with the respective steady-state performance. Namely, the higher temperature levels observed throughout the cycle result in decreased volumetric efficiency, increased NO_x emissions and slightly improved indicated efficiency; and
- no significant distinction is noticed between cast iron and aluminum walls as regards engine transient response.

Further aspects of LHR operation on transient diesel engine response will be discussed in Section 8.5, in terms of second-law balance of the engine processes.

7.3.2 Short Term Temperature Oscillations

An important aspect of engine heat transfer is the (transient) temperature and heat flux variations in the combustion chamber walls. Even under steady-state conditions, these produce considerable temperature gradients inside the cylinder wall components. During a load or speed increase transient event yet greater heat flux rates towards the cylinder walls are noticed, as was depicted in Figure 7.23; as a result, a thermal transient propagates through the walls into the components of the engine, known as thermal shock. The latter is characterized by sharp temperature gradients that may result in high local stress and even material fatigue. The temperature variations in the combustion chamber walls can be divided into two main categories:

- the long term response variations, resulting from the large time scale (of the order of seconds), non-periodic variations of engine speed and/or load; and
- the short term response variations, which are the result of the fluctuations of gas pressure and temperature during an engine cycle, having a time period of the order of ms.

Unlike their long term counterparts, the short term temperature variations develop fully during each engine cycle. This was the result of the work by Assanis and

Heywood [36], who studied the development of the short term response temperature variations in engine cylinder walls that were partially insulated with plasma spray zirconia (PSZ); they highlighted the increased magnitude (of the order of more than 100 K) of the developed temperature oscillations during a steady-state engine cycle. Wong *et al.* [37] investigated the interdependence between engine efficiency and temperature swings amplitude, whereas Rakopoulos *et al.* [38, 39] expanded the analysis by including the effect of silicon nitride insulation and provided comparative results for the various insulation schemes applied. Particularly so during transient operation, these temperature oscillations can assume even greater values, since fueling hence gas temperatures increase significantly within a few cycles after a load or speed increase.

The heat transfer rate from the gas to the walls can be assumed to be a harmonic function of time with a period of one engine cycle. As a result, periodic temperature waves propagate into the wall structure, which nonetheless die out already at a small distance (of very few mm) from the wall inside surface, beyond which the temperature distribution is at steady-state.

In order to calculate the heat transfer rate through a certain location of the combustion chamber wall during a complete engine cycle, the unsteady heat conduction equation must be solved with the appropriate boundary conditions as illustrated in Figure 7.25.

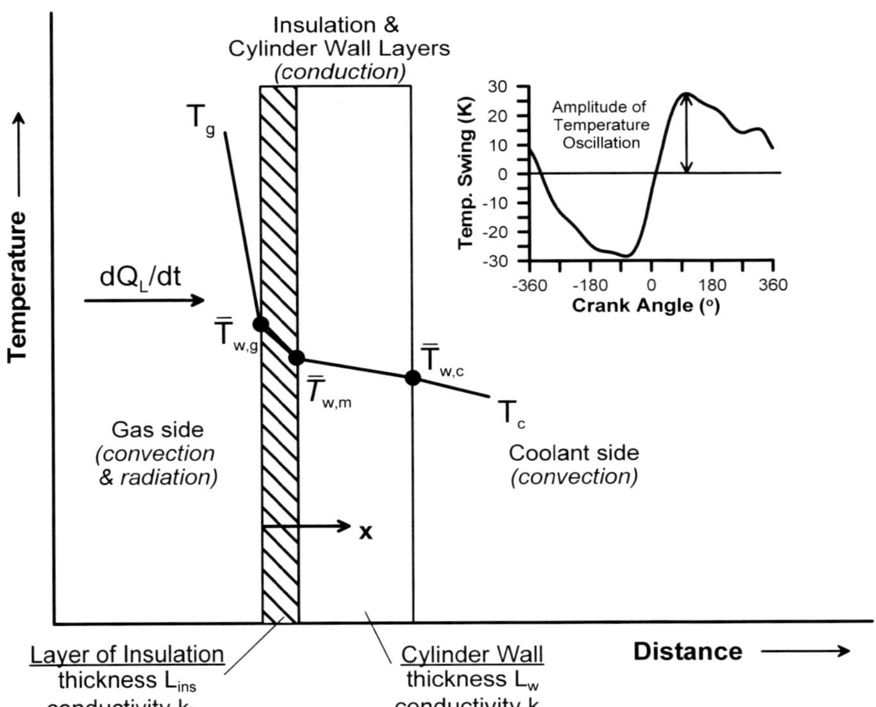

Figure 7.25. Engine cylinder - coolant heat transfer flow, including definition of temperature swing amplitude

Total temperature, *i.e.*, steady-state $\overline{T}_w$ plus time-periodic T_p at any point x within the wall and at any time t, should satisfy the unsteady one-dimensional heat conduction equation

$$\frac{\partial T}{\partial t} = \alpha \frac{\partial^2 T}{\partial x^2} \tag{7.4}$$

where α is the wall (material) thermal diffusivity. The solution of this equation is accomplished by decomposing the problem into its steady-state and time-periodic components.

7.3.2.1 Steady-state Heat Conduction Problem

Applying the boundary conditions to all wall sides (gas-side, coolant side and end of insulation side) of a four-stroke diesel engine, the following equation is obtained for the general case where also a layer of insulation is present with reference to Figure 7.25 (this is an expansion of the analysis presented in Section 2.3.1):

$$\frac{1}{4\pi}\int_0^{4\pi} \frac{dQ_L}{d\varphi} d\varphi = A\,\frac{k_{ins}}{S_{ins}}(\overline{T}_{w,g} - \overline{T}_{w,m}) = A\,\frac{k_w}{S_w}(\overline{T}_{w,m} - \overline{T}_{w,c}) = A\,h_c(\overline{T}_{w,c} - T_c) \tag{7.5}$$

where $dQ_L/d\varphi$ is the heat flux to the cylinder walls (analyzed in Section 9.3.2.2), S_{ins} the thickness of the insulation layer with k_{ins} its thermal conductivity, S_w the cylinder wall thickness with k_w its thermal conductivity, and h_c the heat transfer coefficient from the external wall side (respective temperature $\overline{T}_{w,c}$) to the coolant; the overbar denotes mean temperatures over an engine cycle. Equations 7.5 are solved for the three unknown variables, *i.e.*, the wall temperatures $\overline{T}_{w,g}$, $\overline{T}_{w,m}$ and $\overline{T}_{w,c}$, which change from cycle to cycle during the transient event but can be considered to remain constant throughout each cycle, hence the 'steady-state' definition.

7.3.2.2 Time-periodic Heat Conduction Problem

The time-periodic part $T_p(x,t)$, at any point x within the wall and at any time t, will satisfy the unsteady one-dimensional heat conduction equation within a parallel slab having a thermal diffusivity α:

$$\frac{\partial T_p}{\partial t} = \alpha \frac{\partial^2 T_p}{\partial x^2} \tag{7.6}$$

This continuous partial differential equation can be solved analytically using Fourier analysis techniques, even though during transients $T_p(x,t)$ is not a periodic function. Equation 7.6 is subjected to the boundary condition of the inside wall surface, at x=0, to be exposed to the gas temperature that varies periodically in time,

$$-k_{ins}\left.\frac{\partial T_p}{\partial x}\right|_{x=0} = \bar{h}_g (T_p - T_{pg}) \tag{7.7}$$

with the respective mean heat transfer coefficient from gas to cylinder wall given by

$$\bar{h}_g = \frac{1}{4\pi}\int_0^{4\pi} h_g d\varphi \tag{7.8}$$

The time-periodic part of gas temperature $T_{pg}(t)$ can be expressed as a Fourier series, in the following form:

$$T_{pg}(t) = \sum_{n=1}^{\infty}\left[A_n \cos\left(\frac{2\pi n}{\tau_o} t\right) + B_n \sin\left(\frac{2\pi n}{\tau_o} t\right)\right] \tag{7.9}$$

where τ_o is the time period of the temperature oscillation, which for a four-stroke engine corresponds to a frequency that is half the engine speed. For the wall temperature we get

$$T_{pw} = \sum_{n=1}^{\infty}\frac{\exp(-\xi_n x)}{\sqrt{1+2\zeta_n + 2\zeta_n^2}}\left[C_n \cos\left(\frac{2\pi n}{\tau_o} t - \xi_n x - \theta_n - \delta_n\right)\right] \tag{7.10}$$

where the following quantities are valid for Equations 7.9 and 7.10 ($k_{ins} \equiv k_w$ for the non-insulated wall case):

$$C_n = \sqrt{A_n^2 + B_n^2}$$

$$A_n = \frac{2}{\tau_o}\int_0^{\tau_o} T_g(t)\cos\left(\frac{2\pi n}{\tau_o} t\right) dt \quad , \quad B_n = \frac{2}{\tau_o}\int_0^{\tau_o} T_g(t)\sin\left(\frac{2\pi n}{\tau_o} t\right) dt$$

$$A_o = \bar{T}_g = \frac{1}{\tau_o}\int_0^{\tau_o} T_g(t) dt \tag{7.11}$$

$$\delta_n = \tan^{-1}\left(B_n / A_n\right) \quad , \quad \xi_n = \sqrt{\pi n / \alpha\tau_o} \quad , \quad \theta_n = \tan^{-1}\left(\frac{1}{1+(1/\zeta_n)}\right)$$

$$\zeta_n = \xi_n \left(k_{ins} / h\right) = \sqrt{\left(\pi n k_{ins}^2\right) / \left(\alpha\tau_o h^2\right)}$$

Term $C_n \cos(\xi_n x) = C_n \cos\left[\left(\sqrt{\pi n/\alpha\tau_o}\right)x\right]$ from Equation 7.10 represents a cosine wave of amplitude C_n and wavelength x_o, calculated from the condition $\left(\sqrt{\pi n/\alpha\tau_o}\right)x_o = 2\pi$ and n=1, *i.e.*,

$$x_o = 2\sqrt{\pi\alpha\tau_o} \tag{7.12}$$

Figure 7.26 illustrates the short term temperature oscillations during the same 10–75% load increase transient event of Figure 7.21, showing the response of the amplitude of the swing (defined in Figure 7.25). Whereas in the base, non-insulated, case the amplitude of the oscillation is about 2–3 K, values of up to 64 K at cycle 23 are experienced for the 10–95% load change when the engine is coated with 1.0 mm PSZ. Even higher amplitude is expected for higher load changes or thicker insulation. From Figure 7.26 it is also concluded that there is no practical difference between cast iron and aluminum wall. What should be mentioned here is the high rate of change of the swing amplitude between cycles 5 and 25, where the main portion of the increase in fueling takes place. This rate is of the order of 4 K per cycle for the 10–95% load change or 2 K per cycle for the 10–75% one, and is responsible for the increased thermal loading that is experienced by the engine structure during the transient event.

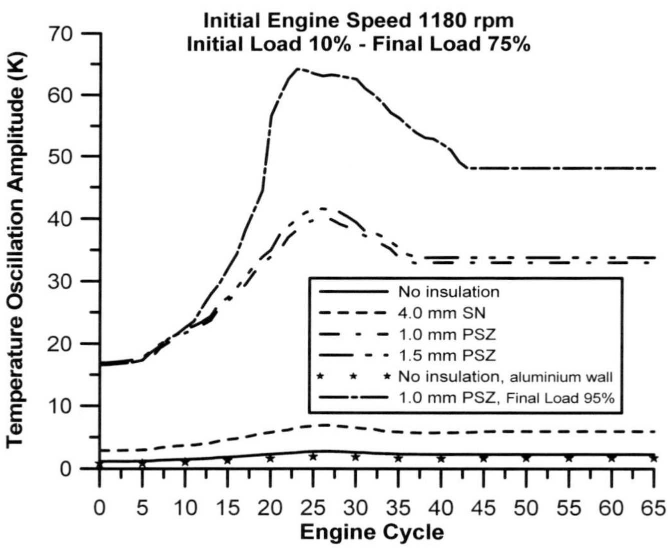

Figure 7.26. Response of the temperature swing amplitude to an increase in load (unless otherwise noted, the cylinder wall is cast iron and the load change 10–75%)

Figure 7.27 provides in more detail the cylinder inside wall temperature oscillation swings against crank angle, for an intermediate cycle of the transient event. A first observation is that the shape of all these curves is similar, a fact that is explained as follows. During the intake stroke, the wall surface is cooled down due to heat transfer from the higher temperature wall to the cooler incoming air. The wall temperature rises during the last stages of the compression stroke, because of heat transfer to the wall from the then higher temperature compressed air. During the middle and last stages of the expansion stroke, the wall temperature decreases because of the decreasing gas temperatures. These wall temperature oscillations have been found to increase with fueling. They also increase with the increasing degree of insulation, however at a decreasing rate.

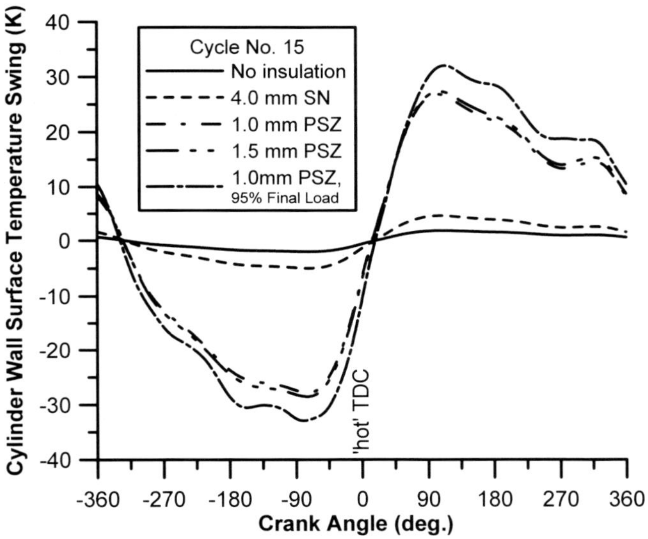

Figure 7.27. Cylinder wall inside temperature oscillation swings against crank angle, during an intermediate cycle of a transient load increase of 10–75%

From Equation 7.10 one can see that the wall temperature swing is proportional to the in-cylinder gas temperature levels (A_n, B_n) and a decreasing function of the parameter $\zeta^2 \propto k_{ins}^2/\alpha h^2$ (note that a constant engine speed is assumed, *i.e.*, constant $N=1/\tau_o$, during each transient cycle). However, the variation of heat transfer coefficient with the degree of insulation is very mild, so that effectively $\zeta^2 \propto k_{ins}^2/\alpha$. Parameter k_{ins}^2/α for cast iron, silicon nitride and plasma spray zirconia, assumes the following values, respectively: 2.1×10^8, 0.36×10^8 and 0.11×10^8 (in SI units). Thus, ζ^2 is a decreasing function of the degree of insulation. The strong increase of the wall temperature swing with the degree of insulation can then be explained by both the higher gas temperatures (*cf.* Figure 7.23) and the lower values of ζ^2. Nonetheless, this increase with the degree of insulation is forwarded at a decreasing rate. This is explained by the fact that, at high degrees of insulation (*e.g.*, for the zirconia cases), the value of ζ^2 becomes very small with respect to 1. Then, the influence on the wall temperature swing value derives effectively only from the relatively milder increase of the gas temperature swing with the degree of insulation.

Figure 7.28 expands on the temperature swing development, depicting the response of the crank angle where the maximum value is observed as well as the depth where the swing dies out. As the transient event develops, the governor responds to the initial speed drop by increasing fueling considerably in order for the engine to cope with the higher loading. This leads to higher values of flame temperatures and an increased duration of combustion, which gradually shifts the crank angle at which the maximum swing is observed later in the expansion stroke. Notice that the respective crank angle at the point of maximum heat transfer coefficient is around 'hot' TDC, whereas the crank angle where the maximum cylinder temperature occurs is 40°CA after TDC. The type of insulation is of much lesser importance here, as all the graphs exhibit the same profile. Similarly, the increased fueling and gas temperatures lead to greater penetration depth of the observed temperature swings. Consistent with the results reached for steady-state engine operation, the heavy insulated cases exhibit a smaller depth of penetration; the latter, combined with the fact that the higher insulated cases lead also to greater temperature amplitudes, is responsible for the much sharper temperature gradients, hence thermal shock experienced by the combustion chamber walls.

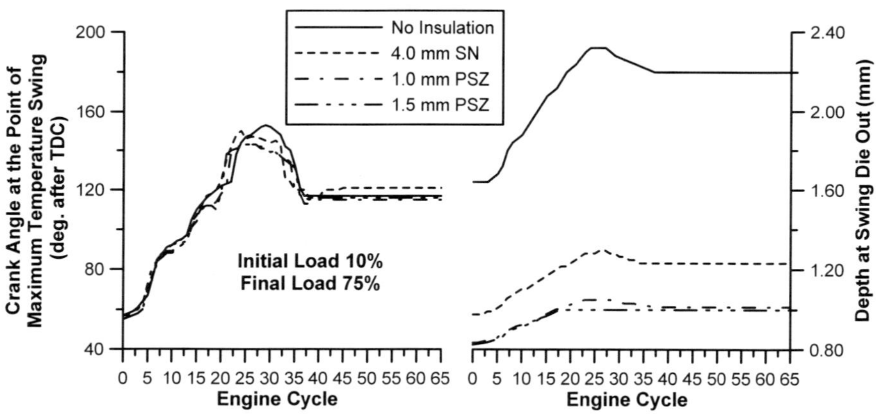

Figure 7.28. Response of crank angle at the point of maximum temperature swing and depth where swing dies out after a ramp increase in load, for various insulation schemes

The latter arguments are illustrated in more detail in Figure 7.29, which shows the variation of wall temperature swings with depth x (inside the wall) extending over a wavelength x_o, at various crank angles (instants of time) extending over the current engine cycle period τ_o (corresponding to 720°CA), at cycle No. 15 of the same 10–75% transient event. Similar type profiles develop during all cycles of the transient event. The strong damping of the wall temperature swings with depth x is apparent. Effectively, the wall swings disappear at a depth of the order of 0.7–3 mm depending on the degree of insulation. It can clearly be seen that when increasing the degree of insulation the amplitude of the wall temperature swings increase highly, while at the same time the corresponding depths inside which they disappear decrease. As a result thermal shock increases too.

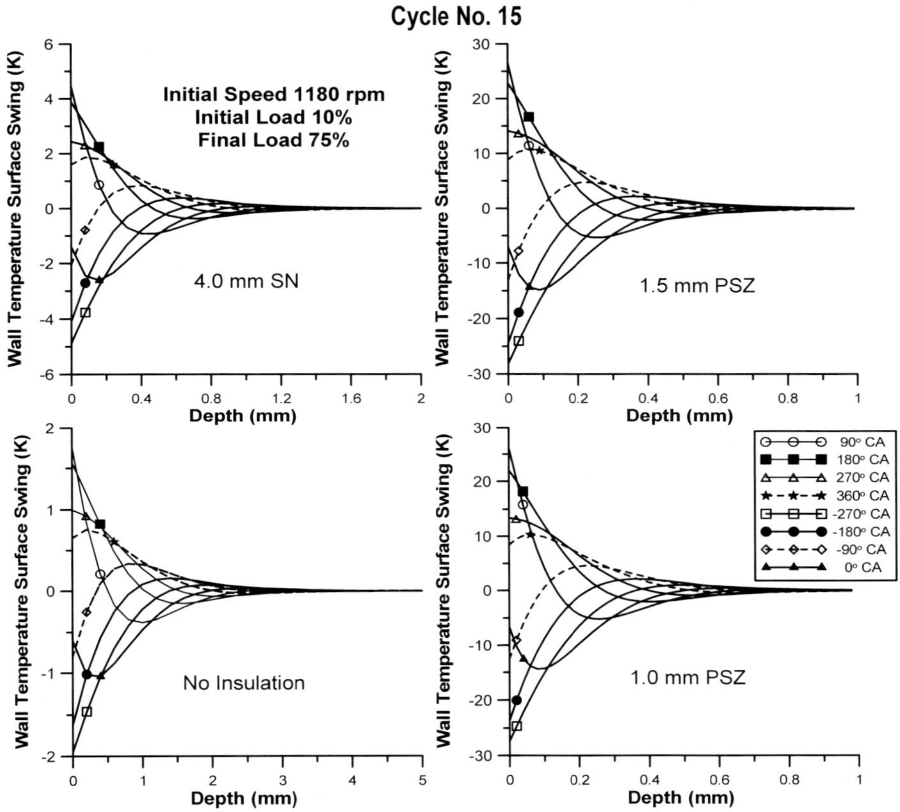

Figure 7.29. Variation of wall temperature swings with depth x (inside the wall) extending over a wavelength, at various crank angles, for the various insulation schemes during an intermediate cycle (No. 15) of a 10–75% load increase transient event [40]

From Equation 7.9 one can see that the wall temperature swing is proportional to a decreasing function of parameter $\zeta^2 \propto 1/h^2$ (for a specific speed, *i.e.*, constant $N \propto 1/\tau_o$, and for the same wall material, *i.e.*, constant k_{ins} and α). From the same equation it can also be deducted that the wall temperature swing phase shifting is an increasing function of parameter ζ. The strong increase of the wall temperature swing with fueling, as the transient event develops, can then be explained by both the higher gas temperatures and gas heat transfer coefficients, *i.e.*, lower ζ. The decrease of depth inside which these oscillations disappear, when increasing the degree of insulation, is explained by noting from Equation 7.12 that the related wavelength $x_o \propto \alpha^{1/2}$ increases with 'α' decreasing in value with the degree of insulation. The combination of large temperature gradients with short penetration lengths, as for example in the zirconia insulation cases, results in high thermal gradients, thus leading to high cyclic thermal loading. Therefore, the material concerned should possess high fatigue durability and thermal shock resistance as well as adequate temperature strength.

References

[1] Schulmeister R. Improvement of load acceptance characteristics of turbocharged diesel engines. MTZ 1980;41:225–32 (in German).

[2] Meyer WE, DeCarolis JJ. Compression temperatures in diesel engines under starting conditions. SAE Paper No. 620521, 1962.

[3] Phatak R, Nakamura T. Cold startability of open chamber direct injection diesel engines. Part I: measurement technique and effects of compression ratio. SAE Paper No. 831335, 1983.

[4] Tsunemoto H, Yamada T, Ishitani H. The transient performance during acceleration in a passenger car diesel engine at the lower temperature operation. SAE Paper No. 850113, 1985.

[5] Gonzalez D. MA, Borman GL, Reitz RD. A study of diesel cold starting using both cycle analysis and multi dimensional calculations. SAE Paper No. 910180, 1991.

[6] Henein NA, Zahdeh AR, Yassine MK, Bryzik W. Diesel engine cold starting: combustion instability. SAE Paper No. 920005, 1992.

[7] Kobayashi A, Kurashima A, Endo S. Analysis of cold start combustion in a direct injection diesel engine. SAE Paper No. 840106, 1984.

[8] Liu H, Henein NA, Bryzik W. Simulation of diesel engines cold-start. SAE Paper No. 2003-01-0080, 2003.

[9] Osuka I, Nishimura M, Tanaka Y, Miyaki M. Benefits of new fuel injection system technology on cold startability of diesel engines – improvement of cold startability and white smoke reduction by means of multi injection with common rail fuel system (ECD-U2). SAE Paper No. 940586, 1994.

[10] Taylor RI. Engine friction: the influence of lubricant rheology. Proc Inst Mech Eng, Part J, J Eng Tribol 1997;211:235–46.

[11] Shayler PJ, Baylis WS, Murphy M. Main bearing friction and thermal interaction during the early seconds of cold engine operation. ASME Trans, J Eng Gas Turbines Power, 2005;127:197–205.

[12] Marek SL, Henein NA. Transient engine and piston friction during starting. SAE Paper No. 922197, 1992.

[13] Arcoumanis C, Yao XG. Transient smoke and unburnt hydrocarbon emissions during cold-start in a turbo-charged DI diesel engine. Institution of Mechanical Engineers, Seminar on 'Transient Performance of Engines', pp. 43–60, 1994.

[14] Peng H-Y, Cui Y, Deng K-Y, Shi L, Li L-G. Combustion and emissions of a direct-injection diesel engine during cold start under different exhaust valve closing timing conditions. Proc Inst Mech Eng, Part D, J Automobile Eng, 2008;222:119–29.

[15] Xiao G-F, Qiao X-Q, Huang Z, Chen Z-P. Improvement of startability of direct-injection diesel engines by oxygen-enriched intake air. Proc Inst Mech Eng, Part D, J Automobile Eng, 2007;221:1453–65.

[16] Armas O, Hernández JJ, Cárdenas MD. Reduction of diesel smoke opacity from vegetable oil methyl esters during transient operation. Fuel 2006;85:2427–38.

[17] Mathis U, Mohr M, Forss A-M. Comprehensive particle characterization of modern gasoline and diesel passenger cars at low ambient temperatures. Atmos Environ 2005;39:107–17.

[18] Graham L. Chemical characterization of emissions from advanced technology light-duty vehicles. Atmos Environ 2005;39:2385–98.
[19] Bielaczyc P, Merkisz L, Pielecha J. Investigation of exhaust emissions from DI diesel engine during cold and warm start. SAE Paper No. 2001-01-1260, 2001.
[20] Gullett BK, Touati A, Oudejans L, Ryan SP. Real-time emission characterization of organic air toxic pollutants during steady state and transient operation of a medium duty diesel engine. Atmos Environ 2006;40:4037–47.
[21] Watson N, Janota MS. Turbocharging the internal combustion engine. London: MacMillan, 1982.
[22] Skopil M, Bulaty T. Calculation of compressor surge with unsteady flow model. Institution of Mechanical Engineers, 6th International Conference on 'Turbocharging and Air-Management Systems', Paper C554/018, pp. 219–25, 1998.
[23] Theotokatos G, Kyrtatos NP. Investigation of a large high-speed diesel engine transient behavior including compressor surging and emergency shut down. ASME Trans, J Eng Gas Turbines Power 2003;125:580–9.
[24] Rakopoulos CD, Michos CN, Giakoumis EG. Study of the transient behavior of turbocharged diesel engines including compressor surging using a linearized quasi-steady analysis. SAE Paper No. 2005-01-0225, 2005.
[25] Fink DA, Cumpsty NA, Greitzer EM. Surge dynamics in a free-spool centrifugal compressor system. ASME Trans, J Turbomachinery 1992;114:321–32.
[26] Borman G, Nishiwaki K. Internal-combustion engine heat transfer. Prog Energy Combust Sci, 1987;13:1–46.
[27] Bryzik W, Kamo R. Tacom/Cummins adiabatic engine program. SAE Paper No. 830314, 1983.
[28] Jennings MJ, Morel T. A computational study of wall temperature effects on engine heat transfer. SAE Paper No. 910459, 1991.
[29] Alkidas AC. Performance and emissions achievements with an uncooled heavy-duty, single-cylinder diesel engine. SAE Paper No. 890144, 1989.
[30] Assanis D, Wiese K, Schwartz E, Bryzik W. The effects of ceramic coatings on diesel engine performance and exhaust emissions. SAE Paper No. 910460, 1991.
[31] Keribar R, Morel T. Thermal shock calculations in I.C. engines. SAE Paper No. 870162, 1987.
[32] Rakopoulos CD, Giakoumis EG. Study of the transient operation of low-heat rejection turbocharged diesel engine including wall temperature oscillations. SAE Paper No. 2007-01-1091, 2007.
[33] Watson N. Dynamic turbocharged diesel engine simulator for electronic control system development. ASME Trans, J Dynam Syst Measurement Control 1984;106:27–45.
[34] Schorn N, Pischinger F, Schulte H. Computer simulation of turbocharged diesel engines under transient conditions. SAE Paper No. 870723, 1987.
[35] Schwartz E, Reid M, Bryzik W, Danielson E. Combustion and performance characteristics of a low-heat rejection engine. SAE Paper No. 930988, 1993.
[36] Assanis DN, Heywood JB. Development and use of a computer simulation of the turbocompounded diesel system for engine performance and component heat transfer studies. SAE Paper No. 860329, 1986.
[37] Wong VW, Bauer W, Kamo R, Bryzik W, Reid M. Assessment of thin thermal barrier coatings for I.C. engines. SAE Paper No. 950980, 1995.
[38] Rakopoulos CD, Antonopoulos KA, Rakopoulos DC, Giakoumis EG. Investigation of the temperature oscillations in the cylinder walls of a diesel engine with special reference to the limited cooled case. Int J Energy Res 2004;28:977–1002.
[39] Rakopoulos CD, Rakopoulos DC, Mavropoulos GC, Giakoumis EG. Experimental and theoretical study of the short-term response temperature transients in the cylinder walls of a diesel engine at various operating conditions. Appl Thermal Eng 2004;24:679–702.

[40] Rakopoulos CD, Giakoumis EG, Rakopoulos DC. Study of the short-term cylinder wall temperature oscillations during transient operation of a turbocharged diesel engine with various insulation schemes. Int J Engine Res 2008;9:177–93.

8

Second-law Analysis

8.1 Introduction

In view of the decreasing energy sources, optimization of thermal processes has initiated vigorous research worldwide. In order to analyze engine performance – that is evaluate the inefficiencies associated with the engine sub-systems – second-law analysis can be a useful alternative to first law, being able to shed more light and provide better insight into the engine processes. For second-law analysis, the key concept is 'availability' (or exergy)[1]; availability is a special case of the more fundamental concept, available energy, introduced by Gibbs [1]. The availability content of a process represents its potential to do useful work being, thus, a quantitative measure of the quality or usefulness of this process. Unlike energy, availability can be destroyed, which is the result of such phenomena as combustion, friction, mixing or throttling. Details about the relevant relationships can be found in [2–6].

The destruction of availability – usually termed irreversibility – is the source for the defective exploitation of fuel into useful mechanical work in a compression (or spark) ignition engine. The reduction of irreversibilities can lead to better engine performance through more efficient exploitation of fuel. To reduce the irreversibilities, we need first to quantify them. That is, the availability destructions through second-law analysis should be evaluated.

Objectives of second law application to internal combustion engines are

- to weigh the various processes and devices, calculating the ability of each one of these to produce work;
- to identify those processes in which destruction or loss of availability occurs, to detect the sources for these destructions, and quantify the losses and destructions;

---

[1] The term availability is preferred in the US; exergy, on the other hand, is more popular in Europe. Both terms will be used interchangeably in this chapter.

- to analyze the effect of various design and thermodynamic parameters on the exergy destruction and losses;
- to propose measures and techniques for the minimization of destruction and losses in order to increase overall efficiency;
- to propose methods for exploitation of losses – most notably exhaust gas to ambient and heat transfer to cylinder walls – now lost or ignored; and
- to define efficiencies so that different applications can be studied and compared, and possible improvements can be measured and quantified.

8.2 Basic Concepts of Availability

8.2.1 Availability of a System

Availability of a system in a given state is defined as the maximum useful work that can be produced through interaction of the system with its surroundings, as it reaches thermal, mechanical and chemical equilibrium. Usually, the terms associated with thermo-mechanical and chemical equilibration are calculated separately, as will be described later in the chapter.

For a closed system experiencing heat and work interactions with the environment, thermo-mechanical availability is given by

$$A^{tm} = (E - U_o) + p_o(V - V_o) - T_o(S - S_o) \tag{8.1a}$$

where $E = E_{kin} + E_{pot} + U$, with E_{kin} the kinetic and E_{pot} the potential energy and p_o and T_o are the fixed pressure and temperature of the environment; U_o, V_o and S_o are the internal energy, volume and entropy of the contents were they brought to p_o and T_o.

Availability is an extensive property with a value greater than or equal to zero. It is obvious that availability is a property the value of which depends not only on the state of the system, but also on the ambient properties. There is no availability in a system when thermal, mechanical and chemical equilibrium exists with the environment. Thermal equilibrium is achieved when the temperature of the system is equal to the temperature of the surrounding environment. In the same way, mechanical equilibrium is achieved when there is no pressure difference between the working medium and the environment.

Chemical equilibrium is established only when there are no components of the working medium, which could interact with those of the environment to produce work. In the case of engines, all the components of the working medium must be either oxidized (*e.g.*, fuel, CO, H), or reduced (*e.g.*, NO, OH) in a reversible way as the system reaches the dead state (see following section for the dead state definition). The only components of the system, which cannot react chemically with the environment and, therefore constitute the components of the mixture at the dead state are O_2, N_2, CO_2 and H_2O. In addition to the work that could be obtained by reversible reactions, some researchers propose a more general definition of chemical availability, which would also take into account the capacity to produce work because of the difference between the partial pressures of the components (when in thermal and mechanical equilibrium with the environment) and the partial pressures of the same components in the environment. This work could be extracted by the use of semi-permeable membranes and efficient low input pressure, high-pressure ratio expansion devices (*e.g.*, Van't Hoff's equilibrium box).

Chemical availability terms are practically useless as regards internal combustion engine units owing to the inability of their recovery, at least for mobile applications [6–8].

8.2.2 Dead State

The choice of reference dead state is of paramount importance when dealing with availability calculations, since this will determine the equilibrium that will be established with the environment and, consequently, the calculated values of availability. This subject has been treated in detail in [3, 7–10].

In general, a system is considered to be at the so-called 'restricted' dead state, when no work potential exists between the system and the environment due to temperature or pressure differences. This is the dead state reached when calculating the thermo-mechanical availability. Usually, the restricted dead state (for any given state) is defined for engine applications to have the same chemical composition as the given state, therefore no work potential exists due to compositional differences; this is most usually true when studying compression ignition engines operation, which involve combustion of lean mixtures.

On the other hand, if chemical equilibrium with the environment is of concern, then we refer to the 'true' or 'unrestricted' dead state, where the chemical potentials of the system equal those of the environment. In such a case, the detailed composition of the environment is needed for the analysis. The (environmental) pressure and temperature conditions of the dead state are usually taken to be p_o=1.01325 bar and T_o=283.15$\div$298.15 K, and if chemical availability is of concern, then the molar composition of the environment is typically taken as: 20.35% O_2, 75.67% N_2, 0.03% CO_2, 3.03% H_2O and 0.92% of various other substances. Changes in the dead state conditions reflect changes in the value of the system availability.

In a closed system, the thermo-mechanical availability given by Equation 8.1a can be re-arranged as follows

$$A^{tm} = E + p_o V - T_o S - G_o = E + p_o V - T_o S - \sum_i m_i \mu_{io} \tag{8.1b}$$

where G_o is the working medium's Gibbs free enthalpy and μ_{io} is the respective chemical potential of species i, both calculated at the restricted dead state conditions, and m_i is the mass of species i. At the restricted dead state the system is in thermal and mechanical equilibrium with the environment. However, no chemical equilibrium exists, which means that some work recovery is possible due to the difference between the composition of the system at the restricted dead state and that of the environment. If the system at the restricted dead state is also permitted to pass into but not react chemically with the surrounding environment, then for perfect gas mixtures chemical exergy is defined as

$$A^{ch} = \sum_i m_i \left(\mu_{io} - \mu_i^o\right) = T_o \sum_i R_{si}\, m_i \ln\left(\frac{x_i}{x_i^o}\right) \tag{8.1c}$$

with μ_i^o the chemical potential of species i at the true dead state, and x_i, x_i^o the mole fractions of species i in the mixture (restricted dead state) and the environment (true dead state), respectively. This chemical availability is a measure of the maximum work when the system comes to equilibrium with the environmental composition.

Total, *i.e.,* thermo-mechanical plus chemical exergy is then given by

$$A = A^{tm} + A^{ch} \quad = E + p_o V - T_o S - \sum_i m_i \mu_i^o \tag{8.1d}$$

8.2.3 General Availability Balance Equation

For an open system experiencing mass exchange with the surrounding environment, the following equation, with reference to Figure 8.1, holds for the total availability on a time rate basis [2–6]:

$$\frac{dA_{cv}}{dt} = \int_j \left(1 - \frac{T_o}{T_j}\right) d\dot{Q}_j - \left(\dot{W}_{cv} - p_o \frac{dV_{cv}}{dt}\right) + \sum_{in} \dot{m}_{in} \psi_{in} - \sum_{out} \dot{m}_{out} \psi_{out} - \dot{I} \tag{8.2}$$

where:

1. dA_{cv}/dt is the time rate of change of the control volume exergy (*i.e.,* engine cylinder, exhaust manifold *etc.*).
2. $\int_j \left(1 - \frac{T_o}{T_j}\right) d\dot{Q}_j$ is the availability term for heat transfer, with T_j the temperature at the boundary of the system, which, in general, is different from the temperature level of a process, and $d\dot{Q}_j$ represents the differential time rate of heat transfer at the boundary of the control volume. This relation shows that increasing the temperature of an energy stream also increases its ability to produce work. This statement is very useful when studying internal combustion engines, since here an increase in the fuel–air equivalence ratio Φ results in an increase in exhaust gas temperatures due to the lean mixtures involved, and thus their potential for work production. Moreover, the above relation denotes that there is actually a limitation imposed by the second law of thermodynamics as regards operation and efficiency of thermal engines.

3. $\left(\dot{W}_{cv} - p_o \frac{dV_{cv}}{dt}\right)$ is the availability term associated with (mechanical) work transfer.
4. $\dot{m}_{in}\psi_{in}$ and $\dot{m}_{out}\psi_{out}$ are the availability terms associated with inflow and outflow of masses, respectively. For a diesel engine, the terms ψ_{in} and ψ_{out} in Equation 8.2 refer to the flow or stream availability (or exergy) of the incoming and the outgoing cylinder mass flow-rates, respectively, given by (neglecting kinetic and potential energy contribution)

$$\psi = \psi^{tm} + \psi^{ch} = h - T_o s - \sum_i x_i \mu_i^o \tag{8.3}$$

with s_o the entropy of the flow-rate were it brought to p_o and T_o. Flow availability is defined as the maximum work that can be obtained as the fluid passes reversibly from the given state to the dead state, while exchanging heat solely with the environment. This property can actually take negative values in case the partial pressure is lower than that of the environment.
5. $\dot{I}$ is the rate of irreversibility production inside the control volume as will be analyzed in more detail later in the chapter.

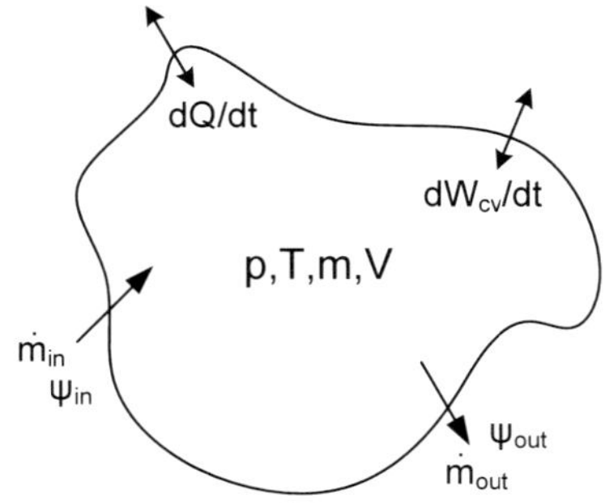

Figure 8.1. Availability balance in an open system

8.2.4 Fuel Availability

The chemical exergy of a substance not present in the environment (*e.g.,* fuel, sulfur, combustion products such as NO or OH *etc.*) can be evaluated by considering an idealized reaction of the substance with other substances for which the chemical exergies are known. This chemical exergy of the fuel can be expressed as follows on a molar basis, in case the fuel exists alone at the restricted dead state [2–6]:

$$\bar{a}_{fch}(T_o, p_o) = \bar{g}_f(T_o, p_o) - \left(\sum_p x_p \bar{\mu}_p^o - \sum_r x_r \bar{\mu}_r^o\right) \tag{8.4a}$$

where index p denotes products (CO_2, H_2O, CO, *etc.*) and index r the reactants (fuel and O_2) of the (stoichiometric) combustion process, while the overbar denotes properties on a per mole basis. Hydrocarbon fuels C_zH_y, which are of special interest to internal combustion engines applications, react with oxygen to produce carbon dioxide and water vapor according to the following chemical reaction:

$$C_zH_y + \left(z + \frac{y}{4} \right) O_2 \rightarrow zCO_2 + \frac{y}{2} H_2O_{(v)} \tag{8.4 b$_1$}$$

Equation 8.4a then becomes [3]

$$a_{fch}\Big|_{T_o, p_o} = -\Delta G_{(T_o, p_o)} + R_{mol} T_o \ln \left[\frac{(x^o_{O_2})^{z+y/4}}{(x^o_{CO_2})^z (x^o_{H_2O})^{y/2}} \right] \tag{8.4 b$_2$}$$

where

$$\Delta G_{(T_o, p_o)} = \left[z\bar{g}_{CO_2} + \left(\frac{y}{2}\right)\bar{g}_{H_2O,(v)} - \bar{g}_{C_zH_y} - \left(z + \frac{y}{4}\right)\bar{g}_{O_2} \right]_{(T_o, p_o)} \tag{8.4 b$_3$}$$

For practical calculations, the above relation is approximated as regards liquid hydrocarbons (on a kg basis) by

$$a_{fch} = LHV \left(1.04224 + 0.011925 \frac{y}{z} - \frac{0.042}{z} \right) \tag{8.4c}$$

with LHV the fuel's lower heating value. For example, for n-dodecane, which closely resembles automotive diesel fuel, Equation 8.4c yields a_{fch}=1.0645 LHV.

Various approximations for the chemical exergy of fossil, liquid and gaseous fuels were reached by Szargut and Styrylska [11], and summarized by Rodriguez [12]. One such approximation for liquid fuels of the general type $C_zH_yO_pS_q$, applicable in compression ignition engines, is

$$a_{fch} = LHV \left[1.0374 + 0.0159 \frac{y}{z} + 0.0567 \frac{p}{z} + 0.5985 \frac{q}{z} \left(1 - 0.1737 \frac{y}{z} \right) \right] \tag{8.4d}$$

8.3 Application of Exergy Balance to the Diesel Engine

8.3.1 Engine Cylinder Exergy Balance

For the engine cylinder, the availability balance equation on a °CA basis is formulated as follows with reference also to Figure 8.2:

$$\overbrace{\frac{dA_{cyl}}{d\varphi}}^{\text{Control Volume}} = \overbrace{\frac{\dot{m}_4\psi_4}{6N}}^{\text{Inflow}} - \overbrace{\frac{\dot{m}_5\psi_5}{6N}}^{\text{Outflow}} - \overbrace{\frac{dA_w}{d\varphi}}^{\text{Work}} - \overbrace{\frac{dA_L}{d\varphi}}^{\text{Heat Loss}} + \overbrace{\frac{dA_f}{d\varphi}}^{\text{Fuel}} - \overbrace{\frac{dI_{cyl}}{d\varphi}}^{\text{Irreversibilities}} \tag{8.5}$$

In the above equation, $\dot{m}_4$ is the incoming flow-rate from the inlet manifold, which consists of air or air plus exhaust gas (in the case of operation with exhaust gas recirculation), whereas $\dot{m}_5$ is the outgoing flow-rate to the exhaust manifold.

$$\frac{dA_w}{d\varphi} = (p_g - p_o)\,\frac{dV}{d\varphi} \tag{8.6}$$

is the (indicated) work transfer, where $dV/d\varphi$ is the rate of change of cylinder volume with crank angle described in Equation 9.31 and p_g the instantaneous cylinder pressure.

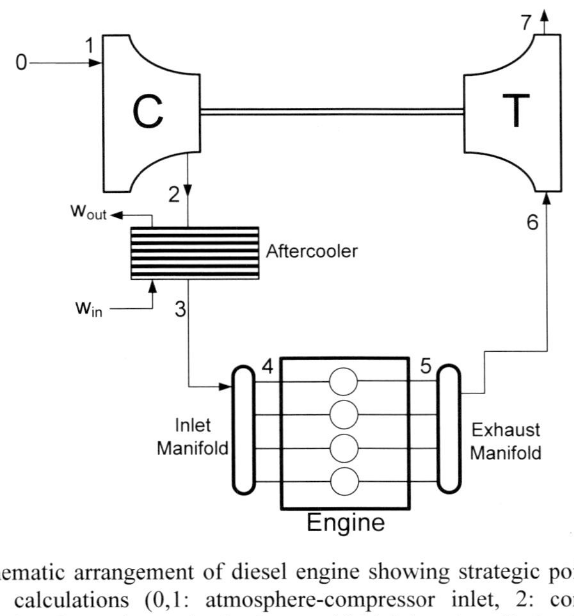

Figure 8.2. Schematic arrangement of diesel engine showing strategic points locations for exergy balance calculations (0,1: atmosphere-compressor inlet, 2: compressor outlet-aftercooler inlet 3: aftercooler outlet-intake manifold inlet, 4: cylinder inlet, 5: cylinder exhaust, 6: exhaust manifold - turbine inlet, 7: turbine outlet)

$$\frac{dA_L}{d\varphi} = \frac{|dQ_L|}{d\varphi}\left(1 - \frac{T_o}{T_g}\right) \tag{8.7}$$

is the heat transfer to the cylinder walls availability, with $dQ_L/d\varphi$ the respective heat flux (Section 9.3.2.2), and T_g the instantaneous cylinder gas temperature. This equation shows that increasing the temperature of the exhaust gas increases its availability too, or the ability of the gas to produce work. This availability loss is usually considered as external to the cylinder control volume, but there are some researchers (for example, Alkidas [13]) who treat the heat losses as another source of irreversibility by defining an individual open thermodynamic system for the water-cooling circuit. By so doing, they usually sum up the combustion irreversibilities and the exergy term for heat transfer (Equation 8.7), and also calculate the availability increase in both the water and oil coolant circuits. In fact, heat loss from gas to the cylinder walls contains a significant amount of availability, which is almost completely destroyed only after this is transferred to the cooling medium.[2]

$$\frac{dA_f}{d\varphi} = \frac{dm_{fb}}{d\varphi} a_{fch} \tag{8.8}$$

is the burned fuel availability, with a_{fch} being the fuel (chemical) availability and $dm_{fb}/d\varphi$ the fuel burning rate (Section 9.3.2.3).

Finally, the term on the left-hand side of Equation 8.5 can be expressed alternatively applying Equation 8.1d, as follows

$$\frac{dA_{cyl}}{d\varphi} = \frac{dU}{d\varphi} + p_o \frac{dV}{d\varphi} - T_o \frac{dS}{d\varphi} - \sum_i \frac{dm_i}{d\varphi} \mu_i^o \tag{8.9}$$

representing the rate of change in the total availability of the cylinder contents. Equation 8.5 can then be solved for the term of irreversibility production $dI_{cyl}/d\varphi$.

8.3.2 In-cylinder Irreversibilities

Hydrocarbon combustion increases entropy; therefore it produces irreversibilities by converting molecules of complicated structure to a mixture of relatively light gaseous fragments. Irreversibility production within the cylinder is due to combustion, mixing of gases at different temperatures, viscous dissipation, turbulence, fluid flow losses, inlet-valve throttling and mixing of the incoming air with the cylinder residuals. An alternative approach entails application of an entropy balance in order to calculate irreversibilities. By so doing, the rate of irreversibility production in the cylinder is linked to the respective entropy generation ΔS_{irr}, *i.e.*, $\dot{I} = T_o \cdot \Delta \dot{S}_{irr}$.

Typical values for in-cylinder irreversibilities are of the order of 20–25% for full-load, turbocharged, diesel engine operation. Greater values are expected for compression ignition engine operating at low loads. For example, for lower than

[2] According to Equation 8.7 (assuming a mean gas temperature of 800 K and an atmospheric temperature of 20°C) the work potential of the heat transfer stream from the exhaust gas to the cylinder walls equals 63% of the respective energy term Q_L. This work potential decreases to 35% after it has been transferred to the cylinder walls (T_{wall}=180°C) and to 17% of the respective heat flux when it is subsequently transferred to the low temperature cooling medium ($T_{coolant}$=80°C).

20% engine loads, in-cylinder irreversibilities waste 40% or even more of the fuel chemical exergy. This is mainly due to the lower gas temperatures involved.

The contribution of combustion to the total in-cylinder irreversibilities in a diesel engine is dominant. Actually, it has been computed being more than 90%. Primus and Flynn [14] calculated in-cylinder, non-combustion irreversibilities at 4.96% of the total, for a six-cylinder, turbocharged and aftercooled, diesel engine operating at 2100 rpm and producing 224 kW. These included thermal mixing of the incoming air with the cylinder residuals and intake valve throttling. Alkidas [13] working on a single-cylinder, naturally aspirated, DI diesel engine of 2.0 L displacement volume, and using a simplified mixing model, estimated air–fuel mixing irreversibilities to be 3% of the total. Rakopoulos and Giakoumis [15] calculated non-combustion, in-cylinder irreversibilities during transient conditions at 5% (maximum) of the total for a six-cylinder, turbocharged and aftercooled diesel engine (decreasing magnitude with increasing load – see also Figure 8.7 later in the chapter). Some fundamental aspects of the mechanism of availability destruction due to combustion in an internal combustion engine are summarized below.

- About 80% of the combustion irreversibilities occur during the heat transfer process between the reacting gas and the yet unburned mixture. The product constituent mixing process assumes most of the remaining percentage. Although Dunbar and Lior [16] reached the above conclusion studying constant volume combustion, their results can be projected for reciprocating engines too, indicating the main mechanism of availability destruction during combustion.
- An increasing combustion temperature (as, for example, is the case with increasing fuel–air equivalence ratio Φ, in compression ignition engine lean operation) decreases the percentage combustion irreversibilities. This conclusion is interrelated to the previous point, since an increasing cylinder charge gas temperature decreases the relative amount of heat transfer from the reacting gas to the yet unburned mixture. Varying the equivalence ratio at which combustion takes place has also a significant effect on the distribution of (thermo-mechanical and chemical) availabilities inside the cylinder.
- The effect of combustion duration and heat release shape is modest on the total amount of (combustion) irreversibilities. On the other hand, their influence on the rate of irreversibility production during an engine cycle is significant.
- The effect of pressure changes on the irreversibilities during the combustion process (other parameters being the same) is modest.
- The amount of combustion irreversibilities can be correlated to the differential change in mixture composition, and notably nothing else. This was the result from the work conducted by Rakopoulos and Andritsakis [17] on a turbocharged diesel engine; they calculated the combustion irreversibility production rate as a function of fuel reaction rate only, and reached the following very interesting equation

$$dI = -\frac{T_o}{T}\sum_j \mu_j \, dm_j \tag{8.10}$$

where index j includes all reactants and products. For perfect gases $\mu_j = g_j$, and for the fuel $\mu_f = a_{fch}$. The above equation reveals that, after all, both heat transfer and work production inside the cylinder only indirectly influence the irreversibilities accumulation.

- Research in the field of alternative fuels has indicated some very interesting results. For example, a decrease in combustion irreversibility can be realized when using lighter than dodecane (*e.g.,* methane) or oxygenated (*e.g.,* methanol) fuels. This is due to the combustion characteristics of these fuels, which involve lower entropy of mixing in the combustion products than the larger n-dodecane molecules [18]. From the second-law analysis point of view, hydrogen combustion was found qualitatively different than the combustion of hydrocarbon fuels since hydrogen oxidation is the combination of two simple diatomic molecules, which yields a triatomic one with significantly more structure than any of the reactants. For this reason, a monotonic decrease in combustion irreversibilities with increasing hydrogen component was calculated for the combustion of CH_4-H_2 mixtures burning in an engine cylinder; this decrease in combustion irreversibilities translates also into an increase in second-law efficiency [19].

8.3.3 Exergy Balance of the Engine Sub-systems

8.3.3.1 Turbocharger

For the compressor, quasi-state conditions are assumed so that there is no accumulation term. Then, the availability balance equation with reference to Figure 8.2 reads

$$\frac{\dot{m}_1\psi_1 - \dot{m}_2\psi_2}{6N} + \frac{\left|\dot{W}_C\right|}{6N} = \frac{dI_C}{d\varphi} \tag{8.11}$$

with $\dot{m}_1 = \dot{m}_2$ the charge air-flow-rate. For the turbine, accordingly

$$\frac{\dot{m}_6\psi_6 - \dot{m}_7\psi_7}{6N} - \frac{\dot{W}_T}{6N} = \frac{dI_T}{d\varphi} \tag{8.12}$$

Both processes are usually assumed adiabatic. In these equations, the terms $\dot{W}_C$ and $\dot{W}_T$ are the power terms for compressor and turbine, respectively.

Irreversibilities in the turbocharger are mainly fluid flow losses due to fluid shear and throttling accounting for around 10% of the total engine irreversibilities

8.3.3.2 Aftercooler or Intercooler

Similarly for the aftercooler, the availability balance equation is given by

$$\frac{\dot{m}_2\psi_2 - \dot{m}_3\psi_3}{6N} - \Delta A_w = \frac{dI_{AC}}{d\varphi} \tag{8.13}$$

where ψ_2 is the flow availability at the compressor outlet – aftercooler inlet, ψ_3 the flow availability at the aftercooler outlet – inlet manifold inlet, and

$$\Delta A_w = \frac{1}{6N}\,\dot{m}_w\,(\psi^w_{out} - \psi^w_{in}) \tag{8.14}$$

where

$$\psi^w_{out} - \psi^w_{in} = c_{pw}\left[T_{w-out} - T_{w-in} - T_o\,\ln\left(\frac{T_{w-out}}{T_{w-in}}\right)\right] \tag{8.15}$$

is the increase in the availability of the cooling medium having mass flow-rate $\dot{m}_w$, specific (mass) heat c_{pw}, initial temperature entering the aftercooler $T_{w\text{-}in}$ and final temperature leaving the aftercooler $T_{w\text{-}out}$. Here, the irreversibilities account for loss of availability due to heat transfer to a cooler medium. The transfer of heat to a cooler medium is a procedure not desirable from the second law of thermodynamics point of view. The particular one is responsible for the loss of around 0.5–1% of the fuel's chemical availability in compression ignition engine applications.

8.3.3.3 Inlet Manifold

For the inlet manifold, the availability balance equation is formulated as follows

$$\frac{dA_{im}}{d\varphi} = \frac{\dot{m}_3\psi_3 - \sum_{j=1}^{n_{cyl}} \dot{m}_{4j}\psi_4}{6N} - \frac{dI_{im}}{d\varphi} \tag{8.16}$$

where ψ_4 is the flow availability at the intake manifold and $j=1–n_{cyl}$ is the cylinder exchanging mass with the inlet manifold. No heat losses are taken into account in most of the cases. The term for irreversibilities $dI_{im}/d\varphi$ accounts mainly for mixing of incoming air with the intake manifold contents, and is, usually, less than 1% of the fuel's chemical availability owing to the low pressures and temperatures involved.

8.3.3.4 Exhaust Manifold

For the exhaust manifold, the availability balance equation is given by

$$\frac{dA_{em}}{d\varphi}=\frac{\sum_{j=1}^{n_{cyl}}\dot{m}_{5j}\psi_{5j}-\dot{m}_6\psi_6}{6N}-\frac{dI_{em}}{d\varphi}+\frac{dA_{Lem}}{d\varphi} \tag{8.17}$$

where index 6 identifies the exhaust manifold state.

The term

$$\frac{dA_{Lem}}{d\varphi}=\frac{|dQ_{Lem}|}{d\varphi}\left(1-\frac{T_o}{T_6}\right) \tag{8.18}$$

accounts for the heat loss in the exhaust manifold (considered as external to the manifold control volume, thus not included in the respective irreversibilities), with T_6 the instantaneous temperature of the manifold contents.

The term $dI_{em}/d\varphi$ is the irreversibility rate in the exhaust manifold, which arises from throttling across the exhaust valve, mixing of cylinder exhaust gas with manifold contents and gas friction along the manifold length. It assumes values of around 1.5–3% of the fuel's chemical availability (greater values correspond to turbocharged engines). Primus and Flynn [14] identified the individual losses in the exhaust manifold of a six-cylinder, turbocharged and aftercooled, diesel engine as follows: exhaust throttling losses accounted for 1.66% of the fuel's chemical availability, exhaust manifold heat loss 0.25%, fluid flow losses 0.57%, and turbine irreversibilities were of the order of 1.69% of the fuel availability.

8.3.4 Second law or Exergy or Exergetic Efficiency

Efficiency is generally used to describe the degree to which the ideal is approached by a specific process or task. An efficiency is defined in order to be able to compare different engine size applications or evaluate various improvement effects, either from the first- or the second-law perspective. The second-law (or exergy or availability) efficiency, also found in the literature as effectiveness or exergetic efficiency, measures how effectively the input (fuel) is converted into product; it is usually of the following form:

$$\varepsilon=\frac{\text{Availability out in product}}{\text{Availability in}}=1-\frac{\text{loss}+\text{destruction}}{\text{Input}} \tag{8.19}$$

Unlike first-law efficiencies, their second-law counterparts weigh the various energy terms according to their capability for work production. Second-law efficiency includes, in addition to exergy losses (*e.g.*, in exhaust gas), the exergy destructions (irreversibilities) too. On the other hand, since energy is conserved, first-law efficiency reflects only energy losses. Further, energy losses are not representative of (and typically overestimate) the usefulness of loss; first-law efficiencies do not explicitly penalize the system for internal irreversibilities being, thus, misleading since they do not actually measure the approach to ideal.

For the cylinder alone, the following second-law efficiency is often applied

$$\varepsilon_1 = \frac{W_{ind}}{M_{fi} a_{fch}} \quad \text{or} \quad \varepsilon_1 = \frac{W_{br}}{M_{fi} a_{fch}} \tag{8.20}$$

with W_{ind} the indicated and W_{br} the brake work production and M_{fi} the total mass of fuel entering the cylinder per cycle. The left-hand side expression of Equation 8.20 is generally preferred, with a subsequent calculation of friction, since the right-hand side expression penalizes the cylinder for the friction irreversibilities.

One could also take into account the difference between outgoing, from the cylinder, A_{out} and incoming, to the cylinder, A_{in} thermo-mechanical availability flows (J/cycle) for their ability to produce extra work, and define the following second-law efficiency:

$$\varepsilon_2 = \frac{W_{ind} + A_{out} - A_{in}}{M_{fi} a_{fch}} \tag{8.21}$$

where (for four-stroke engine operation)

$$A_{out} - A_{in} = \int_0^{720} \left(\dot{m}_5 \psi_5 - \dot{m}_4 \psi_4 \right) \frac{d\varphi}{6N} \tag{8.22}$$

For the compressor and turbine, respectively, the exergy efficiency or effectiveness can be defined as

$$\varepsilon_C = \frac{\dot{m}_C (\psi_2 - \psi_1)}{\left| \dot{W}_C \right|} \qquad \varepsilon_T = \frac{\dot{W}_T}{\dot{m}_T (\psi_6 - \psi_7)} \tag{8.23}$$

The turbine exergy efficiency is a measure of how well the exhaust gas exergy is converted into shaft work. It differs from the isentropic efficiency in as much as the latter compares the actual work developed with the work that would be developed in an isentropic expansion. In any case, both can be classified as second-law efficiencies. The same holds true for the compressor.

For the whole diesel engine plant, we can define

$$\varepsilon_{tot1} = \frac{W_{ind}}{M_{fi} a_{fch}} \quad \text{or} \quad \varepsilon_{tot2} = \frac{W_{ind} + A_{out}^{tot}}{M_{fi} a_{fch}} \tag{8.24}$$

where the exhaust gas to ambient flow availability is $A_{out}^{tot} = \int_0^{720} (dm_7 / d\varphi) \psi_7 d\varphi$.

8.4 Exergy Balance Application to Steady-state Operation

Application of Equations 8.5–8.9 on a diesel engine cylinder is depicted in Figures 8.3 and 8.4, showing the development of both rate and cumulative in-cylinder availability terms (for one cylinder) during an engine cycle. The cumulative terms are defined after integration of the respective rate terms over the engine cycle. Particularly so for steady-state operation, the cumulative value for the cylinder contents availability sums up to zero at the end of the cycle, *i.e.*, $\int \left(dA_{cyl} / d\varphi \right) d\varphi = 0$. Chemical availability is usually neglected in such case studies.[3]

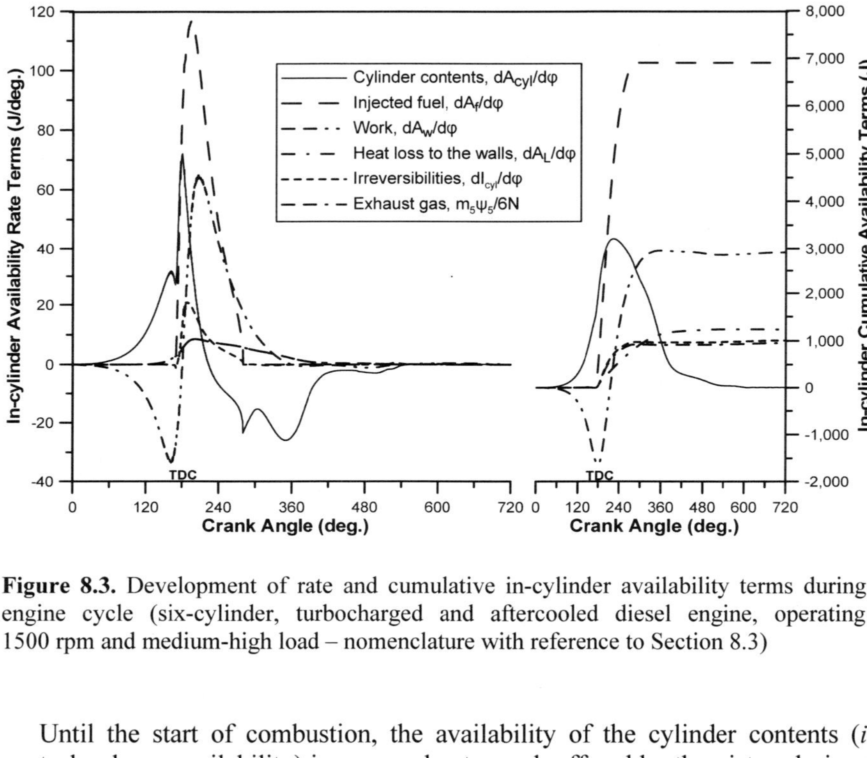

Figure 8.3. Development of rate and cumulative in-cylinder availability terms during an engine cycle (six-cylinder, turbocharged and aftercooled diesel engine, operating at 1500 rpm and medium-high load – nomenclature with reference to Section 8.3)

Until the start of combustion, the availability of the cylinder contents (*i.e.*, control volume availability) increases due to work offered by the piston during the compression process. As the working medium is trapped at a temperature lower

---

[3] The same holds true for the dissociation of the combustion products, as the findings presented concern simulation results derived from a single-zone thermodynamic model. Since the combustion process in a diesel engine does not develop in a uniform way, as the single-zone modeling suggests, it is understood that an error due to underestimation of the local fuel-rich zones' temperatures exists that affects the irreversibility rate production due to heat transfer between the zones. On the other hand, dissociation of the combustion products in the fuel-rich zones results in an amount of chemical availability, which, as has been argued by Shapiro and van Gerpen [20], although unattainable, it should be taken into account in order to obtain an accurate exergy balance from Equation 8.5.

than that of the cylinder walls, availability is transferred through heat to the working medium for the early part of the cycle. Then, the heat transfer direction is reversed as the working medium temperature rises (see Figure 8.4 for a more clear representation of the in-cylinder exergy rate terms during the closed part of the cycle).

As the availability transfer is low during compression, it is obvious that the change of the working medium availability is almost equal to the work transfer, the irreversibility rate being essentially zero. At the point fuel injection commences, a reduction is observed in the control volume availability pattern. This is due to the ignition delay period and the simultaneous loss of heat for evaporation of the injected fuel. After the start of combustion, things change drastically. The burning of fuel causes a considerable increase in pressure and temperature and, consequently, in-cylinder availability and heat loss. As a result, the irreversibility rate increases remarkably too.

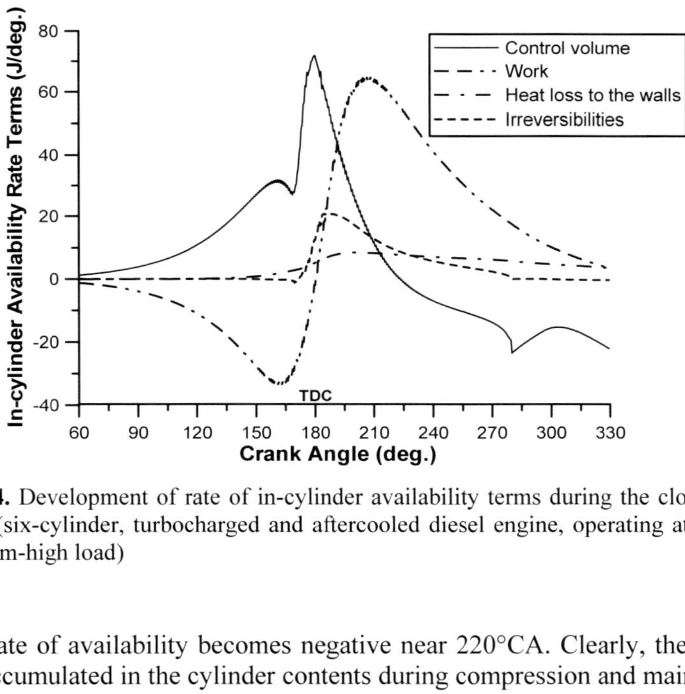

Figure 8.4. Development of rate of in-cylinder availability terms during the closed part of the cycle (six-cylinder, turbocharged and aftercooled diesel engine, operating at 1500 rpm and medium-high load)

The rate of availability becomes negative near 220°CA. Clearly, the available energy accumulated in the cylinder contents during compression and mainly during combustion is being returned in the form of (indicated) work production, which causes the decrease in the availability of the working medium. After the opening of the exhaust valve, the control volume availability rate reaches a second minimum, due to the exhaust gas leaving the cylinder during the blow-down period. At this point (see also Figure 8.6 later in the chapter), a second (much smaller) peak in the cycle irreversibility rate is also observed owing to the pressure and temperature difference between cylinder contents and exhaust manifold. The cumulative availability term continues to decrease so that, at the end of the cycle (720°CA), its value is zero again, as the working medium has returned to its initial state. The main conclusions are, in general, applicable for spark ignition engine operation too.

For the present engine operating point of Figure 8.3, 40.31% (efficiency ε_1) of the fuel's availability is converted into (indicated) work (compared with 42.89% of the fuel's lower heating value); this amount increases to 58.38% (efficiency ε_2) if the difference between outgoing and incoming flows is considered for their ability to produce work. The compressor and turbine efficiencies are 76.38 and 87.42%, respectively. The higher isentropic efficiency of the turbine compared with the compressor, leads to a better exergy efficiency too (Equations 8.23). For the whole diesel engine plant ε_{tot1} = 40.31%, which increases to ε_{tot2} = 53.76% if we also take into account the potential for useful work production of the exhaust gas.

Table 8.1 expands on the previous remarks by tabulating typical first- and second-law balances over an engine cycle for a similar engine configuration [14]. Since second-law analysis assigns different magnitude to each energy streams' ability to produce work, the results presented in Table 8.1 include destructions and losses not to be found in the first-law balance. Perhaps, the most obvious difference lies in the quantification of the exhaust gas to ambient term. Second law shows that this is 'only' 13% of the incoming fuel availability in contrast to the 31% magnitude assigned by the first law, a fact that has to be taken into account if one is considering the addition of a bottoming cycle (see also Section 8.5). The quantification of all losses is also presented (numbers in parentheses reduce the irreversibilities to the total), highlighting the dominance of combustion irreversibilities (21.20% of the fuel's availability, or 75% of the total irreversibilities). Fuel availability was estimated at 1.0338 times the LHV, consequently brake and mechanical efficiency were slightly different when comparing the results from the two laws.

Table 8.1. Comparison of results between first- and second-law balance, and quantification of irreversibilities (six-cylinder, turbocharged and aftercooled, DI diesel engine – reprinted from Primus and Flynn [14], with permission from the American Society of Mechanical Engineers) (numbers in parentheses denote % of total irreversibilities)

	First law (% of fuel energy)	**Second law** (% of fuel exergy)
Work	40.54	39.21
Friction	4.67	4.52
Heat transfer to the walls	17.23	13.98
Aftercooler heat transfer	5.86	1.16
Exh. manifold heat transfer	.39	.25
Exhaust gas to ambient	31.31	12.73
Irreversibilities		**28.15**
Combustion	-	21.20 (75.3)
Thermal mixing	-	0.81 (2.9)
Intake throttling	-	0.58 (2.1)
Exhaust throttling	-	1.66 (5.9)
Fluid flow	-	0.57 (2.0)
Compressor	-	1.64 (5.8)
Turbine	-	1.69 (6.0)

8.5 Exergy Balance Application to Transient Operation

For unsteady operations, the dA/dφ terms given in Section 8.3, for the cylinder and the manifolds, do not sum up to zero (as they actually do for steady-state operation) at the end of a full cycle of the working medium. Their respective cumulative values $\int_0^{4\pi}\left(dA_{cyl}/d\varphi\right)d\varphi$ (four-stroke operation) are, however, very small compared with the other availability terms, especially for naturally aspirated engines or slow response turbocharged engine transients.

For the turbocharger during transient operation, the following availability balance equation is now valid

$$\frac{\dot{m}_T(\psi_6-\psi_7)}{6N}-\frac{\dot{m}_C(\psi_2-\psi_1)}{6N}=\left(\frac{dI_T}{d\varphi}+\frac{dI_C}{d\varphi}+\frac{dE_{kin,TC}}{d\varphi}\right) \tag{8.25}$$

where

$$\frac{dE_{kin,TC}}{d\varphi}=\frac{dW_T-dW_C}{d\varphi}=\frac{1}{2}G_{TC}\frac{d}{d\varphi}(\omega_{TC}^2) \tag{8.26}$$

is the increase in the turbocharger shaft kinetic energy.

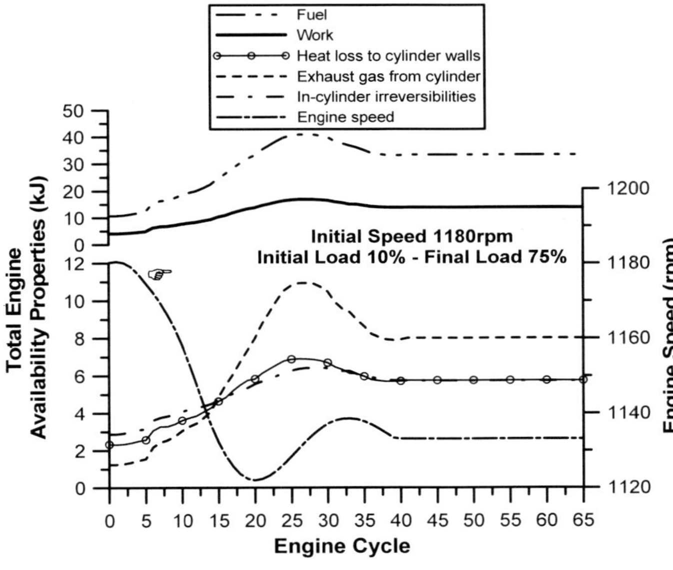

Figure 8.5. Response of total in-cylinder availability terms of a four stroke, turbocharged and aftercooled diesel engine to an increase in load

In Figure 8.5 the response of the in-cylinder availability terms, *i.e.*, work, heat loss to the walls, exhaust gas and irreversibilities is demonstrated as a function of the engine cycles for a 10–75% load increase transient. All of these terms are cumulative values (in 'Joules') over each cycle (for all six cylinders of the engine). The availability term for work and heat loss to the walls increase with fueling, as a function of the engine cycles, because of increases in the charge temperature resulting from increases of the injected fuel quantity and accompanying fuel–air equivalence ratios Φ, following the governor response to the initial speed drop after the load increase. Similar results hold for the exhaust gas from the cylinder term and for the irreversibilities term. All availability values are characterized by a time delay relative to the engine speed, while the in-cylinder irreversibilities follow closely the profile of the injected fuel availability.

For a diesel engine, *i.e.,* lean operation, an increase in equivalence ratio Φ raises the temperatures inside the cylinder, therefore reducing the degradation of the fuel chemical exergy, as this is now transferred to 'hotter' exhaust gases (recall that the greatest part of combustion irreversibilities occurs during the heat transfer process between the reacting gas and the yet unburned mixture). The lower amount of combustion irreversibilities is mainly reflected into increased amount of heat loss or exhaust gas availability; the exploitation of the latter being a key aspect of second-law application to internal combustion engines. Varying the equivalence ratio at which combustion occurs, has also a significant effect on the distribution of thermo-mechanical and chemical availabilities inside the cylinder. It has been shown that the percentage of chemical to total availability increases significantly with relatively rich mixtures pinpointing its significance, mainly, for spark ignition engines operation [8].

An important fact here is that the transient irreversibilities develop in a different way from their 'steady-state' counterparts (*i.e.,* irreversibilities at steady-state operation for the rotational speed and fuel pump rack position of the respective transient cycles); this is highlighted in Figure 8.6. As was analyzed in the previous chapters, during transients, various off-design phenomena occur mainly during the early cycles where the turbocharger lag is prominent.

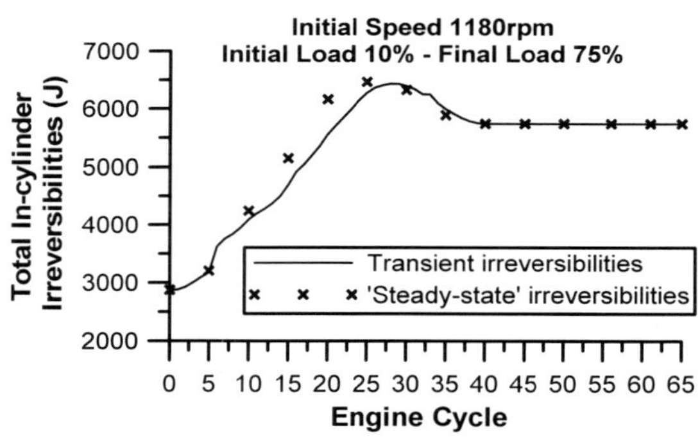

Figure 8.6. Comparison between total in-cylinder irreversibilities of a turbocharged diesel engine during a load increase, with some intermediate 'quasi-state' irreversibilities [21]

In Figure 8.6, a difference up to 11% at the 20th cycle (this is indeed an 'early' cycle of the transient event, as the particular engine–load configuration has a very large value of mass moment of inertia) is observed when comparing the transient with the respective 'steady-state' irreversibilities. The difference is attributed to:

- the differentiated (higher than steady-state conditions, but never greater than stoichiometric for the present engine transient event) fuel–air equivalence ratios, hence lower combustion irreversibilities, experienced during (load increase) transients owing to the turbocharger lag, which significantly affects the air-mass flow-rate;
- the transient operation of the fuel pump that differentiates from the steady-state fuel pump curves; and
- the fact that, during transients, integration of the left-hand side of Equation 8.9 over an engine cycle does not sum up to zero, as the initial conditions of the new cycle differ from the ones of the previous cycle.

In Section 8.3.2, it was mentioned that the contribution of combustion irreversibilities is dominant for in-cylinder processes. Figure 8.7 highlights the contribution of each process involved, *i.e.,* inlet, combustion and exhaust, expanding on the irreversibilities terms (J) inside the cylinder. Combustion irreversibilities assume at least 95% of the total in-cylinder and show increasing importance as the load increase transient event develops. Their greatest part is produced at the early stages of combustion, where the gas temperature is still low. Inlet irreversibilities are overall small due to the relatively low pressure and temperature of the inlet gas (irreversibilities due to throttling across the exhaust valve are not shown here, as they are accounted for in the exhaust manifold irreversibilities).

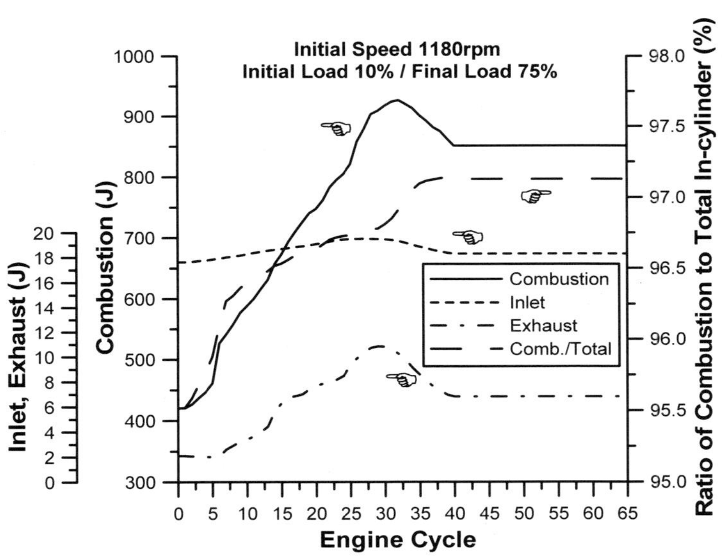

Figure 8.7. Response of various in-cylinder irreversibilities to an increase in load

Flynn *et al.* [7] argued that only that part of the combustion irreversibilities, which is associated with the heat release placement and shape, can be affected by engine development and thus can be improved (highlighting the 'inevitable' of the combustion irreversibility). A key remark here is that increasing the level of combustion temperatures, as for example when increasing the compression ratio or insulating the cylinder walls, results in a relative decrease in the combustion irreversibilities, as the fuel chemical availability is transferred to 'hotter' exhaust gases. This fact denotes that such a process is, in principle, a favorable one from the second-law perspective, highlighting a part of the path that has to be followed for improving engine performance.

The first and the last cycle of a transient event are, actually, steady-state cycles and their development as regards the rate of the various irreversibilities (J/°CA), *i.e.,* in-cylinder (combustion, inlet and exhaust), compressor, turbine, inlet manifold and exhaust manifold as well as injected fuel exergy is quite revealing in terms of engine operation; they are illustrated in Figure 8.8. Both the fuel and the combustion irreversibility profiles develop in a similar way, with the combustion irreversibility term reaching a maximum during the early stage of the (still premixed) combustion phase.

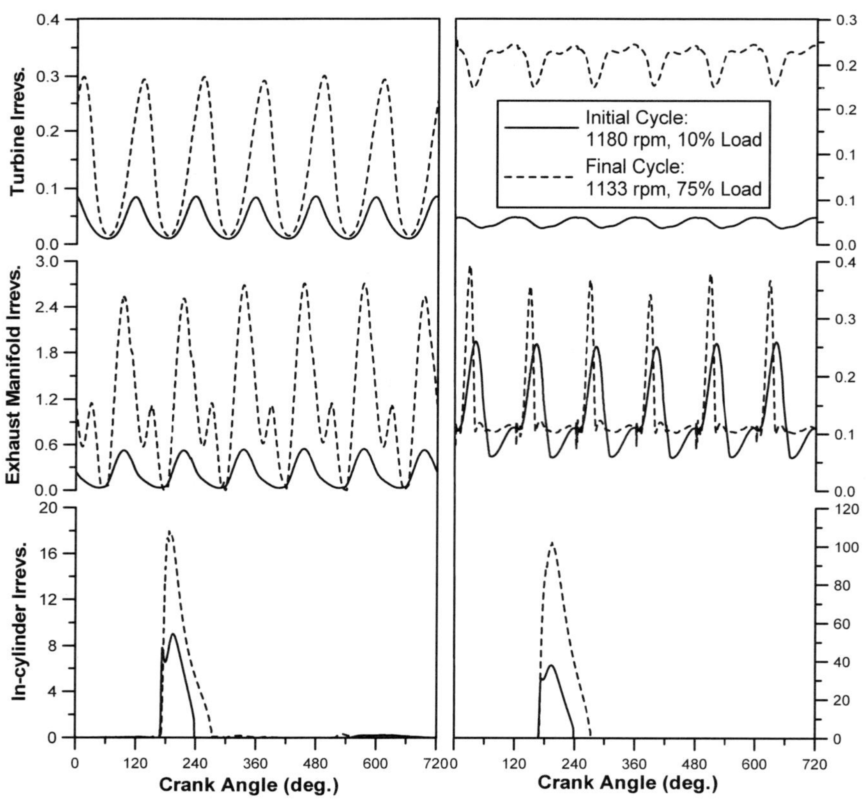

Figure 8.8. Development in the rate (J/°CA) of fuel exergy and irreversibility terms for all diesel engine sub-systems, at the initial and final steady-state conditions of a transient event

On the other hand, the manifolds' and turbocharger's irreversibilities (as well as their other second-law values) incorporate the oscillatory nature of the six-cylinder operation. This is more pronounced for the exhaust manifold and the turbine operation, owing to the pulse turbocharging scheme involved. For the turbine and for the final cycle of the transient event, the ratio of the maximum to minimum irreversibility rate is almost 20 for the particular engine setup.

What is more revealing in terms of second-law analysis is usually the *relative* importance of the various terms, since greater differences in the magnitude are observed when comparing the contribution of each term on the same basis, *i.e.*, fuel exergy or total irreversibilities. This is attempted in Figure 8.9, which shows how cylinder as well as inlet manifold, exhaust manifold, compressor, turbine, and aftercooler percentage irreversibilities develop during the transient event. The main finding is that the relative importance of the in-cylinder irreversibilities decreases as the load increase transient event develops, owing to the increase in loading and thus fueling (*cf.* Figure 8.5). Inlet manifold irreversibilities constitute only a small percentage of the total (not greater than 3% and with decreasing importance during the transient event), whereas those of the exhaust manifold increase substantially from cycle 10 where the main increase in the injected fuel quantity (and thus gas pressures and temperatures) occurs; they reach as much as 15% of the total irreversibilities at cycle 31.

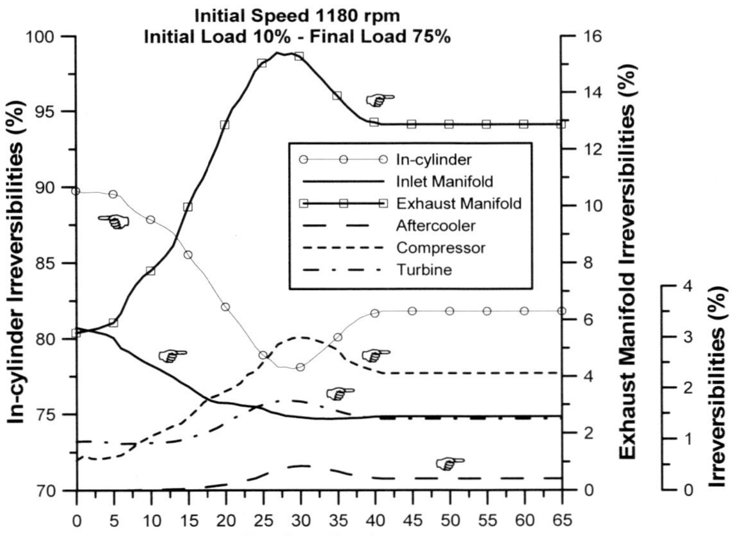

Figure 8.9. Response of various diesel engine and its sub-systems percentage irreversibility terms, after an increase in load

Turbocharger irreversibilities increase during the transient event due to the increase in both compressor and turbine pressures and temperatures, accounting for 4.7% of the total irreversibilities at cycle 28. The compressor irreversibilities outweigh the turbine ones, except for the first cycles where the turbine isentropic efficiency is low. Aftercooler irreversibilities never exceed 0.47% of the total (cycle 30), revealing the low importance of this process as well as the very low potential for work recovery. It is also interesting to note, that the cycle where the maximum (or minimum) percentage occurs differs for each sub-system. This is attributed to the different 'inertia' of each sub-system, which affects its transient response.

The recovery period and the general profile of the various second-law values transient response for a turbocharged diesel engine depend on their respective first-law counterparts, since second-law terms are evaluated using first-law data. It has been shown [6, 22] that

- All the parameters that lead to slow engine speed recovery, such as large exhaust manifold volume or high turbocharger or engine mass moment of inertia, result in similarly slow turbocharger recovery, higher quantities of injected fuel and equivalence ratios Φ and, thus, decreased in-cylinder irreversibilities and increased exhaust gas from cylinder or to ambient availability (when reduced to the fuel exergy). Consequently, an increase in the potential for work recovery during the transient event is established, for example if a bottoming cycle is applied.
- The effect of the aftercooler effectiveness in the engine first-law transient response is minimal, whereas the exergy terms are significantly affected. Cooling the charge air increases the combustion irreversibilities (due to the decrease in the level of cylinder gas temperatures), a fact that is largely counter-balanced by the profits obtained from reduced NO_x emissions.
- In an IDI engine fitted with a prechamber, the main chamber contributes mostly to the total combustion irreversibilities, ranging from approximately 60–70% at low loads to almost 90–95% at full-load conditions, aided by the higher level of pressures and temperatures in the prechamber; the latter's very small volume permits only a small fraction of the fuel to be burned. Lower contribution of the main chamber is experienced in swirl chamber IDI engines.
- In principle, turbocharging has been acknowledged as a very successful way to improve engine efficiency and, mainly, output, as (part of) the available work in the exhaust gas is exploited, although a slight increase in the manifolds and turbocharger irreversibilities is noticed. Turbo-compounding was found to increase the level of pressures, therefore the associated irreversibilities in the exhaust manifold increase too relative to their inlet manifold counterparts; however, a reduction of the cylinders' brake work is experienced when increasing the turbine power output owing to the elevated pumping work [23].

In the previous figures load acceptance transient events were studied from the second-law perspective. In general, similar results apply to acceleration cases; the latter can be considered a much easier operation as regards engine response, since the fuel pump rack moves quickly to the desired final state. Typical results are illustrated in Figure 8.10, showing the response of various exergy terms during a 2500–4500 rpm speed increase transient of a passenger vehicle turbocharged diesel engine.

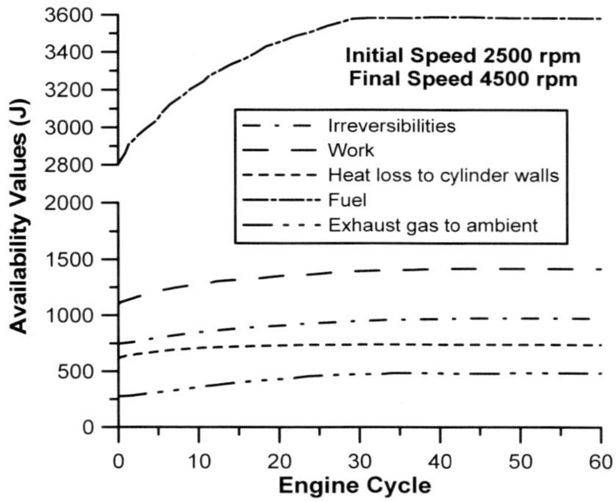

Figure 8.10. Response of various availability terms during 2500–4500 rpm acceleration (four-stroke, four-cylinder, IDI, turbocharged diesel engine of 1.366 L displacement volume; adapted from Bozza *et al.* [24])

Low-heat Rejection Operation

Perhaps the most prosperous field for application of second-law balance to compression ignitions engines is the low-heat rejection (LHR) operation, owing to its considerably differentiated results compared with the first-law analysis (*cf.* Section 7.3).

Background on Steady-state, LHR, Second-law Operation

The analysis of Alkidas [13, 25] for steady-state engine operation showed that the higher insulation of the cylinder walls raises the flame and gas temperature and decreases the percentage combustion irreversibilities. At the same time, the amount of the availability term of the exhaust gas leaving the cylinder is increased. Higher insulation significantly limits the availability destruction associated with heat transfer from the gas to the cylinder walls. This availability potential could then be extracted with the use of a heat transfer device. It is imperative that the engine cooling fluid should not be used for such device, since its low temperature level would make a very poor work recovery, *i.e.,* with the heat transfer from the cylinder walls to the cooling water the majority of the work potential is destroyed. The second-law efficiency ε_2 (Equation 8.21) is also expected to increase with insulation. This is primarily attributed to the increased percentage exhaust gas exergy resulting from the previously mentioned increase in flame temperature.

Transient, LHR, Second-law Engine Operation

Figure 8.11 investigates various insulation schemes focusing on the second-law balance during a load increase transient event, and illustrates the evolution of four basic engine availability values *i.e.,* piston work, heat loss to the cylinder walls, exhaust gas to ambient, and combustion irreversibilities; they are all reduced to the injected fuel chemical exergy. The important finding here is that an increasing degree of insulation proves favorable from the second-law perspective, since the corresponding combustion irreversibilities decrease as a percentage or even as absolute values. This happens due to the fact that an increasing (due to insulation) wall temperature raises the charge temperatures and, consequently, as the ideal Carnot cycle suggests, lowers the degradation of the fuel availability since the fuel's chemical exergy is transferred to 'hotter' gases. For example, for the 1.5 mm PSZ case, the combustion irreversibilities decrease by up to 23.5% compared with the non-insulated configuration, which is indeed a considerable gain; this finding confirms the steady-state results of previous research.

Ideally, 23.5% reduction in combustion irreversibilities would form a sound basis for a remarkable increase in engine efficiency. Unfortunately, although combustion proceeded in a more efficient way, this was not transformed accordingly into increased piston work (Figure 8.11, lower left sub-diagram). Hence, the work term is only marginally improved (+2.3% at cycle No. 25 of the 1.5 mm PSZ case). This finding goes along with the (first-law) isfc results, discussed in Chapter 7.3, and with the conclusions reached for steady-state operation. The reduction of combustion irreversibilities was rather reflected into an increase in the exhaust gas to ambient exergy. Both the absolute (in 'Joules') and percentage values of this term increased with higher degree of insulation owing to the higher temperature of the working medium. This availability amount could be afterwards recovered only if a bottoming cycle is applied, *e.g.,* Rankine cycle or turbo-compound device.

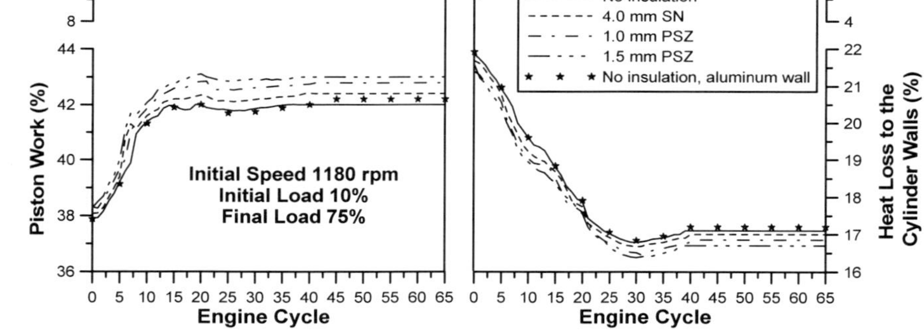

Figure 8.11. Response of various percentage second-law values to an increase in load (unless otherwise noted, the cylinder wall is cast iron) [26]

On the other hand, a higher degree of insulation slightly decreases the exergy of heat loss to the cylinder walls when reduced to the fuel chemical availability.

Equation 8.7 shows that an increasing gas temperature T_g increases the 'quality' (second term in the right-hand side of Equation 8.7) of the respective exergy term. This is contradicted by the fact that a higher insulation scheme reduces the heat flux $dQ_L/d\varphi$ to the cylinder walls (see also Figure 7.23). The synergistic effect of the above factors determines the trend for the percentage exergy term of heat loss depicted in Figure 8.11. However, it should be highlighted that this exergy term of heat loss to the walls possesses now a higher work potential as the wall temperature has increased with insulation. A heat recovery device is needed in order for this availability to be exploited. At the moment, the subsequent heat transfer to the cooling medium practically forces its elimination.

Figure 8.12 illustrates the response of the percentage combustion irreversibilities *i.e.,* combustion irreversibilities reduced to the total engine and sub-systems irreversibilities, during the same transient event. Again, the higher the degree of insulation the lower the percentage combustion irreversibilities, with this trend enhanced the higher the loading. It is interesting to note that even the absolute values (in 'Joules') of the combustion irreversibilities decrease with higher insulations. At the same time, other irreversibility terms such as exhaust manifold (depicted only for the 1.0 mm PSZ coating case in Figure 8.12) or turbine were found to increase. This, rather conflicting, behavior can be explained if we look deeper into the respective irreversibility production mechanisms. For the combustion term, the main mechanism comprises of heat transfer from the burning of fuel to the gas. Consequently, the higher the temperature of the receiving gas (as is the case with increasing insulation) the lower the degradation of the fuel's chemical exergy, hence combustion irreversibilities. On the other hand, turbocharger, aftercooler and manifolds irreversibilities are attributed to throttling, mixing and friction. These processes depend mainly on the level of pressures and temperatures of the working medium; thus, they increase with higher insulations.

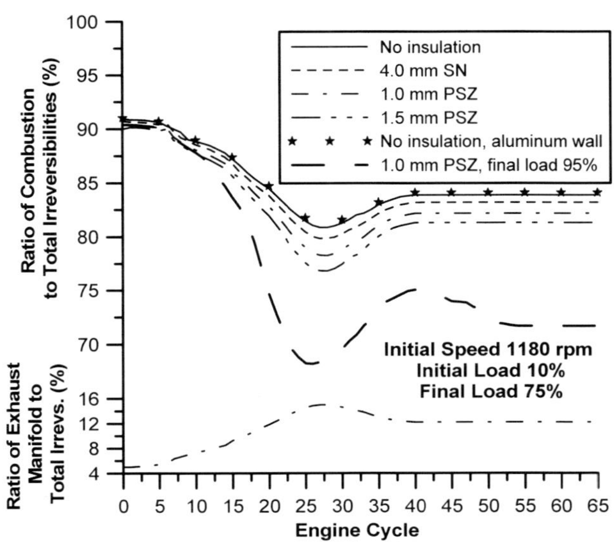

Figure 8.12. Response of percentage irreversibilities to an increase in load (unless otherwise noted, the cylinder wall is cast iron and the load increase is 10–75%)

At the beginning of this chapter it was mentioned that one of the main objectives of second-law application to internal combustion engines is the ability to identify and weigh those processes in which destruction or loss of availability occurs. Heat transfer to the cylinder walls and exhaust gas to ambient have been identified as the two most important ones for diesel engine operation, apart from the predominant combustion irreversibilities. It is useful to quantify the possible gains were those losses exploited in some kind of recovery device, so as to highlight the practical aspect of second-law analysis. This is attempted in Figures 8.13 and 8.14 for the low-heat rejection cases described previously. Only the exhaust gas to ambient stream is considered, however, since this can indeed be exploited by practical means, unlike the heat transfer loss to the walls, which requires a rather unviable heat recovery device for its exploitation. The gain in the work potential of the exhaust gas to ambient (Figure 8.13) reaches 8 kJ or 2 kW during the 4 s of the transient event, for the 1.5 mm PSZ coating compared with the non-insulated operation. This would correspond to a 2.4% increase in the engine brake power. This is the maximum extra power that can be obtained during the specific transient event from the increased insulation, if a bottoming cycle is applied to the particular engine configuration. Of course, larger gains are expected when higher load changes or thicker insulations are applied.

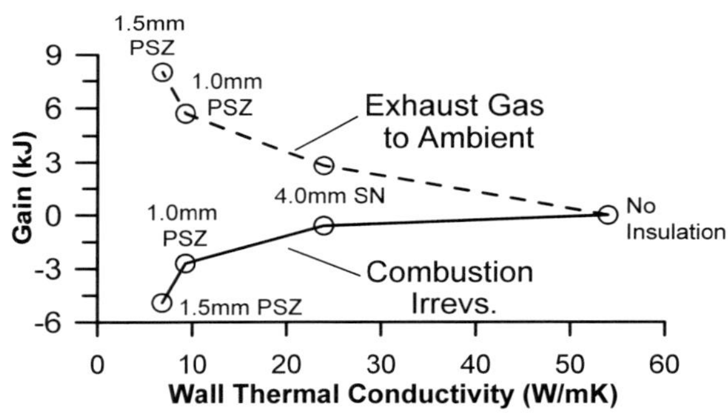

Figure 8.13. Gain from insulation compared with the non-insulated operation during a 10–75% load increase transient event

The respective engine efficiency ε_{tot2} from Equation 8.24 (taking into account both indicated work and exhaust gas to ambient exergy) is plotted in Figure 8.14. Here, it is made obvious that, in work's terms, the gain from the insulation increases the engine efficiency by almost 6.5% (1.5 mm PSZ coating compared with the non-insulated operation). Any thoughts to further exploit this extra amount of work potential must be realized by taking into account the respective steady-state gains (that, hopefully, go along with the transient results) and the financial burden induced from the more complicated engine configuration.

Summarizing, the findings from application of second-law balance to LHR engine operation are the following:

- The second-law balance results are much more coherent than their first-law counterparts, with the second-law values depending strongly on the applied insulation scheme.

- The dominant combustion irreversibilities decrease considerably with higher degrees of insulation both in 'Joules' and as a percentage of either fuel availability or total irreversibilities.
- Unfortunately, this decrease in exergy destruction cannot be recovered as an increase in piston work. It is primarily transformed into an increase in the availability content of the exhaust gas to ambient; the latter can contribute a few percentage points to the total engine plant efficiency if a bottoming cycle is applied.
- The insight gained from such studies, as regards first- *vs.* second-law balance results, is believed to provide a sound basis for an alternative view of internal combustion engine thermodynamic processes possibly to be applied in future works.

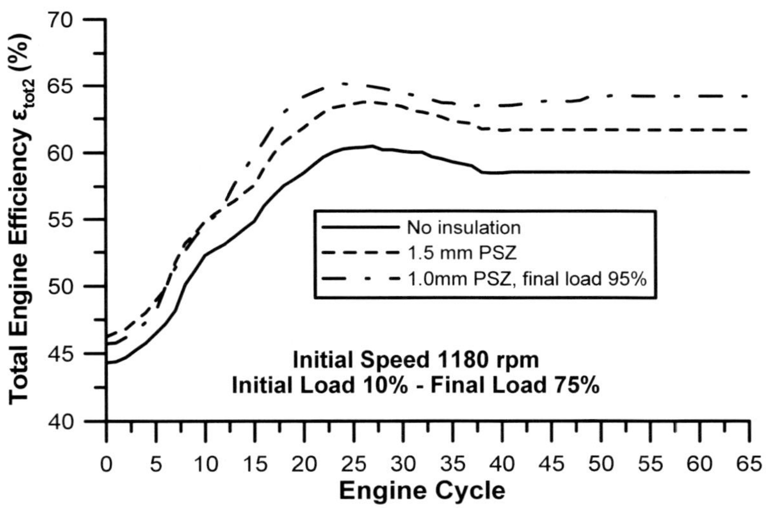

Figure 8.14. Response of total (*i.e.*, indicated work plus exhaust gas to ambient exergy) engine efficiency ε_{tot2} (from Equation 8.24 right-hand side expression) to an increase in load (unless otherwise noted, the load increase is 10–75%)

References

[1] Gibbs JW. The scientific papers, Vol. 1. New York: Dover, 1961.
[2] Haywood RW. Equilibrium thermodynamics for engineers and scientists. New York: John Wiley and Sons, 1980.
[3] Moran MJ. Availability analysis: a guide to efficient energy use. New Jersey: Prentice-Hall, 1982.
[4] Bejan A. Advanced engineering thermodynamics, 3rd edition. New York: John Wiley and Sons, 2006.
[5] Obert EF, Gaggioli RA. Thermodynamics, 2nd edition. New York: McGraw-Hill, 1963.

[6] Rakopoulos CD, Giakoumis EG. Second-law analyses applied to internal combustion engines operation. Prog Energy Combust Sci 2006;32:2–47.
[7] Flynn PF, Hoag KL, Kamel MM, Primus RJ. A new perspective on diesel engine evaluation based on second law analysis. SAE paper No. 840032, 1984.
[8] Van Gerpen JH, Shapiro HN. Second-law analysis of diesel engine combustion. ASME Trans, J Eng Gas Turbines Power 1990;112:129–37.
[9] Wepfer WJ, Gaggioli RA. Reference datums for available energy. In: Gaggioli RA, editor. Thermodynamics: second law analysis. Washington DC: American Chemical Society Symposium Series No. 122, 1980, pp. 72–92.
[10] Gallo WLR, Milanez LF. Choice of a reference state for exergetic analysis. Energy 1990;15:113–21.
[11] Szargut J, Styrylska T. Angenäherte Bestimmung der Exergie von Brennstoffen. Brennstoff-Wärme-Kraft 1964;16:589–96.
[12] Rodriguez L. Calculation of available-energy quantities. In: Gaggioli RA, editor. Thermodynamics: second law analysis. Washington DC: American Chemical Society Symposium Series No. 122, 1980, pp. 39–59.
[13] Alkidas AC. The application of availability and energy balances to a diesel engine. ASME Trans, J Eng Gas Turbines Power 1988;110:462–9.
[14] Primus RJ, Flynn PF. The assessment of losses in diesel engines using second law analysis. ASME WA-Meeting, Anaheim CA, Proc. AES 1986:61–8.
[15] Rakopoulos CD, Giakoumis EG. Availability analysis of a turbocharged diesel engine operating under transient load conditions. Energy 2004;29:1085–104.
[16] Dunbar WR, Lior N. Sources of combustion irreversibility. Combust Sci Technol 1994;103:41–61.
[17] Rakopoulos CD, Andritsakis EC. DI and IDI diesel engines combustion irreversibility analysis. ASME-WA Meeting, New Orleans LA, Proc. AES 1993;30:17–32.
[18] Kyritsis DC, Rakopoulos CD. Parametric study of the availability balance in an internal combustion engine cylinder. SAE paper No. 2001-01-1263, 2001.
[19] Rakopoulos CD, Kyritsis DC. Hydrogen enrichment effects on the second law analysis of natural and landfill gas combustion in engine cylinders. Int J Hydrogen Energy 2006;31:1384–93.
[20] Shapiro HN, Van Gerpen JH. Two zone combustion models for second law analysis of internal combustion engines. SAE paper No. 890823, 1989.
[21] Giakoumis EG, Andritsakis EC. Irreversibility production during transient operation of a turbocharged diesel engine. Int J Vehicle Design 2007;45:128–49.
[22] Rakopoulos CD, Giakoumis EG. Parametric study of transient turbocharged diesel engine operation from the second-law perspective. SAE paper No. 2004-01-1679, 2004.
[23] Primus RJ, Hoag KL, Flynn PF, Brands MC. An appraisal of advanced engine concepts using second law analysis techniques. SAE paper No. 841287, 1984.
[24] Bozza F, Nocera R, Senatore A, Tuccilo R. Second law analysis of turbocharged engine operation. SAE paper No. 910418, 1991.
[25] Alkidas AC. The use of availability and energy balances in diesel engines. SAE paper No. 890822, 1989.
[26] Giakoumis EG. Cylinder wall insulation effects on the first- and second-law balances of a turbocharged diesel engine operating under transient load conditions. Energy Convers Manage 2007;48:2925–33.

9

Modeling

9.1 Introduction

Starting from the late 1960s, computer simulation has contributed enormously towards new evaluation and development of internal combustion engines [see for example, 1–6]. Mathematical tools have become very popular in recent years owing to the continuously increasing improvement in computational power. In order to construct a simulation model, careful observation of the problem and its side effects is initially required (Figure 9.1).

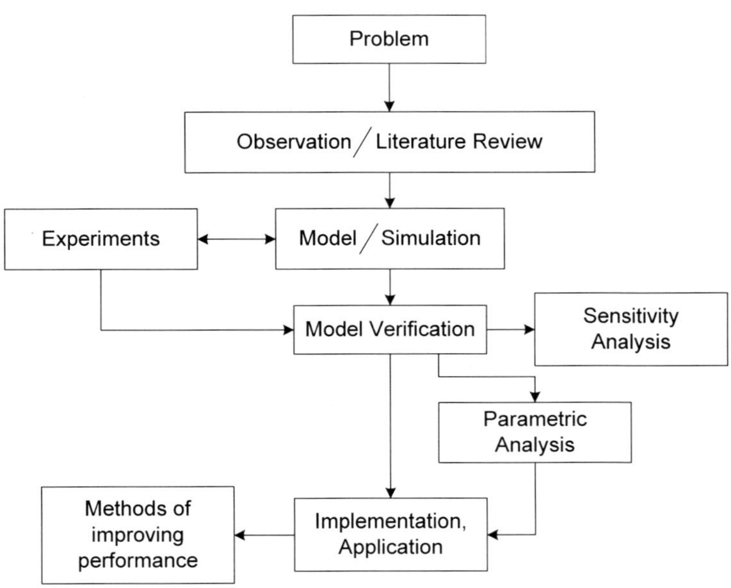

Figure 9.1. Block diagram of modeling process

As regards transient diesel engine operation, it was, mainly, the observation of engine manufacturers in the 1960s that when highly-rated, medium-speed diesel engines are employed in sudden, 0–100% step load changes, severe difficulties are encountered even leading to engine stall, that has initiated research on the subject. The thermodynamic cycle simulations developed over the years have given access to data that otherwise could only be measured with complicated and costly techniques. Moreover, they provided the opportunity to investigate different design solutions early in the development process, optimize key parameters of the system, and perform sensitivity analyses that proved valuable to the engine designer and manufacturer.

In general, a simulation model for (transient) diesel engine operation should

1. be composed of high fidelity sub-models covering the operation of all engine subsystems;
2. give adequate insight into the various engine processes, identifying the effect of key operating parameters;
3. be flexible and with the minimum need for empirical data;
4. have limited requirements in execution time;
5. be modular in order to be able to adopt new subroutines; and
6. be easily adaptable to different engines/operating conditions.

The studies concerning transient diesel engine simulations are based on a preceding mathematical-thermodynamic modeling of the steady-state processes inside the cylinder and its sub-systems. Such models use a combination of analytical and empirical methods to represent the engine components, and they can be classified according to their degree of complication. Analytically based models tend to focus on a °CA variation of engine variables in considerable detail. On the other hand, empirically based models tend to take a wider view, predicting the mean values or trends of the major engine variables with a subsequent loss of resolution and much shorter run times.

The models used for transient simulations are (see also Table 9.1)

- linear or sampled data models applying transfer functions;[1]
- quasi-linear. Mean value models are practically quasi-linear but with the main difference that they are usually built on physically based sub-models of the engine processes rather than empirical correlations;
- filling and emptying models of the zero-dimensional type. These are also termed phenomenological since they incorporate scientific principles for the simulation of the various processes.

Ideally, a complete diesel engine transient simulation code should include true 'transient' sub-models for

1. in-cylinder processes including heat transfer, spray penetration combustion, mass exchange with the manifolds, valve opening schedule, blow-by losses;
2. friction;

---

[1] Such models do not include in-cylinder thermodynamic calculations.

3. inlet and exhaust manifold operation including heat transfer effects;
4. exhaust gas recirculation;
5. engine dynamics in the form of the crankshaft torque balance;
6. turbocharger operation, including compressor and turbine operation, and turbocharger dynamics;
7. waste-gate and/or VGT operation;
8. aftercooler;
9. fuel pump operation and fuel injection;
10. governor or ECU operation with fuel limiting function; and
11. load performance.

Table 9.1. Comparison of specifications and requirements of the various modeling approaches

	Linear	Quasi-linear	Non-linear		
			Filling & Emptying	Wave-action	Multi-dimensional
Mathematical complexity	Very low	Low	High	High	Extremely high
Insight into the relevant phenomena	Very low	Low	Adequate	Adequate	High
Program execution time	Negligible	Negligible	Limited	Increased	Very large
Adaptability	Medium	Low	Medium	Medium-low	Medium

Incorporating detailed sub-models for all the above-mentioned processes, however, would increase considerably the computational time, cost and effort to simulate transient operation. It is not surprising then that most of the modeling techniques developed over the years rely heavily on experimental data or they are actually based on empirical or semi-empirical formulae; this holds particularly true for the 'delicate' processes of combustion and heat transfer but also for the turbocharger simulation.

The various computer models developed over the years have paved the way for an in-depth study of transient operation. However, the excursion to off-design and off-steady-state conditions experienced during transients cannot be dealt with using conventional modeling techniques based on steady-state philosophy. This holds true especially for combustion and friction, but also for the operation of various engine sub-systems such as the turbocharger. Exhaust emissions predictions, although a prosperous field of research during steady-state operation, are extremely limited when transient conditions are involved. This is mainly due to the conflict between detailed in-cylinder thermodynamic codes and the necessity to run a large number of engine cycles.

Figure 9.2 illustrates a simplified simulation flow-chart for transient diesel engine operation studies for a turbocharged and aftercooled diesel engine incorporating exhaust gas recirculation, variable geometry turbocharger/waste-gate valve, and common rail injection; this will be used as the general engine layout for the analysis that follows.

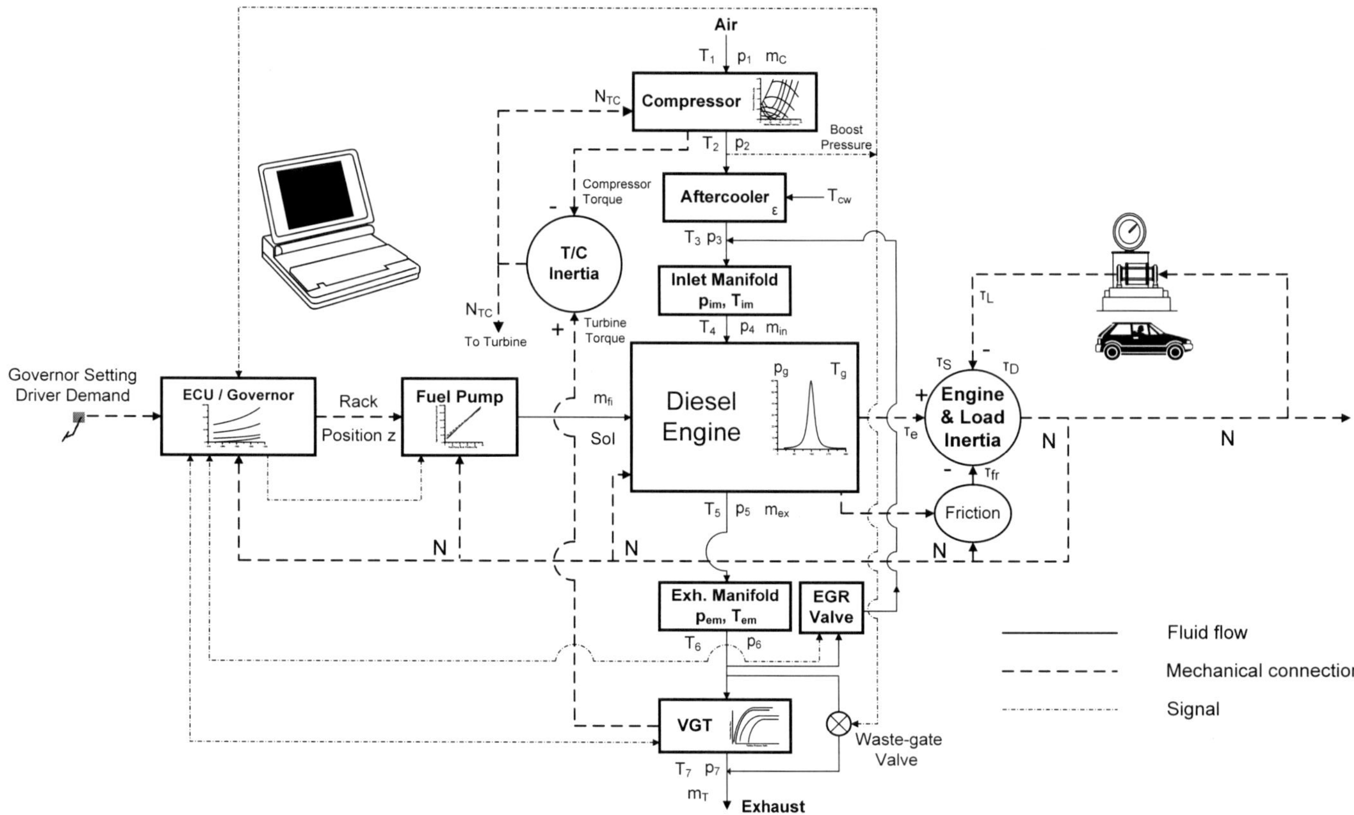

Figure 9.2. Block diagram of simulation code for turbocharged diesel engine transient operation, incorporating fuel limit function, VGT, EGR and waste-gate valve

9.2 Quasi-linear or Mean Value Approach

The basic advantage of quasi-linear modeling lies in its simplicity combined with limited computational requirements. In this approach, the elements of the system, *i.e.,* engine cylinder, manifolds *etc.,* are modeled in terms of steady-state or mean value characteristics using empirical correction coefficients, and neglecting the intermittent and non-linear nature of the engine operating cycle. Simplifications usually adopted are as below.

1. The working medium is assumed either pure air or combustion products, both treated as perfect gases with constant specific heat capacities. Since gas velocities are relatively low, dynamic pressures and temperatures are considered equal to the corresponding static ones.
2. Accumulation volumes in the system have negligible effect.
3. Mean values are used for engine and manifolds pressures and temperatures.
4. The engine output is related to fuel input via empirical equations or tabulated steady-state data.
5. An empirical relation is applied for estimating the exhaust gas temperature, at turbine inlet. This is defined from the inlet manifold temperature, by adding a temperature rise factor to it, to account for combustion effects.
6. A pulse factor is used for correction of the turbine expansion ratio in order to account for pulse turbocharging.

9.2.1 Engine Output

One of the most common approaches regarding engine output prediction was proposed by Ledger and Walmsley [7]; they applied a linear relation for the calculation of indicated mean effective pressure (imep) based on total injected fuel quantity M_{fi}, provided that the air-supply was sufficient for satisfactory combustion:

$$\text{imep} = a_1 M_{fi} \tag{9.1}$$

This approach is based on the thermodynamic analysis of the dual combustion cycle. For air–fuel ratios lower than a critical value, a correction coefficient (termed relative thermal efficiency η_R) is introduced to account for non-linearity between imep and injected fuel quantity [8]. Relative thermal efficiency is defined as the ratio of the actual thermal efficiency η_{th} of the dual combustion cycle (depending on the actual trapped air–fuel ratio) to that at the 30:1 trapped air–fuel ratio limit, *i.e.,*

$$\eta_R = \frac{\eta_{th}}{\left(\eta_{th}\right)_{AFR_{tr}=30}} \tag{9.2}$$

The 30:1 trapped air–fuel ratio limit was selected to be the value above which combustion is supposed to be complete (there is no unburned fuel at the end of combustion); Benson *et al.* [9] preferred a 20:1 air–fuel ratio limit.

Consequently, for the indicated mean effective pressure, the following relation can be applied based on Equation 9.1:

$$\text{imep} = a_1\, M_{fi}\, \eta_R + a_2 \tag{9.3}$$

with a_2 a constant used for engine speed compensation. Brake mean effective pressure bmep can then be calculated by subtracting fmep (Equation 9.95) from imep.

Jensen *et al.* [10] studying a small turbocharged diesel engine, followed an alternative procedure; they developed an expression for the indicated efficiency based on engine speed and fuel–air ratio, namely

$$\eta_i = (a_3 + a_4 N + a_5 N^2)(1 - a_6 \Phi^{a_7}) \tag{9.4a}$$

Indicated work can then be calculated as the product of total injected fuel mass, fuel lower heating value and indicated efficiency

$$W_i = \eta_i\, M_{fi}\, \text{LHV} \tag{9.4b}$$

A completely different approach was followed by Berglund [11], who used steady-state engine maps. He measured brake torque *vs.* speed and fueling at various steady-state conditions and tabulated the obtained values in the form of a look-up table. Other parameters can be mapped with this technique too, *e.g.*, exhaust emissions, fuel consumption or EGR rate. In a more advanced approach, multi-dimensional engine maps can be formed taking into account the effect of other influential engine parameters, *e.g.*, air–fuel ratio or injection timing, or their derivatives on the interesting property. The tabulated engine variables can then be interpolated using the instantaneous speed, fueling, AFR *etc.*, during the load or speed change to estimate their transient values. The major assumption here is that quasi-steady behavior is supposed during transients, which is way beyond reality; to account for the latter non-linearity, some researchers applied correction coefficients to the steady-state maps, as will be discussed later in the section.

9.2.2 Exhaust Gas Temperature

For the estimation of the exhaust gas temperature during steady-state operation, a mean gas temperature rise is firstly applied that represents the average value during steady-state conditions. This is derived from the energy conservation equation applied to a control volume containing the engine block [1]:

$$\Delta T_{e_{st}} = (T_5 - T_4)_{st} = \frac{c}{1 + \text{AFR}} \cdot \frac{\eta_R \text{LHV}}{c_{p_g}} - \frac{a_8}{N} + a_9 \tag{9.5a}$$

The last two terms on the right-hand side of Equation 9.5a are used for engine speed compensation, with constants a_8 and a_9 evaluated after experimental steady-state matching. Term c represents the fraction of the fuel energy contained in the exhaust gases, and it varies with the air–fuel ratio as shown in Figure 9.3.

Use of the above (steady-state) expression during transients involving sudden load increases gives unrealistically high exhaust temperatures; for transient operation, the following relation is applied for the temperature rise [1, 9]

$$\Delta T_{e_{trans}} = \Delta T_{e_{st}} - \Delta T_M \exp(-t/\tau_{em}) \tag{9.5b}$$

Terms ΔT_M and τ_{em} (exhaust manifold time constant representing a warming up delay period) are, mainly, functions of the exhaust gas energy and exhaust manifold size and configuration.

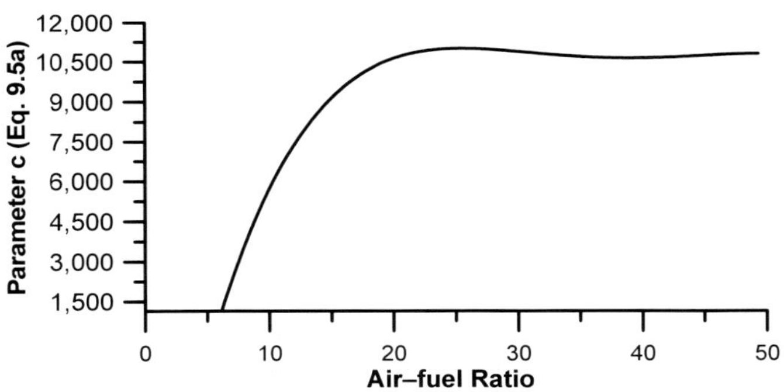

Figure 9.3. Variation of parameter c of Equation 9.5a (in K) with air–fuel ratio

9.2.3 Engine Air-flow

The engine air-flow consists of two parts; the trapped air-flow, which is used for charging of cylinders, and the scavenge air-flow, which consists of air-flowing through the engine during the valve overlap period (four-stroke engine) [9]:

$$\dot{m}_{in} = \dot{m}_{sc} + \dot{m}_{tr} \tag{9.6a}$$

It is assumed that the scavenge mass flow-rate (depending on valve overlap, engine speed and pressure differential across the inlet and exhaust valves) is a fixed percentage of the trapped mass flow-rate, *i.e.*,

$$\dot{m}_{sc} = a_{10}\, \dot{m}_{tr} \tag{9.6b}$$

It was found that coefficient a_{10} ranged between 5 and 10% for engines operating at high speed and having small valve overlap period.

The trapped air-mass flow-rate can be calculated assuming that the trapped volume equals the cylinder volume at inlet valve closure (IVC), using the following equation

$$\dot{m}_{in} = (1 + a_{10}) \frac{p_{IVC}}{R_s T_{IVC}} V_{tr} \, n_{cyl} \frac{N}{120} \tag{9.7a}$$

It is often assumed that the trapped pressure at IVC is 5–10% ($=a_{11}$) higher than the inlet manifold pressure (due to compression effects caused by the upward movement of the piston) and the trapped temperature is 50–70°C ($=\Delta T_{tr}$) higher than the inlet manifold temperature (due to heat transfer effects from the warm engine parts to the cooler inducted air). Consequently, Equation 9.7a is transformed into

$$\dot{m}_{in} = (1 + a_{10}) \frac{(1 + a_{11}) p_{im}}{R_s (T_{im} + \Delta T_{tr})} V_{tr} \, n_{cyl} \frac{N}{120} \tag{9.7b}$$

with p_{im}, T_{im} the inlet manifold pressure and temperature.

A different approach was followed by Younes *et al.* [12], who derived an empirical relation for the volumetric efficiency (based on atmospheric conditions):

$$\eta_{vol} = a_{12} + a_{13} N + a_{14} N^2 \tag{9.8}$$

which can then be used for the calculation of the air-mass flow-rate into the engine cylinders applying the following equation, similar to 9.7b (V_h is the swept volume):

$$\dot{m}_{in} = \eta_{vol} \frac{p_{IVC}}{R_s T_{IVC}} n_{cyl} \, V_h \frac{N}{120} \tag{9.9}$$

A more detailed approximation of η_{vol} incorporates the effect of inlet manifold pressure and temperature, and exhaust manifold pressure.

9.2.4 Transient Discrepancies

It order to improve the agreement between measured and predicted engine performance, it is admitted that the assumption of quasi-steady operation during transients is unrealistic. The following discrepancies during a step load increase have been identified:

- extra heat loss to the 'colder' cylinder walls;
- higher frictional losses, owing to the still low oil temperature and the increased gas pressures;
- system thermal storage effects.

In order to account for the above non-linearities, Jennings *et al.* [13] and Rackmil *et al.* [14] applied a correction coefficient to their tabulated steady-state engine maps, and a time lag, of the order of 100°CA to account for system delays. Winterbone [1] estimated this delay at 120°CA.

A different approach was followed by Benson *et al.* [9], who arbitrarily reduced both engine power and temperature rise by a further 5% in order to account for the above-mentioned discrepancies.

It is made obvious that quasi-linear models are heavily dependent on empirical data in order to adjust Equations 9.1–9.9, or to formulate the steady-state engine maps. However, the non-linear nature of the engine characteristics makes linear interpolation of data unreliable, especially in regions outside of those where the experimental data is available. Moreover, such experimental data are unavailable for low loads/engine speeds, where the most demanding transient tests commence. Quasi-linear modeling plausibly fails to fulfill the second, third and sixth modeling specification mentioned in Section 9.1; unsurprisingly, its use has been limited to studies where the main concern is the speed of execution, *e.g.*, control oriented transient models.

9.3 Filling and Emptying Approach

In filling and emptying models, the engine and its sub-systems are represented as a series of control volumes, linked by valves or orifices that interchange mass, heat and work. The perfect gas law and the principles of mass and energy conservation are applied to predict engine performance. Gas properties are computed using polynomial relations of temperature and pressure as will be discussed in Section 9.3.1.1. The computational step is quite small, of the order of 1°CA or even less, during which quasi-steady conditions are assumed for the thermodynamic processes. The filling and emptying, zero-dimensional models that are used as a basis for the simulation of diesel engine transient performance are usually single-zone, *i.e.*, treating the cylinder contents as a uniform mixture. Such models are termed 'zero-dimensional', in the sense that they do not involve any consideration of the flow field dimensions. Apart from the single-zone models, the urgent need to control pollutant emissions from internal combustion engines has led to the development of other more complicated models, such as two-zone and multi-zone models, which furnish increased accuracy and flexibility for such complex phenomena as the formation of nitric oxide and soot in engine cylinders.

In a single-zone model, the working fluid in the engine is assumed to be a thermodynamic system that undergoes energy and mass exchange with the surroundings, where the energy released during the combustion process is obtained by applying the first law of thermodynamics to the system. Usual assumptions of single-zone modeling are

- spatial homogeneity of pressure (for two- and multi-zone models too);
- spatial homogeneity of temperature (for the whole cylinder or for each zone considered in two-zone or multi-zone models);

- working fluid considered as perfect gas;
- gas properties (enthalpy, internal energy *etc.*) are modeled using polynomial relations with temperature (and pressure);
- heat released from combustion is distributed evenly throughout the cylinder;
- enthalpy associated with the injected fuel is usually not significant, hence ignored;
- spatially averaged, instantaneous (time resolved) heat transfer rates are used, to estimate heat transfer to the cylinder walls (Section 9.3.2.2); and
- dissociation is usually, but not always, neglected.

In two-zone models, on the other hand, the working fluid is imagined to consist of two zones, a burned and an unburned zone. These zones are actually two distinct thermodynamic systems with energy and mass interactions between themselves and their common surroundings, the cylinder walls; further to the previous assumptions the following are also usually applied:

- no heat transfer takes place between burned and unburned zones; and
- the work required to transfer fluid from the unburned zone to the burned zone is negligible.

9.3.1 Thermodynamics Fundamentals

9.3.1.1 State Properties

In computer simulations following the filling and emptying approach, use is made of polynomial relations for the thermodynamic data of each species in the mixture needed for the analysis of the engine processes. One such, widely adopted, approach is based on the JANAF Table thermodynamic data [3, 15, 16]. An alternative approach is provided by Benson and Whitehouse [4]; for the evaluation of absolute molar (index 'm') specific enthalpy of species i (O_2, N_2, CO_2, H_2O, N, NO, OH, H, O, *etc.*) it holds that

$$h_{mi}(T) = R_{mol} \sum_{j=1}^{5} a_{ij} T^{j} + h_{moi} \tag{9.10}$$

where h_{moi} is the molar enthalpy at absolute zero. Values of constants a_j for each species i are provided in Table 9.2.

Assuming that the gases obey the perfect gas relation, the molar heat capacity under constant pressure c_{pmi} of species i is then given by

$$c_{pmi}(T) = \frac{dh_{mi}}{dT} = R_{mol} \sum_{j=1}^{5} j a_{i,j} T^{j-1} \tag{9.11}$$

For the specific internal energy it holds accordingly that

$$u_{mi}(T) = h_{mi}(T) - R_{mol}T \tag{9.12}$$

hence,

$$u_{mi}(T) = R_{mol}\left(\sum_{j=1}^{5} a_{ij}T^{j} - T\right) + h_{moi} \tag{9.13}$$

and the respective molar heat capacity under constant volume c_{vmi} is given by

$$c_{vmi}(T) = R_{mol}\left(\sum_{j=1}^{5} j a_{i,j}T^{j-1} - 1\right) \tag{9.14}$$

The molar entropy of each species, primarily needed for the second-law analysis of Chapter 8, can be computed from the following relation:

$$s_{mi} = s_{moi} + \int_{T_o}^{T} c_{pmi}\frac{dT}{T} - R_{mol}\ln\left(\frac{p_i}{p_o}\right) \tag{9.15}$$

Table 9.2. Polynomial coefficients a_j (corresponding temperature range 500–3000 K and reference pressure 1 atm)

Gas	j=1	j=2	j=3	j=4	j=5	j=6	$h_{mo}=u_{mo}$
CO_2	3.10	2.73×10^{-3}	-7.89×10^{-7}	8.66×10^{-11}	0	6.58	-3.94×10^{8}
H_2O	3.74	5.66×10^{-4}	4.95×10^{-8}	-1.82×10^{-11}	0	0.96	-2.39×10^{8}
N_2	3.34	2.94×10^{-4}	1.95×10^{-9}	-6.57×10^{-12}	0	3.76	0
O_2	3.25	6.52×10^{-4}	-1.50×10^{-7}	1.54×10^{-11}	0	5.71	0
CO	3.32	3.77×10^{-4}	-3.22×10^{-8}	-2.19×10^{-12}	0	4.63	-1.14×10^{8}
H_2	3.43	-8.18×10^{-6}	9.67×10^{-8}	-1.44×10^{-11}	0	−3.84	0
OH	3.48	1.36×10^{-4}	-4.55×10^{-9}	2.93×10^{-12}	0	2.13	3.87×10^{7}
H	2.50	0.00	0.00	0.00	0	−0.46	2.16×10^{8}
O	2.76	-2.51×10^{-4}	1.00×10^{-7}	-1.39×10^{-11}	0	3.73	2.47×10^{8}
NO	3.50	2.99×10^{-4}	-9.59×10^{-9}	-4.90×10^{-12}	0	5.11	8.99×10^{7}
N	2.50	2.87×10^{-6}	-2.45×10^{-9}	6.15×10^{-13}	0	4.19	4.71×10^{8}
CH_4	1.94	4.96×10^{-3}	-1.24×10^{-6}	1.62×10^{-10}	0	8.15	-6.69×10^{7}
C_3H_8	1.14	1.46×10^{-2}	-2.96×10^{-6}	0.00	0	-	-0.91×10^{8}
C_8H_{18}	−0.72	4.64×10^{-2}	-1.68×10^{-5}	2.67×10^{-9}	0	-	-4.93×10^{8}
$C_{12}H_{26}$	6.40	5.30×10^{-2}	-1.27×10^{-5}	1.06×10^{-9}	0	-	-2.90×10^{8}

Assuming that the reference temperature is again 0 K (hence $s_{moi} = 0$ according to the third law of thermodynamics), we get

$$s_{mi} = \int_0^T c_{pmi} \frac{dT}{T} - R_{mol} \ln\left(\frac{x_i p}{p_o}\right) = s'_{mi}(T, p_o) - R_{mol} \ln\left(\frac{x_i p}{p_o}\right) \tag{9.16}$$

with x_i the molar fraction of species i in the mixture; the standard state entropy $s'_{mi}(T, p_o)$ of species i at pressure of 1 atm, which is a function of temperature only, is then given by the following property relation:

$$s'_{mi}(T, p_o) = R_{mol}\left(a_{i1} \ln T + 2a_{i2} T + 3a_{i3} \frac{T^2}{2} + 4a_{i4} \frac{T^3}{3} + 5a_{i5} \frac{T^4}{4} + a_{i6}\right) \tag{9.17}$$

Finally, the molar Gibbs free enthalpy is defined as

$$g_{mi} = h_{mi} - T s_{mi} \tag{9.18}$$

The latter can be calculated by combining Equations 9.10 and 9.16.

Further, it is widely accepted that for the range of pressures and temperatures experienced in internal combustion engine applications, the mixture can be assumed to obey the perfect gas relation, namely

$$pV = mR_s T \tag{9.19}$$

9.3.1.2 Hydrocarbon–Oxygen Reaction and Dissociation Effects

Since combustion is the most delicate process occurring in an internal combustion engine, basic equations for its analysis are incorporated in every detailed computer simulation code. The overall reaction of one kmol of hydrocarbon fuel C_nH_m, assuming chemical equilibrium and neglecting dissociation, is described by the following expression for lean mixtures $(\Phi \leq 1)$:

$$\begin{aligned} &C_n H_m + \overbrace{\frac{1}{\Phi}\left(n + \frac{m}{4}\right) O_2 + 3.76 \frac{1}{\Phi}\left(n + \frac{m}{4}\right) N_2}^{\text{Air: } 79\% N_2 \,/\, 21\% O_2 \text{ v/v}} \rightarrow \\ &n\, CO_2 + \frac{m}{2} H_2O + \left(\frac{1}{\Phi} - 1\right)\left(n + \frac{m}{4}\right) O_2 + 3.76 \frac{1}{\Phi}\left(n + \frac{m}{4}\right) N_2 \end{aligned} \tag{9.20}$$

The number of kmoles of the constituents (CO_2, H_2O, O_2, N_2 and C_nH_m, respectively, if no dissociation is taken into account), at any time t, can be calculated as a function of the number of kmoles of fuel burned up to that time and the equivalence ratio Φ. Therefore, the internal energy of the mixture is then computed from Equation 9.13 by knowing the instantaneous composition and the specific internal energies of its constituents.

Knowledge of the exact gas composition inside the combustion chamber, at each instant of time, is critical for accurate calculation of thermodynamic cycle models. For this purpose, dissociation occurring at high temperatures (typically above 1500 K) should also be taken into consideration, at least when two-zone analysis is involved. To this aim, a chemical equilibrium scheme can be employed considering a certain number of species present inside the combustion chamber. Such an equilibrium scheme can also form the baseline for predicting the exhaust nitric oxide (NO) and carbon monoxide (CO), which are produced in a kinetically controlled fashion (Section 9.11.2).

A two-way reaction between four species A, B, C and D, is described by the following expression:

$$y_a A + y_b B \rightleftarrows y_c C + y_d D \tag{9.21}$$

with y_i, i=a,b,c,d the stoichiometric coefficients of the reaction.

The respective equilibrium (or dissociation) constant K_p is given by

$$K_p = \frac{(x_c)^{y_c}(x_d)^{y_d}}{(x_a)^{y_a}(x_b)^{y_b}} p^{y_c + y_d - y_a - y_b} \tag{9.22}$$

with x_i, i=a,b,c,d the molecular concentrations of species A, B, C and D, and p the combustion chamber (uniform) pressure.

For the dissociation constant K_p, mainly a function of temperature, the following relation is also valid [4]:

$$\ln(K_p) = \ln\left(\frac{(p_c)^{y_c}(p_d)^{y_d}}{(p_a)^{y_a}(p_b)^{y_b}}\right) = \frac{-\Delta G_T^o}{R_{mol}T} =$$
$$\left\{\sum_{n=1}^{R} \frac{y_n g_{mTn}}{R_{mol}T} - \sum_{k=1}^{P} \frac{y_k g_{mTk}}{R_{mol}T}\right\} - \frac{1}{R_{mol}T}\left(\sum_{k=1}^{P} y_k h_{mok} - \sum_{n=1}^{R} y_n h_{mon}\right) \tag{9.23}$$

with 'P' denoting products and 'R' denoting reactants for reaction 9.21. The last term on the right-hand side of Equation 9.23 is the enthalpy of reaction at absolute zero. Term $g_{mT}/(R_{mol}T)$ needed in Equation 9.23 can be computed based on the polynomial relations of the previous section, for each reactant 'n' and product 'k', *i.e.*,

$$\frac{g_{mT}}{R_{mol}T} = a_1(1 - \ln T) - \left(a_2 T + a_3\frac{T^2}{2} + a_4\frac{T^3}{3} + a_5\frac{T^4}{4}\right) - a_6 \tag{9.24}$$

A typical dissociation scheme applied in diesel engine computer simulations, assumes that the working medium is a mixture of 11 species (O_2, N_2, CO_2, H_2O, H_2, OH, NO, CO, O, H, N) and fuel vapor; the following seven reactions among the above 11 species take place:

$$\begin{aligned}
&\frac{1}{2}H_2 \rightleftarrows H && H_2 + \frac{1}{2}O_2 \rightleftarrows H_2O \\
&\frac{1}{2}O_2 \rightleftarrows O && \frac{1}{2}N_2 + \frac{1}{2}O_2 \rightleftarrows NO \\
&\frac{1}{2}N_2 \rightleftarrows N && \frac{1}{2}H_2 + \frac{1}{2}O_2 \rightleftarrows OH \\
&CO + \frac{1}{2}O_2 \rightleftarrows CO_2 &&
\end{aligned} \tag{9.25}$$

For each one of the above reactions, Equation 9.23 can be formulated. The remaining four equations that are needed in order to solve the dissociation scheme are provided by the mass balance between species C, H, O and N from hydrocarbon–oxygen reaction (Equation 9.20). The resulting scheme of 11 algebraic equations can then be solved using typical numerical analysis techniques, to find the concentration of each species at each computational step in the computer simulation.

9.3.2 In-cylinder Calculations

9.3.2.1 First law of Thermodynamics Applied to the Engine Cylinder

A. Single-zone Modeling

The basis for the thermodynamic analysis of engine processes is the first law of thermodynamics; its application to the engine cylinder reads (see also Figure 9.4)

$$\frac{dQ_L}{d\varphi} - p_g \frac{dV}{d\varphi} = \frac{dU}{d\varphi} - \sum_j \frac{dm_j}{d\varphi} h_j \tag{9.26}$$

where $dQ_L/d\varphi$ is the rate of heat loss to the cylinder walls (liner, piston crown, cylinder head – analyzed in the next sub-section), p_g is the uniform pressure of the cylinder contents, dm_j is the mass exchanged (positive when entering) in the time step $d\varphi$, and h_j is the specific enthalpy of it. Subscript j denotes fuel injection (index 'fi'), exchange with the exhaust manifold (index 'ex'), inlet manifold ('in') and crankcase blow-by losses ('bl').

The second fundamental equation applied describes the mass conservation

$$\frac{dm}{d\varphi} = \sum_j \frac{dm_j}{d\varphi} h_j = \sum \frac{dm_{inlet}}{d\varphi} h_{inlet} - \sum \frac{dm_{out}}{d\varphi} h_{out} \tag{9.27}$$

By expressing the perfect gas relation (Equation 9.19) in differential form, we get

$$p_g \frac{dV}{d\varphi} + V\frac{dp_g}{d\varphi} = mR_s \frac{dT_g}{d\varphi} + R_s T_g \frac{dm}{d\varphi} \tag{9.28}$$

The application of the first law of thermodynamics to the engine cylinder (open cycle) can then be formulated as follows, using the previous equations:

$$\sum_i m_i c_{vi} \frac{dT_g}{d\varphi} = \frac{dQ_L}{d\varphi} - \frac{mR_s T_g}{V}\frac{dV}{d\varphi} + \sum_j \frac{dm_j}{d\varphi} h_j - \sum_i u_i \frac{dm_i}{d\varphi} \tag{9.29}$$

with i representing each species considered.

With reference to Figure 9.4, the volume V of the engine cylinder, needed in the previous equations, is given by

$$V = V_{cl} + \pi(D^2/4)\, x + V_{bowl} \tag{9.30}$$

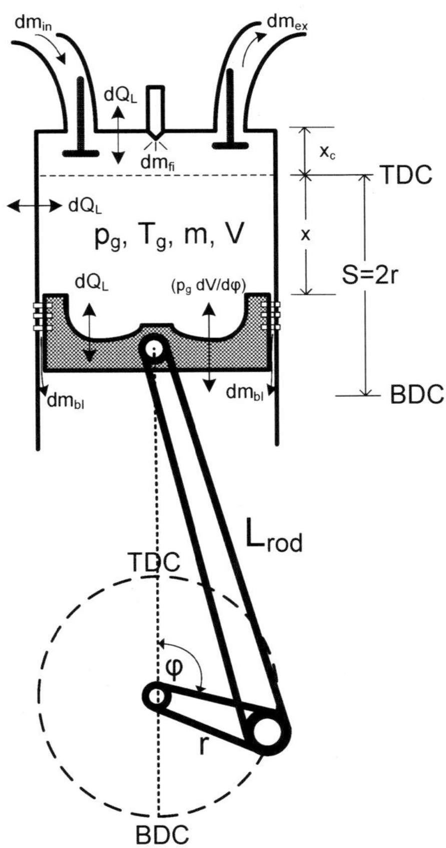

Figure 9.4. Schematic diagram of engine cylinder identifying main dimensions, gas properties, and mass and heat exchange

Differentiating Equation 9.30 with respect to crank angle φ, we get

$$\frac{dV}{d\varphi}=\frac{\pi D^2}{4}\left[r\sin\varphi\left(1+\frac{\lambda\cos\varphi}{\sqrt{1-\lambda^2\sin^2(\varphi)}}\right)\right] \tag{9.31}$$

with V_{bowl} the volume of the piston bowl, $V_{cl}=\pi(D^2/4)x_c$ the clearance volume and x the piston displacement measured from the TDC position, which is given by

$$x=r\left(1-\cos\varphi\right)+L_{rod}\left(1-\sqrt{1-\lambda^2\sin^2\varphi}\right) \tag{9.32}$$

where L_{rod} is the connecting rod length, r the crank radius, and $\lambda=r/L_{rod}$ the ratio of crank radius to connecting rod length. Another useful relation is

$$d\varphi=6\,N\,dt \tag{9.33}$$

with N the engine speed expressed in rpm; Equation 9.33 is employed for transforming the various terms from time to degree crank angle basis. Moreover, the mean piston velocity is given by

$$\bar{u}_{pist}=\frac{SN}{30} \tag{9.34}$$

B. Two-zone Modeling

A typical, comprehensive two-zone model that can be used in transient simulation analyses has been proposed by Rakopoulos *et al.* [17]. Such models should take into account all the processes taking place in the cylinder, *i.e.*, in-cylinder air-motion, fuel spray development and mixing, spray impingement on the walls, turbulent heat transfer and combustion chemistry. Droplets evaporation and fuel ignition delay can be taken into account through the combustion sub-model (Section 9.3.2.3).

During combustion and expansion, depending on the number of injector nozzle holes, an equal number of sprays constitute the entire burning section. In this case, apart from the perfect gas state equation, the first law of thermodynamics for an open system is applied for each zone. For the surrounding air-zone (index 'a'), which only loses mass (air) to the burning zone, the first law of thermodynamics is written as

$$dQ_L=dU+p_g dV+h_a dm_a \tag{9.35}$$

while for the burning zone, which gains mass (air) from the air-zone and also an enthalpic flow from the fuel ready to be burned in the time step, the first law of thermodynamics is given by

$$dQ_L = dU + p_g dV - h_a dm_a - h_f dm_f \tag{9.36}$$

In the above equations, the internal energy U of the mixture is computed by knowing the instantaneous composition and the specific internal energies of the constituents as described in Section 9.3.1.1.

Fuel Spray Penetration and Air Entrainment Rate

The correlation of Arai *et al.* [18] can be used for the spray tip location as a function of time, based on relevant experimental data and turbulent gas jet theory (Figure 9.5). The fuel injected inside the combustion chamber breaks up into globules, forming a cone-shaped spray corresponding to each one of the 'z' nozzle holes. If the kmoles of air trapped in the cylinder (in one cycle) are $N_{a_{tot}} = m_{a_{tot}} / M_a$ (with M_a the molecular weight of air), then the phenomenon will continue until each spray penetration reaches a value (D/2+πD/z), or until it entrains a maximum quantity of air equal to $N_{a_{tot}} / z$.

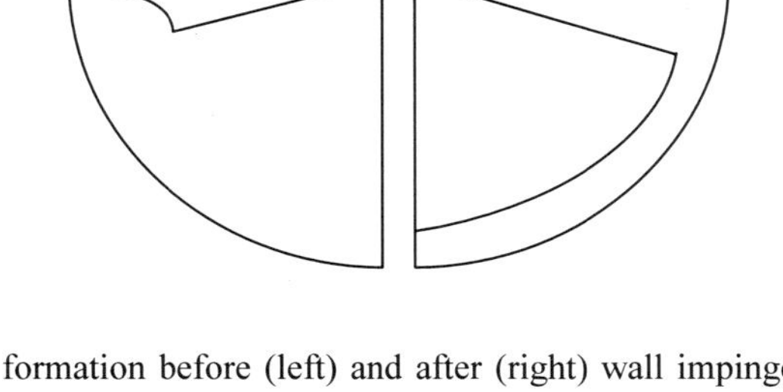

Figure 9.5. Spray formation before (left) and after (right) wall impingement for two-zone analysis

First, the parameters related to the fuel jet break-up point should be calculated. The break-up time t_{br} is found by equating the two expressions of spray penetration correlation before and after t_{br}, corresponding to the break-up length $x=x_{br}$:

$$x = 0.39\sqrt{2\,\overline{\Delta p_{inj}} / \rho_f} \cdot t \quad \text{for} \quad 0 < t \le t_{br} \tag{9.37a}$$

$$x = 2.95\,(\overline{\Delta p_{inj}} / \rho_a)^{0.25} \sqrt{D_N\, t} \quad \text{for} \quad t \ge t_{br} \tag{9.37b}$$

$$t_{br} = \frac{28.61\,\rho_f\,D_N}{\sqrt{\rho_a\,\overline{\Delta p_{inj}}}}, \qquad x_{br} = \overline{u}_{inj}\,t_{br} \tag{9.37c}$$

where $\overline{\Delta p_{inj}}$ is the mean pressure drop across the injector nozzle, ρ_f the liquid fuel density, and ρ_a the density of air inside the cylinder; the mean spray velocity from each nozzle hole of diameter D_N is given by

$$\overline{u}_{inj} = \frac{\dfrac{M_{fi}}{z\,\Delta\varphi_{inj}}\,6\,N}{\rho_f\left(\dfrac{\pi D_N^2}{4}\right)} \tag{9.38}$$

In Equation 9.38, $\Delta\varphi_{inj}$ is the total duration of fuel injection in degrees CA and M_{fi} the total amount of injected fuel per cycle. Therefore, the mean pressure drop in the nozzle is given by

$$\overline{\Delta p_{inj}} = \frac{1}{2}\rho_f\left(\frac{\overline{u}_{inj}}{c_D}\right)^2 \tag{9.39}$$

Accounting for air swirl, by introducing the air swirl ratio $R_s=N_s/N$ with N_s the cylinder charge rotational speed, the above correlations are modified as follows:

$$x_{brs} = x_{br}\left(1+\frac{\pi\,R_s\,N\,x_{br}}{30\cdot u_{inj}}\right)^{-1}, \qquad t_{brs} = \frac{x_{brs}}{u_{inj}} = \left(\frac{x_{brs}}{x_{br}}\right)t_{br} \tag{9.40}$$

Afterwards we can proceed to the main calculations. At each crank angle step, the fuel injection rate and the cumulative fuel injected in each spray are, respectively,

$$\dot{m}_{fi_j} = \dot{m}_{fi}/z \qquad \text{and} \qquad m_{fi_j} = m_{fi}/z \tag{9.41}$$

where index 'j' denotes each one of the z sprays.

Similarly, the spray tip velocity and the pressure drop across the nozzle are, respectively,

$$u_{inj} = \dot{m}_{fi_j}\,6\,N/(\rho_f A_N) \qquad \text{and} \qquad \Delta p_{inj} = \frac{1}{2}\rho_f\,(u_{inj}/c_D)^2 \tag{9.42}$$

After the calculation of the fuel jet break-up point characteristics, the following cases are discriminated for the spray penetration x and the spray angle θ (for no swirl, or with swirl), according to the value of time t:

For $t \le t_{brs}$

$$x = u_{inj}t, \quad x_s = x\left(1+\frac{\pi R_s N x}{30 u_{inj}}\right)^{-1}, \quad \theta = 0, \quad \theta_s = 0 \tag{9.43}$$

For $t > t_{brs}$

$$\begin{aligned} x &= 2.95\left(\frac{\Delta p_{inj}}{\rho_a}\right)^{0.25}\sqrt{D_N}\, t^{\beta}, \quad x_s = x\left(1+\frac{\pi R_s N x}{30 u_{inj}}\right)^{-1} \\ \theta &= 2\tan^{-1}\left(\frac{1}{A'}4\pi\sqrt{\frac{\rho_a}{\rho_f}}\frac{\sqrt{3}}{6}\right), \quad \theta_s = \theta\left(\frac{x}{x_s}\right)^2 \end{aligned} \tag{9.44}$$

where $A' = 3 + 0.28\left(L_N / D_N\right)$, with L_N the nozzle hole length. In the above relations, the exponent of time β=0.50 if $x_s \le D/2$, that is before wall impingement, while it is taken equal to 0.48 if $x_s \ge D/2$, that is after wall impingement for a single-cylinder, naturally aspirated engine. It is seen that the same correlations are opted to be used both before and after wall impingement, using different values for β. Thus, the complexities involved due to uncertainties in modeling the real physical mechanisms existing in the spray picture after wall impingement are avoided.

The mass of entrained air in the spray (of conical shape) is

$$\begin{aligned} m_{a_j} &= \frac{\pi}{3}\left(\tan\frac{\theta}{2}\right)^2 \rho_a (x - x_{br})^3 \\ m_{a_{js}} &= \frac{\pi}{3}\left(\tan\frac{\theta_s}{2}\right)^2 \rho_a (x_s - x_{brs})^3 \end{aligned} \tag{9.45}$$

where in the above relations it is considered that the volume taken by the fuel is negligible against that of the air, as well as the paraboloid part at the base of the cone against that of the pure cone.

The above relations for the air-mass entrained into the fuel spray apply only in the case where $t_{brs} < t \le (\Delta\varphi_{inj} / 6N) + t_{brs}$. In the case where $t > (\Delta\varphi_{inj} / 6N) + t_{brs}$, a conical part at the tail (index t) of the spray has to be subtracted. Then, the above equations become

$$m_{a_j} = \frac{\pi}{3}\rho_a\left(\tan\frac{\theta}{2}\right)^2 (x - x_{br})^3 - \frac{\pi}{3}\rho_a\left(\tan\frac{\theta_t}{2}\right)^2 (x_t - x_{br})^3 \tag{9.46a}$$

$$m_{a_{js}} = \frac{\pi}{3}\rho_a \left(\tan\frac{\theta_s}{2}\right)^2 (x_s - x_{brs})^3 - \frac{\pi}{3}\rho_a \left(\tan\frac{\theta_{ts}}{2}\right)^2 (x_{ts} - x_{brs})^3 \tag{9.46b}$$

where $x_t = 2.95\ (\Delta p_{inj}/\rho_a)^{0.25}\sqrt{D_N}\ [t-(\Delta\varphi_{inj}/6N)-t_{br}]^{\beta_t}$ is the cone length to be subtracted in the case of no-swirl, and $x_{ts} = x_t\ (1+\pi R_s N x/30 u_{inj})^{-1}$ is the corresponding cone length to be subtracted when swirl exists.

Finally, the corresponding spray angles for the tail part of the cone to be subtracted are, respectively,

$$\theta_t = \theta, \qquad \theta_{ts} = \theta_t \left(\frac{x_t}{x_{ts}}\right)^2 \tag{9.47}$$

For the fuel mass that is effectively inside the spray (the fuel starts to be considered inside the spray only after the break-up length), we have

$$m_{f_j} = m_{fi_j}(t - t_{brs})/t \quad \text{for} \quad t \le \Delta\varphi_{inj}/6N \tag{9.48}$$

$$m_{f_j} = \frac{m_{fi_j}}{(\Delta\varphi_{inj}/6N)}(t - t_{brs}) \quad \text{for} \quad \Delta\varphi_{inj}/6N < t \le (\Delta\varphi_{inj}/6N) + t_{brs} \tag{9.49}$$

$$m_{f_j} = m_{fi_j} \quad \text{for} \quad t > (\Delta\varphi_{inj}/6N) + t_{brs} \tag{9.50}$$

with j the spray number.

Lastly, the air–fuel ratio (by mass) AFR is, respectively

$$AFR_{spr} = m_{a_j}/m_{fj} \qquad \text{and} \qquad AFR_{spr_s} = m_{a_{js}}/m_{f_j} \tag{9.51}$$

9.3.2.2 Heat Transfer

Heat transfer between gas and cylinder walls is due to convection and radiation, with the latter primarily originating from hot soot particles. The intrinsic delicate task of modeling the heat transfer rate dQ_L/dt to the cylinder walls, needed in Equation 9.26, is usually accomplished via relations of the following type:

$$\frac{dQ_L}{dt} = A\,h_g (T_g - T_w) \tag{9.52}$$

where the corresponding surface area is given by

$$A = 2\left(\pi D^2/4\right) + A_p \tag{9.52a}$$

with $A_p=\pi Dx$ (x defined in Figure 9.4 and Equation 9.32), and h_g the instantaneous (convective) heat transfer coefficient. For the latter, global approximations are applied based on assumptions of overall, empirical, instantaneous spatially average heat transfer. Coefficient h_g is, generally, assumed the same for all surfaces and all operating conditions in the engine cylinder since, in single-zone modeling, T_g stands for the bulk mean gas temperature. For the estimation of h_g the following three models are usually applied; they all originate from dimensional analyses that correlate h_g to gas thermal conductivity, density, specific heat capacity and a representative length.

1. The Annand correlation [19]

$$\frac{dQ_L}{dt} = A\left[a\,\frac{k_g}{D}\,Re^b\left(T_g - T_w\right) + c\left(T_g^4 - T_w^4\right)\right] = A\,h_g\left(T_g - T_w\right) \tag{9.53}$$

where a, b and c are constants evaluated after experimental matching at steady-state conditions, k_g is the gas thermal conductivity, and the Reynolds number

$$Re = \rho_g\, u_{char}\, L_{char} / \mu_g \tag{9.54}$$

is calculated with a characteristic length, usually, equal to the piston diameter D and a characteristic speed equal to the mean piston velocity (Equation 9.34), or derived from a k-ε turbulence sub-model.

A significant advantage of the Annand correlation is that it includes a separate radiation term; however, the use of mean bulk gas temperature in single-zone modeling (instead of the actual non-homogeneous temperature distribution in the cylinder) limits the benefits from this advantage (notice that the radiation term in Equation 9.53 varies with the fourth power of temperature). Typical values for constants a, b and c, identified in Benson and Whitehouse [4] for a wide range of engines, are

- $0.06 < a < 0.76$
- $0.54 < b < 0.90$
- $c = 0.576\sigma$

with σ the Stefan–Boltzmann constant (5.67×10^{-8} W/m^{2}K^4); during compression and gas exchange c=0.

One of the weak points of this model is the relatively broad range of constant 'a', which rather suggests that important parameters may have been ignored in formulating Equation 9.53.

Further, one can calculate the gas thermal conductivity $k_g=3.17\times10^{-4}\,T_g^{0.772}$, and dynamic viscosity $\mu_g=3.3\times10^{-7}\,T_g^{0.7}/(1+0.027\,\Phi)$ [3].

2. The Woschni correlation [20]

In this approach, the heat transfer coefficient h_g needed in Equation 9.52 is approximated by

$$h_g = c\, D^{d-1}\, p_g^d\, u_g^d\, T_g^{0.75-1.62d} \tag{9.55}$$

with u_g the local average gas velocity, considered proportional to the mean piston speed during intake, compression and exhaust; during combustion and expansion a rise due to combustion is added:

$$u_g = \left[c_1 \bar{u}_{pist} + c_2 \frac{V_h T_r}{p_r V_r} (p_g - p_m) \right] \tag{9.56}$$

with p_r, T_r, V_r, the pressure, temperature and volume at reference conditions (*e.g.*, at inlet valve closure), p_m the corresponding motored cylinder pressure assuming a polytropic process, *i.e.*, $p_m = p_r \left(V_r / V \right)^{1.3}$, and c, c_1, c_2 constants depending on the particular engine characteristics (*e.g.*, swirl), and varying according to the particular process of the cycle. Typical values for the constants in Equations 9.55 and 9.56 are

- $c = 129.8$
- $d = 0.8$
- $c_1 = 6.18 + 0.417R_s$ (gas exchange) or $c_1 = 2.28 + 0.308R_s$ (closed part); R_s stands for the swirl ratio (0…3)
- $c_2 = 0$ (compression and gas exchange)
- $c_2 = 3.24\times10^{-3}$ for DI engines, or $c_2 = 6.22\times10^{-3}$ for IDI engines (closed part of the cycle).

3. The Hohenberg correlation [21], which is actually a simplification of the Woschni relation

$$h_g = \left(129.8\, p^{0.8}\, V_h^{-0.06}\, T_g^{-0.4} \right) \left(\bar{u}_{pist} + 1.4 \right)^{0.8} \tag{9.57}$$

For the latter equation, the corresponding piston surface area in Equation 9.52a is A_{pist}=A(piston crown) + $A^{0.3}$ (piston top land).

During transients, the improved model of Annand [22] can prove quite useful, since it takes into account the variation of gas temperature. In this case, Equation 9.53 is transformed into

$$\frac{dQ_L}{dt} = A \left\{ \frac{k_g}{D} Re^b \left[a \left(T_g - T_w \right) + \frac{a'}{\omega} \frac{dT_g}{dt} \right] + c \left(T_g^4 - T_w^4 \right) \right\} \tag{9.58}$$

Even for the simple case where a uniform wall temperature is assumed for all cylinder surfaces, the thermal inertia of the cylinder wall has to be taken into account during transient operation. This can be accomplished with the use of a heat transfer scheme, based on electrical circuit analogy that models the temperature distribution from the gas to the cylinder wall up to the coolant. By so doing the cylinder wall thickness, thermal conductivity and thermal diffusivity are taken into account. For the heat transfer inside the cylinder wall, one-dimensional unsteady heat conduction equation reads

$$\frac{\partial T}{\partial t} = \alpha_w \frac{\partial^2 T}{\partial x^2} \tag{9.59}$$

with $\alpha_W = k_W/(\rho_W c_W)$ the wall thermal diffusivity, ρ_W the wall density and c_W its specific heat capacity.

Applying the boundary conditions to all wall sides (gas side and coolant side) of a four-stroke diesel engine, the following equation is obtained with reference to Figure 9.6:

$$\frac{1}{4\pi}\int_0^{4\pi} \frac{dQ_L}{d\varphi} d\varphi = A\,\frac{k_w}{L_w}\left(\overline{T}_w - \overline{T}_{w,c}\right) = A\,h_c\left(\overline{T}_{w,c} - T_c\right) \tag{9.60}$$

where L_W is the cylinder wall thickness with k_W its thermal conductivity and h_c the heat transfer coefficient from the external wall side (respective temperature $\overline{T}_{w,c}$) to the coolant. Equations 9.60 are solved for the two unknown mean wall temperatures $\overline{T}_w$ (needed in Equation 9.52) and $\overline{T}_{w,c}$, which change from cycle to cycle during the transient event but can be considered to remain constant throughout the cycle.

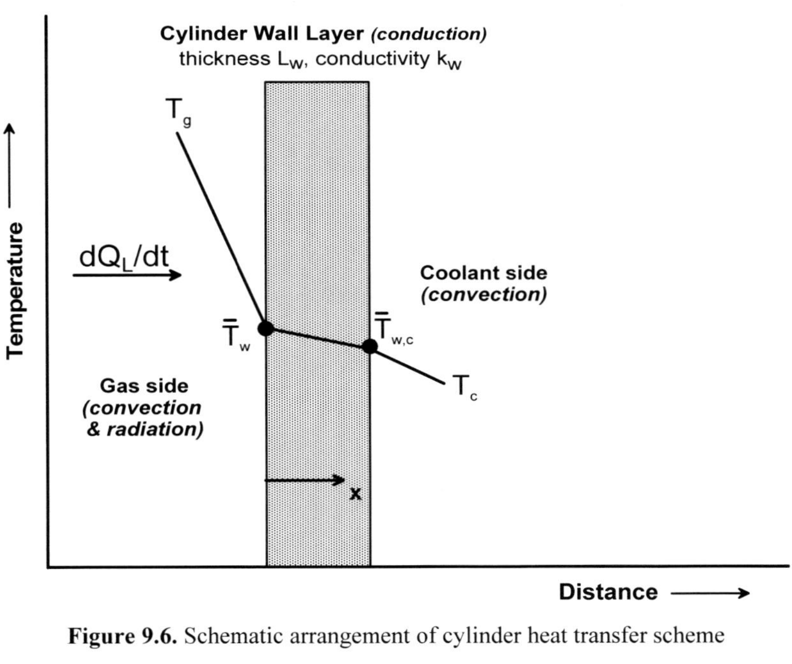

Figure 9.6. Schematic arrangement of cylinder heat transfer scheme

An alternative approach is to use a 'hysteresis' (time lag due to thermal inertia) expression in order to update the wall temperature at each consecutive cycle, which changes as a result of the increase in fueling and/or speed [23, 24]. This approach is based on the well known fact that the update of cylinder wall temperature is notably delayed compared with changes in fueling (*cf.* Figures 2.22 and 2.23), a fact that has actually led many researchers to assume constant wall temperature throughout the transient event (*cf.* the sensitivity analysis in Section 9.13)

$$T_w = T_{wo} + \Delta T_w \left(1 - e^{-d_t t}\right) \tag{9.61}$$

In Equation 9.61, d_t is a coefficient determining how 'fast' or how 'slow' the wall temperature responds to the changes in fueling; it depends on the wall construction (material) and insulation used. Further, T_{wo} is the initial (at the point the transient event commences) gas-side wall temperature and $\Delta T_w = T_{wst} - T_{wo}$, with T_{wst} the respective steady-state gas-side wall temperature according to the current fueling (primary dependence) and rotational speed.

9.3.2.3 Combustion

When it comes to combustion modeling, the traditional approach is application of a semi-empirical, physically based heat release rate law, the accuracy of which has been confirmed under steady-state conditions. This is validated by

- matching simulated to measured pressure diagrams at steady-state conditions or
- direct match between predicted heat release rates and elaborated ones from steady-state pressure diagrams.

The combustion process in a compression ignition engine with single injection strategy is a premixed-diffusion one. Combustion models simulate the above procedure employing first an ignition delay approximation, and second a fuel burning law.

For ignition delay calculations, use is typically made of the following Arrhenius type expression providing τ_{id} in ms:

$$\tau_{id} = C_1 \, p_{gm}^{c_2} \, \exp\left(\frac{E_a}{R_{mol} T_{gm}}\right) \tag{9.62}$$

where p_{gm} and T_{gm} are the mean values of gas pressure and temperature in the cylinder during ignition delay, requiring an iterative computational procedure, E_a an apparent activation energy, and C_1, c_2 constants depending primarily on the fuel but also on injection and air-flow characteristics. Heywood [3] summarizes the empirical constants from Equation 9.62 from various researchers, and provides alternative ignition delay correlations.

The fuel burning rate procedure can be simulated using:

1. Wiebe (Vibe) functions [25]

$$m_{fb}(\varphi) = 1 - \exp\left[-c\left(\tilde{\varphi}\right)^{d+1}\right] \tag{9.63a}$$

with m_{fb} the cumulative fuel burning mass, $\tilde{\varphi} = \left(\varphi - \varphi_s\right) / \Delta\varphi_{burn}$ the crank angle measured from ignition, non-dimensionalized by the total combustion duration $\Delta\varphi_{burn}$, φ_s the crank angle at the start of combustion, 'c' a combustion efficiency coefficient, and 'd' a parameter, which adjusts the shape of the curve. Differentiating with respect to non-dimensional angle $\tilde{\varphi}$, we get the fuel mass burning rate

$$\dot{m}_{fb}(\varphi) = c\,(d+1)\,(\tilde{\varphi})^{d} \exp\left[-c(\tilde{\varphi})^{d+1}\right] \tag{9.63b}$$

2. Watson model [26]

This is a two-part expression derived after extended calculation and curve-fitting of experimental fuel burning rate diagrams for swirl-type DI engines, very popular in transient simulation codes; it is described by the following relations:

$$\begin{aligned} m_{fb}(\tilde{t}) &= \beta\, f_{pre}(\tilde{t}) + (1-\beta)\, f_{diff}(\tilde{t}) \\ \beta &= 1 - a\,\Phi^{b} / \tau_{id}^{c} \\ f_{pre}(\tilde{t}) &= 1 - \left(1 - \tilde{t}^{k_1}\right)^{k_2} \\ f_{diff}(\tilde{t}) &= 1 - \exp\left(-k_3 \tilde{t}^{k_4}\right) \end{aligned} \tag{9.64}$$

In Equations 9.64, $\tilde{t}$ is the time measured from ignition, non-dimensionalized by the total combustion time, β is the fraction of fuel burned in the premixed phase, f_{pre} and f_{diff} are the premixed and diffusion burning functions (the latter is based on the Wiebe function Equation 9.63a), and k_i i=1–4 are constants, specific for each engine and depending on factors having strong influence on combustion rate, *i.e.*, ignition delay period τ_{id}, overall fuel–air equivalence ratio Φ and engine speed N.

The best-fit values of these parameters for a range of DI diesel engines, based on nominal combustion duration of 125°CA, were found to be

- $k_1 = 2.0 + 1.25\times10^{-8}(\tau_{id}N)^{2.4}$
- $k_2 = 5{,}000$
- $k_3 = 14.2/\Phi^{0.644}$
- $k_4 = 0.79 k_3^{0.25}$
- $0.8 < a < 0.95$
- $0.25 < b < 0.45$
- $0.25 < c < 0.50$.

3. Whitehouse–Way model [27]

Again, this model concerns a two-part expression; for the preparation rate (kg/°CA) it holds that

$$P = K\, m_{fi}^{1-x}\, m_u^x\, p_{O_2}^y \tag{9.65a}$$

This relation was initially derived on the assumption that the rate of preparation is proportional to the surface area of all the droplets having uniform diameter, while proportionality with respect to the droplets diameter has also been proposed. A variable value of the exponent x makes allowance for this fact and the usually existing non-uniform droplets diameter distribution. The last term on the right-hand side of Equation 9.65a allows for the effect of oxygen availability on mixing.

For the reaction rate, an Arrhenius type equation is used incorporating also the ignition delay period

$$R = \frac{K' p_{O_2}}{N\sqrt{T_g}} e^{\frac{-E_a}{T}} \int (P-R)d\varphi \qquad (kg/^{o}\,CA) \tag{9.65b}$$

with p_{O_2} the partial pressure of oxygen in the mixture, N the engine speed in rps, E_a the activation energy, $m_{fi} = \int (dm_{fi}/d\varphi)d\varphi$ the cumulative mass of injected fuel up to the crank angle φ considered, with injection rate $dm_{fi}/d\varphi$ discussed in Section 9.8, $m_u = m_{fi} - \int P d\varphi$ the mass of unprepared fuel, and K, K', x and y constants according to the engine under study. Typical values for the constants identified by Benson and Whitehouse [4] are

- $0.01 < x < 1$
- $y = 0.4$
- $0.008 < K < 0.020$ bar^{-y}
- $E_a = 15{,}000$ K
- $1.2\times10^{10} < K' < 65\times10^{10}$ (K$^{0.5}$/bar s).

Finally, the fuel burning rate is given by the relations

$$\frac{dm_{fb}}{d\varphi} = R \quad \text{if} \quad R<P \qquad \text{and} \qquad \frac{dm_{fb}}{d\varphi} = P \quad \text{if} \quad R>P \tag{9.65c}$$

The second of Equations 9.65c expresses the fact for diesel engine combustion that at the high temperatures corresponding to the main period of combustion, the time taken for burning of the prepared fuel is negligible compared with the preparation time. Then, preparation and combustion rates are effectively indistinguishable, so that the mechanism of diffusion combustion prevails. On the other hand, the first of Equations 9.65c expresses the fact that the chemical kinetics are important only at the beginning of the burning period, corresponding to the ignition delay period. Then, the temperature is usually very low for rapid burning, so that the mechanism of premixed combustion prevails. As the temperature rises, the burning rate increases until it is controlled by the lack of the 'reservoir' of the prepared fuel, which is exhausted, so that then the above described diffusion (preparation) mechanism takes over.

The three above-mentioned models are especially applicable for use in single- or two-zone simulations of the engine processes, where the combustion details are not of primary concern.

4. In the form of Arrhenius equations in the very rare case of transient multi-zone models [28]; for the latter, detailed sub-models for spray penetration, droplet evaporation, air–fuel mixing *etc.* are required as discussed earlier:

$$m_{fb} = A\ \rho_g^2\ x_{fv}^a\ x_{o_2}^b\ \exp(\frac{-E_a}{R_{mol}T_g}) \tag{9.66}$$

with x_{fv} the fuel vapor and x_{o_2} the oxygen concentration, and ρ_g the gas density in the specific zone.

Unfortunately, transient thermodynamic modeling is not always capable of capturing the significant deteriorations in combustion discussed in Section 2.3.2. One reason may be the possible inability of the empirical combustion models to predict engine performance at higher than unity fuel–air equivalence ratios. Another one is their inability to take into account phenomena of fuel accumulation that are usually experienced during the initial cycles of a turbocharged diesel engine transient cycle. For the latter quasi- or multi-dimensional modeling would be required, limited at the moment owing to the need to simulate a large number of engine cycles. One major inaccuracy lies in a usually adopted simplification, namely the assumption of constant combustion coefficients during transients.[2] To overcome these burdens, many empirical approaches have been proposed.

- Woschni and Anisits [29] proposed a correction of the parameter 'd' and burn duration $\Delta\varphi_{burn}$ with operating conditions for the Wiebe function 9.63

$$d = d_o\left(\frac{\tau_{id_o}}{\tau_{id}}\right)^{0.5}\left(\frac{p_{IVC}}{p_{IVC_o}}\frac{T_{IVC_o}}{T_{IVC}}\right)\left(\frac{N_o}{N}\right)^{0.3} \tag{9.67a}$$

$$\Delta\varphi_{burn} = \Delta\varphi_{burn_o}\left(\frac{\Phi}{\Phi_o}\right)^{0.6}\left(\frac{N}{N_o}\right)^{0.5} \tag{9.67b}$$

[2] For example, constant burn duration equal to 125°CA for each transient cycle is usually assumed when applying the Watson combustion model.

where subscript 'o' denotes reference conditions with known heat release rate shape.

- Betz and Woschni [30] extended Equations 9.67 in order to be applicable in the region of less than unity relative air–fuel ratios, typically encountered during turbocharger lag

$$\Delta\varphi_{burn} = \Delta\varphi_{burn_o} \left(\frac{\Phi}{\Phi_o}\right)^{0.6} \left(\frac{N}{N_o}\right)^{0.5} \eta_u^{0.6} \tag{9.68}$$

with η_u the energy conversion rate that equals, theoretically, $1/\Phi$ for $\Phi>1$, and 1 for $\Phi \leq 1$. This relation was developed for a DI diesel engine of medium swirl, operating at medium engine speed. Experiments on DI and IDI engines made possible an empirical approximation of the 'real' energy conversion rate; the latter was found to depend on actual fuel–air equivalence ratio Φ as well as on the maximum Φ (or minimum $\lambda=1/\Phi$) that gives tolerable smoke emissions, *i.e.*

$$\begin{aligned} \eta_u &= 1.0 && \text{for } \lambda_{smoke} \leq \lambda \\ \eta_u &= \alpha\,\lambda\, e^{c\lambda} - b && \text{for } 1.0 < \lambda < \lambda_{smoke} \\ \eta_u &= 0.95\,\lambda + d && \text{for } \lambda \leq 1 \end{aligned} \tag{9.69a}$$

where λ_{smoke} the minimum (smoke-limited) value of $\lambda=1/\Phi$. Further,

$$\alpha = \frac{0.05 - d}{\lambda_{smoke}\, e^{c\lambda_{smoke}} - e^c}\ , \quad b = \alpha e^c - 0.95 - d$$
$$c = \frac{-1}{\lambda_{smoke}}\ , \quad d = -0.0375 - \frac{\lambda_{smoke} - 1.17}{15} \tag{9.69b}$$

- Assanis *et al.* [31] proposed and validated a new ignition delay correlation to be incorporated in the Watson combustion model during transient conditions. This relation, unlike the usually applied one (Equation 9.62), and following transient engineering intuition, incorporates the effect of fuel–air equivalence ratio, Φ

$$\tau_{id} = C_1\, p_{gm}^{c_2}\, \Phi^{-c_3} \exp\left(\frac{E_a}{R_{mol} T_{gm}}\right) \tag{9.70}$$

with C_1, c_2, and c_3 constants specific for each engine. By including the global equivalence ratio dependence, Assanis *et al.* did not assume that the local fuel–air mixture in a DI diesel engine is perfectly mixed. Instead, it was postulated that the global equivalence ratio is actually a measure of the probability of finding local pockets of fuel–air mixture within flammability limits for auto-ignition sites to be promoted. The adjustable constants in Equation 9.70 were fitted to minimize the least-square error between measured and correlated ignition delay, as well as to ensure that the latter would tend to zero at extremely high load. The ignition delay data were correlated as a function of the overall equivalence ratio and the mean pressure and temperature over the ignition delay interval. This ignition delay formula gave satisfactory results, mainly for medium to high DI engine speeds using the following values for the constants concerned, $C_1 = 2.4,\ c_2 = -1.02$ and $c_3 = -0.2$.

- Benson and Whitehouse [4] proposed that the constant K in the (dominant) preparation rate Equation 9.65a of the Whitehouse–Way model, be correlated with the Sauter mean diameter (SMD) of the fuel droplets (*i.e.*, the mean droplet diameter that has the same surface to volume ratio to that of the total spray) through a formula of the type

$$K \propto \frac{1}{SMD^{c}} \tag{9.71}$$

This approach was adopted, for example, by Rakopoulos *et al.* [32] during transient operation, using a SMD correlation proposed by Hiroyasu *et al.* [33]

$$SMD(\mu m) = C\ \Delta p_{inj}^{-0.135}\ \rho_g^{0.121}\ V_{fi}^{0.131} \tag{9.72}$$

Thus, account was taken of variations in fueling (through the volume of injected fuel V_{fi}), compression pressure (through gas density ρ_g) and injection pressure Δp_{inj} across the nozzle, which typically vary during transients. The initial estimation for the constant 'c' in Equation 9.71 was 2; Rakopoulos and Giakoumis [32] found a value of c=2.5 more appropriate for their six-cylinder, turbocharged diesel engine.

- Incorporation of the effect of fuel–air equivalence ratio has been theoretically investigated by Rakopoulos and Giakoumis [34] concerning the preparation constant K and the activation constant E_a of the Whitehouse–Way model

$$K = K_o \left(\frac{\Delta\Phi}{\Phi_o} \right)^{c_1} \quad \text{and} \quad E_a = (E_a)_o \left(\frac{\Delta\Phi}{\Phi_o} \right)^{c_2} \tag{9.73}$$

A slightly different approach, being in the middle of quasi-linear and filling and emptying modeling, has been also adopted [35, 36]. Here, a °CA based thermodynamic code was used to generate a steady-state, multi-dimensional, look-up engine torque map. The latter is used in the dynamic simulation of the engine or whole vehicle performance. This approach is mainly chosen for limiting computational time, particularly as regards Transient Cycles simulation. However, it fails to fully take into account the fundamental aspects of transient operation differentiation from steady-state conditions.

9.3.2.4 Flow Through Valves

For the evaluation of the mass exchange between cylinder and manifolds needed in Equation 9.26 (index j='in' or j='ex' – see also Figures 9.2 and 9.4), the correlation for gas mass-flow through a restriction can be used, accounting for one-dimensional, quasi-steady, compressible flow:

$$\dot{m}_j = c_D A_V \frac{p_u}{R_u T_u} \sqrt{\frac{2\gamma R_u T_u}{\gamma - 1} \left[\left(\frac{p_d}{p_u}\right)^{\frac{2}{\gamma}} - \left(\frac{p_d}{p_u}\right)^{\frac{\gamma+1}{\gamma}} \right]} \tag{9.74a}$$

This relation holds true when $p_d / p_u > [2/(\gamma+1)]^{\gamma/(\gamma-1)}$, while for the opposite case

$$\dot{m}_j = c_D A_V \frac{p_u}{R_u T_u} \sqrt{\gamma R_u T_u \left(\frac{2}{\gamma+1}\right)^{\frac{\gamma+1}{\gamma-1}}} \tag{9.74b}$$

The discharge coefficient c_D is a function of the instantaneous valve lift (the latter can be simulated, for example, using polynomial expressions with respect to crank angle) to valve head diameter ratio, A_v is the geometric valve flow area and γ the specific heat capacities ratio; index 'd' denotes downstream and 'u' upstream conditions. Similar equations can be applied for the blow-by losses or for the flow between main chamber and prechamber in an IDI engine, by proper account of the effective flow area $c_D A_V$. Details about the geometric valve flow area A_V are provided in [3].

9.4 Manifolds

At the most simplified level of quasi-linear modeling, both inlet and exhaust manifolds are modeled as no volume components. A more advanced version of quasi-linear modeling has also been reported, in which the filling and emptying approach is adopted for manifolds simulation [*e.g.*, 37] as will be described in the following paragraph.

Most of the researchers who simulated in-cylinder processes with the filling and emptying modeling technique have extended this approach to the manifolds operation too. Here, the perfect gas law and the first law of thermodynamics are applied, taking into account the manifold volume and the heat loss from the (exhaust) manifold walls. Equation 9.26 holds true for each of the two manifolds but now term $dV/d\varphi=0$:

$$\frac{dQ_{Lim}}{d\varphi} = \frac{dU_{im}}{d\varphi} + \frac{dm_{a-im}}{d\varphi} h_{a-im} - \sum_{cyl=1}^{n_{cyl}} \frac{dm_{im-cyl}}{d\varphi} h_{im-cyl} \tag{9.75}$$

$$\frac{dQ_{Lem}}{d\varphi} = \frac{dU_{em}}{d\varphi} + \sum_{cyl=1}^{n_{cyl}} \frac{dm_{cyl-em}}{d\varphi} h_{cyl-em} - \frac{dm_{em-a}}{d\varphi} h_{em-a} \tag{9.76}$$

where index 'a-im' denotes mass exchange between inlet manifold and intercooler or compressor or environment for naturally aspirated engine, index 'im-cyl' denotes mass exchange between inlet manifold and each cylinder, index 'cyl-em' denotes mass exchange between exhaust manifold and each cylinder (Equations 9.74), and index 'em-a' denotes mass exchange between exhaust manifold and turbine, or environment for naturally aspirated engine. Care has to be taken for heat loss effects as regards the exhaust manifold:

$$\frac{dQ_{Lem}}{dt} = A_{em} h_{em} (T_{em} - T_{w-em}) = A_{em} \left(a \frac{k_{w-em}}{\bar{D}_{em}} Re^b \right) (T_{em} - T_{w-em}) \tag{9.77}$$

whereas for the inlet manifold, owing to the low gas temperatures involved, $dQ_{Lim} / dt \approx 0$.

It is well known that the gas flow in the exhaust manifold of a diesel engine is markedly unsteady, resulting in the propagation of pressure waves that may affect engine performance if the exhaust pipes are long enough and the valve overlap period substantial. The influence of the unsteady flow waves' propagation is more intense in the case of pulse turbocharged engines, which form the majority in vehicular applications. Here, the issue of the inlet and, mainly, exhaust gas interactions with the engine turbocharger is prominent.

The travel time of pressure wave in the exhaust manifold pipe is given by [2]

$$\Delta\varphi = \frac{12 \, L N}{\alpha} \tag{9.78}$$

where α is the sonic velocity and L the equivalent pipe length. For long enough exhaust pipes, *e.g.*, L= 0.80 m, $\Delta\varphi = 35°$CA at 2000 rpm, which means that a wave action simulation should be applied. However, for shorter lengths, typical in automotive applications, or for lower engine speeds, the filling and emptying approach seems to be sufficient.

Wave action models, also termed gas-dynamic, are labeled as non-steady and one-dimensional; they provide the solution of the unsteady flow equations in the engine ducts. In the more general case, wave action models are a set of first order non-linear, non-homogenous and hyperbolic partial differential equations of mass, energy, and momentum conservation using the perfect gas law. They are solved by the method of characteristics or by finite difference techniques. Special care has to be taken for accurate modeling of the necessary boundary conditions at the duct ends, *e.g.*, boundary conditions at the cylinder end, at the pipe open end, and at the exhaust system turbine end. Gas-dynamic models can provide an adequate evaluation of the pulsating flow phenomena, which are of great importance for pulse turbocharged diesel engines; nonetheless, a much higher computational time is required. A detailed coverage of the subject together with the formulation of the respective equations can be found in [1, 2].

9.5 Multi-cylinder Engine Transient Operation

At steady-state operation, the performance of each cylinder is essentially the same, owing to the constant position of the governor clutch resulting in the same amount of fuel being injected per cycle. Under transient operation, however, and particularly for engines with mechanical injection systems, each cylinder experiences different fueling and air-mass flow-rates during the same engine cycle. This happens due to the combined effect of

- continuous movement of the fuel pump rack that is initiated by the load or speed change; and
- continuous change of the turbocharger compressor operating point.

As regards speed changes only the first cycles are practically affected. However, when load changes are investigated, significant variations can be experienced throughout the whole transient event. The usual approach, here, is the solution of the governing equations for one cylinder and the subsequent use of suitable phasing images of this cylinder's behavior. This approach was first introduced by Watson [38] and was widely adopted for limiting computational time. Unlike this, Rakopoulos and Giakoumis [34, 39] developed a true multi-cylinder engine model. In this model, all the governing differential and algebraic equations presented in the previous sections are solved individually for every cylinder of the engine; this resulted in (significant) differentiations in both fueling and air-mass flow-rates for each cylinder during the *same* cycle of a transient event. The drawback of this approach is that it increases the computational time almost linearly with the number of cylinders involved.

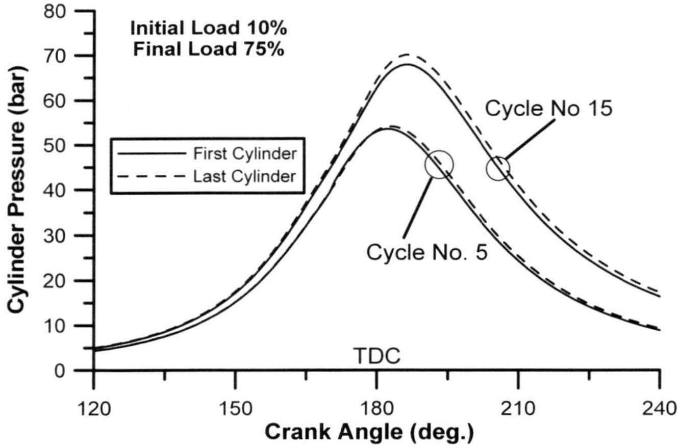

Figure 9.7. Effect of multi-cylinder modeling on the prediction of transient engine operation

Figure 9.7 is a typical illustration of the results obtained by comparing the 'single-cylinder' with the 'multi-cylinder' engine modeling approach, on a six-cylinder, turbocharged and aftercooled diesel engine. For the examined transient case of 10–75% load increase, the pressure diagrams of the first and the last in firing order cylinders are depicted for cycle No. 5 and No. 15. Cylinder pressures assume up to 5% higher values when the first and the last in firing order cylinders are under investigation. This differentiates accordingly the performance of each cylinder. Similarly, all other properties that depend on the fueling rate are also differentiated from cylinder to cylinder during the same transient cycle, thus affecting the whole engine transient response.

9.6 Turbocharger

The mathematical representation of compressor and turbine characteristics is necessary to calculate the interaction between turbocharger and (diesel) engine. The majority of diesel engine transient simulations, either quasi-linear or filling and emptying, have adopted the use of compressor and turbine performance characteristics or 'maps' as provided by the manufacturer or measured in-house. These give the interdependence between isentropic efficiency η_{is}, pressure ratio r, mass flow-rate $\dot{m}$ and rotational speed N_{TC}, *i.e.*,

$$\eta_{isC} = f_1(r_C, N_{TC}) \qquad \text{and} \qquad \dot{m}_C = f_2(r_C, N_{TC}) \tag{9.79a}$$

for the compressor, and

$$\eta_{isT} = g_1\left(\frac{u}{c_o}, N_{TC}\right) \qquad \text{and} \qquad \dot{m}_T = g_2(r_T, N_{TC}) \tag{9.79b}$$

for the turbine, with $(u/c_o)_T$ the turbine blade to speed ratio. Usually, N_{TC} and $\dot{m}$ are corrected for standard inlet conditions.

In the conventional interpolation approach, turbo-machinery maps are digitized and entered into the simulation code in a tabular form or in the form of (for example) second-order polynomials. At a particular step in the cycle simulation, the tables or polynomials are interpolated in a two dimensional fashion to calculate two unknown map variables from two known variables. Rakopoulos *et al.* [32] reported that no difference in the predictions from a single-zone model was observed, if this computational step was less than 20°CA.

While this interpolation technique is very straightforward to implement, there are some difficulties associated with its use in transient engine simulations [38, 40].

- the non-linear nature of compressor and turbine characteristics makes linear interpolation of data in look-up tables somewhat unreliable in regions outside of those where the experimental data are available;

- experimental data are usually unavailable for low rotor speed/low-pressure ratios;
- commencement of a transient test from a zero or very small load means that the compressor practically operates with a pressure ratio less than unity, a fact complicating the situation even more;
- small changes in pressure correspond to much greater changes in mass flow-rate in the compressor map;
- obtaining accurate turbine efficiency data is a well acknowledged problem, with general correlations usually applied that are based on the turbine blade to speed ratio; and
- the assumption of quasi-steady flow inside the turbo-machinery is perhaps the major simplification.

Overall, it is highly uncertain that the operating points of the compressor and turbine during a transient event follow their steady-state map values. To this aim, Fink *et al.* [41] proposed a 'flattening' of compressor curves during transient operation by using a differential equation, which takes into account the compressor 'through-flow' time τ_C:

$$\frac{dr_C}{dt} = \frac{1}{\tau_C}(r_{C_{st}} - r_C) \tag{9.80}$$

with r_C the current (transient) pressure ratio and $r_{C_{st}}$ the respective steady-state pressure ratio according to the compressor map.

Use of 1-D turbocharger modeling would be helpful in assessing 'real' transient turbocharger operation, but at the same time it would significantly increase the complexity and the PC execution time of the developed code.

In recent years, an increased interest in variable geometry turbines is observed, since these offer scope for improving transient response and fuel consumption, and regulate EGR flow as was discussed in Sections 2.4 and 6.4.4. At the most simplified level, a variable geometry turbine can be modeled as a series of fixed geometry units; interpolation between the various units maps is then undertaken in order to compute the instantaneous turbine characteristics according to VGT vane position. However, the variations between the maps corresponding to different vane positions in the VGT are non-linear, owing to the fact that the swing blade tip position relative to the rotor tip changes with the blade angle. The alternative technique that has been applied by some researchers [*e.g.*, 42, 43] is to utilize artificial neural networks to capture the properties of the fixed or variable geometry turbine in the whole operating range. The modeling process includes construction of the network of the appropriate architecture, and training of the network on the digitized turbo-machinery characteristics. Like any interpolation technique, this is more accurate when working within the boundaries of the training data; extrapolation should be avoided. Accuracy is also enhanced by increasing the amount of training data presented.

For the turbocharger transient operation, the following differential equation describes the angular momentum balance on its shaft based on Newton's second law of motion for rotating systems

$$\eta_{mTC}\tau_T - |\tau_C| = G_{TC}\frac{d\omega_{TC}}{dt} \tag{9.81}$$

The respective power terms are given by (constant pressure turbocharging)

$$\dot{W}_C = \dot{m}_C(h_2 - h_1) \approx \dot{m}_C\bar{c}_{pC}(T_2 - T_1) \tag{9.82}$$

for the compressor indicated power, with the temperature T_2 after the compressor computed from

$$T_2 = T_1\left\{1 + \frac{1}{\eta_{isC}}\left[\left(\frac{p_2}{p_1}\right)^{\frac{\gamma_a - 1}{\gamma_a}} - 1\right]\right\} \tag{9.83}$$

and

$$\dot{W}_T = \dot{m}_T(h_6 - h_7) \approx \dot{m}_T\bar{c}_{pT}(T_6 - T_7) \tag{9.84}$$

is the instantaneous value of the turbine indicated power, with the temperature T_7 downstream of the turbine (see also Figure 9.2) given by

$$T_7 = T_6\left\{1 - \eta_{isT}\left[1 - \left(\frac{p_7}{p_6}\right)^{\frac{\gamma_R - 1}{\gamma_R}}\right]\right\} \tag{9.85}$$

Further, η_{mTC} is the turbocharger mechanical efficiency that is mainly a function of its speed.

Mass conservation for the whole diesel engine plant gives

$$\dot{m}_T = \dot{m}_C + \dot{m}_{fi} - \dot{m}_{bl} \tag{9.86}$$

In filling and emptying modeling, the above equations for the turbocharger are solved at each degree crank angle using the current values of inlet manifold and exhaust manifold pressures and temperatures, accounting, thus, for pulse effects in the exhaust manifold. As regards quasi-linear modeling, however, and in order to account for the pulse operation, a 'pulse' factor is introduced [1, 8, 44]. Winterbone [1] working on a six-cylinder, turbocharged DI diesel engine correlated the pulse factor K_T with the cyclic mass-flow parameter $\dot{m}_T \sqrt{T_6}/(p_7\, N)$ via a linear relation:

$$K_T = c\left(\frac{\dot{m}_T \sqrt{T_6}}{p_7\, N}\right) + d \tag{9.87}$$

with 'c' and 'd' constants specific for each engine configuration. The pulse factor is basically a function of the number of cylinders and the exhaust manifold configuration. Consequently, the 'effective' expansion ratio $(p_6/p_7)_{eff}$ needed in Equation 9.85 is given by

$$r_{T_{eff}} = \left(\frac{p_6}{p_7}\right)_{eff} = K_T\left(\frac{p_6}{p_7}\right) = K_T r_T \tag{9.88}$$

When present, the *waste-gate* valve should also be simulated. If the valve is open, a force balance exists described by the following second-order differential equation on a time basis for the spring-mass system displacement x [45]:

$$m_{wgv}\,\ddot{x} = F_{eg} - (p_2 - p_1)\,\bar{A}_d - (F_o + k_{wgv}\, x) - F_{fr} - c_D\,\dot{x} \tag{9.89}$$

with m_{wgv} the valve, diaphragm and spring mass, F_{eg} the exhaust gas force determined as a function of turbine pressure ratio and waste-gate valve lift, F_o the initial spring force of the valve, k_{wgv} the spring stiffness, F_{fr} the friction force, $F_D = c_D\dot{x}$ the damping force, and $\bar{A}_d$ the effective diaphragm area depending on the geometric area and the differential pressure upstream and downstream of the valve.

The *aftercooler* is usually modeled via its steady-state thermal effectiveness, being, mainly, a function of charge air-mass flow, $\varepsilon = f_1(\dot{m}_C)$, for example $\varepsilon = 1 - \dot{m}_C^2$. The steady-state temperature T_3 at the aftercooler exit can then be correlated with its inlet temperature T_2 using the following expression:

$$T_3 = T_2\,(1-\varepsilon) + \varepsilon\, T_{cw} \tag{9.90}$$

with T_{cw} the cooling medium temperature. During transients, the thermal inertia of the aftercooler should be taken into account using a relation similar to Equation 9.61.[3]

---

[3] A similar Equation to 9.90 can be applied to model the EGR cooler with the effectiveness being, mainly, a function of engine speed and exhaust gas mass flow-rate. Moreover, Equations 9.74 describing the flow through a restriction can be used to compute the flow through the EGR valve, with the valve effective area being a quadratic function of valve lift, and the valve pressure drop often neglected.

Pressure drop of the charge air across the aftercooler can be approximated by $\Delta p = p_2 - p_3 = f_2(\dot{m}_C)$, for example $p_2 - p_3 = c\,\dot{m}_C^2$.

Compressor Surge

Simulation of the compressor operation when it experiences surge, requires the representation of the compressor characteristics to the 'left' of the surge line; the latter are not provided by the manufacturer, and can be defined accurately only after costly and time-consuming experimental work on a compressor test bed. A simpler method is described here that has been adopted by many researchers in the field with satisfactory results.

For axial flow or multistage centrifugal compressors, the region of the compressor map to the left of the surge line can be represented by the following cubic polynomial [46, 47]

$$r_C = r_{C_o} + c\left[1 + \frac{3}{2}\left(\frac{\dot{m}_C}{d} - 1\right) - \frac{1}{2}\left(\frac{\dot{m}_C}{d} - 1\right)^3\right] \tag{9.91}$$

Parameters 'c' and 'd' are defined in Figure 9.8, and r_{C_o} is the compressor pressure ratio at zero mass-flow.

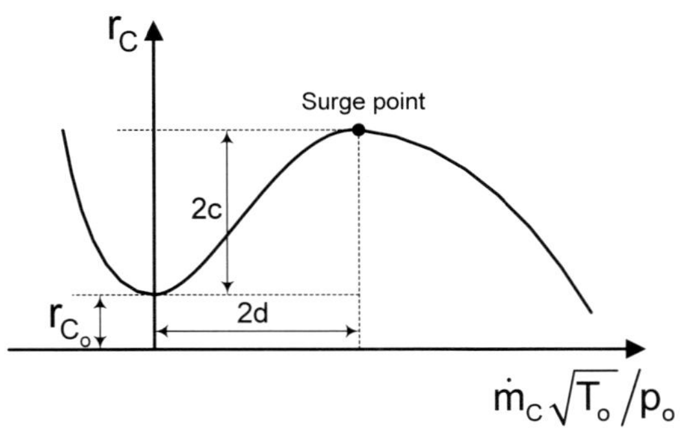

Figure 9.8. Compressor characteristic showing the definition of parameters d and c

In Figure 9.8, the point on a constant speed line on the compressor map where surge occurs depends on whether the compressor diffuser is vaned or vaneless. Surge occurs when the slope of the speed line is zero or slightly positive with vaneless diffusers, or slightly negative with vaned diffusers. Because the compression system operating point during surge passes through this zone quite rapidly, as very abrupt mass-flow excursions are observed at almost constant pressure, this part of the compressor characteristics does not greatly affect the prediction of the compression system transient behavior. For the pressure ratio at compressor zero mass flow, we have

$$r_C = r_{C_o} = \left[1 + \frac{\gamma_a - 1}{2\gamma_a R_s T_o} \omega_C^2 \left(r_2^2 - r_1^2\right)\right]^{\frac{\gamma_a}{\gamma_a - 1}} \tag{9.92}$$

where r_1 and r_2 are the compressor eye and impeller outer radii, respectively. The previous equation can be derived applying radial equilibrium theory for the compressor impeller blades and assuming isentropic flow processes. Referring to the compressor eye radius, this is defined as the compressor impeller eye mean geometric radius, which divides the impeller eye area into two sections of equal area

$$r_1 = \sqrt{\frac{r_{h1}^2 + r_{t1}^2}{2}} \tag{9.93}$$

where r_{h1} and r_{t1} are the impeller eye hub and tip radii, respectively.

For automotive centrifugal compressors, the negative flow branch of the characteristic can be matched with a parabola

$$r_C = r_{C_o} + 2c\left(\dot{m}_c/2d\right)^2 \tag{9.94}$$

whereas a third-order polynomial expression, such as the one of Equation 9.91, can be used for the region between zero mass flow and the surge line [48].

9.7 Friction

9.7.1 Mean fmep Method

For the calculation of friction inside the cylinder, the most fundamental approach is application of *mean fmep* equations, *i.e.*,

$$\text{fmep} = \alpha + \beta p_{max} + \gamma \overline{u}_{pist} \tag{9.95}$$

where p_{max} is the peak cylinder pressure, $\overline{u}_{pist}$ is the corresponding mean piston speed, and α, β and γ are constants derived after calibration against experimental data at steady-state conditions. Consequently, friction torque is given by

$$\tau_{fr}(\varphi) = \overline{\tau}_{fr} = \text{fmep}\ A_{pist} r / 2\pi \tag{9.96}$$

This friction torque remains constant at every degree crank angle during a cycle, with its values differentiating only from engine cycle to cycle, thus, being particularly suitable for use in quasi-linear models. The important part here is the $\gamma\bar{u}_{pist}$ term, indicating the dominant effect of engine speed on friction torque.

9.7.2 Rezeka–Henein Model

One of the main drawbacks of Equation 9.95, apart from its mean, over the engine cycle, value, is that it disregards oil viscosity and temperature effects. Hence, its application in transient engine simulation models should be made with caution, and only for the conditions (oil temperature and type) under which it was derived. During transient diesel engine operation, where the engine speed and fueling change continuously, more detailed, per degree crank angle, friction estimations are required in order for the models to predict accurately the instantaneous engine speed from the crankshaft torque balance (Equation 9.111). One such model is the semi-empirical Rezeka–Henein [49] friction model; it separates friction torque into six terms allowing for modeling at each degree crank angle. Three of these parts are concerned with the cylinder movement, as follows:

$$\tau_{fr1} = c_1 \left[\mu_{oil} (r\omega R_1)(p_r + p_g) w_{ring} \right]^{0.5} D (n_o + 0.4 n_c)\, r \left| R_1 \right| \tag{9.97a}$$

for the ring viscous lubrication,

$$\tau_{fr2} = c_2\, \pi D n_c\, w_{ring} \left(p_r + p_g \right)\left(1 - \left| \sin\varphi \right| \right) r \left| R_1 \right| \tag{9.97b}$$

for the ring mixed lubrication, and

$$\tau_{fr3} = c_3 \left(\mu_{oil} \frac{r\omega R_1}{h_{oil}} \right) D\, L_{ps}\, r \left| R_1 \right| \tag{9.97c}$$

for the piston skirt losses.

The remaining three parts are concerned with the crankshaft, as follows

$$\tau_{fr4} = c_4\, n_v\, F_s\, r \left| R_1 \right| \omega^{-0.5} \tag{9.97d}$$

for the valve train,

$$\tau_{fr5} = c_5\, \mu_{oil}\, \omega \tag{9.97e}$$

for the auxiliaries, while for the loaded bearing it holds

$$\tau_{fr6} = c_6 \frac{\pi D^2}{4} r_{jb}\, p_g \left| \cos\varphi \right| \omega^{-0.5} \tag{9.97f}$$

In the above relations, n_o is the number of oil rings, n_c the number of compression rings, w_{ring} the ring width, p_r the elastic ring pressure, μ_{oil} the oil kinematic viscosity, h_{oil} the oil film thickness, L_{ps} the length of the piston skirt, n_v the number of (inlet and exhaust) valves per cylinder, F_s the valve spring force, r_{jb} the journal bearing radius,. For the quantity R_1 it holds that

$$R_1 = u_{pist} / r\omega = \sin\varphi + (\lambda \sin\varphi \cos\varphi) / \sqrt{1-\lambda^2 \sin^2\varphi} \tag{9.98}$$

Total friction torque at each degree crank angle is the sum of the above six terms, *i.e.*,

$$\tau_{fr}(\varphi) = \sum_{i=1}^{6} \tau_{fr_i}(\varphi) \tag{9.99}$$

Coefficients c_1 to c_6 in Equations 9.97 are derived through calibration with experimental data at steady-state conditions, bearing in mind that the mean friction torque over the 720°CA period of a four-stroke engine cycle, when converted to fmep, should be equal to the result given by Equation 9.95. Figure 9.9 illustrates the development of each friction torque term according to the Rezeka–Henein model during an engine cycle for low-load operation of a six-cylinder turbocharged and aftercooled diesel engine.

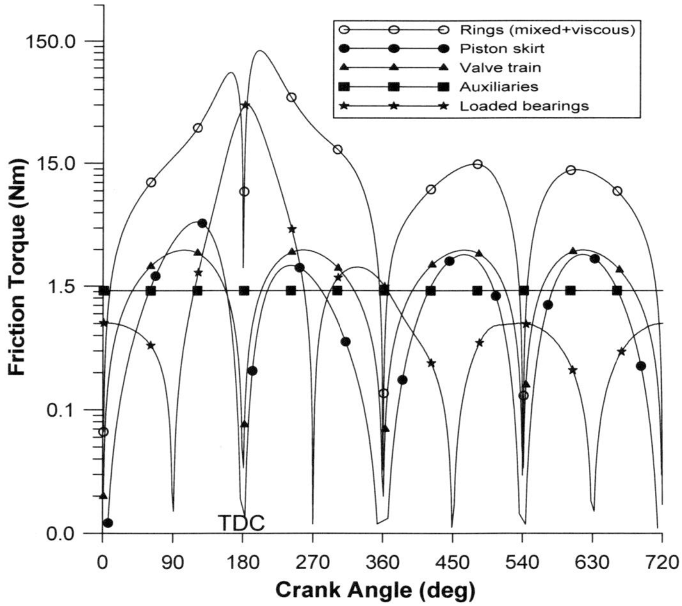

Figure 9.9. Variation of friction torque terms during a cycle using the Rezeka–Henein model

The main drawback of the Rezeka–Henein friction model is the need for many empirical coefficients; Tuccilo *et al.* [50] proposed a relation for the correction of constants c_i (i=1–6) of Equations 9.97 in order to account for engine speed effects:

$$c_i = c_{nom_i} (N / N_{nom})^{d_i} \tag{9.100}$$

where the values of constants c_{nom_i} are already known at a nominal speed N_{nom} and d_i are the correction exponents found after calibration against experimental data at various steady-state conditions.

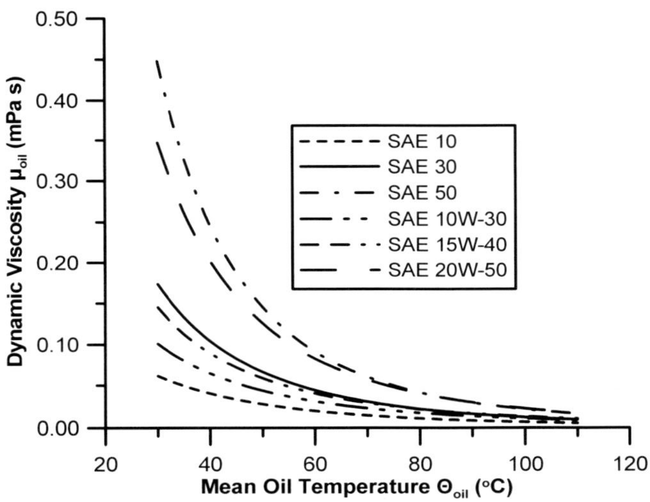

Figure 9.10. Effect of oil temperature and type on viscosity using Equation 9.101

Oil viscosity μ_{oil}, needed in Equations 9.97c and 9.97e, can be approximated by a Vogel equation [51]

$$\mu_{oil} = C_{oil} \cdot e^{\left(\frac{\Theta_1}{\Theta_{oil} + \Theta_2}\right)} \tag{9.101}$$

where C_{oil}, Θ_1 and Θ_2 are constants depending on oil type, and Θ_{oil} is the mean, over an engine cycle, oil temperature in °C. Increase in oil temperature is generally desirable as it reduces the oil film viscosity, thus decreasing the amount of friction; the same holds for lower viscosity oils containing friction modifiers. Figure 9.10 illustrates the effect of oil temperature and type on the viscosity from Equation 9.101; significant differences are observed, mainly at low temperatures highlighting the influence of oil viscosity on engine friction torque during cold starting. Values for coefficients C_{oil}, Θ_1 and Θ_2 are provided in [6, 51] for various oil-types.

9.7.3 Account for Transient Discrepancies

The sensitivity of transient operation predictions to friction modeling errors was investigated by Watson [38]. He showed that a (rather exaggerated) 50% overestimation in friction torque could lead to an almost equal increase in predicted final engine speed drop. He also proposed application of Equation 9.95 at each computational step, rather than at each cycle. By so doing, account is taken of the changing gas pressure profile during a cycle, which, as is depicted in Figure 3.19, plays a significant role in the development of true friction torque.

Winterbone and Loo [52], based on the results by Winterbone and Tennant [53], applied the following equation for the correction of fmep during transient operation of a two-stroke, turbocharged diesel engine:

$$\mathrm{fmep}_{\mathrm{trans}} = \mathrm{fmep}_{\mathrm{st}}\left[1+0.0025\left(\mathrm{N_D}-\mathrm{N}\right)\right] \tag{9.102}$$

where $\mathrm{fmep}_{\mathrm{st}}$ is the corresponding steady-state fmep from Equation 9.95, N is the current engine speed and $\mathrm{N_D}$ the demand speed.

Rakopoulos and Giakoumis [32] proposed a more fundamental correction for the friction torque during transients

$$\tau_{\mathrm{fr}}(\varphi)_{\mathrm{trans}} = \tau_{\mathrm{fr}}(\varphi)\left(1+\varepsilon(\varphi)/\varepsilon_{\max}\right) \tag{9.103}$$

Here, the instantaneous value for the total friction torque $\tau_{\mathrm{fr}}(\varphi)$ (as computed, for example, from Equation 9.96 or 9.99), is corrected according to the current crankshaft angular acceleration $\varepsilon(\varphi)$, thus providing the 'real' transient friction torque $\tau_{\mathrm{fr}}(\varphi)_{\mathrm{trans}}$ to be used in the crankshaft torque balance Equation 9.111. The exact value of coefficient c_{fr} needs experimental investigation and, obviously, depends on the specific engine application.

9.8 Fuel Injection

In the majority of transient simulations, the total amount of fuel injected per cycle and cylinder $\mathrm{M_{fi}}$, is found by applying *steady-state* fuel pump curves at the instantaneous values of engine speed N and fuel pump rack position z of the current transient cycle, *i.e.*,

$$\mathrm{M_{fi}} = \mathrm{f}(\mathrm{z},\mathrm{N}) \tag{9.104}$$

Assuming quasi-steady flow through the injector nozzle, the rate of fuel injection is given by

$$\dot{\mathrm{m}}_{\mathrm{fi}} = \sqrt{2\rho_{\mathrm{f}}\left(\mathrm{p_{inj}}-\mathrm{p_g}\right)}\,\overline{\mathrm{A}}_{\mathrm{N}} \tag{9.105}$$

with ρ_f the fuel density, p_{inj} the fuel pressure across the injection nozzle, and $\bar{A}_N = c_D A_N$ the total effective flow area of the injector nozzles. Ferguson [6] proposes the following equation for a more accurate estimation of the fuel injection rate

$$\frac{\dot{m}_{fi}}{M_{fi}} = \frac{\omega}{\Delta\varphi_d\,\Gamma(n)}\left(\frac{\varphi-\varphi_s}{\Delta\varphi_d}\right)^{n-1}\exp\left(\frac{\varphi_s-\varphi}{\Delta\varphi_d}\right) \tag{9.106}$$

where

- $\ln\Gamma(n)=(n-\frac{1}{2})\ln(n)-n+\frac{1}{2}\ln(2\pi)+\frac{1}{12n}-\frac{1}{360n^3}+\frac{1}{1260n^5}-\frac{1}{1680n^7}$;
- $1<n<2$ for DI engines, and $3<n<5$ for IDI engines;
- φ_s is the crank angle at the start of injection; and
- $\Delta\varphi_d$ is the duration of injection.

Use of the above steady-state approach during transients constitutes a rather coarse simplification since the fuel pump, as was analyzed in Section 2.2, experiences a transient operation of its own during the engine's dynamic conditions, with a very possible differentiation in the amount of injected fuel per cylinder compared with the one under *similar* (engine speed and fuel pump rack position) steady-state conditions.

In a more advanced approach, a mathematical fuel injection model can be applied to simulate the fuel pump-injector lift mechanism for mechanical injection pumps as, for example, the one developed by Rakopoulos and Hountalas [54] and incorporated in the transient diesel engine simulation code by Rakopoulos and Giakoumis [32, 34]; this model takes into account both the delivery valve and the injector needle motion.

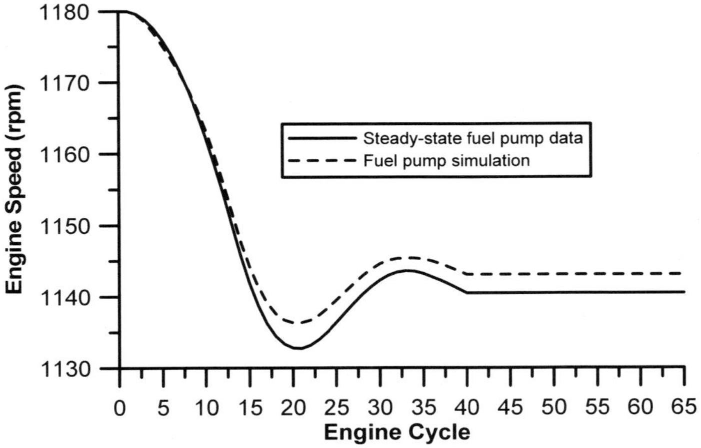

Figure 9.11. Effect of detailed fuel pump modeling on the prediction of transient diesel engine operation after a step load increase

The unsteady gas flow equations are solved using the method of characteristics, providing the dynamic injection timing as well as the duration and the rate of injection for each cylinder at each transient cycle. The obvious advantage is that the transient operation of the fuel pump is also taken into account, through the fuel pump residual pressure, which is built-up together with the other variables during the transient event. Moreover, such an individual fuel injection subroutine can be combined with a true multi-cylinder model, and be called upon once for every cylinder at each cycle with the values of angular velocity, fuel pump rack position and pump residual pressure existing at the point of the individual cylinder's static injection timing. The time burden imposed is not of concern, since it is executed in not more than 0.1 s in a typical PC. Typical results, when applying both fuel injection approaches, are depicted in Figure 9.11, where a difference of the order of 8% is observed in the lowest engine speed (and consequently the other engine variables).

9.9 Mechanical Governor

In Section 3.2.2, the fundamental equations describing the governor clutch movement were given. Simulating the governor based entirely on its physics may appear attractive, but in practice it is very rarely applied, since, in most cases, detailed technical data for the governors are not readily available. In order to find the instantaneous fuel pump rack position, which is initiated by the governor clutch movement during transient operation, a second-order differential equation is usually applied. This incorporates the effect of the various governor elements, such as the centrifugal force of the ball weights, the spring force, damping, *etc.*; one such equation used by Winterbone [1] is

$$\dot{z} = \frac{1}{\tau}\left[c_1\left(N_D - N\right) + c_2 - z\right] \tag{9.107a}$$

The governor of Equation 9.107a is a proportional one; adding an integral term can yield a P-I control that can restore engine speed to its original value after the new load has been applied:

$$z_t = z_{t-1} + \frac{1}{\tau}\left[c_1\left(N_D - N_t\right) + c_2 - z_{t-1}\right]dt + \int_0^\infty c_3\left(N_t - N_{t-1}\right)dt \tag{9.107b}$$

Watson and Marzouk [55] used the following second-order relation for a six-cylinder, truck engine governor:

$$c_4\ddot{z} + c_5\dot{z} + \left(1 + c_6 y\right)\left(z + c_7 y\right) - c_8\omega^2 z = c_9\left(\omega^2 - \omega_D^2\right) \tag{9.108}$$

Rakopoulos and Giakoumis [32, 34] applied the following equation for a heavy-duty, six-cylinder, turbocharged diesel engine:

$$\frac{d^2z}{d\varphi^2} = c_{10}\frac{dz}{d\varphi} + c_{11}z + c_{12}z\omega^2 + c_{13}\omega^2 + c_{14} \quad (9.109)$$

whereas for Qiao *et al.* [56] the appropriate equation for the governor of a passenger vehicle, four-cylinder, naturally aspirated engine was

$$\frac{d^2z}{d\varphi^2} + c_{15}\frac{dz}{d\varphi} + c_{16}(z+y) - c_{17}\,z\omega^2 = c_{18}\,\omega^2 \quad (9.110)$$

In all the above equations, z is the fuel pump rack position, N, ω are the current engine speed and angular velocity, respectively, y is the governor control lever position, and index D corresponds to the demand value. All the c_i (i=1–18) constants in the previous equations are derived after calibration against experimental data under transient conditions applying, for example, least-squares analysis.

9.10 Crankshaft Torque Balance

The instantaneous values for engine speed and angular acceleration needed in the transient engine simulation are derived from the conservation of angular momentum principle applied on the total system (engine-load). Its mathematical formulation, neglecting crankshaft torsional stiffness and damping, is as follows, if G_{tot} represents the total system mass moment of inertia (engine, flywheel and load) reduced to the crankshaft axis

$$\tau_e(\varphi) - \tau_{fr}(\varphi) - \tau_L(\varphi) = G_{tot}\frac{d\omega}{dt} \quad (9.111)$$

where

1. $\tau_e(\varphi) = \sum_{i=1}^{n_{cyl}} \tau_{ei}(\varphi)$ denotes the total engine indicated torque that includes the contribution of gas, inertia and (the negligible) gravitational forces of all cylinders. Individual cylinder indicated torque $\tau_{ei}(\varphi)$ is given explicitly by

$$\tau_e(\varphi) = \overbrace{\tau_g(\varphi)}^{\text{Gas}} + \overbrace{\tau_{in}(\varphi)}^{\text{Inertia}} + \overbrace{\tau_{gr}(\varphi)}^{\text{Gravitational}} =$$

$$\left[\overbrace{\left(p_g(\varphi)\,A_{pist}\frac{u_{pist}(\varphi)}{r\omega}\right)}^{\text{Gas}} - \overbrace{\left(m_l\,b(\varphi)\frac{u_{pist}(\varphi)}{r\omega}\right)}^{\text{Inertia}} + \overbrace{\left(m_l\,g\frac{u_{pist}(\varphi)}{r\omega} + m_r\,g\,\sin\varphi\right)}^{\text{Gravitational}}\right] r \quad (9.112)$$

with u_{pist} the piston linear velocity (Equation 3.2), b the piston linear acceleration (Equation 3.3), m_l the reciprocating masses (Equation 3.6a) and m_r the rotating masses (Equation 3.6b).

2. $\tau_{fr}(\varphi)$ is the total (from all cylinders) friction torque (Equations 9.96 or 9.99).
3. $\tau_L(\varphi)$ is the resistance torque term

$$\tau_L(\varphi) = C_1 + C_2\omega^{C_3} \tag{9.113}$$

Combining the above equations, we get

$$\frac{d\omega}{d\varphi} = \frac{\alpha}{\omega} + \frac{\beta_1}{\omega} + \beta_2\omega^{C_3-1} + \frac{\gamma}{\omega} + \frac{\delta}{\omega} \tag{9.114}$$

where (neglecting the gravitational forces contribution)

$$\alpha = \frac{-\sum_{i=1}^{n_{cyl}}\left(m_l\, b(\varphi_i)\, u_{pist}(\varphi_i)/r\omega\right)}{G_{tot}}$$ is the inertia torque term,

$\beta_1 = -C_1/G_{tot}$ comes from the constant term of the load torque,

$\beta_2 = -C_2/G_{tot}$ comes from the speed dependent term of the load torque,

$$\gamma = \frac{\sum_{i=1}^{n_{cyl}} p_g(\varphi_i)\, A_{pist}\, u_{pist}(\varphi_i)/r\omega}{G_{tot}}$$ is the gas torque term, and

$$\delta = \frac{-\sum_{i=1}^{n_{cyl}} \tau_{fr}(\varphi_i)}{G_{tot}}$$ is the friction torque term

with φ_i the individual cylinder crank angle according to the ignition order.

From Equation 9.114, the crankshaft angular acceleration $\varepsilon(\varphi)$ can then be derived by differentiating with respect to crank angle φ:

$$\varepsilon(\varphi) = \frac{d\omega}{dt} = \alpha + \beta_1 + \beta_2\omega^{C_3} + \gamma + \delta \tag{9.115}$$

9.11 Exhaust Emissions

9.11.1 Global Approximations

Owing to the obvious limitations of quasi-linear and single-zone modeling as regards exhaust emissions prediction, many researchers incorporated empirical approaches in their models. Such a simple relation for transient soot emissions, on a Celesco smokemeter, was given in [1]. This gave satisfactory results for the acceleration of a six-cylinder, turbocharged diesel engine:

$$\text{Smoke} = b_1\,\text{AFR}^{b_2} \tag{9.116}$$

Watson [23] used the following correlation for a six-cylinder, turbocharged, DI diesel engine:

$$\frac{\text{Smoke}}{(\text{Smoke})_{\text{ref}}} = b_3\left(\frac{N}{N_{\text{ref}}}\right)^{b_4}\left(\frac{\Phi}{\Phi_{\text{ref}}}\right)^{b_5}\exp\left[b_6\frac{(1-\beta)}{(1-\beta_{\text{ref}})}\right] \tag{9.117}$$

with β the premixed burning fraction (Equation 9.64).

Younes *et al.* [12] incorporated in their quasi-linear code the following relation for the Bosch smoke number-BSN of a four-cylinder, turbocharged, diesel engine:

$$\text{BSN} = b_7\left(\frac{\dot{m}_{\text{air}}}{\dot{m}_{\text{fi}}}\right)^{b_8N^3+b_9N^2+b_{10}N+b_{11}\dot{m}_{\text{air}}+b_{12}} \tag{9.118}$$

and a similar one for HC emissions.

Rakopoulos and Giakoumis [32] found the following BSN relation more appropriate for their six-cylinder, medium duty, turbocharged diesel engine:

$$\text{BSN} = b_{13}\,e^{b_{14}\text{AFR}} \tag{9.119}$$

Jiang and Van Gerpen [57] developed the following equation for estimating the rate of particulate emissions during the EPA transient cycle

$$P_{\text{rate}} = b_{15} + b_{16}\Phi + b_{17}\Phi^2 + b_{18}\Phi^3 + b_{19}\Phi^4 \tag{9.120}$$

In Equations 9.116–9.120, constants b_i (i=1–19) are derived using experimental data at steady-state conditions for various engine speeds and loads; all approximations gave overall moderately satisfactory results.

A different perspective entails use of the steady-state maps and further definition of appropriate weighting factors to account for transient discrepancies; such an approach was followed by Brace *et al.* [58], who applied neural network techniques on steady-state data to predict transient exhaust emissions from a quasi-linear model. Ericson *et al.* [59] also used steady-state maps and accounted for transient non-linearities, mainly through global relative air–fuel ratio corrections, albeit with moderate success.

9.11.2 Nitric Oxide Formation Model

Since it is well established that NO formation is controlled by chemical kinetics rather than chemical equilibrium, the generally accepted kinetics model proposed by Lavoie *et al.* [60] can be adopted in two- or multi-zone transient simulation models; this describes the extended Zel'dovich kinetics scheme. According to this model, the governing chemical reactions for NO formation and their respective forward reaction rate constants (m^3/kmol/s) are as follows:

$$\begin{aligned} &N + NO \rightleftarrows N_2 + O, \quad k_{1f}=3.1\times 10^{10}\, e^{(-160/T)} \\ &N + O_2 \rightleftarrows NO + O, \quad k_{2f}=6.4\times 10^{6}\, T \cdot e^{(-3125/T)} \\ &N + OH \rightleftarrows NO + H, \quad k_{3f}=4.2\times 10^{10} \end{aligned} \tag{9.121}$$

The one way equilibrium rates for the above three reactions are defined as

$$R_1=k_{1f}(N)_e(NO)_e, \quad R_2=k_{2f}(N)_e(O_2)_e, \quad R_3=k_{3f}(N)_e(OH)_e \tag{9.122}$$

where index 'e' denotes equilibrium concentrations and term $\alpha=(NO)/(NO)_e$. Finally, the rate of change of NO concentration is expressed as

$$\frac{1}{V}\frac{d\left((NO)V\right)}{dt}=2\left(1-\alpha^2\right)\frac{R_1}{1+\alpha\dfrac{R_1}{R_2+R_3}} \tag{9.123}$$

9.11.3 Soot Formation Model

For the calculation of the net soot that is formed inside the cylinder, the model of Hiroyasu *et al.* [33] can be used, for example, as modified be Lipkea and DeJoode [61]. The net soot formation rate is expressed as the difference between the formation and oxidation rates, which are given by

$$\frac{dm_{sf}}{dt}=A_{sf}\, dm_f^{0.8}\, p_g^{0.5}\, e^{\left(-E_{sf}/\left(R_{mol}T_g\right)\right)} \tag{9.124}$$

$$\frac{dm_{sc}}{dt}=A_{sc}m_{sn}\left(p_{O_2}/p_g\right)p_g^{n}\, e^{\left(-E_{sc}/\left(R_{mol}T_g\right)\right)} \tag{9.125}$$

where pressures are in bar and masses in kg, dm_f is the fuel vapor mass to be burned in the current time step, and p_{O_2} is the partial pressure of oxygen in the zone. Also, constants A_{sf}, A_{sc}, activation energies E_{sf}, E_{sc} and the exponent n are evaluated after experimental matching at steady-state conditions. Finally, the net soot formation rate is expressed as follows:

$$\frac{dm_{sn}}{dt} = \frac{dm_{sf}}{dt} - \frac{dm_{sc}}{dt} \tag{9.126}$$

9.12 Solution of Equations

The calculation of all the above given differential and algebraic equations in a filling and emptying simulation model starts at the inlet valve closure position. At the initial operating point (steady-state conditions), an estimation of temperature, pressure and composition (including residual gas and exhaust gas recirculation) is made for all control volumes as well as for the turbocharger (compressor pressure ratio and turbocharger speed). Thereafter, a series of cycles (3–5) is performed for the current initial engine speed and loading/fueling up to the moment when the thermodynamic variables do not change from cycle to cycle, beyond a certain limit set for accuracy (*e.g.,* 0.1%). All first-law equations (combustion, heat transfer, blow-by losses, manifolds, turbocharger, first law of thermodynamics *etc.*) are solved during this running process. Especially as regards turbocharging, the procedure is as follows: first, an estimation is made concerning compressor boost pressure and isentropic efficiency. These are considered constant during the whole engine cycle. Using the previous estimations and the values from manifolds/intercooler operation, the turbine inlet conditions are derived that define the turbine, hence, compressor work. If the initial estimations are right, Equation 9.81 (with $d\omega_{TC}/dt=0$) is confirmed. If not, the new compressor operating point is reached, which leads to a new test cycle until final convergence, where also the turbocharger rotational speed is determined from the compressor map.

After *thermodynamic* equilibrium is achieved at steady-state conditions, the simulation program proceeds with the transient subroutines (*e.g.,* governor, crankshaft torque balance) now 'turned on' but still without the new applied load or speed change. This is made so that a *dynamic* equilibrium is also established at the initial operating conditions. This process typically requires one to two engine cycles. After that, the load or speed change is applied according to its schedule, and the transient response evolves. The crankshaft torque balance (Equation 9.114) is solved at each degree crank angle, providing the new values for angular velocity and acceleration. The updated value of angular velocity is used for the thermodynamic calculations in the next step, which, in turn, provide new values for engine torque, used in Equation 9.112, and so on. At each degree crank angle and after the new angular velocity has been found, the governor equation (Section 9.9) is solved, which provides new values for the fuel pump rack position.

At static injection timing position and with the current values of fuel pump rack and engine speed, the fuel injection map or fuel injection model (Section 9.8) subroutine is called for the evaluation of the amount and profile of fuel injected for the particular cylinder. The latter initiates the heat release law subroutines (Section 9.3.2.3), which specify the profile and values of cylinder pressure and temperature. At each degree crank angle, and with the current values of engine speed and gas pressure, the friction torque sub-routines (Section 9.7) are also called.

The respective manifolds first-law analysis (Section 9.4) provides the gas properties at turbine inlet; the latter are used for the turbine expansion and through Equations 9.81–9.83 the respective compressor operating point is established that specifies the engine air-supply. Turbocharger equations need not be called at each (crankshaft) degree crank angle. For better convergence, during combustion and expansion the computational step can be reduced to one half or even one fourth of a degree crank angle. Dynamic equations, however, can still be called at each degree crank angle.

At the end of the cycle, mean values of all important thermodynamic and dynamic properties are computed incl. torque and exhaust emissions. The model is assumed to finally converge when certain engine variables such as engine speed, governor position, turbocharger operating point and engine and load torques (mean values over the engine cycle) present no change in the current cycle compared with their values in the previous one beyond a certain limit of accuracy. This limit can be set at 0.02% for the speed and torques and 0.1% for the governor/fuel pump rack position, *i.e.*,

$$\left|\frac{x - x_o}{x_o}\right| < 0.1\% \tag{9.127}$$

with x the current value of the selected typical engine or turbocharger variable. The above-mentioned computational procedure is schematically illustrated in the flow-chart of Figure 9.12.

For the solution of differential equations, various methods proposed in numerical analysis handbooks can be applied; one such method is the predictor-corrector. We assume a differential equation of the following general form:

$$y'' = f(x, y, y') \tag{9.128}$$

and also

$$y_1 = y \text{ and } y_2 = y' = y_1' \tag{9.129}$$

Knowing the values at computational step k, we estimate the new values for the unknown point (k+1) according to

$$\begin{aligned} y_{1,k+1}^{(0)} &= y_{1,k} + h\, y_{2,k} \\ y_{2,k+1}^{(0)} &= y_{2,k} + h\, f(x_{1,k}, y_{1,k}, y_{2,k}) \end{aligned} \tag{9.130}$$

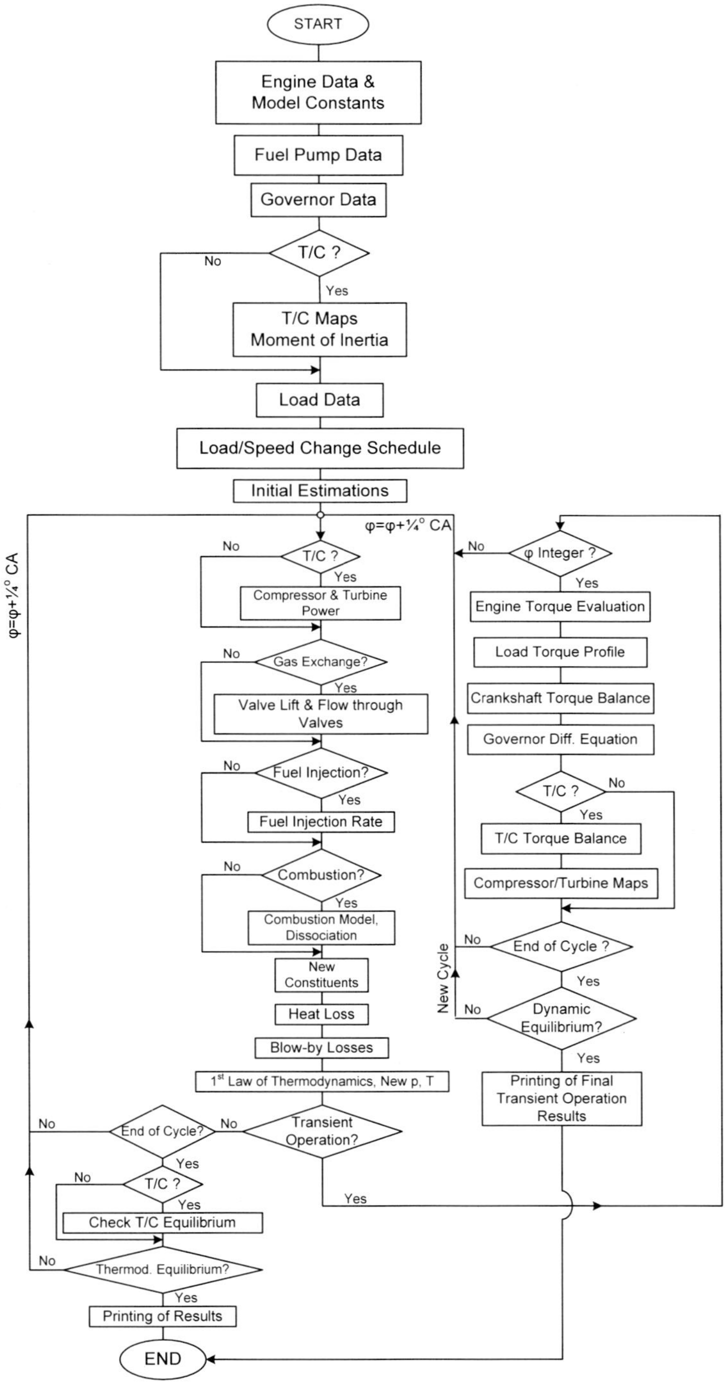

Figure 9.12. Simplified flow-chart of transient simulation code (solution of thermodynamic equations every 1/4°CA and of the dynamic equations every 1°CA)

and also

$$\begin{aligned} y_{1,k+1}^{(i+1)} &= y_{1,k} + \frac{h}{2}\left(y_{2,k} + y_{2,k+1}^{i}\right) \\ y_{2,k+1}^{(i+1)} &= y_{2,k} + \frac{h}{2}\left(f(x_{1,k}, y_{1,k}, y_{2,k}) + f(x_{1,k+1}^{(i)}, y_{1,k+1}^{(i)}, y_{2,k+1}^{(i)})\right) \end{aligned} \tag{9.131}$$

where i is the number of iterations required for convergence and h the time step between two computational points (*e.g.*, 1°CA). When the accuracy limit (set by a similar to Equation 9.127) is met, the solution at the specific computational point has then been reached.

A typical transient case of load acceptance (30–60 engine cycles) requires approximately one second on a modern PC when quasi-linear modeling is used, while it needs one minute via single-zone modeling and more than 20 hours when quasi-dimensional analysis is adopted. The above times are increased proportionally if separate analyses for each specific cylinder are applied. Moreover, as regards zero-dimensional studies, a multi-zone model would increase the required time almost linearly with the number of zones analyzed.

9.13 Sensitivity Analysis

Transient simulation codes suffer from the need to run a large number of engine cycles, particularly when Transient Cycles analysis is involved. This often results in ignoring or simplifying specific sub-models for the sake of speeding up program execution. These simplifications can be justified if the goal is to identify key parameters that influence transient response or just general trends. However, they can prove quite limiting if the scope is an in depth study of transient phenomena, especially, nowadays, when stringent regulations concerning engine exhaust emissions dominate the industry. Typical modeling simplifications include

- single-zone modeling;
- assumption of constant cylinder wall temperature throughout the whole transient event;
- solution of one cylinder's equations by assuming that all the others behave in the same way for each transient cycle;
- use of steady-state fuel injection data;
- use of steady-state turbocharger data;
- use of steady-state dynamometer/load curves;
- filling and emptying modeling of exhaust manifold operation instead of wave action simulation;
- simulation of friction torque with the use of mean fmep relations; and
- assumption that both combustion and friction evolve during transients in the same way as they do under steady-state operation.

A number of these simplifications have been studied by Rakopoulos and Giakoumis [39] on a high moment of inertia, turbocharged and aftercooled, medium duty diesel engine. It was shown that only the assumptions of constant cylinder wall temperature and steady-state dynamometer/load curves (for load changes or for small speed changes only) are justified. All the other assumptions can induce a modeling error, which ranges from 5–20% according to the examined transient schedule and engine–load setup. Typical results are presented in Table 9.3, which also includes comments on the imposed burden in typical PC execution time.

Table 9.3. Comparison between usual and more advanced approaches for filling and emptying transient diesel engine modeling: summary of the differentiation in results and of the computational time burden induced (data are valid for the following case: six-cylinder, turbocharged and aftercooled diesel engine, with a speed range of 1000–1500 rpm, coupled to a hydraulic brake, with relatively high total mass moment of inertia and tight governing, operating on load increase transients)

Simulation section	Usual approach	Examined-more detailed approach	Differentiation in results for current engine set up	Computational time burden
Cylinder wall temperature	Constant value	Heat convection-conduction scheme (Equation 9.60)	Negligible	Negligible
Cylinder-manifolds interdependence	Single-cylinder approach	'True' multi-cylinder approach (Section 9.5)	5–10%	Considerable
Fuel injection	Steady-state fuel pump curves	Fuel pump - injector mechanism model (Section 9.8)	8%	Very small
Dynamometer	Steady-state curve	Transient modeling of brake/load	None *	Small
Exhaust manifold	Filling and emptying	Method of characteristics (Section 9.4)	< 3% **	Considerable
Friction	'mean' fmep	Modeling per deg. CA (Section 9.7)	> 6%	Negligible
Differentiation of transient from steady-state	No compensation	'Deterioration' of friction-combustion rates (Sections 9.3.2.3 and 9.7.3)	> 10%	None

* Estimation only – no actual simulations carried out

** Based on differentiation of results during steady-state operation

References

[1] Horlock JH, Winterbone DE. The thermodynamics and gas dynamics of internal combustion engines, Vol. II. Oxford: Clarendon Press, 1986.

[2] Watson N, Janota MS. Turbocharging the internal combustion engine. London: MacMillan, 1982.

[3] Heywood JB. Internal combustion engine fundamentals. New York: McGraw-Hill, 1988.
[4] Benson RS, Whitehouse ND. Internal combustion engines. Oxford: Pergamon Press, 1979.
[5] Stone R. Introduction to internal combustion engines, 3rd edition. London: MacMillan, 1999.
[6] Ferguson CR, Kirkpatrick AT. Internal combustion engines: applied thermosciences, 2nd edition. New York: Wiley, 2001.
[7] Ledger JD, Walmsley S. Computer simulation of a turbocharged diesel engine operating under transient load conditions. SAE Paper No. 710177, 1971.
[8] Winterbone DE, Benson RS, Closs GD, Mortimer AG. A comparison between experimental and analytical transient test results for a turbocharged diesel engine. Proc Inst Mech Eng 1976;190:267–76.
[9] Benson RS, Ledger JD, Whitehouse ND, Walmsley S. Comparison of experimental and simulated transient responses of a turbocharged diesel engine. SAE Paper No. 730666, 1973.
[10] Jensen J-P, Kristensen AF, Sorenson SC, Houbak N, Hendricks E. Mean value modeling of a small turbocharged diesel engine. SAE Paper No. 910070, 1991.
[11] Berglund S. A model of turbocharged engines as dynamic drivetrain members. SAE Paper No. 933050, 1993.
[12] Younes R, Champoussin JC, Liazid A. Modeling of turbocharged diesel engine for optimal control. Entropie 1993;174/175:31–42 (in French).
[13] Jennings MJ, Blumberg PN, Amann RW. A dynamic simulation of the Detroit diesel electronic control system in heavy duty truck powertrains. SAE Paper No. 861959, 1986.
[14] Rackmil CI, Blumberg PN, Becker DA, Schuller RR, Garvey DC. A dynamic model of a locomotive diesel engine and electrohydraulic governor. ASME Trans, J Eng Gas Turbines Power 1988;110:405–14.
[15] JANAF thermochemical tables, 2nd edition. NSRDS-NBS37, US National Bureau of Standards, 1971.
[16] Gordon S, McBride BJ. Computer program for calculation of complex chemical equilibrium composition, rocket performance, incident and reflected shocks, and Chapman-Jouguet detonations. NASA publication SP-273, 1971.
[17] Rakopoulos CD, Rakopoulos DC, Giakoumis EG, Kyritsis DC. Validation and sensitivity analysis of a two zone diesel engine model for combustion and emissions prediction. Energy Convers Manage 2004;45:1471–95.
[18] Arai M, Tabata M, Hiroyasu H, Shimizu M. Disintegrating process and spray characterization of fuel jet injected by a Diesel nozzle. SAE Paper No. 840275, 1984.
[19] Annand WJD. Heat transfer in the cylinders of reciprocating internal combustion engines. Proc Inst Mech Eng 1963;177:973–90.
[20] Woschni G. A universally applicable equation for the instantaneous heat transfer coefficient in the internal combustion engine. SAE paper No. 670931, 1967.
[21] Hohenberg GF. Advanced approaches for heat transfer calculations. SAE Paper No. 790825, 1979.
[22] Annand WJD, Ma TH. Instantaneous heat transfer rates to the cylinder head surface of a small compression-ignition engine. Proc Inst Mech Eng 1970-71;185:976–87.
[23] Watson N. Dynamic turbocharged diesel engine simulator for electronic control system development. ASME Trans, J Dynamic Syst Measurement Control 1984;106:27–45.
[24] Rakopoulos CD, Giakoumis EG. Study of the transient operation of low-heat rejection turbocharged diesel engine including wall temperature oscillations. SAE Paper No. 2007-01-1091, 2007.

[25] Wiebe I. Halbempirische formel fuer die Verbrennungsgeschwindigkeit. Moscow: Verlag der Akademie der Wissenschaften der UdSSR, 1967.
[26] Watson N, Pilley AD, Marzouk M. A combustion correlation for diesel engine simulation. SAE paper No. 800029, 1980.
[27] Whitehouse ND, Way RGB. Rate of heat release in diesel engines and its correlation with fuel injection data. Proc Inst Mech Eng 1969-70; 3J(184):17–27.
[28] Bazari Z. A DI diesel combustion and emission predictive capability for use in cycle simulation. SAE Paper No. 920462, 1992.
[29] Woschni G, Anisits F. Eine Methode zur Vorausberechnung der Änderung des Brennverlaufs mittelschnelllaufender Dieselmotoren bei geänderten Betriebsbedingungen. MTZ 1973;34:106–15.
[30] Betz A, Woschni G. Energy conversion rate and rate of heat release of turbocharged diesel engines under transient conditions. MTZ 1986;47:263–7 (in German).
[31] Assanis DN, Filipi ZS, Fiveland SB, Syrimis M. A predictive ignition delay correlation under steady-state and transient operation of a direct injection diesel engine. ASME Trans, J Eng Gas Turbines Power 2003;125:450–7.
[32] Rakopoulos CD, Giakoumis EG, Hountalas DT, Rakopoulos DC. The effect of various dynamic, thermodynamic and design parameters on the performance of a turbocharged diesel engine operating under transient load conditions. SAE Paper No. 2004-01-0926, 2004.
[33] Hiroyasu H, Kadota T, Arai M. Development and use of a spray combustion modeling to predict diesel engine efficiency and pollutant emissions. Bulletin JSME 1983;26:569–75.
[34] Rakopoulos CD, Giakoumis EG. Review of thermodynamic diesel engine simulations under transient operating conditions. SAE Paper No. 2006-01-0884, SAE Trans, J Engines 2006;115:467–505.
[35] Ertl C, Kranawetter E, Stütz W. Simulation of the dynamic behavior of diesel engines with electronic management system. MTZ 1997;58:612–8 (in German).
[36] Lin C-C, Filipi Z, Louca L, Peng H, Assanis D, Stein J. Modelling and control of a medium-duty hybrid electric truck. Int J Heavy Vehicle Syst 2004;11:349–71.
[37] Rakopoulos CD, Michos CN, Giakoumis EG. Study of the transient behavior of turbocharged diesel engines including compressor surging using a linearized quasi-steady analysis. SAE Paper No. 2005-01-0225, 2005.
[38] Watson N. Transient performance simulation and analysis of turbocharged diesel engines. SAE Paper No. 810338, 1981.
[39] Rakopoulos CD, Giakoumis EG. Sensitivity analysis of transient diesel engine simulation. Proc Inst Mech Eng, Part D, J Automobile Eng 2006;220:89–101.
[40] Moraal P, Kolmanovksy IV. Turbocharger modeling for automotive control applications. SAE Paper No. 1999-01-0908, 1999.
[41] Fink DA, Cumpsty NA, Greitzer EM. Surge dynamics in a free-spool centrifugal compressor system. ASME Trans, J Turbomachinery 1992;114:321–32.
[42] Filipi Z, Wang Y, Assanis D. Effect of variable geometry turbine (VGT) on diesel engine and vehicle system transient response. SAE Paper No. 2001-01-1247, 2001.
[43] Jung M, Ford RG, Glover K, Collings N, Christen U, Watts MJ. Parameterization and transient validation of a variable geometry turbocharger for mean-value modeling at low and medium speed-load points. SAE Paper No. 2002-01-2729, 2002.
[44] Shamsi SS. Development of a real-time digital computer simulation of a turbocharged diesel engine. SAE Paper No. 800521, 1980.
[45] Schorn N, Pischinger F, Schulte H. Computer simulation of turbocharged diesel engines under transient conditions. SAE Paper No. 870723, 1987.

[46] Moore FK, Greitzer EM. A theory of post-stall transients in axial compression systems. Part I: development of equations. ASME Trans, J Eng Gas Turbines Power 1986;108:68–76.
[47] Botros KK. Transient phenomena in compressor stations during surge. ASME Trans, J Eng Gas Turbines Power 1994;116:133–42.
[48] Hansen KE, Jørgensen P, Larsen PS. Experimental and theoretical study of surge in a small centrifugal compressor. ASME Trans, J Fluids Eng 1981;103:391–5.
[49] Rezeka SF, Henein NA. A new approach to evaluate instantaneous friction and its components in internal combustion engines. SAE Paper No. 840179, 1984.
[50] Tuccilo R, Arnone L, Bozza F, Nocera R, Senatore A. Experimental correlations for heat release and mechanical losses in turbocharged diesel engines. SAE Paper No. 932459, 1993.
[51] Cameron A, Ettles CCMc. Basic lubrication theory, 3rd edition. Chichester: Ellis Horwood, 1981.
[52] Winterbone DE, Loo WY. A dynamic simulation of a two-stroke turbocharged diesel engine. SAE Paper No. 810337, 1981.
[53] Winterbone DE, Tennant DWH. The variation of friction and combustion rates during diesel engine transients. SAE Paper No. 810339, 1981.
[54] Rakopoulos CD, Hountalas DT. A simulation analysis of a DI diesel engine fuel injection system fitted with a constant pressure valve. Energy Convers Manage 1996;37:135–50.
[55] Watson N, Marzouk M. A non-linear digital simulation of turbocharged diesel engines under transient conditions. SAE Paper No. 770123, 1977.
[56] Qiao J, Dent JC, Garner CP. Diesel engine modelling under steady and transient conditions using a transputer based concurrent computer. SAE Paper No. 922226, 1992.
[57] Jiang Q, Van Gerpen JH. Prediction of diesel engine particulate emission during transient cycles. SAE Paper No. 920466, 1992.
[58] Brace CJ, Deacon M, Vaughan ND, Charlton SJ, Burrows CR. Prediction of emissions from a turbocharged passenger car diesel engine using a neural network. Institution of Mechanical Engineers, 5th International Conference on 'Turbocharging and Turbochargers', Paper C484/046, London, 1994, pp. 83–91.
[59] Ericson C, Westerberg B, Egnell R. Transient emission predictions with quasi stationary models. SAE Paper No. 2005-01-3852, 2005.
[60] Lavoie GA, Heywood JB, Keck JC. Experimental and theoretical study of nitric oxide formation in internal combustion engines. Combust Sci Technol 1970;1:313–26.
[61] Lipkea WH, DeJoode AD. Direct injection diesel engine soot modeling: formulation and results. SAE Paper No. 940670, 1994.

Appendix A

Exhaust Emission Regulations and Transient Cycles

A.1 Introduction

Starting from the 1980s, (diesel-engined) vehicles have been tested for exhaust emissions, prior to type approval, using sophisticated standardized transient tests (Transient Cycles); these are usually characterized by long duration (up to 30 minutes) consisting of both speed and load changes under varying operating schedules. A Transient Test Cycle is a sequence of test points each with a defined vehicle speed to be followed by the vehicle under study, or with a defined rotational speed/torque to be followed by the engine under transient conditions; these test points are divided in time steps, mostly seconds, during which acceleration is assumed constant. Such standardization is necessary as it makes it possible to compare different vehicles/engines that fulfill the same operation. Although the primary use of Transient Cycles is for emission certification of (light-duty or heavy-duty) automotive applications, recently, specialized Test Cycles for non-road mobile engines have been proposed too. In order for the exhaust emission measurements to be representative of real engine operation, Transient Test Cycles incorporate some or all of the following driving conditions

- cold and hot starting;
- frequent accelerations and decelerations;
- changes of load;
- idling conditions typical of urban driving;
- sub-urban or rural driving schedule; and
- motorway driving.

By applying a Transient Cycle for the testing of new vehicles, the complete engine operating range is tested and not just the maximum power or torque operating points. Moreover, the serious discrepancies that are experienced during abrupt transients are taken into account; in fact, as was discussed in Chapter 5, large overshoot of particulate and gaseous emissions are observed under load or speed increase situations, mainly owing to turbocharger lag. It should, however, be pointed out that the primary objective of a Transient Cycle procedure is to establish the *total* amount of exhaust emissions rather than indicate the specific parts or conditions under which these emissions are produced. Further, legislative Test Cycles assume straight roads with zero gradient, thus no account is taken of the respective road-dependent resistance torque.

Transient Cycles require highly sophisticated experimental facilities (a fully automated test-bed with electronically controlled motoring and dissipating (chassis) dynamometer, fast response exhaust gas analyzers, dilution tunnels, *etc.*) in order to be accurately reproduced, and complicated, time consuming computer models for their simulation. Many countries in the world have developed Transient Cycles for emission testing of their vehicles; in the following paragraphs, the Transient Cycles valid in the European Union, the USA and Japan will be summarized together with the respective exhaust emission legislation levels. These Transient Cycles concern the testing of passenger vehicles, light-duty (commercial) vehicles, heavy-duty vehicles, heavy-duty engines, and non-road mobile engines.

Passenger cars and light-duty vehicles usually undergo a vehicle speed *vs.* time Test Cycle on a chassis dynamometer, and the results are expressed in g/km. Since vehicle testing is much more difficult for heavy-duty or non-road vehicles, the exhaust emission certification procedure for the latter usually makes use of an engine rather than a vehicle cycle; this is realized on an engine test bed, where the engine under study follows a prescribed engine speed/torque *vs.* time procedure[1] and the results are usually expressed in g/kWh.

A.2 European Union (EU)

A.2.1 Emission Standards

Emission requirements for light-duty vehicles have existed in the EU since the early 1970s, while the first requirements for heavy-duty vehicles came in at the end of the 1980s. The standards are defined in a series of EU Directives stating the progressive introduction of increasingly stringent standards (see also Figure A.15 later in the chapter) [1, 2].

A.2.1.1 Passenger Vehicles and Light Trucks

European Union emission regulations for new light-duty vehicles (passenger cars and light commercial vehicles) were initially specified in Directive 70/220/EEC, which was amended a number of times; some of the most important amendments are reproduced in Table A.1 for passenger cars (category M_1, *i.e.,* vehicles used for the carriage of passengers and comprising no more than 8 seats in addition to the

---

[1] The engine speed/torque *vs.* time schedule derives from a respective vehicle speed *vs.* time pattern applying a suitable drivetrain model.

driver's seat) and for light commercial vehicles (category N_1, *i.e.*, vehicles used for the carriage of goods having a maximum mass not exceeding 3.5 tonnes).

Table A.1. EU emission standards for passenger cars (category M_1), and light commercial vehicles (category N_1), g/km

		Date	CO	HC+NOx	NOx	PM
M_1 *	**Euro 1**	1992.07	2.72	0.97	-	0.14
	Euro 2, IDI	1996.01	1.00	0.70	-	0.08
	Euro 2, DI	1996.01[a]	1.00	0.90	-	0.10
	Euro 3	2000.01	0.64	0.56	0.50	0.05
	Euro 4	2005.01	0.50	0.30	0.25	0.025
	Euro 5	2009.09[b]	0.50	0.23	0.18	0.005[c]
	Euro 6	2014.09	0.50	0.17	0.08	0.005[c]
N_1,Class I < 1305 kg **	**Euro 1**	1994.10	2.72	0.97	-	0.14
	Euro 2, IDI	1998.01	1.00	0.70	-	0.08
	Euro 2, DI	1998.01[a]	1.00	0.90	-	0.10
	Euro 3	2000.01	0.64	0.56	0.50	0.05
	Euro 4	2005.01	0.50	0.30	0.25	0.025
	Euro 5	2009.09[b]	0.50	0.23	0.18	0.005[c]
	Euro 6	2014.09	0.50	0.17	0.08	0.005[c]
N_1,Class II 1305-1760 kg **	**Euro 1**	1994.10	5.17	1.40	-	0.19
	Euro 2, IDI	1998.01	1.25	1.00	-	0.12
	Euro 2, DI	1998.01[a]	1.25	1.30	-	0.14
	Euro 3	2001.01	0.80	0.72	0.65	0.07
	Euro 4	2006.01	0.63	0.39	0.33	0.04
	Euro 5	2010.09[d]	0.63	0.295	0.235	0.005[c]
	Euro 6	2015.09	0.63	0.195	0.105	0.005[c]
N_1,Class III >1760 kg **	**Euro 1**	1994.10	6.90	1.70	-	0.25
	Euro 2, IDI	1998.01	1.50	1.20	-	0.17
	Euro 2, DI	1998.01[a]	1.50	1.60	-	0.20
	Euro 3	2001.01	0.95	0.86	0.78	0.10
	Euro 4	2006.01	0.74	0.46	0.39	0.06
	Euro 5	2010.09[d]	0.74	0.35	0.28	0.005[c]
	Euro 6	2015.09	0.74	0.215	0.125	0.005[c]

* At the Euro 1–4 stages, passenger vehicles > 2.500 kg were type approved as Category N_1 vehicles

** N_1 ref. mass classes for Euro 1, 2: Class I ≤ 1250 kg, Class II 1250–1700 kg, Class III > 1700 kg

[a] until 1999.09.30 (after that date, DI engines must meet the IDI limits)

[b] 2011.01 for all models

[c] proposed to be changed to 0.003 g/km using the PMP procedure

[d] 2012.01 for all models

A vast improvement in acceptable passenger vehicle PM emissions of the order of 80% is noticed from Euro 4 to Euro 5 standards, which means that a diesel particulate filter is probably required. On the other hand, only a small reduction (28%) in acceptable NO_x emissions has been legislated by the European Commission. This emission limit has been set so that reductions can be achieved by further internal engine measures. Since Euro 5 levels practically require installation of particulate filters in the exhaust stream, the European Commission preferred to avoid an obligation for installing an additional NO_x after-treatment system at the same stage. A further significant change from Euro 4 to Euro 5 standards is the proposal that the durability period over which manufacturers must ensure the functioning of pollution control devices be extended from 80,000 km to 160,000 km. This change will more realistically reflect the actual life of vehicles and ensure that emission control systems continue to function throughout the whole vehicle life.

Under the draft implementing legislation (status 2008), a standard for the acceptable particle number concentration has been set to 6×10^{11}/km, applicable, at the latest, upon entry into force of Euro 6 levels. Concerning CO_2, the European Commission has signed voluntary agreements with the European (ACEA), Japanese and Korean Automobiles Manufacturers Associations to reduce the respective emissions. These agreements are expected to adopt quite ambitious emission targets in the years to come (proposed value of 130 g CO_2/km from 2012, and, possibly, 95 g/km from 2020). New standards may be introduced for emissions that are now unregulated.

A.2.1.2 Heavy-duty Vehicles

Heavy-duty engines emission regulations apply latin numerals instead of the arabic ones used for light-duty vehicle standards. The first EU Directive to regulate emissions from heavy-duty vehicles, *i.e.,* road vehicles with 'technically permissible maximum laden mass' over 3,500 kg, was published in 1988 (88/77/EEC). The latter Directive was followed by a number of amendments (summarized in Table A.2); Euro VI emission levels are still to be definitized.

Table A.2. EU emission standards for heavy-duty diesel engines, g/kWh (smoke in m^{-1})

	Date	Test Cycle	CO	HC	NOx	PM	Smoke
Euro I	1992, <85kW	ECE R-49	4.5	1.1	8.0	0.612	-
	1992, >85kW		4.5	1.1	8.0	0.36	-
Euro II	1996.10		4.0	1.1	7.0	0.25	-
	1998.10		4.0	1.1	7.0	0.15	-
Euro III	2000.10	ESC & ELR	2.1	0.66	5.0	0.10 0.13*	0.8
Euro IV	2005.10		1.5	0.46	3.5	0.02	0.5
Euro V	2008.10		1.5	0.46	2.0	0.02	0.5
Euro VI[1]	2013.01[2]		1.5	0.13	0.4	0.01	-
Euro III	2000.10	ETC	5.45	0.78[3]	5.0	0.16 0.21*	-
Euro IV	2005.10		4.0	0.55[3]	3.5	0.03	-
Euro V	2008.10		4.0	0.55[3]	2.0	0.03	-
Euro VI[1]	2013.01[2]		4.0	0.16	0.4	0.01	-

* for engines of less than 0.75 L swept volume per cylinder and rated power speed > 3000 rpm

[1] Formal proposal 16/12/2008

[2] 2014.10 for all models

[3] Non-methance hydrocarbons (NMHC)

As regards heavy-duty diesel-engined vehicles, in 2005 Euro IV standards were implemented in Europe with very low limit values for particulate emissions. When deciding on this legislation, it was anticipated that manufacturers would need to use particulate traps or technology with corresponding performance to meet those limit values. For heavy-duty vehicles, a further tightening of the NO_x limit values was introduced from 2008. To meet these limit values manufacturers need to use some kind of after-treatment device. For EURO VI emission certification the Worldwide Harmonized Transient Cycle (Figure A.18) will be applied. To prevent the possibility that the Euro VI PM mass limit is met using open filters that would enable a high number of ultra fine particles to pass, it is planned to introduce at a later stage a new particle number standard, in addition to the mass based limit. The particle number standard would be introduced once the final results of the UN/ECE Particulate Measurement Programme (PMP) become available.

A.2.2 Transient Cycles

A.2.2.1 ECE+EUDC Transient Cycle for Passenger Vehicles and Light Trucks
Passenger vehicles and light trucks in the European Union are tested for emissions using the ECE+EUDC Test Cycle (also referred to as the MVEG Cycle); this procedure is performed on a chassis dynamometer (EEC Directive 90/C81/01). The entire testing procedure (Figure A.1) comprises of four ECE segments repeated without interruption, followed by one EUDC (European Urban Driving Cycle) segment.

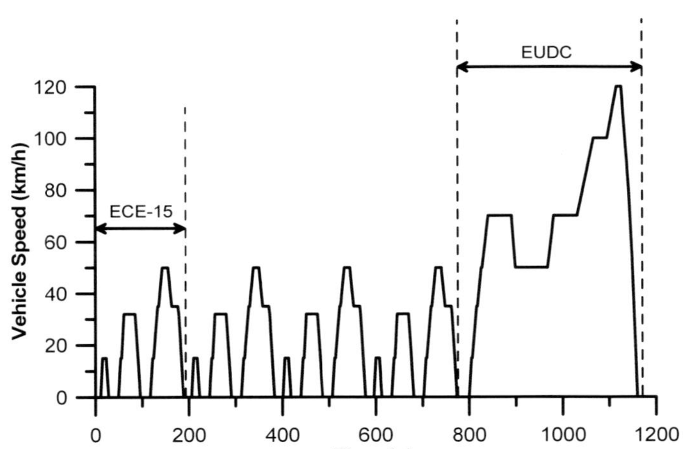

Figure A.1. ECE+EUDC Driving Cycle comprising of four ECE 15 segments (duration 4×195 s = 780 s, distance 4×1.013 km = 4.052 km, average speed 18.7 km/h, maximum speed 50 km/h) and an EUDC Cycle (duration 400 s, distance 6.955 km, average speed 62.6 km/h, maximum speed 120 km/h; max. speed is 90 km/h for low powered vehicles with less than 30 kW maximum power)

The ECE Cycle is an urban driving cycle (based on Paris traffic conditions) with relatively soft accelerations representing typical urban driving conditions in a big European city. It is characterized by relatively low vehicle speeds and engine loads. The EUDC segment has been added after the fourth ECE run to account for more aggressive, high speed (sub-urban or motorway) driving modes. Initially, the vehicle was allowed to soak before the test for at least 6 hours at a test temperature of 20–30°C. It was then started and allowed to idle for 40 s. From year 2000, the idling period has been eliminated, *i.e.*, engine is cold started and the emission sampling process begins immediately. This modified cold start procedure is also referred to as the New European Driving Cycle (NEDC) and is also used for CO_2 emissions measurement. Emissions are sampled during the ECE Cycle according to the constant volume sampling technique (Chapter 4); they are analyzed, and expressed in g/km for each of the pollutants concerned.

A.2.2.2 European Transient Cycle for Heavy-duty Vehicles

The ETC (European Transient Cycle) Test Cycle has been introduced, together with the ESC (European Stationary Cycle consisting of 13 steady-state modes) for emission certification of heavy-duty diesel engines in Europe starting in the year 2000 (EC Directive 1999/96/EC).[2]

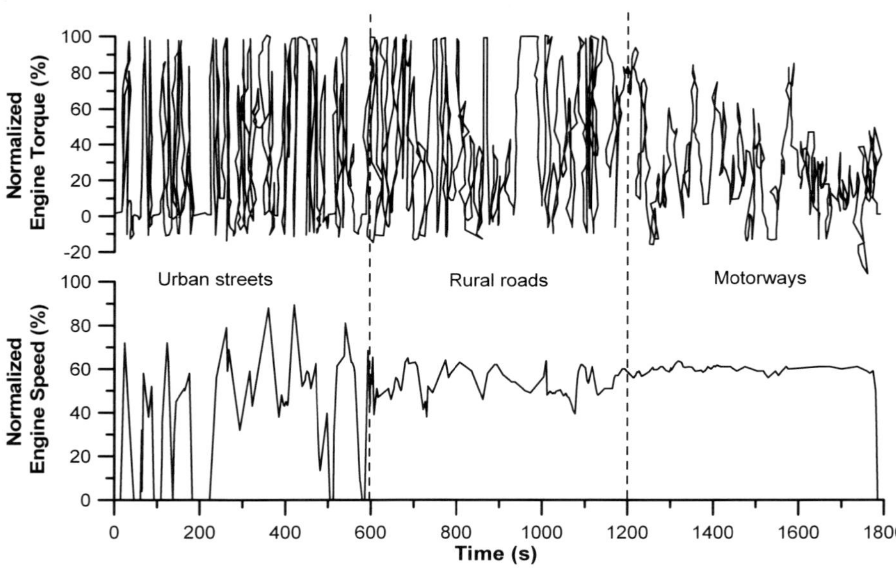

Figure A.2. ETC Transient Cycle for heavy-duty diesel engines

The ESC and ETC Cycles replace the earlier R-49 test (a 13-mode steady-state diesel engine Test Cycle). The ETC Cycle has been developed by the FIGE

[2] Smoke opacity is measured on the ELR (European Load Response) Test, consisting of a sequence of load steps at constant engine speeds, specified in the same Directive.

Institute (Forschungsinstitut Geräusche und Erschütterungen), Aachen, Germany and is based on real road cycle measurements of heavy-duty vehicles. Different driving conditions are represented by three parts (each of 600 s duration) of the ETC Cycle, incorporating urban, rural and motorway driving as well as motoring sections (Figure A.2).

- Part one represents city driving corresponding to a maximum speed of 50 km/h, frequent starts, stops, and idling.
- Part two is rural driving starting with a steep acceleration segment; the corresponding average speed of the vehicle is about 72 km/h.
- Part three is motorway driving with corresponding average vehicle speed of about 88 km/h.

For the purpose of engine certification, the ETC Cycle is performed on an engine dynamometer. To this aim, the engine under study needs first to be mapped for determining the speed *vs.* torque curve. Figure A.2 illustrates normalized engine speed and normalized engine torque *vs.* time for the ETC Cycle; for the particular engine under test, speed is de-normalized using the following equation [2]

$$\text{Actual Speed} = \frac{\%\ \text{speed (reference speed} - \text{idle speed)}}{100} + \text{idle speed} \tag{A.1}$$

with the reference speed N_{ref} corresponding to the 100% speed values specified in the engine dynamometer schedule, defined as follows:

$$N_{ref} = N_{lo} + 95\%\left(N_{hi} - N_{lo}\right) \tag{A.2}$$

with N_{hi} the highest engine speed, where 70% of the declared (by the manufacturer) maximum power occurs, and N_{lo} the lowest engine speed, where 50% of the declared maximum power occurs.

Similarly, engine torque is de-normalized to the maximum torque at the respective speed using the following equation:

$$\text{Actual Torque} = \frac{\%\ \text{torque} \cdot (\text{max.torque})}{100} \tag{A.3}$$

with the maximum torque value found from the respective engine mapping curve. Negative torque values in the upper sub-diagram of Figure A.2 correspond to motoring points, and for the purpose of reference cycle generation these take unnormalized values determined in directive 1999/96/EC.

A.2.2.3 Non-road Mobile Diesel Engines

For non-road mobile diesel engines, the Non-road Transient Cycle (NRTC) has been developed in co-operation with the US Environmental Protection Agency. This Test Cycle is used in parallel with the Non-road Steady Cycle (NRSC). The Test Cycle is an engine dynamometer transient driving schedule of total duration about 1200 s. Engine speed and torque during the NRTC Test are demonstrated in Figure A.3. De-normalization of speed and torque for the actual engine under study is accomplished using Equations A.1–A.3.

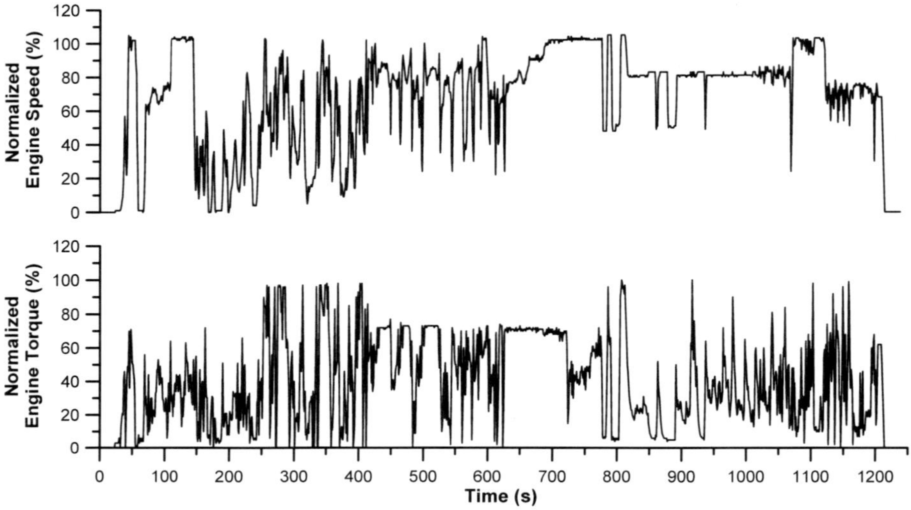

Figure A.3. NRTC dynamometer schedule for non-road mobile diesel engines

A.3 United States of America

A.3.1 Emission Standards

In the US, there are two types of emission standards; Federal and California. The US Environmental Protection Agency (EPA) sets Federal emission standards and the California Air Resources Board (CARB) sets California standards. Both regulation programs have adopted the principle that vehicles doing the same job, must meet the same emissions standards, regardless of the size of the vehicle or the fuel used [1, 3, 4].

A.3.1.1 Passenger, Light and Medium Duty Vehicles – EPA Tier II Standard
Currently (2008), vehicles sold in the United States must meet 'Tier II' standards that went into effect in 2004. Tier II standards are currently being phased in a process that should be complete by 2009. Within the Tier II ranking, there is a subranking ranging from (level) Bin 1–10, with 1 being the cleanest (zero emission vehicle - ZEV) and 10 being the 'dirtiest'. Table A.3 outlines and defines the

vehicle categories used in the EPA Tier II standards. Under the Tier II regulation, the same emission standards apply to all vehicle weight categories, *i.e.,* cars, minivans, light-duty trucks, and SUVs have the same emission limit. Table A.4 illustrates the emission standards for all pollutants (certification bins) when tested on the Federal Test Procedure (FTP), described in Section A.3.2.1. Engines in commercial vehicles above 8,500 lbs GVWR, such as cargo vans or light trucks, continue to certify to heavy-duty engine emission standards. Since light-duty emission standards are expressed in g/mile, vehicles with large engines (such light trucks or SUVs) have to use more advanced emission control technologies than vehicles with smaller engines in order to meet the standards.

Table A.3. Vehicle categories used in EPA Tier II standards

Vehicle Category			**Abbreviation**	**Requirements**
Light-Duty Vehicle			LDV	max. 8.500 lb GVWR
Light-Duty Truck			LDT	max. 8.500 lb GVWR. max. 6.000 lb curb weight. max. 45 ft2 frontal area
	Light light-duty truck		LLDT	max. 6.000 lb GVWR
		Light-duty 1	LDT1	max. 3.750 lb LVW1
		Light-duty 2	LDT2	min. 3.750 lb LVW1
	Heavy light-duty truck		HLDT	min. 6.000 lb GVWR
		Light-duty 3	LDT3	max. 5.750 lb ALVW2
		Light-duty 4	LDT4	min. 5.750 lb ALVW2
Medium-Duty Passenger Vehicle			MDPV	max. 10.000 lb GVWR3

[1] LVW (loaded vehicle weight) = curb weight + 300 lb

[2] ALVW (adjusted loaded vehicle weight) = average of GVWR (Gross Vehicle Weight Rating) and curb weight

[3] Manufacturers may alternatively certify engines for diesel fueled MDPVs through the heavy-duty diesel engine regulations

A.3.1.2 Passenger, Light and Medium Duty Vehicles – California Low Emission Vehicle II (LEV II) Standards

Traditionally, California emission standards have been more stringent than the EPA requirements, but their evolution and structure is similar to that of the Federal legislation. In November 1998, the California ARB adopted a proposal that requires manufacturers to produce vehicles, beginning with the 2004 model year, that meet the LEV II standards. LEV II regulation is reproduced in Table A.5 for passenger cars and light-duty vehicles with gross vehicle weight rating up to 14,000 lb [4].

Table A.4. US EPA Tier II emission standards, g/mile

Bin	Intermediate life (5 years / 50,000 miles)				Bin	Full useful life				
	NMHC	CO	NOx	HCHO*		NMHC	CO	NOx†	PM	HCHO*
Temporary bins										
10[a,b,c,d,g,h]	0.125 (0.160)	3.4 (4.4)	0.4	0.015 (0.018)	10[a,b,c,d]	0.156 (0.230)	4.2 (6.4)	0.6	0.08	0.018 (0.027)
9[a,b,c,e,g,h]	0.075 (0.140)	3.4	0.2	0.015	9[a,b,c,e]	0.090 (0.180)	4.2	0.3	0.06	0.018
Permanent bins										
8[b,f,h]	0.100 (0.125)	3.4	0.14	0.015	8[b,f]	0.125 (0.156)	4.2	0.20	0.02	0.018
7[h]	0.075	3.4	0.11	0.015	7	0.090	4.2	0.15	0.02	0.018
6[h]	0.075	3.4	0.08	0.015	6	0.090	4.2	0.10	0.01	0.018
5[h]	0.075	3.4	0.05	0.015	5	0.090	4.2	0.07	0.01	0.018
4	-	-	-	-	4	0.070	2.1	0.04	0.01	0.011
3	-	-	-	-	3	0.055	2.1	0.03	0.01	0.011
2	-	-	-	-	2	0.010	2.1	0.02	0.01	0.004
1	-	-	-	-	1	0	0	0	0	0

* Formaldehyde
† Average manufacturer fleet NOx standard is 0.07 g/mile for Tier II vehicles
[a] Bin deleted at end of 2006 model year (2008 for HLDTs)
[b] The higher NMHC, CO and HCHO values apply only to HLDTs & MDPVs (expire after 2008)
[c] An additional temporary bin restricted to MDPVs, expires after model year 2008
[d] Optional temporary NMHC standard of 0.195 g/mile (50,000 miles) and 0.280 g/mile (full useful life) applies for qualifying LDT4s and MDPVs only
[e] Optional temporary NMHC standard of 0.100 g/mile (50,000 miles) and 0.130 g/mile (full useful life) applies for qualifying LDT2s only
[f] Higher temporary NMHC standard deleted at end of 2008 model year
[g] Intermediate life standards are optional for diesels certified to bin 10
[h] Intermediate life standards are optional for any test group certified to a 150,000 miles useful life

Table A.5. California LEV II emission standards for passenger cars and light-duty vehicles, GVWR < 8,500 lbs and for medium duty vehicles, GVWR > 8,500 lb, g/mile

Category	Vehicle	50,000 miles / 5 years					120,000 miles / 11 years				
		NMOG	CO	NOx	PM	HCHO	NMOG	CO	NOx	PM	HCHO
LEV	< 8,500 lb	0.075	3.4	0.05	-	0.015	0.090	4.2	0.07	0.01	0.018
ULEV		0.040	1.7	0.05	-	0.008	0.055	2.1	0.07	0.01	0.011
SULEV		-	-	-	-	-	0.010	1.0	0.02	0.01	0.004
LEV	8,501-10,000 lb	-	-	-	-	-	0.195	6.4	0.2	0.12	0.032
ULEV		-	-	-	-	-	0.143	6.4	0.2	0.06	0.016
SULEV		-	-	-	-	-	0.100	3.2	0.1	0.06	0.008
LEV	10,001-14,000 lb	-	-	-	-	-	0.230	7.3	0.4	0.12	0.040
ULEV		-	-	-	-	-	0.167	7.3	0.4	0.06	0.021
SULEV		-	-	-	-	-	0.117	3.7	0.2	0.06	0.010

LEV: Low Emission Vehicles
ULEV: Ultra Low Emission Vehicles
SULEV: Super Ultra Low Emission Vehicles

A.3.1.3 Heavy-duty Truck and Bus Engines

Model year 1988–2004 US Federal (EPA) emission standards for heavy-duty diesel truck engines are summarized in Table A.6. On December 2000 the EPA signed new much more stringent emission standards for model year 2007 and later heavy-duty highway engines.

Table A.6. EPA emission standards for heavy-duty diesel engines, g/bhp-h

Year	HC	CO	NOx	PM
1988	1.3	15.5	10.7	0.6
1990	1.3	15.5	6.0	0.6
1991	1.3	15.5	5.0	0.25
1994	1.3	15.5	5.0	0.1
1998	1.3	15.5	4.0	0.1
2004	0.50*	15.5	2.5	0.1
2007	0.14*	15.5	0.20	0.01
* Non-methane hydrocarbons				

Different sets of emission level requirements have been legislated for bus, non-road, locomotive, marine and stationary diesel engines.

A.3.2 Transient Cycles

A.3.2.1 Passenger Vehicles and Light-duty Trucks

The FTP-75 (Federal Test Procedure) Transient Cycle is used for emission certification of light-duty vehicles in the US.

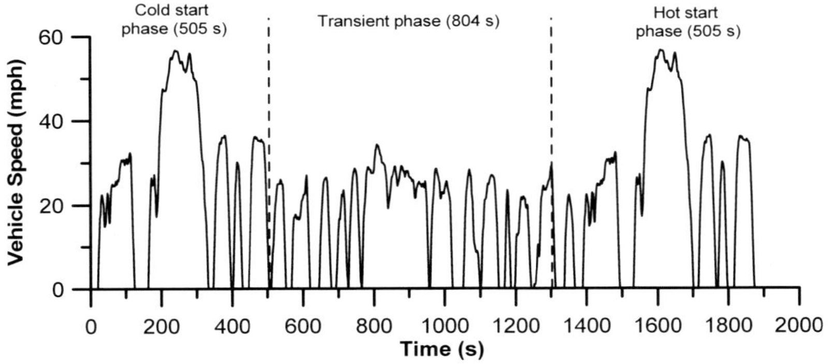

Figure A.4. US FTP-75 Transient Cycle (duration 1874 s, distance 11.04 miles, average speed 21.21 mph, maximum speed 56.7 mph)

The FTP-75 Cycle (Figure A.4) is derived from the earlier FTP-72 Cycle (also known as Urban Dynamometer Driving Schedule) by adding a third phase of 505 s, identical to the first phase of FTP-72 but with a hot start. The third phase starts after the engine is stopped for 10 minutes. Thus, the entire FTP-75 Cycle consists of cold start phase, transient phase and hot start phase. The emissions from each phase are collected in a separate Teflon bag, analyzed and expressed in g/mile. Effective model year 2000, vehicles have to be additionally tested on two Supplemental Federal Test Procedures designed to address shortcomings with the FTP-75 in the representation of: 1. aggressive, high speed and/or high acceleration driving behavior (see also Table A.10), rapid speed fluctuations, and driving behavior following startup (US06), and 2. the engine load and emissions associated with the use of air-conditioning units (SC03).

Another light-duty vehicles Test Cycle used in the US is the Highway Fuel Economy Test Driving Schedule (HWFET), demonstrated in Figure A.5, which simulates highway driving conditions under 60 mph.

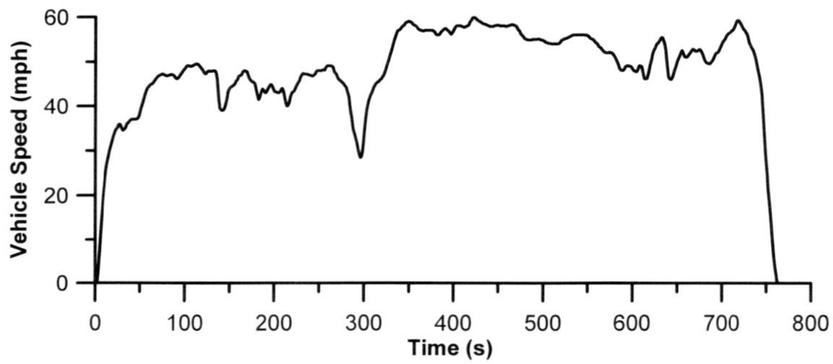

Figure A.5. US Highway Fuel Economy Test Driving Schedule (HWFET) (duration 765 s, distance 10.26 miles, average speed 48.3 mph, maximum speed 59.9 mph)

The California Unified Cycle (UC – also known as LA92 Test) (Figure A.6) is a chassis dynamometer driving schedule for light-duty vehicles developed by the California Air Resources Board. The test is also referred to as the Unified Cycle Driving Schedule (UCDS). The UC test has a similar three-bag structure, but is a more aggressive Driving Cycle than the Federal FTP-75, being characterized by higher speeds and accelerations, fewer stops per mile, and less idle time.

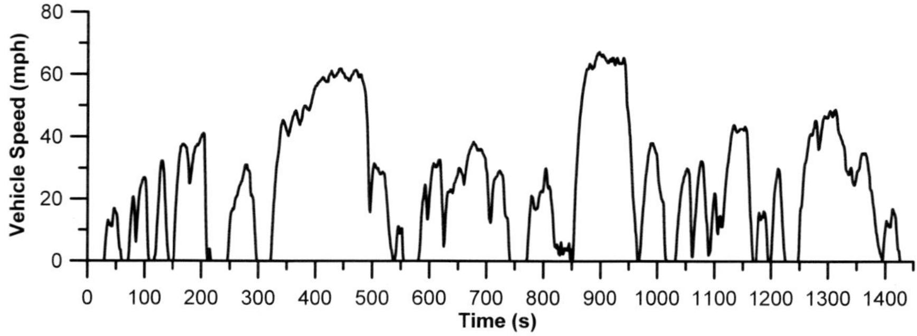

Figure A.6. California Unified (UC) Transient Cycle (duration 1435 s, distance 9.82 miles, average speed 24.63 mph, maximum speed 67.2 mph)

The NYCC Test Cycle (Figure A.7) simulates low speed, congested, urban driving with frequent stops and idling, typical of the state of New York.

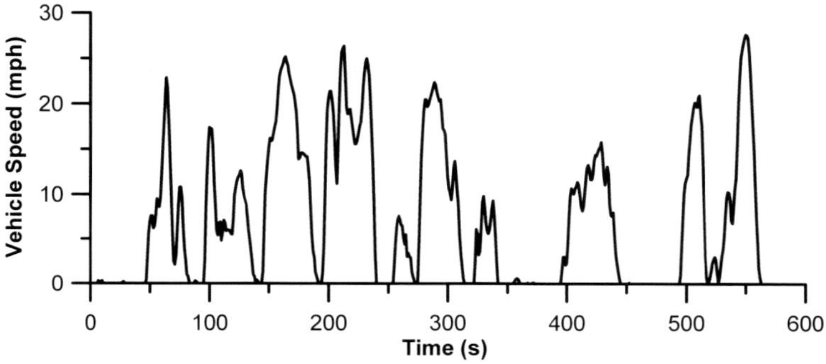

Figure A.7. New York City Cycle (NYCC) (duration 598 s, distance 1.18 miles, average speed 7.1 mph, maximum speed 27.7 mph)

A.3.2.2 Heavy-duty Vehicles

The EPA UDDS (Urban Dynamometer Driving Schedule) Transient Test Cycle is illustrated in Figure A.8; it has been developed for chassis dynamometer testing of heavy-duty diesel-engined vehicles.

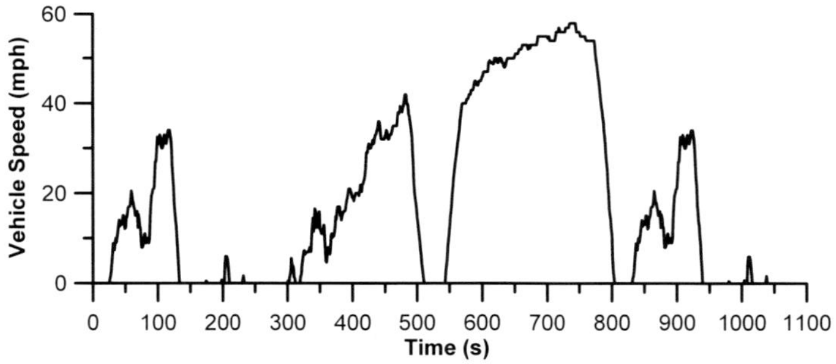

Figure A.8. UDDS Transient Cycle for heavy-duty vehicles (duration 1060 s, distance 5.55 miles, average speed 18.86 mph, maximum speed 58 mph)

Apart from the above-mentioned Transient Cycles, other more specialized Transient Tests are available in the US, as for example the Manhattan Bus Cycle, New York Composite, Central Business District, *etc*.

A.3.2.3 Heavy-duty Engines

FTP Transient Cycle (Engine Dynamometer)

The FTP heavy-duty Transient Cycle is currently used for emission testing of heavy-duty on-road engines in the US. The transient test was developed to take into account the variety of heavy-duty trucks and buses in American cities, including traffic in and around the cities on roads and expressways. The FTP transient test is based on the UDDS chassis dynamometer driving cycle and is demonstrated in Figure A.9. The equivalent average speed is about 30 km/h and the equivalent distance travelled is 10.3 km for a running period of 20 minutes.

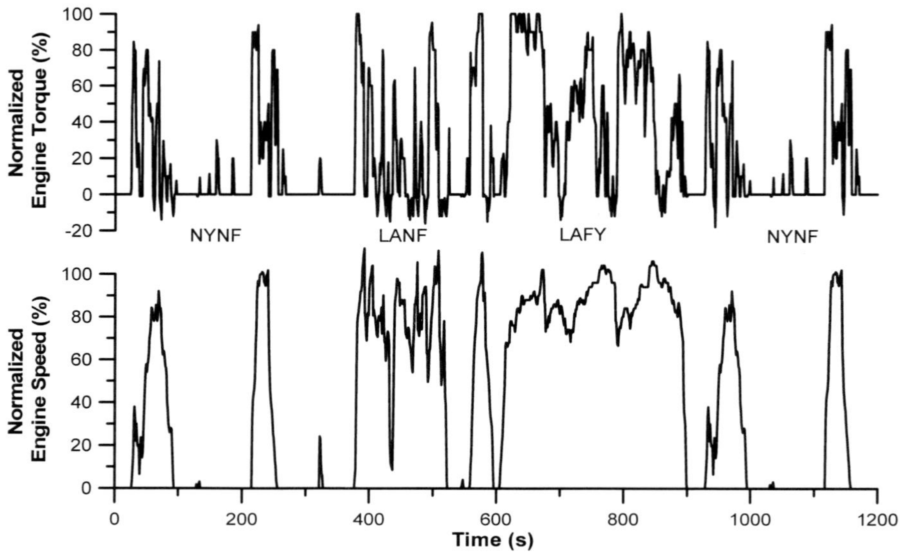

Figure A.9. US Transient Cycle FTP (engine dynamometer)

As is the case with the ETC and NRTC European Transient Cycles for heavy-duty and non-road diesel engines, the FTP Test Cycle includes motoring segments too and, therefore, requires an electric dynamometer. This Transient Cycle consists of four phases:

1. NYNF (New York Non Freeway) phase typical of light urban traffic with frequent starts and stops.
2. LANF (Los Angeles Non Freeway) phase typical of crowded urban traffic with few stops.
3. LAFY (Los Angeles Freeway) phase simulating crowded expressway traffic in Los Angeles.
4. NYNF phase.

The FTP Test Cycle comprises cold start after parking overnight, followed by idling, acceleration and deceleration phases, and a wide variety of different speeds and loads; it is carried out twice and the second repetition is made with a warm start after a stop of 20 min upon completion of the first cycle. A de-normalization procedure similar to that described by Equations A.1–A.3 applies here too.

A.3.2.4 Non-road Mobile Diesel Engines

Figures A.10 and A.11 illustrate two Transient Cycles for non-road mobile diesel engines valid in the US, realized on an engine dynamometer.

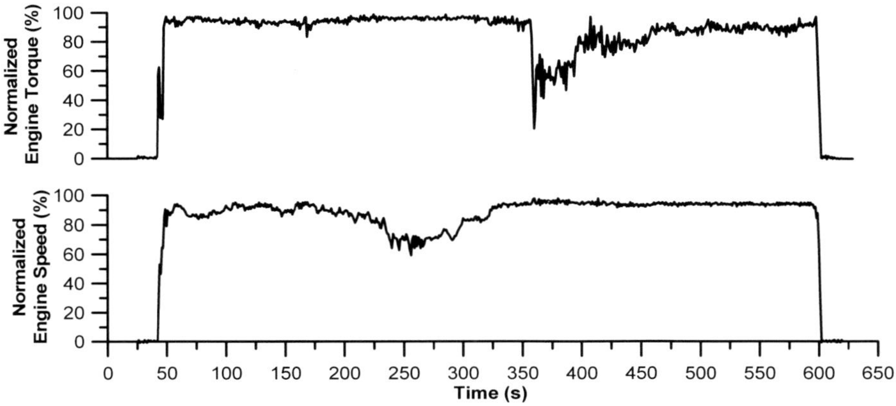

Figure A.10. US EPA agricultural tractor Transient Cycle

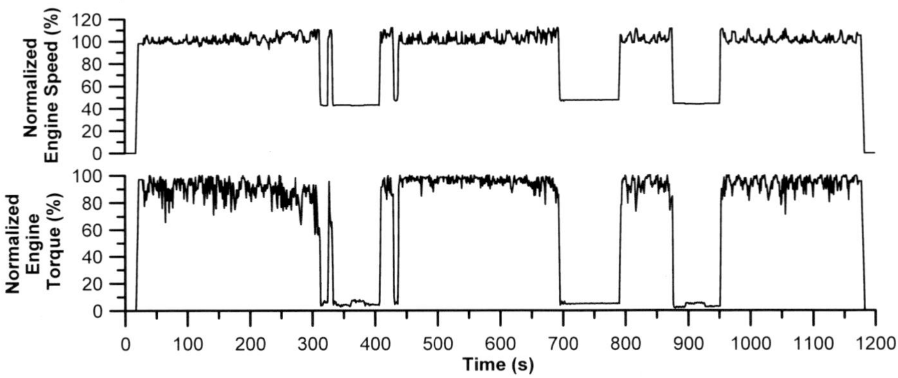

Figure A.11. US EPA excavator Transient Cycle

A.4 Japan

A.4.1 Emission Standards

According to the Japanese emission legislation, two types of standards are used, denoted as 'mean' and 'max'. 'Mean' standards are to be met as a type approval limit and as a production average. 'Max' standards are met as an individual limit in series production vehicles [1, 5].

A.4.1.1 Passenger Vehicles and Light Trucks

Emission standards for new diesel-engined passenger and light commercial vehicles with gross vehicle weight lower than 3,500 kg are summarized in Tables A.7 and A.8, respectively.

Table A.7. Japanese emission standards for diesel-engined passenger vehicles, g/km

Vehicle Weight	Date	Test Cycle	CO		HC		NOx		PM	
			mean	(max)	mean	(max)	mean	(max)	mean	(max)
<1250 kg*	1986	10-15 mode	2.1	(2.7)	0.40	(0.62)	0.70	(0.98)		
	1990		2.1	(2.7)	0.40	(0.62)	0.50	(0.72)		
	1994		2.1	(2.7)	0.40	(0.62)	0.50	(0.72)	0.20	(0.34)
	1997		2.1	(2.7)	0.40	(0.62)	0.40	(0.55)	0.08	(0.14)
	2002		0.63		0.12		0.28		0.052	
	2005[a]	JC08[b]	0.63		0.024[c]		0.14		0.013	
	2009		0.63		0.024[c]		0.08		0.005	
>1250 kg*	1986	10-15 mode	2.1	(2.7)	0.40	(0.62)	0.90	(1.26)		
	1992		2.1	(2.7)	0.40	(0.62)	0.60	(0.84)		
	1994		2.1	(2.7)	0.40	(0.62)	0.60	(0.84)	0.20	(0.34)
	1998		2.1	(2.7)	0.40	(0.62)	0.40	(0.55)	0.08	(0.14)
	2002		0.63		0.12		0.30		0.056	
	2005[a]	JC08[b]	0.63		0.024[c]		0.15		0.014	
	2009		0.63		0.024[c]		0.08		0.005	

* equivalent inertia weight (EIW); vehicle weight of 1265 kg
[a] full implementation by the end of 2005
[b] full phase-in by 2011
[c] non-methane hydrocarbons

Table A.8. Japanese emission standards for diesel-engined light commercial vehicles with gross vehicle weight ≤ 3,500 kg (≤ 2,500 kg before 2005), g/km

Gross Vehicle Weight	Date	Test	CO		HC		NOx		PM	
			mean	(max)	mean	(max)	mean	(max)	mean	(max)
≤1700 kg	1988	10-15 mode	2.1	(2.7)	0.40	(0.62)	0.90	(1.26)		
	1993		2.1	(2.7)	0.40	(0.62)	0.60	(0.84)	0.20	(0.34)
	1997		2.1	(2.7)	0.40	(0.62)	0.40	(0.55)	0.08	(0.14)
	2002		0.63		0.12		0.28		0.052	
	2005[a]	JC08[c]	0.63		0.024[d]		0.14		0.013	
	2009		0.63		0.024[d]		0.08		0.005	
>1700 kg	1988	6 mode	790*	(980)*	510*	(670)*	DI:380* IDI:260*	(500*) (350*)		
	1993	10-15 mode	2.1	(2.7)	0.40	(0.62)	1.30	(1.82)	0.25	(0.43)
	1997[b]		2.1	(2.7)	0.40	(0.62)	0.70	(0.97)	0.09	(0.18)
	2003		0.63		0.12		0.49		0.06	
	2005[a]	JC08[c]	0.63		0.024[d]		0.25		0.015	
	2009		0.63		0.024[d]		0.15		0.007	

* ppm
[a] full implementation by the end of 2005
[b] 1997: manual transmission vehicles; 1998: automatic transmission vehicles
[c] full phase-in by 2011
[d] non-methane hydrocarbons

A.4.1.2 Heavy-duty Vehicles

Emission legislation for diesel engines on heavy-duty vehicles has traditionally been less stringent in Japan than the corresponding legislation in the US and the EU. However, because of the deterioration in air quality, Japan has decided to implement far-reaching legislation in stages to reduce the emission of both NO_x and PM. A first stage, which is of the same order of magnitude as the Euro IV standards, has been implemented in 2005 and a further, much more stringent stage has been planned for 2009. Emission standards for diesel-engined heavy-duty vehicles in Japan are reproduced in Table A.9.

Table A.9. Japanese emission standards for diesel-engined heavy-duty vehicles with gross vehicle weight > 3,500 kg (> 2,500 kg before 2005), g/kWh

Date	Test	CO		HC		NOx		PM	
		mean	**(max)**	**mean**	**(max)**	**mean**	**(max)**	**mean**	**(max)**
1988/89	6 mode	790*	(980*)	510*	(670*)	DI: 400* IDI: 260*	(520*) (350*)		
1994	13 mode	7.40	(9.20)	2.90	(3.80)	DI: 6.00 IDI: 5.00	(7.80) (6.80)	0.70	(0.96)
1997[a]	13 mode	7.40	(9.20)	2.90	(3.80)	4.50	(5.80)	0.25	(0.49)
2003[b]	13 mode	2.22		0.87		3.38		0.18	
2005[c]	JE05	2.22		0.17[d]		2.0		0.027	
2009	JE05	2.22		0.17[d]		0.7		0.01	

* ppm

[a] 1997: GVW ≤ 3,500 kg; 1998: 3,500 < GVW ≤ 12,000 kg;1999: GVW > 12,000 kg

[b] 2003: GVW ≤ 12,000 kg; 2004: GVW > 12,000 kg

[c] full implementation by the end of 2005

[d] non-methane hydrocarbons

A.4.2 Transient Cycles

In 2005 the Japanese Ministry of Environment introduced two new Test Cycles for emission certification of light and heavy-duty vehicles, *i.e.*,

- JC08 test mode for light-duty vehicles of gross vehicle weight < 3,500 kg,
- JE05 test mode for heavy-duty vehicles of gross vehicle weight > 3,500 kg.

For heavy-duty vehicles, the new test mode JE05 has become effective in 2005. For light-duty vehicles, the new Test Cycle is expected to be fully phased-in by 2011. In the 2005–2011 period, emissions will be determined using weighted averages from the new test mode, earlier 10–15 mode and 11 mode Cycles. The Cycles can be run as either cold- or warm-start tests, depending on the application.

A.4.2.1 Light-duty Vehicles

10–15 Mode Transient Cycle
During the transition period 2005–2011 the 10–15 mode Transient Cycle, illustrated in Figure A.12, is valid.

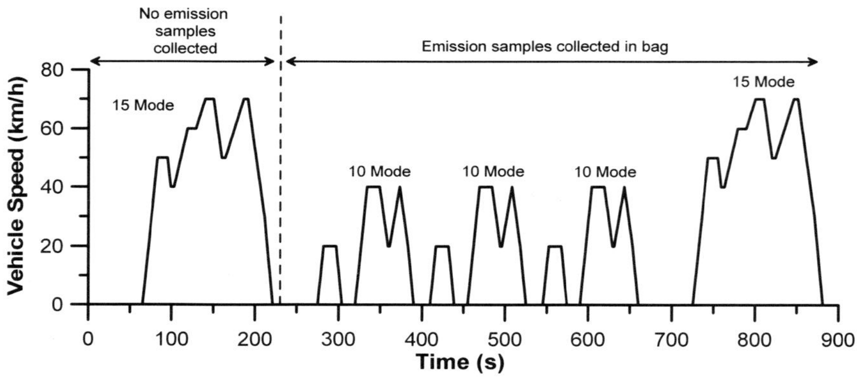

Figure A.12. Japanese 10–15 mode Transient Cycle (duration 891 s, distance 6.34 km, average speed 25.61 km/h, maximum speed 70 km/h including the initial 15 mode segment)

Transient Cycle JC08
Figure A.13 illustrates the proposed light-duty JC08 Transient Cycle, which will be fully implemented by 2011; this simulates typical big-city congested traffic conditions with idling and frequent accelerations/decelarations.

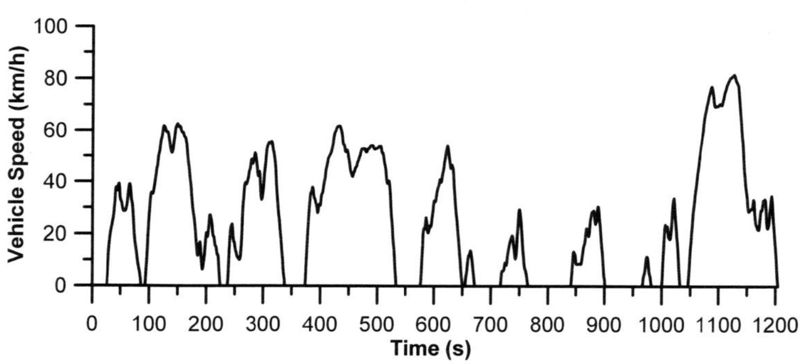

Figure A.13. Japanese JC08 Transient Cycle for light-duty vehicles (duration 1204 s, distance 8.159 km, average speed 24.4 km/h, maximum speed 81.6 km/h)

A.4.2.2 Heavy-duty Vehicles Transient Cycle JE05

The JE05 cycle is a transient driving schedule of approximately 1800 s total duration. It is defined through vehicle speed *vs*. time points, as depicted in Figure A.14, and is based on typical Tokyo driving conditions.

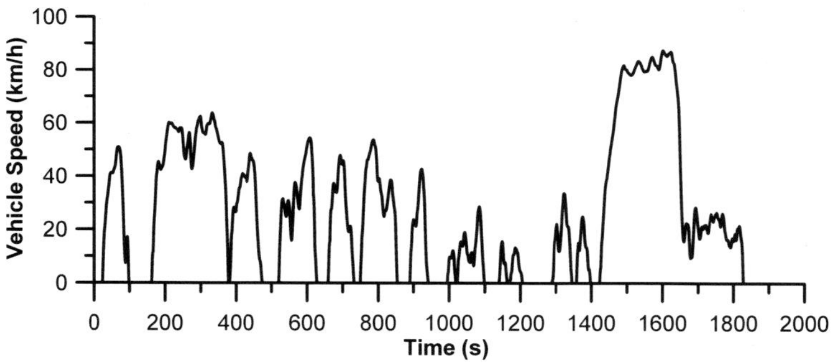

Figure A.14. Japanese JE05 Transient Cycle for vehicles with GVW > 3,500 kg (duration 1829 s, distance 13.89 km, average speed 27.34 km/h, maximum speed 87.6 km/h)

For engine dynamometer testing, engine speed and torque *vs*. time data must be generated based on the vehicle speed points applying a suitable drivetrain model.

A.5 Overall: Comparative Data

Figures A.15–A.17 summarize the progressive tightening of acceptable NO_x and PM emissions for European passenger vehicles (Figure A.15, based on Table A.1), US passenger vehicles (Figure A.16, based on Table A.4) and heavy-duty engines (Figure A.17).

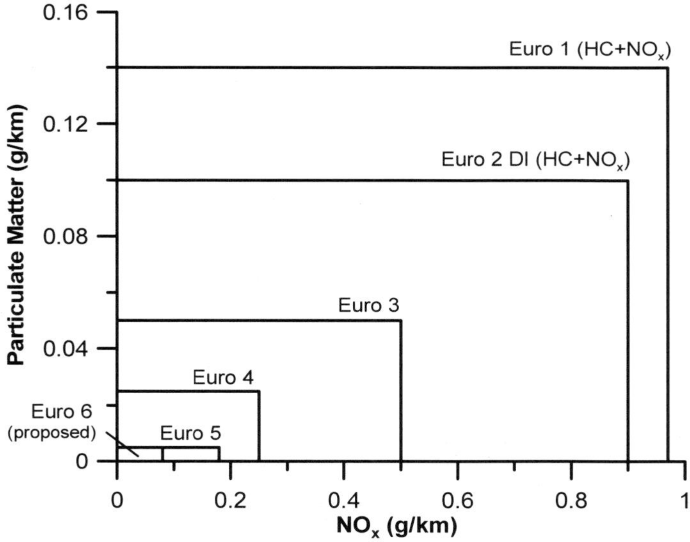

Figure A.15. Gradual tightening of European emission standards for DI diesel-engined passenger cars (for Euro 1 and Euro 2 emission levels, 'NO_x' values correspond to the sum of NO_x and HC emissions, see also Table A.1)

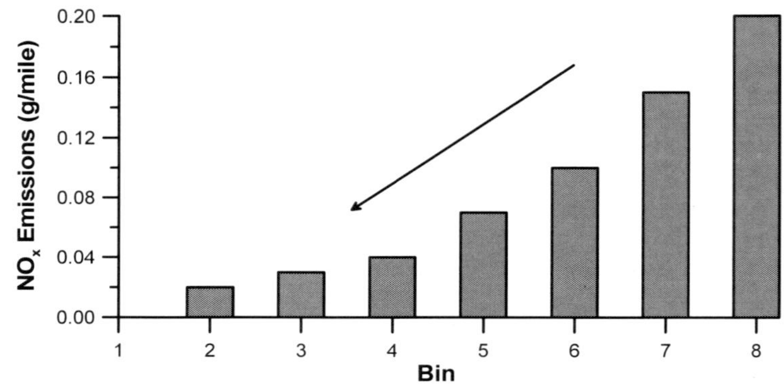

Figure A.16. Gradual tightening of acceptable NO_x emission limits within the US Tier II standards

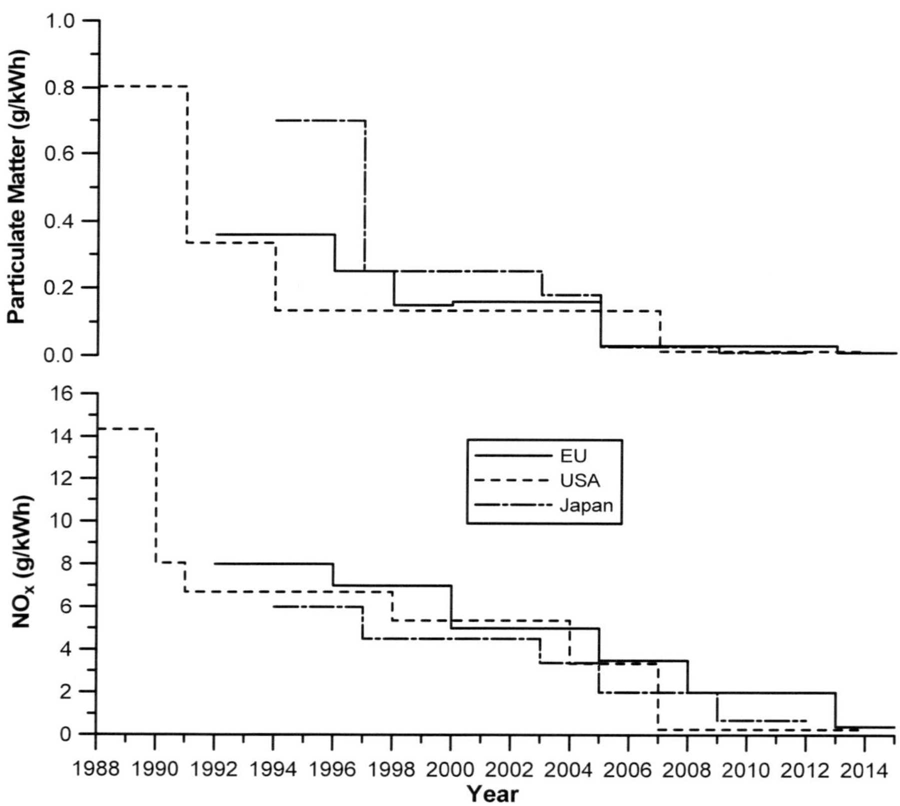

Figure A.17. Comparative heavy-duty truck DI diesel engines NO_x and PM emission levels over the years in the EU, the US and Japan (European PM levels after EURO III (2000) correspond to ETC Cycle limits)

Finally, Table A.10 summarizes the different parameters (distance, duration, average and maximum speeds, mean and maximum (positive) accelerations) of the various light-duty and heavy-duty Transient Cycles discussed in the previous paragraphs.

Table A.10. Summarization of various Transient Cycle parameters

	Country	**Vehicle type**	**Distance**	**Duration**	**Average speed**	**Maximum speed**	**Maximum acceleration**	**Mean acceleration *****	**Idling time**
		-	km	s	km/h	km/h	m/s^2	m/s^2	%
ECE 15	EU	LD	4×1.013= 4.052	4×195= 780	18.70	50	1.04	**0.748**	32.7
EUDC	EU	LD	6.955	400	62.59	120	0.83	0.378	10.3
ECE+EUDC	EU	LD	11.007	1180	33.58	120	1.04	0.623	25.1
FTP-75	USA	LD	**17.76**	**1874**	34.13	91.23	1.48	0.51	19.1
FTP US06	USA	LD	12.89	596	**77.71**	**129.23**	**3.76**	0.67	6.9
FTP SC03	USA	LD	5.76	596	34.79	88.17	2.28	0.502	18.8
HWFET	USA	LD	16.51	765	77.58	96.38	1.43	0.194	**0.8**
NYCC	USA	LD	1.897	598	11.42	44.57	2.68	0.66	37.9
CU	USA	LD	15.79	1435	39.62	108.12	3.08	0.672	16.3
UDDS	USA	HD	8.93	1060	30.35	93.32	1.96	0.4827	33.2
J10-15*	Japan	LD	6.34	891	25.61	70	0.81	0.5245	32.6
J10-15**	Japan	LD	4.165	660	22.72	70	0.81	0.5689	32.4
JC08	Japan	LD	8.16	1204	24.40	81.6	1.69	0.426	27.1
JE05	Japan	HD	13.89	1829	27.34	87.6	1.59	0.314	25.2

LD: passenger & light duty / HD: heavy duty
* incl. warm-up phase
** excl. warm-up phase
*** only positive values of acceleration considered

Clearly, the supplemental FTP US06 and the California Unified Cycle are the most aggressive in terms of both mean and maximum acceleration, a fact that is also expected to accordingly affect total exhaust emissions. In contrast, European Cycles are 'softer', as was the case with the Japanese 10–15 Mode.

A.6 Worldwide Heavy-duty Transient Cycle

At its 34th session in June 1997, The UNECE Group of Experts on Pollution and Energy (GRPE), under the guidance of Working Party 29, mandated the *ad hoc* group WHDC with the development of a 'Worldwide harmonised Heavy-duty Certification' procedure [6]. The objective of the research program was to develop a worldwide harmonized engine Test Cycle for the emissions certification procedure of heavy-duty vehicles/engines that would

- become a uniform global basis for engine certification regarding exhaust emissions;
- be representative of worldwide real-life heavy-duty engine operation;
- give the highest potential for the control of real-life emissions;
- be applicable in the future to state-of-the-art technology; and

- match emissions in relative terms for accurate ranking of different engines/technologies.

In order to proceed to the development of the worldwide harmonized engine Test Cycle, the United Nations research group conducted a collection and analysis of driving behavior data as well as a statistical investigation of heavy-duty vehicle usage in different regions of the world. From this database, a representative worldwide Transient Vehicle Cycle (WTVC) of 1800 s duration and 40 km/h average speed has been derived (illustrated in Figure A.18, upper sub-diagram), as well as a normalized engine speed (Figure A.18 lower sub-diagram) and engine torque *vs*. time Transient Cycle for heavy-duty diesel engines.

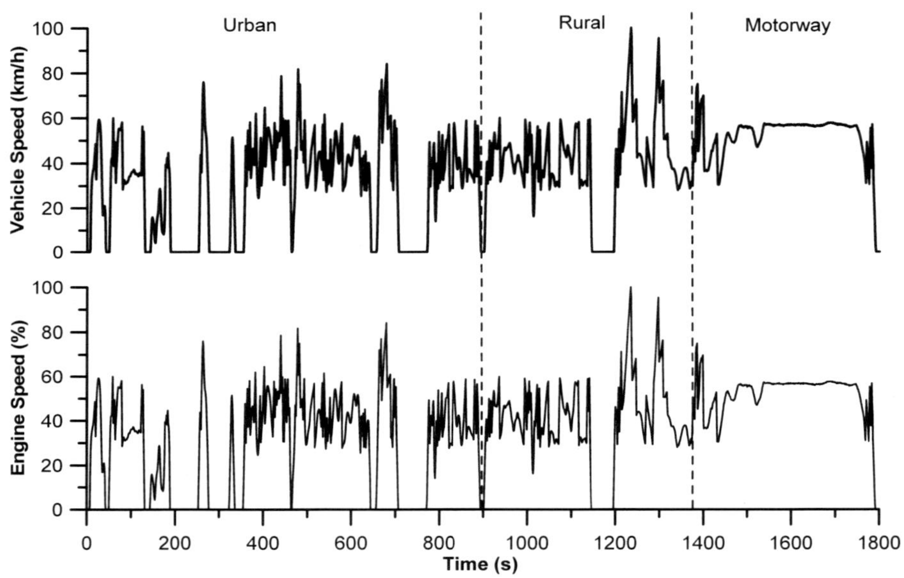

Figure A.18. The proposed worldwide heavy-duty vehicle Transient Cycle [6]

References

[1] http://www.dieselnet.com, Ecopoint Inc.

[2] http://ec.europa.eu/enterprise/automotive/directives/vehicles/index.htm

[3] http://www.epa.gov/nvfel/testing/dynamometer.htm#engcycles

[4] http://www.arb.ca.gov

[5] http://www.env.go.jp/en/

[6] TRANS/WP29/GRPE/2001/2, 'Development of a Worldwide Harmonised Heavy-duty Engine Emissions Test Cycle', Final Report, Informal document No. 2, GRPE 42nd session, 28. May–1. June 2001, ECE-GRPE WHDC Working Group, Convenor: Dr. Cornelis Havenith, Author: Heinz Steven, April 2001.

Appendix B

Fundamentals of Control Theory

The purpose of an engine control system is to keep the performance of the engine and its sub-systems within specified acceptable levels. Perhaps, the most well established controller is the speed governor. To achieve the control function, the control system incorporates sensors for measuring the interesting properties, a processor, nowadays most usually electronic, that gathers the signals from the sensors and calculates the optimum control action, and actuators that perform the required action or function. On the basis of their operating principle, control systems are classified as

- open-loop or
- closed-loop systems.

Usually the control systems applied in (diesel) engines incorporate the closed-loop strategy. A closed-loop control system calculates the control action based on the error between measured (feedback) and demand values (Figure B.1).

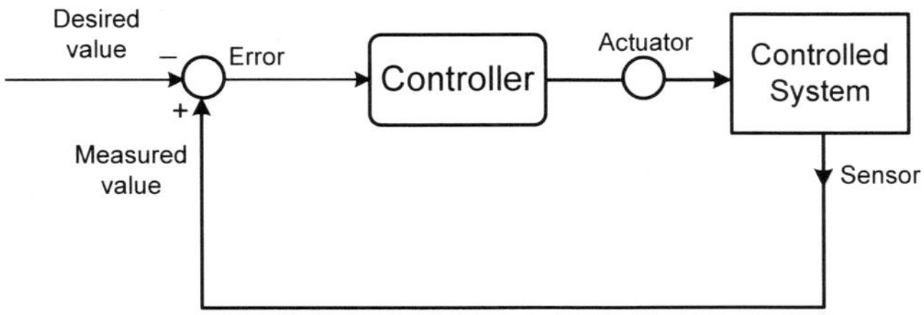

Figure B.1. Simplified closed-loop control system

Open-loop controllers, on the other hand, do not incorporate any feedback action (Figure B.2) and can only provide sluggish response.

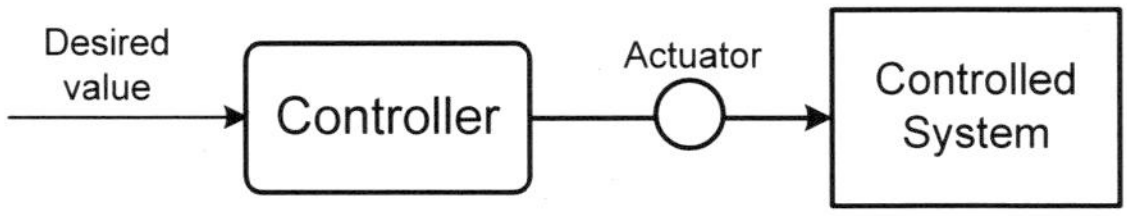

Figure B.2. Typical open-loop control system

The desired control behavior in an open- or closed-loop system is usually plotted against engine speed and load and stored in a look-up table in the form of a 3-D map. Correction coefficients are also applied to compensate for non-ideal conditions, *e.g.,* cold-starting, operation at high altitude or transients. More advanced techniques make use of model-based control, where the desired value is determined via a linear or, better still, quasi-linear system model (such as the ones described in Section 9.2) simulating the basic thermodynamic and dynamic aspects of the system.

An interesting alternative to the classic closed-loop system is feed-forward, namely the technique of adding to the control action to compensate for a measured disturbance (Figure B.3). A typical feed-forward control system is demonstrated in Figure 2.41 (EGR control).

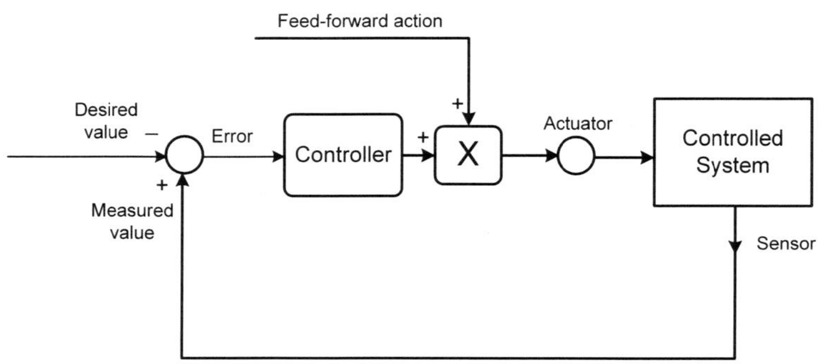

Figure B.3. Closed-loop control system with feed-forward action

The most popular controller, particularly for systems with small transport delays, handles the error between measured and demand values via P-I-D action (Figure B.4). The proportional term (P) offers speed of response, the integral term (I) ensures that there is no control error under steady-state conditions and the derivative term (D) ensures that the control signal is modified if the error changes quickly.

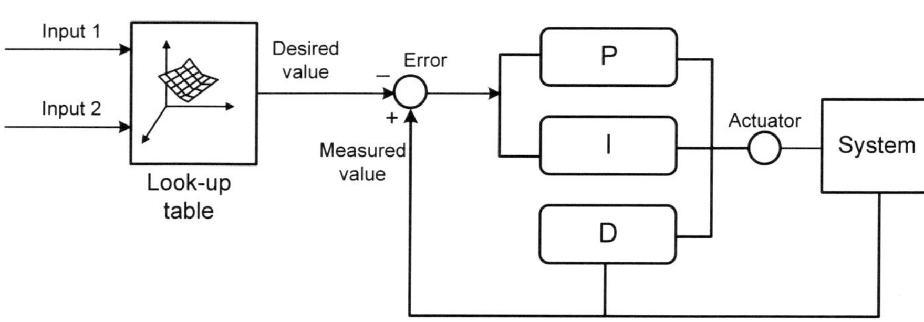

Figure B.4. Closed-loop P-I-D controller incorporating a look-up table

In recent years, the evolution of electronics has facilitated the use of alternative control actions, namely

- multivariable techniques (handling several input and output variables);
- adaptive control; this is able to change its structure to adapt to new conditions;
- predictive control, applying a knowledge of the system to be controlled to make predictions for future control actions;
- robust control; and
- neural/fuzzy logic control.

Advanced controllers communicate with the various system parts via a Controller Area Network (CAN) bus and may incorporate sophisticated features such as self-diagnosis.

Index